Lecture Notes in Mechanical Engineering

Lecture Notes in Mechanical Engineering (LNME) publishes the latest developments in Mechanical Engineering - quickly, informally and with high quality. Original research reported in proceedings and post-proceedings represents the core of LNME. Volumes published in LNME embrace all aspects, subfields and new challenges of mechanical engineering. Topics in the series include:

- Engineering Design
- Machinery and Machine Elements
- Mechanical Structures and Stress Analysis
- Automotive Engineering
- Engine Technology
- Aerospace Technology and Astronautics
- Nanotechnology and Microengineering
- Control, Robotics, Mechatronics
- MEMS
- Theoretical and Applied Mechanics
- Dynamical Systems, Control
- Fluid Mechanics
- Engineering Thermodynamics, Heat and Mass Transfer
- Manufacturing
- Precision Engineering, Instrumentation, Measurement
- Materials Engineering
- Tribology and Surface Technology

To submit a proposal or request further information, please contact the Springer Editor in your country:

China: Li Shen at li.shen@springer.com
India: Dr. Akash Chakraborty at akash.chakraborty@springernature.com
Rest of Asia, Australia, New Zealand: Swati Meherishi at swati.meherishi@springer.com
All other countries: Dr. Leontina Di Cecco at Leontina.dicecco@springer.com

To submit a proposal for a monograph, please check our Springer Tracts in Mechanical Engineering at http://www.springer.com/series/11693 or contact Leontina.dicecco@springer.com

Indexed by SCOPUS. The books of the series are submitted for indexing to Web of Science.

More information about this series at http://www.springer.com/series/11236

Štefan Medvecký · Slavomír Hrček ·
Róbert Kohár · František Brumerčík ·
Viera Konstantová
Editors

Current Methods of Construction Design

Proceedings of the ICMD 2018

Springer

Editors
Štefan Medvecký
Department of Design and Mechanical Elements, Faculty of Mechanical Engineering
University of Žilina
Žilina, Slovakia

Slavomír Hrček
Department of Design and Mechanical Elements, Faculty of Mechanical Engineering
University of Žilina
Žilina, Slovakia

Róbert Kohár
Department of Design and Mechanical Elements, Faculty of Mechanical Engineering
University of Žilina
Žilina, Slovakia

František Brumerčík
Department of Design and Mechanical Elements, Faculty of Mechanical Engineering
University of Žilina
Žilina, Slovakia

Viera Konstantová
Department of Design and Mechanical Elements, Faculty of Mechanical Engineering
University of Žilina
Žilina, Slovakia

ISSN 2195-4356 ISSN 2195-4364 (electronic)
Lecture Notes in Mechanical Engineering
ISBN 978-3-030-33145-0 ISBN 978-3-030-33146-7 (eBook)
https://doi.org/10.1007/978-3-030-33146-7

This Springer imprint is published by the registered company Springer Nature Switzerland AG
The registered company address is: Gewerbestrasse 11, 6330 Cham, Switzerland

Organize Committee

International Scientific Committee of ICMD 2018

Bošanský Miroslav, Slovak University of Technology in Bratislava (SK)
Dynybyl Vojtěch, Czech Technical University in Prague (CZ)
Dzimko Marián, University of Žilina (SK)
Folta Zdeněk, Technical University of Ostrava (CZ)
Harlecki Andrej, University of Bielsko-Biała (PL)
Hartl Martin, Brno University of Technology (CZ)
Herák David, Czech University of Agriculture in Prague (CZ)
Homišin Jaroslav, Technical University of Košice (SK)
Hosnedl Stanislav, University of West Bohemia (CZ)
Hrček Slavomír, University of Žilina (SK)
Juraszek Janusz, University of Bielsko-Biała (PL)
Kučera Ĺuboš, University of Žilina (SK)
Krzysiak Zbigniew, University of Life Sciences in Lublin (PL)
Lašová Václava, University of West Bohemia (CZ)
Medvecký Štefan, University of Žilina (SK)
Mazal Pavel, Brno University of Technology (CZ)
Němček Miloš, Technical University of Ostrava (CZ)
Němčeková Miroslava, Slovak University of Technology in Bratislava (SK)
Nicoletto Gianni, University of Parma (IT)
Pešík Lubomír, Technical University of Liberec (CZ)
Petrů Michal, Technical University of Liberec (CZ)
Radek Norbert, Technical University of Kielce (PL)
Rusnák Juraj, Slovak University of Agriculture in Nitra (SK)
Ševčík Ladislav, Ph.D., Technical University of Liberec (CZ)
Takeichi Yoshinori, Toyohashi University of Technology (JPN)
Vaňek Václav, University of West Bohemia (CZ)
Vereš Miroslav, Slovak University of Technology in Bratislava (SK)
Živčák Jozef, Technical University of Košice (SK)

Steering Committee Members, University of Žilina (SK)

Slavomír Hrček, email: slavomir.hrcek@fstroj.uniza.sk
Viera Konstantová, email: viera.konstantova@fstroj.uniza.sk
Anna Závodská, email: anna.zavodska@rc.uniza.sk
Milan Benko, email: milan.benko@fstroj.uniza.sk

Review Committee of ICMD 2018

Bašťovanský Ronald, University of Žilina (SK)
Brončeк Jozef, University of Žilina (SK)
Brumerčík František, University of Žilina (SK)
Dekýš Vladimír, University of Žilina (SK)
Dynybyl Vojtěch, Czech Technical University in Prague (CZ)
Dzimko Marián, University of Žilina (SK)
Folta Zdenek, Technical University of Ostrava (CZ)
Gajdáč Igor, University of Žilina (SK)
Grega Róbet, Technical University of Košice (SK)
Herák Dávid, Czech University of Agriculture in Prague (CZ)
Hrček Slavomír, University of Žilina (SK)
Kohár Robert, University of Žilina (SK)
Konstantová Viera, University of Žilina (SK)
Kučera Ĺuboš, University of Žilina (SK)
Kaššay Peter, Technical University of Košice (SK)
Lukáč Michal, University of Žilina (SK)
Madaj Rudolf, University of Žilina (SK)
Medvecká B. Silvia, Technical University of Košice (SK)
Mičian Miloš, University of Žilina (SK)
Němček Miloš, Technical University of Ostrava (CZ)
Nový František, University of Žilina (SK)
Pešík Lubomír, Technical University of Liberec (CZ)
Petrů Michal, Technical University of Liberec (CZ)
Radek Norbert, Technical University of Kielce (PL)
Sládek Augustín, University of Žilina (SK)
Smetanka Lukáš, University of Žilina (SK)
Ševčík Ladislav, Technical University of Liberec (CZ)
Takeichi Yoshinori, Toyohashi University of Technology (JPN)
Vaňek Václav, University of West Bohemia (CZ)
Weis Perer, University of Žilina (SK)

Preface

The book *Current Methods of Construction Design* has been created on the basis of contributions of the 59th International Conference of Machine Design Departments (ICMD 2018). It is the 6th conference in history organized by the University of Žilina, Faculty of Mechanical Engineering, Departments of Design and Machine Elements.

This book is a follow-up to earlier successful books: *Modern Methods of Construction Design*, 2014, and *The Latest Methods of Construction Design*, 2016.

This conference is mainly focused on Machine Design and Optimization, Engineering Analyses, Tribology and Nanotechnology, Additive Technologies, Hydraulics and Fluid Mechanisms, Modern Material and Technology, Biomechanics, Biomimicry and Innovation. The main objective of the conference is to present new results of research and development, to improve the quality of teaching at the universities, as well as to transfer new innovative solutions into practice, in the mentioned topics of the conference. The conference provides space for engineers, Ph.D. students and university educators to discuss their work results and find new inspiration.

The conference was organized under the auspices of Dean of the Faculty of Mechanical Engineering of the University of Žilina, Prof. Dr. Ing. Milan Sága. The Demänovská value, in Slovak Republic, was chosen as a venue for the 59th conference.

Conference History

1960 Brno University of Technology, Brno (CZ)
1962 University of West Bohemia, Pilsen (CZ)
1963 Slovak University of Technology in Bratislava, Bratislava (SK)
1964 Technical University of Liberec, Liberec (CZ)
1965 Technical University of Košice, Košice (SK)
1966 Czech Technical University in Prague, Prague (CZ)

1967 Technical University of Ostrava, Ostrava (CZ)
1968 University of Žilina, Žilina (SK)
1969 University of West Bohemia, Pilsen (CZ)
1970 Brno University of Technology, Brno (CZ)
1971 Technical University of Zvolen, Zvolen (SK)
1972 Slovak University of Technology in Bratislava, Bratislava (SK)
1973 Technical University of Liberec, Liberec (CZ)
1974 Technical University of Košice, Košice (SK)
1975 Brno University of Technology, Brno (CZ)
1976 Czech Technical University in Prague, Prague (CZ)
1977 Czech University of Life Sciences Prague, Prague (CZ)
1978 University of Žilina, Kysucké Nove Mesto (SK)
1979 Technical University of Liberec, Liberec (CZ)
1980 Slovak University of Technology in Bratislava, Smolenice (SK)
1981 Brno University of Technology, Brno (CZ)
1982 Technical University in Zvolen, Šachtičky (CZ)
1983 Technical University of Košice, Michalovce (SK)
1984 Technical University of Ostrava, Podolánky (CZ)
1985 Czech Technical University in Prague, Kladno (CZ)
1986 University of West Bohemia, Pilsen, Pilsen (CZ)
1987 Slovak University of Technology in Bratislava, Bratislava (SK)
1988 Brno University of Technology, Brno (CZ)
1989 Technical University of Liberec, Liberec (CZ)
1990 University of Žilina, Papradno (SK)
1991 Technical University of Košice, Herľany (SK)
1992 Technical University of Ostrava, Tešinske Beskydy (CZ)
1993 Technical University in Zvolen, Zvolen (SK)
1994 Czech Technical University in Prague, Kladno (CZ)
1995 Brno University of Technology Brno, Brno (CZ)
1996 University of West Bohemia, Pilsen, Pilsen (CZ)
1997 Slovak University of Technology in Bratislava, Gabčíkovo (SK)
1998 Technical University of Liberec, Liberec (CZ)
1999 University of Žilina, Beluškė Slatiny (SK)
2000 Technical University of Košice, Herľany (SK)
2001 Technical University of Ostrava, Ostrava (CZ)
2002 Technical University of Zvolen, Bienska dolina (SK)
2003 Czech Technical University in Prague, Prague (CZ)
2004 Brno University of Technology, Blansko (CZ)
2005 Technical University of Liberec, Sedmihorky (CZ)
2006 Czech University of Life Sciences Prague, Prague (CZ)
2007 Slovak University of Technology in Bratislava, Smolenice (SK)
2008 University of West Bohemia, Šumava, Srní (CZ)
2009 University of Žilina, Terchová (SK)
2010 Technical University of Košice, Slovenský Raj (SK)
2011 Brno University of Technology, Mikulov (CZ)

2012 Technical University of Ostrava, Sepetná (CZ)
2013 Technical University of Liberec, Hejnice (CZ)
2014 Technical University in Prague, Beroun (CZ)
2015 Slovak University of Technology in Bratislava
2015 Slovak University of Agriculture in Nitra, Beladice (SK)
2016 University of West Bohemia, Železná Ruda (CZ)
2017 Czech University of Life Sciences Prague, Prague (CZ)
2018 University of Žilina, Demänovská dolina (SK)

Žilina, Slovakia

Štefan Medvecký
Slavomír Hrček
Róbert Kohár
František Brumerčík
Viera Konstantová

Contents

Hydraulics and Fluid Mechanisms

Modern Material and Technology

Biomechanics, Biomimicry and Innovation

Contributors

Pavel Bartonicek Technical University of Liberec, Liberec, Czech Republic

Ronald Bašt'ovanský University of Žilina, Žilina, Slovak Republic

Yunus Özcan Baz Ondokuz Mayis University, Atakum/Samsun, Turkey

Michal Belorit University of Žilina, Žilina, Slovak Republic

Milan Benko University of Žilina, Žilina, Slovak Republic

Peter Bezák Institute of Competitiveness and Innovations, University of Žilina, Žilina, Slovak Republic

Marek Bisták University of Žilina, Žilina, Slovak Republic

Miroslav Bošanský Slovak University of Technology in Bratislava, Bratislava, Slovak Republic

Jozef Brončck University of Žilina, Žilina, Slovak Republic

František Brumerčík University of Žilina, Žilina, Slovak Republic

Eliška Cézová Czech Technical University in Prague, Prague, Dejvice, Czech Republic

Martin Chára Technical University of Liberec, Liberec 1, Czech Republic

Jakub Chmelar Czech Technical University in Prague, Praha 6, Czech Republic

Bahadır Demirel Ondokuz Mayis University, Atakum/Samsun, Turkey

Martin Dolanský Technical University of Liberec, Liberec 1, Czech Republic

Martin Dub Czech Technical University in Prague, Prague 6, Czech Republic

Ján Ďungel CEIT Engineering Services, s.r.o, Žilina, Slovak Republic

Stanislav Ďuriš Slovak University of Technology in Bratislava, Bratislava 1, Slovak Republic

Filip Dušek Technical University of Ostrava, Ostrava, Poruba, Czech Republic

Vojtěch Dynybyl Czech Technical University in Prague, Praha 6, Czech Republic

Marián Dzimko University of Žilina, Žilina, Slovak Republic

Peter Fabian University of Žilina, Žilina, Slovak Republic

Zdeněk Folta Technical University of Ostrava, Ostrava, Poruba, Czech Republic

Adam Furstenzeller Slovak University of Agriculture in Nitra, Nitra, Slovak Republic

Igor Gajdáč University of Žilina, Žilina, Slovak Republic

Tomáš Gajdošík University of Žilina, Žilina, Slovak Republic

Eva Gajdošová University of Žilina, Žilina, Slovak Republic

Martin Gorschenek University of West Bohemia in Pilsen, Pilsen, Czech Republic

Stanislav Grambličká University of Žilina, Žilina, Slovak Republic

Robert Grega Technical University of Košice, Košice, Slovak Republic

Ladislav Gulan Slovak University of Technology in Bratislava, Bratislava, Slovak Republic

Gurkan Alp Kagan Gürdil Ondokuz Mayis University, Atakum/Samsun, Turkey

Andrzej Harlecki University of Bielsko-Biala, Bielsko-Biala, Poland

Jiří Havlík Technical University of Ostrava, Ostrava, Poruba, Czech Republic

David Herak Czech University of Life Sciences Prague, Prague, Czech Republic

Michal Herda University of Žilina, Žilina, Slovak Republic

Michal Hoč University of Žilina, Žilina, Slovak Republic

Juraj Hodál Slovak University of Technology in Bratislava, Bratislava, Slovak Republic

Josef Hodek COMTES FHT a.s., Dobřany, Czech Republic

Jan Hoidekr Czech Technical University in Prague, Prague, Czech Republic

Slavomír Hrček University of Žilina, Žilina, Slovak Republic

Branislav Hučko Slovak University of Technology in Bratislava, Bratislava, Slovak Republic

Lukáš Jančar Technical University of Ostrava, Ostrava, Poruba, Czech Republic

Radka Jírová Technical University of Liberec, Liberec, Czech Republic

Jan Kanaval Czech Technical University in Prague, Prague, Dejvice, Czech Republic

Peter Kaššay Technical University of Košice, Košice, Slovak Republic

Róbert Kohár University of Žilina, Žilina, Slovak Republic

Viera Konstantová University of Žilina, Žilina, Slovak Republic

Daniel Koutný Brno University of Technology, Brno, Czech Republic

Matúš Kovaliček University of Žilina, Žilina, Slovak Republic

Ivo Křístek Technical University of Ostrava, Ostrava, Poruba, Czech Republic

Zbigniew Krzysiak University of Life Sciences in Lublin, Lublin, Poland

Václav Kubec COMTES FHT a.s., Dobřany, Czech Republic

Ĺuboš Kučera University of Žilina, Žilina, Slovak Republic

Petr Kulhavý Technical University of Liberec, Liberec 1, Czech Republic

Petr Lepšík Technical University of Liberec, Liberec 1, Czech Republic

Frantisek Lopot Czech Technical University in Prague, Prague 6, Czech Republic

Aleš Lufinka Technical University of Liberec, Liberec 1, Czech Republic

Michal Lukáč University of Žilina, Žilina, Slovak Republic

Rudolf Madaj University of Žilina, Žilina, Slovak Republic

Maroš Majchrák University of Žilina, Žilina, Slovak Republic

Miroslav Malý Technical University of Liberec, Liberec, Czech Republic

Pavel Maly Czech Technical University in Prague, Prague 6, Czech Republic

Jaromír Markovič Slovak Legal Metrology n. o, Banská Bystrica, Slovak Republic

Rudolf Martonka Technical University of Liberec, Liberec, Czech Republic

Ivan Mašín Technical University of Liberec, Liberec, Czech Republic

Zbigniew Matuszak Maritime Academy in Szczecin, Szczecin, Poland

Martin Mazač Technical University of Liberec, Liberec, Czech Republic

Štefan Medvecký University of Žilina, Žilina, Slovak Republic

Markus Melicher Slovak University of Technology in Bratislava, Bratislava 1, Slovak Republic

Erik Mikuláš Slovak University of Technology in Bratislava, Bratislava 1, Slovak Republic

Václav Mosler Technical University of Ostrava, Ostrava, Poruba, Czech Republic

Pavel Mossoczy Czech Technical University in Prague, Praha 6, Czech Republic

Milos Nemcek Technical University of Ostrava, Ostrava, Poruba, Czech Republic

Aleksander Nieoczym Lublin University of Technology, Lublin, Poland

Ondřej Novák Technical University of Liberec, Liberec, Czech Republic

Michael Oeljeklaus Škoda Auto a.s. Mladá Boleslav, Mladá Boleslav, Czech Republic

Łukasz J. Orman Kielce University of Technology, Kielce, Poland

Jakub Palenčár Slovak University of Technology in Bratislava, Bratislava 1, Slovak Republic

Rudolf Palenčár Slovak University of Technology in Bratislava, Bratislava 1, Slovak Republic

David Paloušek Brno University of Technology, Brno, Czech Republic

Libor Pantělejev Brno University of Technology, Brno, Czech Republic

Peter Pavlásek Slovak University of Technology in Bratislava, Bratislava, Slovak Republic

Jan Pavlík Technical University of Ostrava, Ostrava, Poruba, Czech Republic

Lubomír Pešík Technical University of Liberec, Liberec 1, Czech Republic

Karel Petr Czech Technical University in Prague, Prague 6, Czech Republic

Michal Petrů Technical University of Liberec, Liberec 1, Czech Republic

Vítězslav Pfliegel Technical University of Liberec, Liberec, Czech Republic

Jacek Pietraszek Cracow University of Technology, Cracow, Poland

Pavol Podhora Žilinská Univerzita v Žiline, Žilina, Slovak Republic

Roman Polák University of West Bohemia in Pilsen, Pilsen, Czech Republic

Silvester Poljak Žilinská Univerzita v Žiline, Žilina, Slovak Republic

Antonín Prantl COMTES FHT a.s., Dobřany, Czech Republic

Adam Przemyk University of Bielsko-Biala, Bielsko-Biala, Poland

Norbert Radek Kielce University of Technology, Kielce, Poland

Tomáš Riegr Technical University of Liberec, Liberec, Czech Republic

Juraj Rusnák Slovak University of Agriculture in Nitra, Nitra, Slovak Republic

Jan Rybář Slovak University of Technology in Bratislava, Bratislava 1, Slovak Republic

Martina Ryvolová Technical University of Liberec, Liberec, Czech Republic

Milan Sapieta University of Žilina, Žilina, Slovak Republic

Robert Sásik University of Žilina, Žilina, Slovak Republic

Ladislav Ševčík Technical University of Liberec, Liberec, Czech Republic

Dana Šišmišová Slovak University of Technology in Bratislava, Bratislava 1, Slovak Republic

Rudolf Skyba University of Žilina, Žilina, Slovak Republic

Juraj Slovák Slovak University of Technology in Bratislava, Bratislava 1, Slovak Republic

Lukáš Smetanka University of Žilina, Žilina, Slovak Republic

Patrik Sniehotta Technical University of Ostrava, Ostrava, Poruba, Czech Republic

Peter Spišák University of Žilina, Žilina, Slovak Republic

Pavel Srb Technical University of Liberec, Liberec 1, Czech Republic

František Starý Czech Technical University in Prague, Prague, Dejvice, Czech Republic

Ján Šteininger University of Žilina, Žilina, Slovak Republic

Marián Stopka University of Žilina, Žilina, Slovak Republic

Michal Stupavský University of Žilina, Žilina, Slovak Republic

Roman Svoboda BAG DEVELOPMENT s. r. o, Mladá Boleslav, Czech Republic

Yoshinori Takeichi Toyohashi University of Technology, Hibarigaoka, Tempakucho, Toyohashi-shi, Aichi, Japan

Dariusz Tarnapowicz Maritime University of Szczecin, Szczecin, Poland

Szymon Tengler University of Bielsko-Biala, Bielsko-Biala, Poland

Radek Teply University of West Bohemia, Pilsen, Czech Republic

František Tóth Slovak University of Agriculture in Nitra, Nitra, Slovak Republic

Miroslav Trochta Technical University of Ostrava, Ostrava-Poruba, Czech Republic

Michal Tropp University of Žilina, Žilina, Slovak Republic

Matej Urbanský Technical University of Košice, Košice, Slovak Republic

Tomáš Vach Technical University of Liberec, Liberec, Czech Republic

Ján Vachálek Slovak University of Technology in Bratislava, Bratislava 1, Slovak Republic

Ha Van Nguyen Technical University of Liberec, Liberec, Czech Republic

Václav Vaněk University of West Bohemia in Pilsen, Pilsen, Czech Republic

Pavol Vašek Slovak University of Technology in Bratislava, Bratislava 1, Slovak Republic

Martin Vicen University of Žilina, Žilina, Slovak Republic

Josef Vosáhlo Technical University of Liberec, Liberec, Czech Republic

Petr Votápek University of West Bohemia, Plzeň, Czech Republic

Michal Vrabec University of Žilina, Žilina, Slovak Republic

Peter Weis University of Žilina, Žilina, Slovak Republic

Jiří Začal Technical University of Ostrava, Ostrava-Poruba, Czech Republic

Stanislav Zeman Slovak University of Technology in Bratislava, Bratislava 1, Slovak Republic

Josef Zvoníček Brno University of Technology, Brno, Czech Republic

Machine Design and Optimization

Design Optimization of the Octavia SK37x Driver's Seat—DFMA Principle Application

Pavel Bartonicek and Ivan Mašín

Abstract The article describes the optimization of the driver's seat adjustment mechanism. Technical, functional and customer requirements for the seat were laid. The original seat was analyzed using: DFA analysis, complexity calculation, theoretical efficiency calculation, process flow diagram. An analysis of existing production technology and a rated optimization potential has been done. TRIZ analysis was performed. We looked for system elements that could be substituted by a supersystem and then replaced. The goal was to simplify the mechanism, to reduce the number of elements while preserving the function. Low-pressure casting and sheet forming technologies and welding were compared. Four optimization options have been developed using the FEM. The load data measured by the real crash test and provided by the manufacturer were used as input data. Innovated parts had to meet these load criteria. It is a condition of the innovated driver's seat mechanism to show the same or better resistance to the existing mechanism in the case of a passenger car accident. Customer safety has to be preserved or improved. An economic assessment has been made. Based on this, the best optimization option was selected and presented. The economic return, the complexity of production, the most suitable technology, the potential for further optimization were taken into account.

Keywords DFMA · FEM · DFA analysis · TRIZ · Process flow diagram · Complexity · Economic evaluation · Driver's seat · Optimization · Customer ratings

1 Introduction

The passenger car seat has undergone permanent changes since the early nineteenth century when the first car was invented. As time passes, it adds additional features to meet the needs of increasingly demanding customers and ever-tougher legislation. The primary feature of the seat was to provide users with driving support and now performs many functions, namely the passenger cools in the summer season, heats

---

P. Bartonicek (✉) · I. Mašín
Technical University of Liberec, Studentská 2, 46117 Liberec, Czech Republic
e-mail: pavel.bartonicek@gmail.com

Š. Medvecký et al. (eds.), *Current Methods of Construction Design*, Lecture Notes in Mechanical Engineering, https://doi.org/10.1007/978-3-030-33146-7_1

up in the winter, protects against lateral impact, allows ergonomic seating and, last but not least, has the ability to dampen vibrations. The contemporary seat is made up of a supporting frame with a foamed polyurethane foam and a wire reinforcement. Comfort increases height adjustment, lumbar support and headrest. The parts of the seat are fastened to each other by welds or clips, forming a body of different weights and interacting with each other. From the point of view of the designer, the seat is a challenge because in the product life management process periodic innovations to maintain competitiveness compared to changes in market orientation and ever-increasing technical demands are being introduced as legislation in place for destination sales. The current trend in the automotive industry is the shift from robustness to relieving, which is the reason behind efforts to save weight and thus reduce emissions and CO_2 content, in an effort to maintain both the functionality and benefit of the design. The parts are thinner, the materials change, "unnecessary" value that the customer does not appreciate. As a sign of this trend, we have optimized the selected part of the SK37x Octavia car seat, namely the feed mechanism, because it has a defined number of parts and has the potential to innovate when applying DFMA principles.

2 Description of the Original Longitudinal Displacement of the Driver's Seat Octavia Car

The seat (see Fig. 1) consists of a seat and a backrest. The pipe construction is coupled to the tipper by means of a tilting mechanism for tilting the backrest. The seat is attached to the passenger car floor by a height adjustment mechanism. The lowermost part is the longitudinal displacement of the seat, which provides a longitudinal

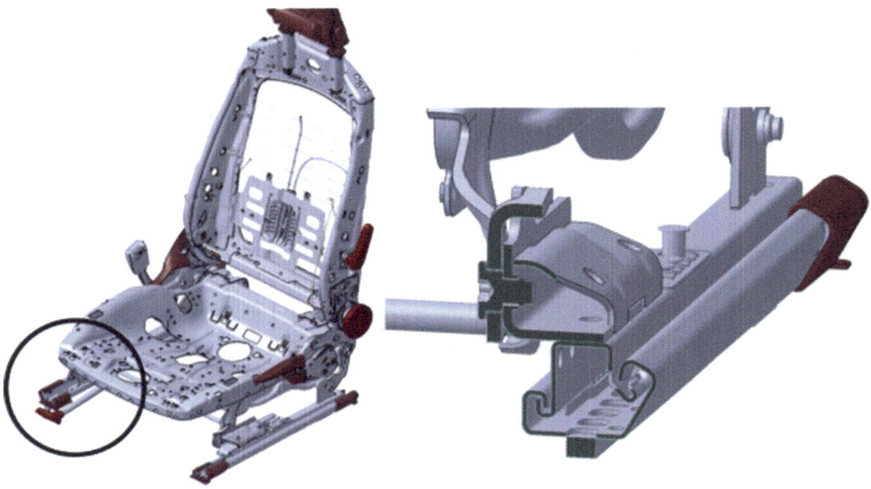

Fig. 1 A driver seat frame without PUR foam, detail of the undercarriage mechanism

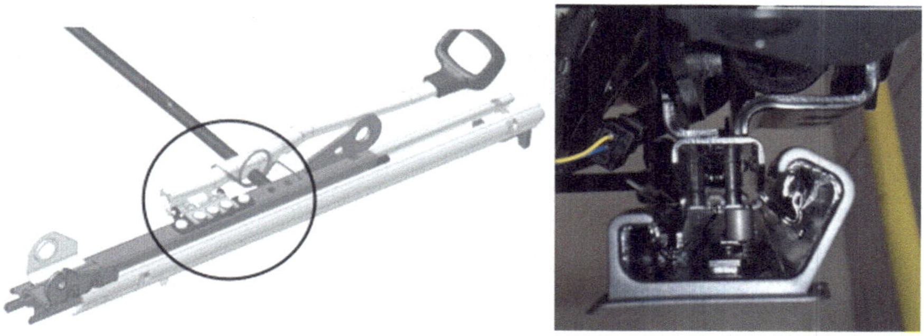

Fig. 2 A partial shift mechanism, detail of locking pins

displacement. The seat is attached to the body by a screw connection (4 pcs). The seat and backrest further consist of PUR foam laid on a support consisting of a perforated steel support. This extends the driver's weight to the entire area and creates comfort for the driver's seat. Tubular structure and perforated steel sheets are fixed to the seat frame by a welded joint.

The base of the passenger car seat (see Fig. 2) is made of metal profiles. The lower profile fulfills two functions: it fixes the driver's seat to the car body and at the same time provides a longitudinal displacement (sliding element + ball rolling). A complete saddle of the driver's seat and the seat belt attachment are attached to the top profile of the undercarriage mechanism. The undercarriage mechanism is controlled by a lever with an ergonomic handle. When moving upwards, the locking pins are unlocked and the top bar can move. Locking is active when the control lever is released, the locking is possible in small steps, depending on the position of the locking pins and the opposing holes in the lower bar. The control torque from the lever with the connecting rod is transmitted to the other half of the mechanism. This second half of the seat travel mechanism is the same, mirror-inverted.

Functional analysis [1] was carried out with the help of data provided by the car manufacturer Škoda Auto a.s. The individual parts were marked and numbered. Basic parts and parts that are not primary, i.e., subparts, have been identified and can be modified, replaced and reduced. These are fasteners, connectors, wires, washers. Liquids such as lubricants, adhesives, sealants and the like have not been included in the disintegration. The number of part interactions has been identified with each other. The goal is to increase complexity (see A TRIZ analysis [2]) has been carried out which has shown the simplification potential and seat parts that can be replaced by replacing parts in the over-assembly with the function of parts in the subassembly. However, the function of the longitudinal sliding mechanism of the seat has been retained.

$$\mathrm{CF} = \sqrt{\mathrm{TNP} \times \mathrm{TNI}} = \sqrt{26 \times 69} = 42.3 \tag{1}$$

The DFA calculation of complexity factor (1), where CF is complexity factor, TNP is total number of parts and TNI is total number of mutual interactions, introduces a parameter that has been used to evaluate another new, upgraded variant of the undercarriage.

$$\mathrm{TE} = \frac{\mathrm{TMNP}}{\mathrm{TNP}} * 100 = \frac{3}{26} * 100 = 12\% \tag{2}$$

Theoretical efficiency of the original variant (2) is calculated by using the above-mentioned formula, where TE is theoretical efficiency, TMNP is the theoretical minimum number of parts and TNP is the total number of parts introduced by the second evaluation parameter.

Figure 3 shows that there are three essential parts in the assembly, without the assembly being a must and serving as a platform. The rest of the parts are theoretically replaceable, or their function can take a larger whole. There are 25 parts in total. By computing from Eqs. (1) and (2), we concluded that the theoretical calculation of efficiency based on the number of parts is 12%. This efficiency should be higher than 60% according to the recommendation [3, 4]. Here, in our case, we are limiting restrictions, i.e., we must not reduce passive safety and we must maintain the overall cross section of the locking pins that are calculated to absorb the impact in the event of a passenger car accident. These are values calculated from the mathematical model of the seat and verified with real vehicle crash tests.

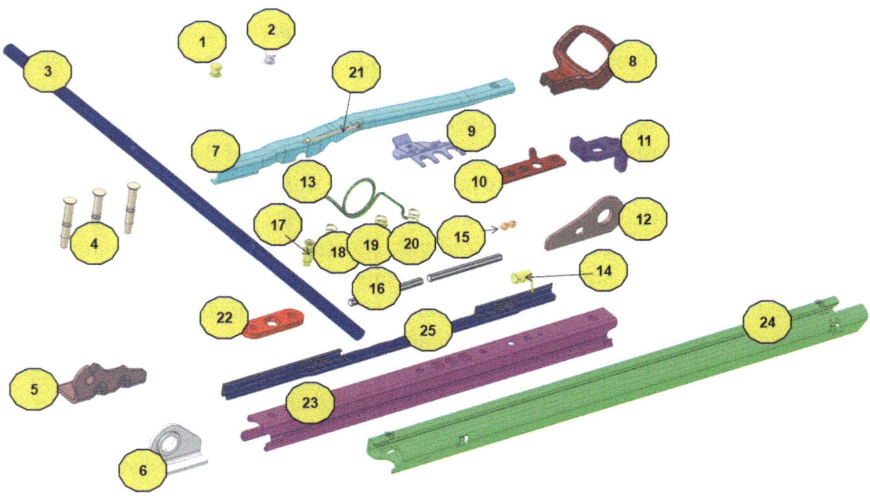

Fig. 3 A falling parts of the seat travel mechanism

3 Innovation of the Longitudinal Displacement of the Driver's Seat

3.1 Increasing the Complexity of the Seat Mechanism

The first step of optimization was to increase the complexity of the seat mounts and the safety belt holder, by reducing the number of parts No. 6 and No. 5 (safety belt holder). The parts were incorporated together, the number of parts was reduced, the assembly operations were simplified, the production flow accelerated, the material economy was simplified. The seat was simplified and the complexity index increased.

3.2 Innovation of the Overall Aretation Mechanism

The locking mechanism (see Fig. 4) has been structurally simplified. The locking pins (No. 4) have been replaced by a locking profile that takes over the function. Three arresting springs (No. 4) were reduced and the control function took over the single, central spring of the "ridge" part (No. 9).

The part (No. 9) was created by bending and shearing technology from steel sheet metal. By optimizing, the material was saved and the production was abducted. The locking pins (see Fig. 6) proved unnecessarily complicated. The mounting support

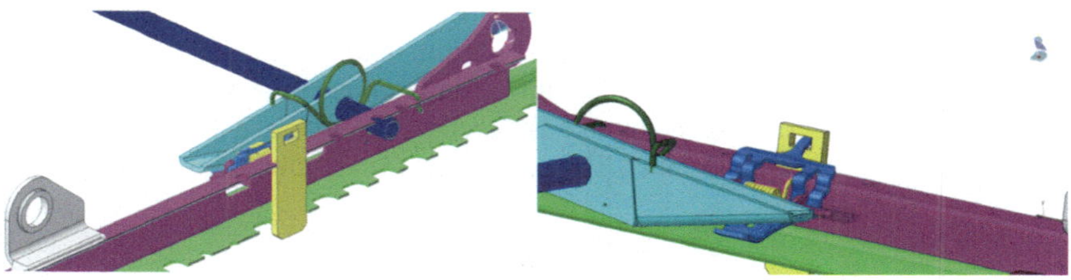

Fig. 4 A cut through the innovative lock mechanism, detail of the control of the comb part (No. 9)

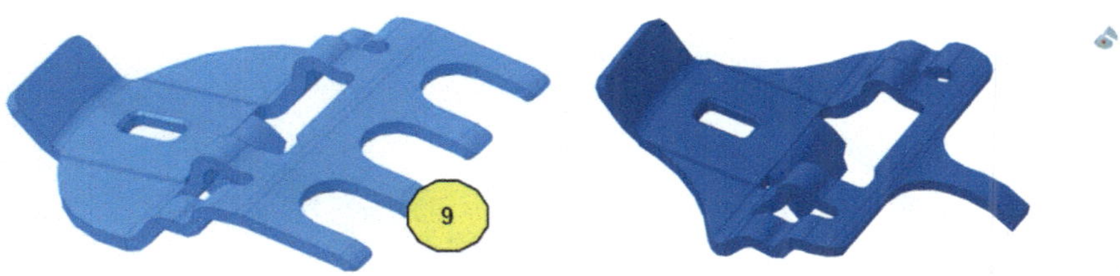

Fig. 5 Original part of the ridge (No. 9) and the new innovated part

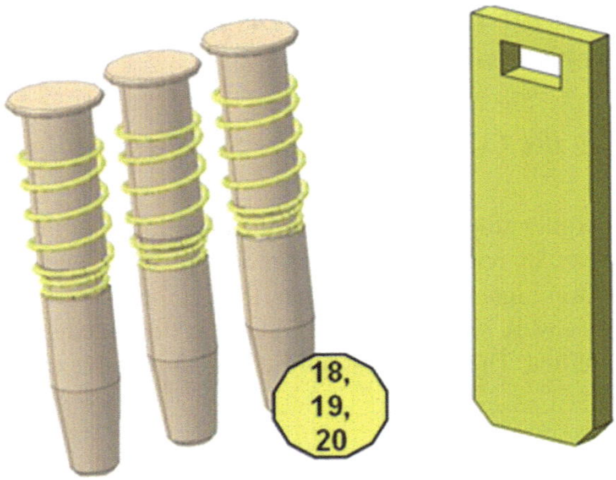

Fig. 6 Original parts and the new innovated one

consisted of manipulating the pins, applying the springs and pressing into the undercarriage profile of the seat. The springs filled the holding function and facilitated control. Overall, the surgery was assessed as difficult and lengthy. In the first step, the number of locking pins was reduced. In the second step, the number of springs was reduced. By using the TRIZ method, three central springs functioned as one central spring, which had to be reinforced (No. 14) (Fig. 5).

If we consider that the material of the optimized pin will be identical to the material of the locking pins, then we can proceed with the consideration that the cross section of the new part must be the same as the cross section of the original pins. The pin has a diameter $d = 8$ mm, i.e., cut with a 50 mm^2 area. For three pins, the cross section is 150 mm^2. The new stud designed by us has a rectangular shape of 30×6 mm, i.e., a cut with a surface area of 175 mm^2. The conditions (3) of safety must be respected.

$$S_{\text{original}}\ (\text{mm}^2) \leq S_{\text{innovated}}\ (\text{mm}^2) \tag{3}$$

Simplify the construction of the lock lever. The original locking lever (see Fig. 7)

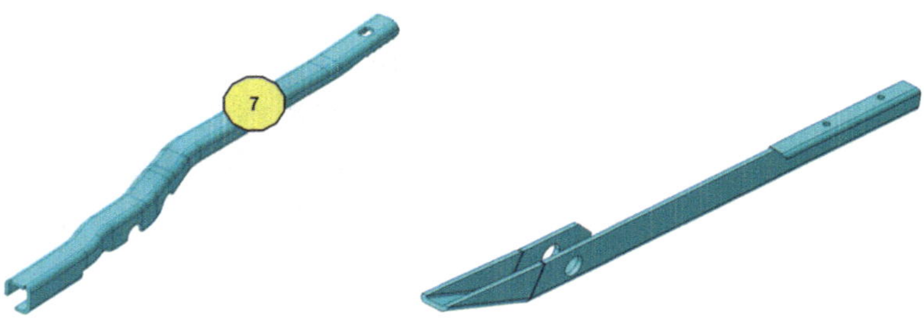

Fig. 7 Original lever lock control lever, the new innovative lever

is formed by pressing a sheet of 2.5 mm and then bent so as to get past the drive tracks. The shaft (No. 3) is secured by welding. In the final operation, the handle is attached to it (No. 8). The original lever is a complicated production and the bending is a result of the shape of the handle the designer tried to get optically into the center of the seat, away from the undercarriage profile. The original lever is a molding with a bending operation in several planes, with the operation of the cuts and the need to fix the finishing of the assembly and subsequent welding to the workpiece. Our innovative solution is based on the finding that the arresting lever is only stressed in one direction, not dynamic with a high overload, no greater than $F_{max} = 57$ N (Octavia SK37x Lock Release Act). A new design is a simpler, reinforced construction at the point of transferring the cleansing force, on the contrary weakened in places that are not stressed.

Punching operations have been minimized. Poka Yoke solution, a notch to prevent misalignment of the part insertion (No. 3), while fixing the part for the next operation, joining the workpiece by welding, thus preventing clamping and welding costs. Surface parts Ra = 1.6 because it is a visible part, surface treated by painting with car surface finish, the preferred appearance here. The cost of molding is therefore lower than the original solution and the number of cuts has been reduced, in particular, the multi-plane bending has been removed so that the locking lever handle extends along the longitudinal *Y*-axis from the right side of the drive mechanism. The new design has eliminated this handicap, and the design has been simplified. Installation is simpler, orientation of insertion in one direction. Reliability and lifetime presumed throughout the life of the car are not stressed by dynamic forces or other significant flying forces. Disassembly is not due to weld joints. The overall balance of savings is high, the potential for further increasing the degree of optimization.

We used the MKP method to compare the two parts, the original lever (No. 7) and the newly modeled levers. The material is structural steel, weldable, unalloyed steel of class 11. These are designed for less stressed parts, construction, tensile strength Rm is 470–630 MPa, yield strength Re about 355 MPa. As a stress, we consider double the force used to unlock the seat, i.e., $F_{max} = 50$ N (source of Škoda Auto a.s.). Higher stress has no justifiable reason, the normally loaded 75-kg driver seat is unlocked with the force of $F \approx 40$ N. Using finite element method (FEM) (see Fig. 8), we evaluated both parts according to the HMH hypothesis and maximum deformation.

The stress of both models (original, new) is therefore assumed by the force $F_1 = 100$ N, which on the longer arm causes the moment $Mk_1 = 23.6$ Nm and shorter arm, after the pin $Mk_2 = 6.9$ Nm, This applies to the original lever. The new lever (see Fig. 9) is shorter, the U-shape is replaced by L-shape due to material savings and production time. Here, the force $F_1 = 100$ N causes torque load $Mk_1 = 25.9$ Nm and $Mk_2 = 6.4$ Nm.

The TRIZ methodology has shown that the next transition to a higher level (theory of "S" innovation curves) is the abandonment of mechanical control and the replacement of the lock, for example, by the principle of electromagnetism (see Fig. 10).

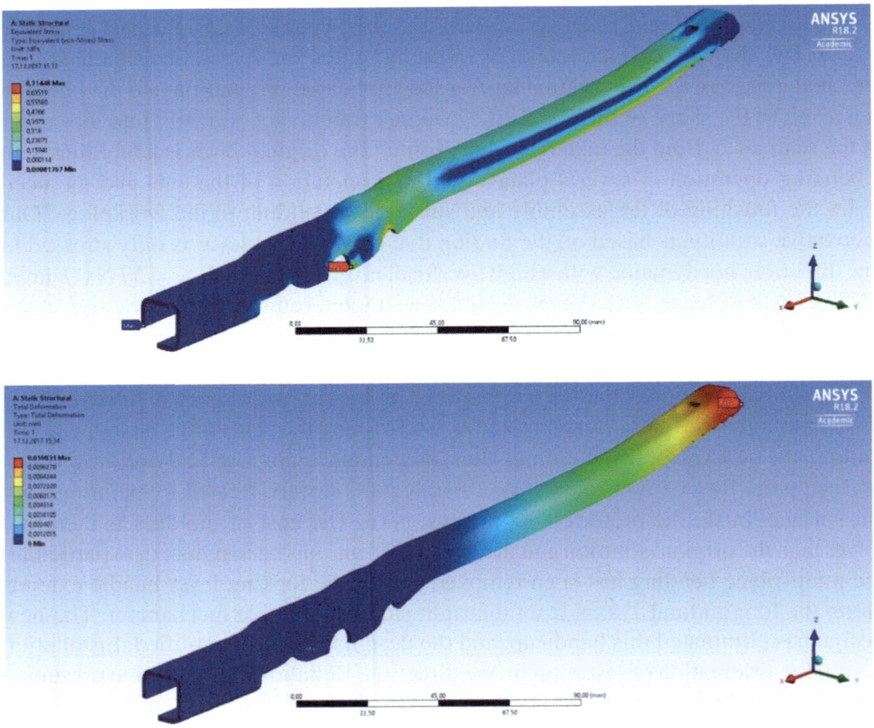

Fig. 8 Original lever, HMH hypothesis and max. deformation

4 Conclusion

We have carried out a study of the optimization of parts of the undercarriage mechanism of the passenger car seat. We conducted the application in line with the DFMA

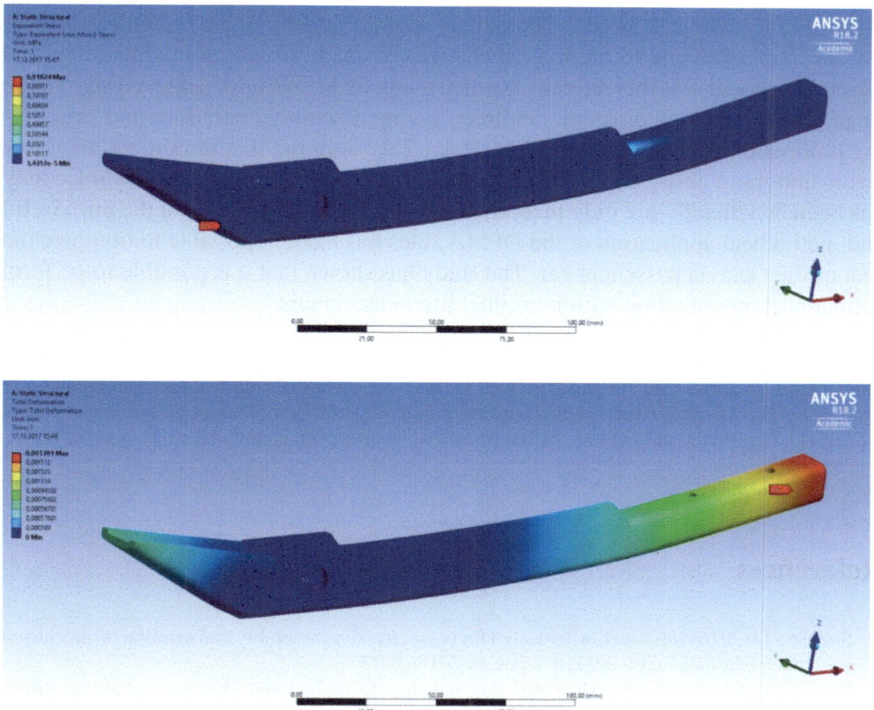

Fig. 9 New innovated lever, HMH hypothesis and max. deformation

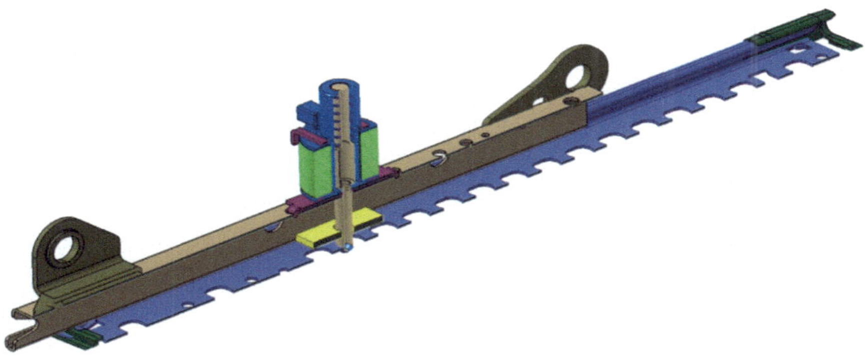

Fig. 10 Electromagnetic principle of arresting solution

methodology, which showed the potential for optimization. At places with the potential for innovation and technological change, we made structural modifications and the design model was theoretically verified using FEM. We have achieved significant financial savings both direct and indirect, saving assembly operations and simplifying production. We have introduced a Poka Yoke solution that minimizes assembly errors and more easily defines workflows. From our point of view, the whole study has been beneficial—not only theoretical—because it has shown that the purposeful and methodical application of the DFMA rules has made it possible to optimize the seat of the Octavia passenger car. The study has shown that it is possible to perform further seat optimization within product life management.

Acknowledgements I would like to thank the development department of a car manufacturer Škoda Auto a.s. for the data provided as well as thank the technical collaborators The Institute for Nanomaterials, Advanced Technology and Innovation (CxI) TUL, for technical assistance, namely Doc. Ph.D. Ing. Petrů, and Institute of Industrial Engineering IPI, namely Doc. Dr. Ing. Mašín.

References

1. Stienstra D (2016) Introduction to design for (cost effective) assembly and manufacturing. http://me.gatech.edu/files/capstone, last accessed 2018/07/13
2. Princip návrhu výrobku. Katedra částí a mechanismů strojů, Technická Univerzita Liberec, http://www.kst.tul.cz/podklady/principy_navrhu/Preklad_upraveny.pdf. Last accessed 2018/03/19
3. Boothroyd G, Dewhurst P, Knight WA (1989) Dewhurst product design for manufacture and assembly. Marcell Dekker, Inc.
4. Giddings, Lewis (1997) Simultaneous engineering study of phase II injector assembly line

Description of the Methodology of the Automated Optimization Process

Michal Belorit, Slavomír Hrček and Marek Bisták

Abstract The article deals with optimization methods, their basic properties and the division. Particular emphasis is placed on numerical methods that are suitable to optimize the problem with a large number of design variables and are relatively robust. The methodology by which a construction can be optimized by the automated process is also described. The process consists of linking commercial programs, namely Creo Parametric, ANSYS Workbench and MATLAB. The program Creo secures cad data; ANSYS Workbench serves to calculate FEM simulations; and MATLAB manages an optimization process based on inputs and outputs and creates the interface between programs and provides the flow of information. Also, the article describes how to create the link between these programs. Thanks to this connection, it is possible to automate the optimization process by creating a program in MATLAB environment and also to select and implement a suitable optimization algorithm according to the optimization task.

Keywords Automated optimization · Optimization methods · FEM analysis

1 Introduction

The technical problems of the present are becoming very complex and demanding to find suitable solutions, especially due to limitations and requirements resulting from ecology, production efficiency, energy sources, etc. The solution often becomes more difficult also because of frequent conflicts of requirements, such as increasing the energy efficiency and reliability of technical devices while simultaneously reducing weight [1]. Along with it, an optimal solution is necessary to be obtained in the shortest possible time. Therefore, it is important to make adequate use of all available computing capabilities.

An important part of every field of development and innovation is the optimization process. It is a step toward achieving a workable solution that is the best of all possible solutions for the goal.

M. Belorit · S. Hrček (✉) · M. Bisták
University of Žilina, Univerzitná 8215/1, 01026 Žilina, Slovak Republic
e-mail: slavomir.hrcek@fstroj.uniza.sk

Š. Medvecký et al. (eds.), *Current Methods of Construction Design*, Lecture Notes in Mechanical Engineering, https://doi.org/10.1007/978-3-030-33146-7_2

From a mathematical point of view, optimization can be considered as the discipline in which the extreme, a minimum or a maximum, of the given function $F(x)$ is searched in the given set M. This function can be called an optimization or purpose function. And the set M means the area for which the correct solution can be found [2].

Optimization is a concept that is beginning to emerge very frequently, especially in the last two centuries, but the roots of finding the optimal solution go into the deep past. And virtually every area of human activity and decision-making includes optimization problems and tasks. An optimization task can be compiled from these problems and consequently obtain a sufficient solution using suitable optimization methods and algorithms [3].

Many of the optimization methods were mathematically described by personalities such as Isaac Newton, Leonard Euler, or Joseph Luis Lagrange. However, the emergence of relatively complex but effective optimization methods and their justifiable use was made possible by the introduction of computing technique. It has brought powerful tools into every field of technical development. Modern engineering has also adapted a rational design approach, called structural optimization, which is based on the application of modern computing equipment and appropriate software [4].

2 Methods and Materials

2.1 Basic Types of Optimization Tasks

According to Rosinova [5], optimization tasks are possible to divide from different perspectives into the following basic types.

According to the character of the system model:

- Static optimization tasks—the model deals with steady states, no time dependencies.
- Dynamic optimization tasks—the model includes time dependencies, the dynamic system is optimized, and the role of this type is also referred to as the role of the optimal management.

Depending on the type of variable and purpose function:

- Continuous tasks—in general, tasks whose variables are from an infinite set and the purpose function is continuous. They are mostly less demanding for calculation, due to the course of the respective functions and boundaries at a particular point where the behavior of nearby functions can be assumed.
- Discrete tasks—the variables get values from the final set, and the task solution can be complicated by the behavior of the purpose function and the boundary when passing from one point of the set of permissible solutions to the other point.

According to the character of the set of permissible solutions:

- Tasks without boundaries.
- Tasks with boundaries—the set of permissible solutions are defined by the given boundaries and the variables must meet them.

Depending on the character of the solution (extremum):

- Global optimization—searches for global optimum or global extremum.
- Local optimization—the local extreme of the purpose function is searched.

2.2 Classification of Optimization Methods

Depending on the type of an optimization task, a suitable method is chosen for its solution. In some cases, the optimization task can be solved by analytical methods. Those can achieve the exact solution. However, the complexity of a task does not allow such methods to be used to obtain a solution. Therefore, numerical methods are used in practical optimization tasks.

In the literature, optimization methods are classified differently, especially depending on the focus of the particular work. And oftentimes, it is possible to find that one method is in two different works included in uneven classification groups. But the basic classification is:

Analytical methods. They are suitable for tasks defined in the analytical form because searching for extremes by these methods is based on the derivation calculation. They deal with one-dimensional or multidimensional tasks with or without boundaries.
Numerical methods. The iterative methods in which the optimal value of the respective purpose function F searches around the starting point by successively moving in individual iterative steps to points with a more advantageous value of function $F(x)$ until the determined tolerance of the solution is reached. The tolerance means the value defining the criterion for termination of the optimization algorithm.
Deterministic methods. These methods are characterized in particular by the fact that the search procedure or the method of selecting the direction and the length of the step is not random. On the contrary, there are mathematically defined precise procedures for the given method on how to approach these two parameters. Gauss-Seidel, Hooke-Jeeves method, the method of Nelder and Mead, and Newton method are the most known deterministic methods.
Stochastic methods. These are non-deterministic methods that are characterized by the unrepeatability of the extremal searching process due to the application of the randomly generated vector in the iteration cycle. So they are able to solve multidimensional optimization tasks and search the global optimum, but sometimes because of a large number of analyzes, reanalysis, the length of the solution reaches unrealistic time. They can be labeled as universal methods applicable to any optimization tasks with or without boundaries. In many stochastic methods, inspiration is given

to living or non-living nature [6]. For example, the following methods belong to the stochastic optimization category and do not require whole optimization function just its values:

- *Monte Carlo method*—is method that only requires information about the functional values of the target function and on restrictive conditions for the search for optima. It is possible to get global optimum with this method, and it is relatively easy to program its algorithm. It is characterized by the fact that each self-performed optimization process with the same starting point can be represented by a sequence of other points, and even in some special cases, it can proceed to a different optimum [4].
- *Blind algorithm* is a basic stochastic algorithm that repeatedly generates a random solution from a given domain and remembers it only if it was better than the solution already recorded in the previous history of the algorithm. In general, we can say that the blind algorithm does not contain any strategy of solution design based on the previous algorithm history. Each solution is built independently of previous solutions. It records a solution that provides the lowest functional value for the time the procedure is activated. When the procedure is terminated, this solution is an output parameter.
- *Hill climbing algorithm* the blind algorithm can be simply generalized to the hill climbing algorithm where the best local solution is iteratively searched for in a certain area, and this solution is used as the center of the new area in the next step. The basic idea is that, given a particular solution, we build a random number of new solutions so that variables are randomly changed in the chosen solution (the chosen solution is the center of the randomly generated solution). From this area, the best solution is chosen and will be used in the next iteration step as the center of the new area. This process is repeated as many times as is prescribed at the beginning and the best solution that occurred during the algorithm history is recorded.
- *Tabu search*—eliminates the main drawback of the climbing algorithm. After a certain number of iterative steps, the climbing algorithm often returns to the local optimal solution that has already occurred in its previous course. The "short-term memory" is introduced into the climbing algorithm that remembers for a short interval of the previous history of the algorithm solutions that provide local optimal solutions. Hence, the algorithm is able to avoid stuck in local extremum. This modified climbing algorithm systematically scans the entire area of the solution in which we look for a global optimum of a function.
- *Simulated annealing*—a global optimization technique that goes through the search area by testing random mutations on a single solution. A mutation that enhances fitness is always accepted. Mutations that reduce fitness are probabilistically accepted on the basis of the fitness value difference and the descending stage parameter. In jargon, simulated annealing is the search for the lowest energy instead of maximum fitness. This technique can also be used in standard genetic algorithms by simply starting with a relatively high degree of mutation that reduces excess time beyond the intended range.

- *Genetic algorithm*—this is a search method used in computer science to find approximate solutions for optimization and search issues. Genetic algorithms are a special set of evolutionary algorithms that use evolutionary biology-inspired processes such as inheritance, mutation, natural selection, and recombination. They are mostly implemented as a computer simulation in which a population of abstract images, called chromosomes, of candidate solutions, called individuals, of the optimization problem, lead to better solutions. The solution is traditionally represented as binary strings, but a different coding method is also possible. The evolution starts with a population of completely random individuals (generation). In each generation, the fitness of the entire population is evaluated, multiple individuals are randomly selected from the current population (based on their fitness) and modified (mutated or crossed) to create a new population that becomes the starting point of the next iterative step of the algorithm. In genetic algorithms, an individual or partial solution to the problem, called a chromosome (also called a genotype), is represented by a set of parameters that define the proposed solution to the problem the genetic algorithm is trying to solve. Chromosomes are commonly defined as simple data and instruction strings, but a large number of other data structures can also be used to store chromosomes [7].

3 Results

The methodology requires to determine conditions and requirements of optimization problem that will define an optimization task. An overview and basic characterization of various optimization techniques, especially those that are suitable for design use and potentially can be applied to solve the optimization task, is essential. The optimization process is schematically depicted in Fig. 1 and consists of next three stages:

1. The basic part, which is a 3D virtual geometric model, can be processed in any commercial CAD system that has support for Workbench Associative Interface in ANSYS Workbench. Creo Parametric, Inventor, NX, and SolidWorks are all suitable. The CAD model is necessary to appropriately modify to avoid interference between the individual parts or the geometry itself that can be caused by altering the parameters and thus to fail the model. Model Simplification is necessary to do before FEM software calculation. This will accelerate calculation process but calculating reliability will be maintained and that is essential.
2. An interactive link between the CAD system and the FEM software is created. This means that the FEM software is able to change the values of the geometry parameters and adapt the geometry to allow for different combinations of design variables to be investigated in the optimization task. All boundary conditions, contacts, the appropriate model network, and other necessary features to obtain individual solutions are defined. As FEM program, ANSYS Workbench is

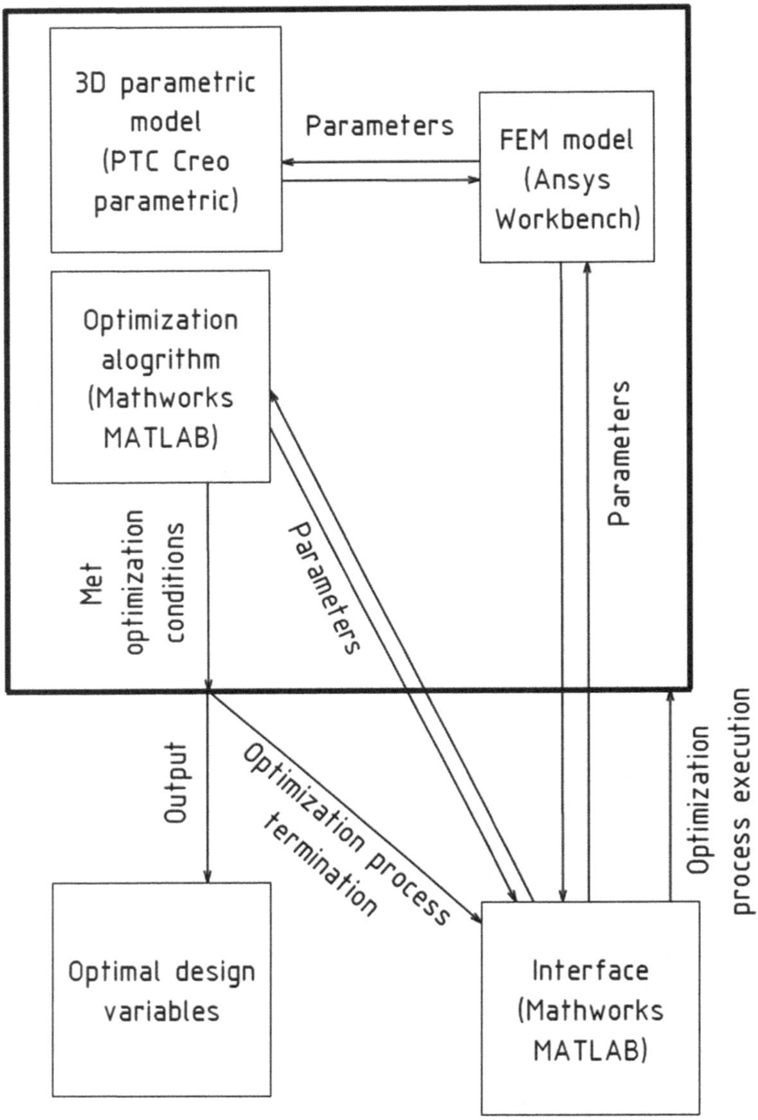

Fig. 1 Schematic of optimization process

chosen, because of its mentioned linkage with CAD systems and scripting possibilities. It uses Python as scripting language. And, the commands can be sent to the Workbench, thanks to MATLAB toolbox called ANSYS aaS, via MATLAB environment. Because of that, it is possible to programmatically control design points and reach the outputs of the FEM analysis. FEM software is actually the purpose function of the optimization task.

3. The optimization task is not defined analytically. Hence, it is necessary to choose a suitable optimization method that uses only the functional values of the individual design points. The optimization itself is implemented in the MATLAB computing environment of MathWorks. This program has its own optimization package. However, if none of offered built-in optimization methods conform requirements, optimization algorithm still can be programmed manually.

The important part is the interface, created in MATLAB too, that serves as a control unit. It is in charge of optimization process execution with estimated design variable values. When FEM solver is done, information and solution are passed via the interface to the optimization algorithm. The algorithm evaluates the solution. If the conditions for termination of the optimization process are met, an output with information about optimal design variables is created and the process is terminated. If not, the optimization algorithm determines new design variables and the interface sends them back to FEM software and runs the solver again with new values. This repeats until the global optimum is not found or a number of iterations reach its limitation or another criterion is met.

4 Discussion

There is a lot of software for static structural analysis. Even many CAD software already have built-in modules for this type of analysis and are often associated with optimizing the design. However, they are commonly limited and do not offer the user the required options. Hence, in such a situation, it is appropriate to choose a combination of suitable software available to take advantage of them. But it is important to ensure that these programs can be linked together. The above-mentioned ANSYS Workbench is guaranteed to provide interactive geometry reading connections with the most commercially known CAD systems. And MATLAB software, because of the toolbox, makes it possible to control the ANSYS Workbench from the environment of MATLAB itself. And it is also a powerful data processing and visualization tool, as well as programming and controlling the entire optimization algorithm. Therefore, by creating a robust program, it is possible to automate such an optimization process and customize it.

5 Conclusion

The paper in the introduction describes the need to use optimization techniques to mechanical design. In the following section, it deals with the description of basic types of optimization tasks and the basic division of optimization methods. Subsequently, the selected numerical methods are described. And later, the methodology of an automated optimization design process, based on the linkage of commercial software, is described. Where Creo Parametric provides CAD data, ANSYS Workbench serves to calculate FEM analyzes and MATLAB serves as an interface and simultaneously managing optimization. Thanks to this connection, it is possible to automate the optimization process by creating a program in MATLAB environment and also to select and implement an optimization algorithm according to the optimization task.

Acknowledgements This study was supported by The Ministry of Education, Science, Research and Sport of the Slovak Republic under the Contract No. VEGA 1/0595/18—optimizing the internal geometry of roller bearings with line contact in order to increase their durability and reduce their structural weight.

References

1. Medvecká I, Biňasová V, Kubinec L (2017) Planning and performance evaluation of the manufacturing organizations. Procedia Eng 192:46–51
2. Buša J (2011) Mini minimalizácia. FEI, TU, Košice
3. Lukáč M, Brumerčík F, Krzywonos L, Krzysiak Z (2017) Transmission system power FLOW model. Commun Sci Lett Univ Žilina 19(2):27–31
4. Sága M, Vaško M, Kocúr R, Tóth Ľ, Kohár R (2006) Aplikácia optimalizačných algoritmov v mechanike. VTS, University of Žilina, Žilina
5. Rosinová D, Dúbravská M (2008) Optimalizácia. Vydavateľstvo STU, Bratislava
6. Bäckryd RD, Ryberg AB, Nilsson L (2017) Multidisciplinary design optimisation methods for automotive structures. Int J Automot Mech Eng 14(1):4050–4067
7. Garg A, Juneja D (2012) a comparison and analysis of various extended techniques of query optimization. Int J Adv Technol 3(3):184–194

Innovation of the Cutting Tools of the Modular Production Line for the Production of Sophisticated Adhesive Wound Covers

Martin Dolanský

Abstract This article deals with the innovation and construction design of the cutting module, which is part of a modular production line for the production of sophisticated adhesive wound covers with respect to the Cleanroom ISO Class 8 standard and chlorine resistance. The start of the project is based on innovative methods. First, patent survey and market research were conducted. In addition, the identification and analysis of innovation opportunities were carried out, and finally, the innovation statement was formulated. The QFD method and the morphological matrix were chosen to generate the concepts. Subsequently, one of the concepts was selected using multi-criteria decision-making provided by the AHP tool. The chosen variant was further designed and processed, including strength calculations of important parts. Deformation stress tests were performed in finite element software. Integral part of this work is the technical documentation of the individual components from which the product can be realized.

Keywords Cutting module innovation · Adhesive cover · Concept · Cutting tool

1 Introduction

At present, the demands and requirements in the various technical sectors are increasing and their development is still accelerating. For example, in the healthcare industry today, from the original uniform shapes of the adhesive cover plates, which must be further cut or otherwise adjusted to the desired size by their user, they proceed to the production of various slip-sticks. These measures, however, include requirements for adaptation—the upgrading of existing production lines, especially their drawing tools.

Today, wide ranges of patches are used in healthcare. A patch is a dressing material that serves to cover the wound, fix another dressing material or injure yourself. It is equipped with a pillow that can carry drugs or disinfectants. Plaster material is subject to high demands. The patch must be adherent, but it should not damage the

M. Dolanský (✉)
Technical University of Liberec, Studentská 1402/2, 46117 Liberec 1, Czech Republic
e-mail: dolansky36@seznam.cz

Š. Medvecký et al. (eds.), *Current Methods of Construction Design*, Lecture Notes in Mechanical Engineering, https://doi.org/10.1007/978-3-030-33146-7_3

skin during scanning. It should not be caused by allergies, it should be breathable, durable, supple, acceptable aesthetic and last but not least affordable. The patch may be in a nonsterile or sterile container, which overlaps, in particular, the portion intended for direct contact with the wound. Adhesive surfaces are covered with easily removable covers to protect the sticky surface from drying and dirt. The adhesive layer is based on an acrylic base. Some adhesives may contain latex, which is not very suitable due to the frequent occurrence of allergic reactions in patients. The substrate material for the adhesive depends on the purpose of the patch; for example, plastic substrate (PVC, PE, PU, etc.) is resistant to damp environments. They are also highly absorbent, breathable and elastic (nonwoven, woven, stretched) coatings. Soft patches of soft, nonwoven textiles are used for sensitive skin. As a rule, the cushion is made mainly of treated cotton and other natural or artificial fibers so that it sticks but does not sneak into the wound [1].

1.1 Purpose of Work

The aim of this work is to innovate the cutting tools of the modular production line for the production of sophisticated adhesive wound covers. Designed concepts and their design solutions are based on specified parameters. The entire production line must meet the Cleanroom ISO Class 8 standard [2]. This standard, which is in the production of not only pharmaceutical products, indicates the level of contamination of the production environment by various pollutants such as dust, microbes, aerosol particles and other chemicals. The level of contamination is determined by the number of particles per 1 m^3 and according to ISO 14644-1 and ISO 14698 is divided into 9 levels. Different chlorine disinfectants and cleaners are required to meet these requirements. Therefore, individual parts and components must be made of durable construction materials in relation to these materials.

With the requirement to make a modular production line for the production of sophisticated rubber wound dressings, EMTEX, s.r.o. turned to the Technical University of Liberec based in Dvůr Králové. This company is engaged in the production and trade of piece and meter textiles, bedding and adhesive wound covers. Current hat production in the named company is divided into three steps and is implemented through three production lines. First, on the first device, the adhesive is applied to the backing material. The next step is to attach the pad and overlay to the adhesive backing. The last step is to simply cut the patches into square or rectangular shapes.

The main reason for the innovation of cutting tools is the ability to produce sophisticated shapes of patches with a view to easily changing the sheared shape. Tools must be designed to be able to cut all individual parts of the patch that are made of plastic, textile and paper materials. Another requirement for the cutting module is the possibility of dividing the patch materials perforated or completely along the edges of the cut. The design of the module must be designed so that it is possible to place the device "flush" on the base plate of the production line. The split strip of the blank must be at a distance of 45 mm from this base plate.

The design of the product proceeded according to selected methods from innovative engineering [3]. First, it is necessary to plan the innovation of shearing tools, which consists of the following steps:

- Identification of innovation opportunities and their evaluation
- Develop a timetable
- Innovation statement.

This information will be further used to design 3 concepts. Equally important indicator when designing a concept is the QFD (quality function deployment) method, which ensures that customer voice is already included in the design and product planning phase. Another important tool for creative product concept generation is the morphological matrix, which features partial functions and assigns individual solution options. Consequently, the proposed concepts are evaluated using the Analytical Hierarchy Process (AHP) method, which selects the most appropriate concept variant. After selecting the most suitable variant, it will be elaborated in detail, which will be further used for its realization. Construction will also include strength calculations of key components and parts and possible optimization of the frame.

Finally, the innovative solution will be evaluated and compared with the existing one. The most important indicators for the evaluation will be the possibility of editing of any shapes, the time of changing the tools for changing the shear shape and the cutting speed.

2 Planning Product Innovation

This chapter will outline the progress of innovation from initial planning, project planning, to formulate innovative intent and statements to innovative opportunities that include patent and competitive market research.

2.1 Schedule

First, the timetable for an innovative project in MS Project 2013 was set the problem is divided into multiple subphases, resources are assigned to the tasks, and a time limit is set including other necessary measures for individual tasks [4]. The program from this information will generate a project plan shown in the Gantt chart (see Fig. 1), and calculates the estimated end date.

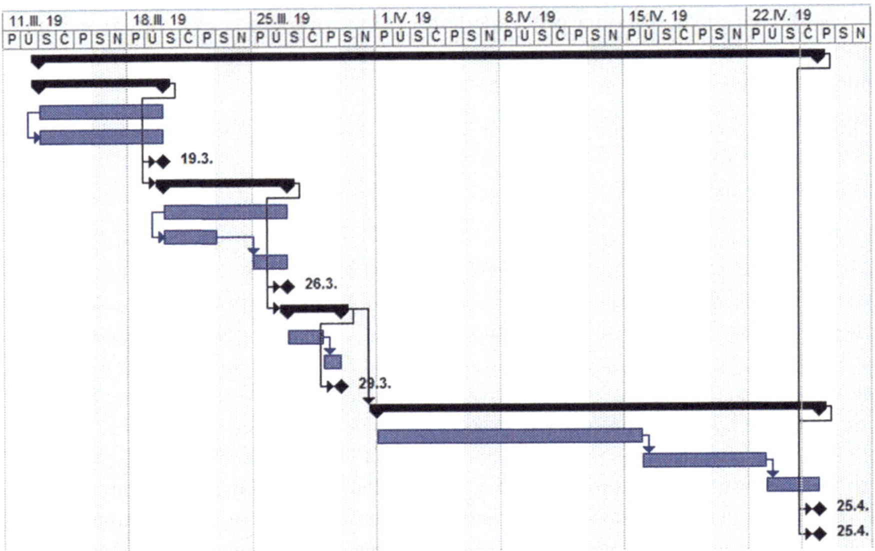

Fig. 1 Gantt chart

2.2 *Innovative Opportunities*

An integral part of the planning process is the identification of opportunities. In this section, it is decided to create a new one or partial improvements existing with respect to the product platform. To analyze opportunities a patent survey, market research,

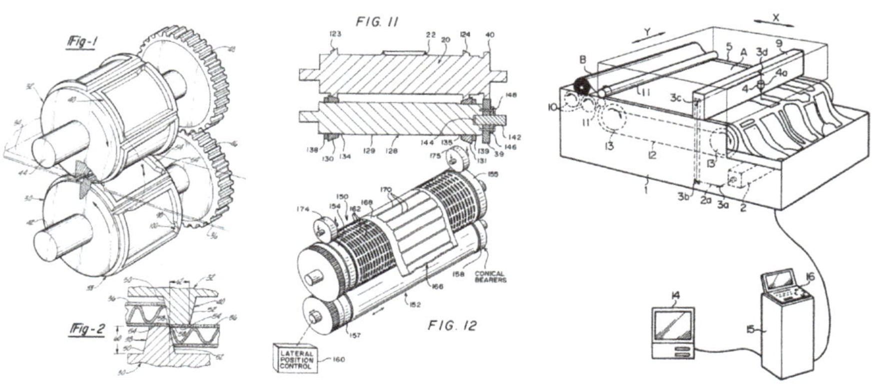

Fig. 2 Patent research, US4608895A Rotary die cutting (1984) (left). US5388490A Rotary die cutting system and method for sheet material (1990) (middle). US5200592A Automatic cutting apparatus for cloth (1990) (right) [5]

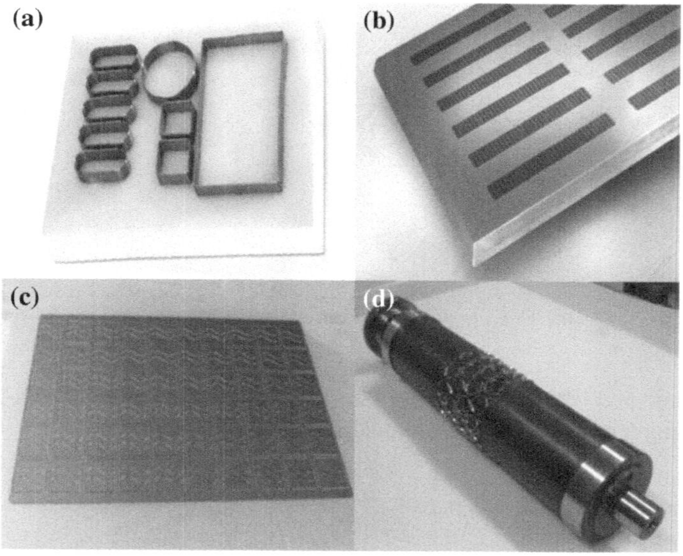

Fig. 3 Market research. Cutting tool with belt bendable blade (**a**). Magnetic plate with cutting plate (**b**). Cutting plate (**c**). Magnetic roller with cutting plate (**d**)

and user and customer consultation took place. The most interesting products found on the market and in the patent survey are shown in Figs. 2 and 3.

3 Generating Concepts

The next stage of the innovation process is the draft concept itself. Using innovative engineering methods, 3 concepts were generated in this chapter. Drafting will be divided into the following subactivities:

- Identification of customer needs.
- Specification of product characteristics using the QFD method.
- Creative concept generation using the morphological matrix.
- Generate concepts themselves.
- Evaluation and selection of the best concept.

The first indicator for creating a new product is customer requirements and ideas. Therefore, it is necessary to obtain the essential information from the customer, interpret them correctly, organize these needs into groups, determine the relative importance of these needs and review them retrospectively.

Now there is a formulation of customer needs. It seeks to accurately and measurably describe what the product must contain, know, etc. from the above information.

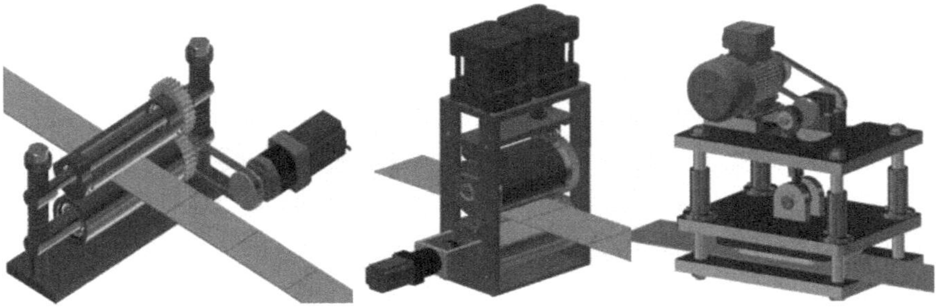

Fig. 4 Designed concepts. Concept A—(left), Concept B—(middle), Concept C—(right)

The QFD method is the dominant tool for this formulation. From the QFD matrix, the following 5 most important parameters were highlighted:

- Maximum shear force.
- Minimum life of functional parts.
- Maximum shear width.
- Time to change the cut shape.

Weight of the device in the synthesis of already known solutions and creative solutions to critical problems, the solution of partial problems has to be assessed and combined. One of the combination methods of creative product concept generation is the morphological matrix. In the matrix, they are placed on the vertical axis of the function, and all possible sketches are assigned to them in the columns. The solution options will be combined to create possible designs and concepts. The following 3 concepts appear in the following Fig. 4, after a combination of partial product solution solutions in the morphological matrix:

- Concept A deals with two spring-loaded rotating cylinders. Material separation occurs at the contact point of the cylinders by means of a cut edge on the top roller. The surface is removable and easily dismountable with 6 screws.
- Concept B is again the use of a pair of cylindrical rotary tools. The lower roller is again smooth and fulfills the function of the pressure roller. A major change over the first proposed concept is that the top roller has a magnetic surface. A so-called cutting plate can be attached to this magnetic surface. In a sheet metal sheet of thickness up to 1 mm, almost any desired shape for cutting material can be cut.
- The third designed C concept is to cut wound covers with a band bendable blade. This division principle is known as a linear cutter. The knife, together with the plate in which it is stored, forms the so-called cutting tool with the stop plate.

In order to select the most appropriate option, a structured technique for organizing and analyzing complex decisions based on the mathematics and psychology, called the AHP abbreviation, was chosen. The chosen tool helps judges find what best fits their goals by dividing difficult issues into smaller issues by assigning the relevant criteria and dividing them into hierarchies. It provides a comprehensive and rational framework for structuring a decision-making problem.

According to AHP analysis, the B concept is the best-considered concept.

4 Developing Final Variant

First, according to the technical specifications given in Table 1, the required cylinder dimensions were determined. The calculations reached the cylinder diameters $d = 130$ mm and the area width for the cutting plate $b = 180$ mm.

Stainless steel pneumatic cylinders according to CRDNG, ISO 15552 standards have been designed to provide a theoretical force of up to 4712 N as a source of compressive force. With this maximum force, the design of the bearings and the pressing rollers was carried out and the lifetime control was carried out. The lowest lifecycle was in rolls where 272 operating hours were loaded with a maximum force of the cylinders. However, in the normal operation of the cylinders, only minor parts of the maximum possible compressive force of the cylinders will be used.

Furthermore, a deflection control of the underpressure back pressure cylinder was performed. The deflection calculation was performed in the ANSYS Workbench 2016 software, which is based on the finite element method (see Fig. 5).

The results show that the roll deflection at the maximum force the pneumatic cylinders are capable of being developed is 0.0055 mm. This deformation value is satisfactory. Consequently, the strength check of the cylinder was performed, including the tough impact strength at which the safety $k = 6.12$.

Another major problem was the design of the cylinder drive. Since no high torque is required for the rotation of the cylinders, an electric stepper motor designated NEMA 23 has been designed. This engine is suitable for our application due to its size, weight,

Table 1 Technical specification

Product parameter	Value
Maximum possible width of material web	150 mm
The maximum possible length of the wound cover	380 mm
The distance of the blank from the base plate Max.	Max. 45 mm
Cutting speed	1 pcs/s with deposition on a conveyor belt at a speed of 200–1500 cm/min
Cutout shape	Arbitrary
Required enclosure dimensions	75/100 mm; 100/100 mm; 150/100 mm; 150/150 mm

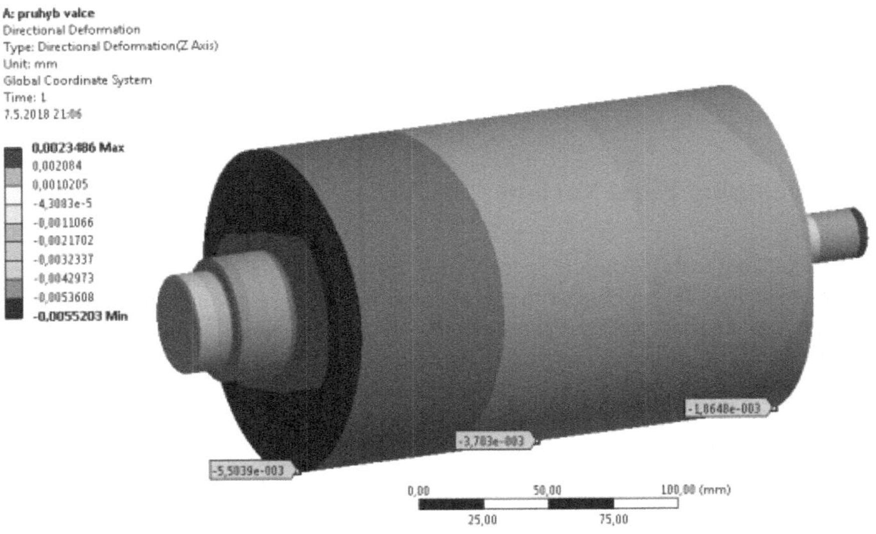

Fig. 5 Deformation of the cylinder

accuracy, and ease of operation. The engine is attached to the planetary GEA-60-3-60-ST-063 stepper gearbox, which reduces the gear ratio by 3 and increases the torque of the engine.

The torque from the gearbox is transmitted to the pressure roller by means of a three-piece clutch marked OLDHAM type OX57G with a through-hole. The last phase of the construction was the design of the frame for all components. The roller frame will be securely fastened to the anchor plate of the production line by means of screws. It has to be manufactured with high precision and is designed to be sufficiently

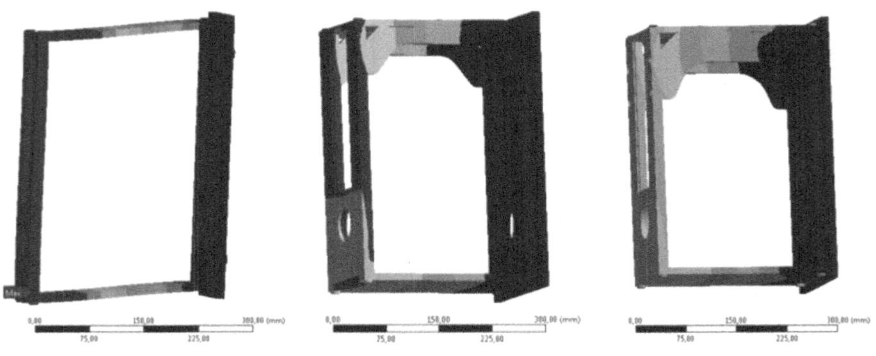

Fig. 6 Phase 1—without reinforcing ribs, deformation is 0.221 mm (left). Phase 2—added reinforcing ribs, deformation is 0.061 mm (middle). Phase 3—added bigger reinforcing ribs, deformation is 0.035 mm (right)

Fig. 7 Final construction

stiff and to avoid the shear and other axial crossing of the rolls and thus the quality of the cut. Therefore, a deformation analysis of the frame was necessary. The static frame deformation calculation was performed in the ANSYS Workbench 2016 software, provided that the amount of deformation is not greater than 0.05 mm (see Figs. 6 and 7).

The optimization resulted in a reduction in deformation from the original 0.221 mm to the final 0.035 mm. The frame designed in this way meets the rigid requirements for proper operation of the device.

5 Results and Conclusion

All specified requirements have been met from the parameters specified in the introductory chapter. The maximum bandwidth of the material was maintained and the maximum possible length of the cut shape was increased from the required minimum 380 to about 405 mm. The design of the device is designed so that the strip of sheared material is at a distance of 45 mm from the production base plate. The NEMA stepper motors with planetary gearbox for stepper motors have been designed to drive cutters, tension rollers and belt conveyors. For both pairs of cylinders, the engine power series 23 was selected and the 34 series was selected for the belt conveyor. Designed motors meet the specified 1 pcs/s cutting speed parameter with speed control capability to cover the desired belt conveyor belt speed range of 200–1500 cm/min.

At work, all the goals were met. From the design, it is possible to implement a modular production line cutting device for the production of sophisticated adhesive wound covers.

Acknowledgements This publication was written at the Technical University of Liberec as part of the project "Innovation of products and equipment in engineering practice" with the support of the Specific University Research Grant, as provided by the Ministry of Education, Youth and Sports of the Czech Republic in the year 2018.

References

1. Mikulová M (2017) Přehled náplastí, které naleznete v lékárně: Od textilních po náplasti ve spreji. Last accessed 2017/06/03
2. Pharmacist Pharma Journal. http://www.pharmacistspharmajournal.org/2010/02/clean-room-classification-aspects-of.html#.Wu87Wuqsa00. Last accessed 2018/01/15
3. Mašín I (2012) Inovační inženýrství - Plánování a návrh inovovaného výrobku. 1. vydání, TUL Liberec, Liberec
4. Lepšík P, Vodička J (2012) Plánování a řízení společných projektů 1. vydání. TUL Liberec, Liberec
5. Google Patents homepage. https://patents.google.com, Rotary die cutting. b.r. US4608895A. g 1984-03-14., Rotary die cutting system and method for sheet material. b.r. US5388490A. Granted 1990-05-10., Automatic cutting apparatus for cloth. b.r. US5200592A. Granted 1990-10-04. Last accessed 2018/01/02

Mobile Positioning Device Powered by an Electromotor

Filip Dušek

Abstract This article deals with the design of positioning device used for easier montage of various parts, especially blowers and compressors. The working desk is part of the motion positioning device which rotates around horizontal axis of positioning device. Working desk is held in pair of bearings and mathematically solved as a refracted beam. Design of the device must be universal for many types of assemblies and cannot be fixed to the ground therefore, is obtained with two pairs of wheels to provide mobility. While assembling pressing machines bearing with electromotor are fit into the housing which is mounted to the positioning device, the device is in horizontal position toward floor and is able to absorb forces from pressing process to the positioning device. FEM analysis is used for stress examination of critical areas. When pressuring is started, positioning device must hold on a stationary position. FEM analysis is used to determine maximum radial ratio of blower fixed to the geometric center/centerline of the positioning device. Positioning device must stay stable in all operations, therefore was created chart of the maximum stable position around horizontal axis for 500 kg load from horizontal to the most critical vertical position turned by 90°.

Keywords Positioning · Mobile device · Electromotor

1 Introduction

Positioning devices (see Figs. 1 and 2) are widely used for many assembling operations, while mobility of device can be very practical during running production operations. The customer submitted request to the production company to create mobile positioning device. The income attributes were complete size of device, mobility, reliability and possibility to remove/change the completely base desk of the rotating table without any needs for difficult assembling.

After proposal of the general parameters and design, the positioning device was prototyped and sent to production. However, after half a year customer requested

F. Dušek (✉)
Technical University of Ostrava, 17. Listopadu 15/2172, 70833 Ostrava, Poruba, Czech Republic
e-mail: filip.dusek@vsb.cz

Š. Medvecký et al. (eds.), *Current Methods of Construction Design*, Lecture Notes in Mechanical Engineering, https://doi.org/10.1007/978-3-030-33146-7_4

Fig. 1 Positioning device before expedition [1]

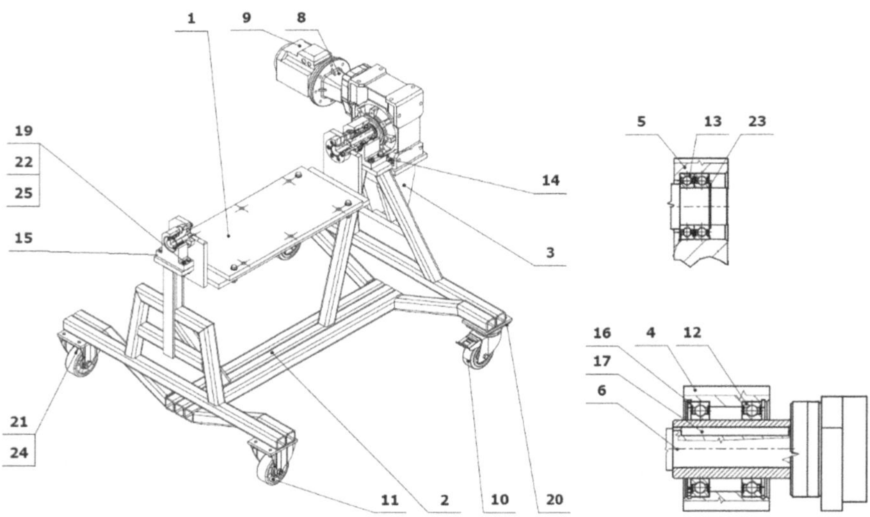

Fig. 2 Complete design of device (Legend: 1—Working table, 2—Frame, 3—Motor support, 4—Driven pivot support, 5—Pivot support, 6—Support pivot, 8—Worm gearbox, 9—Electric motor, 10—Moveable wheel, 11—Fixed wheel, 12—Ball bearing, 13—Ball bearing, 14—assembly of housing A, 15—assembly of housing B)

revision. It was necessary to determine maximal allowed load capacity. After technical analysis, the maximal load capacity was determined as 500 kg. The objective of this work was to verify this analysis and also check stability of positioning table while assembling.

2 Design of Positioning Device

The main part of the table consists of welded frame which is assembled from square profiles. Mobility of device is provided by pair of movable and pair of fixed wheels with additional brake. Driving mechanism is consisting of electromotor with 1.5 kW power and worm gearbox with ratio 320. Output speed of driving mechanism is 4.5 rpm.

The output gearbox shaft is mounted to the coupling located in housing. Housings are mounted to the welded frame using bolts. There is a pair of ball bearings in every housing. Driving and carrying pivots are part of housings. Driving pivot is tubular and therefore provides socket for driving shaft. Outer part of the pivot is put into bearings located in housing. Torque is transferred between parts via parallel key lock. Carrying pivot is put into bearings and mounted to the rotating table via bolts.

The working table is assembled from welded thick sheet metals with holes used to mount assembled parts. Whole working table is mounted by bolts on the pivots. Torque is transferred by two components. Firstly, by friction of bolts and secondly by shape of key locks as seen in Fig. 3.

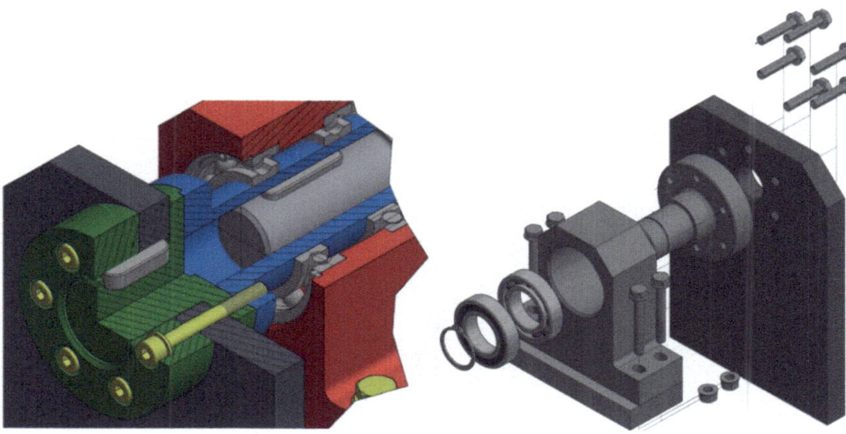

Fig. 3 Model of assembly of housing A (left), assembly of housing B (right)

3 Stability and Maximum Distance from Center of Rotation

One of many problems was to determine maximal dimension of beam which is motor able to turn with without consecutive overload. The assumption is that the center of mass of a maximum weight load is positioned out of positioning device table. Related to this is also stability of whole table which must stay stable during whole operation and avoid overload and collapse. Distance from center of mass is possible count according to Eq. (1), (see Fig. 4).

$$x(\alpha) = \frac{T}{Q \cdot \sin(\alpha)} \tag{1}$$

where x—distance from center of mass, T—nominal torque from gearbox, Q—weight force. Stability positioning device is presented in Fig. 5.

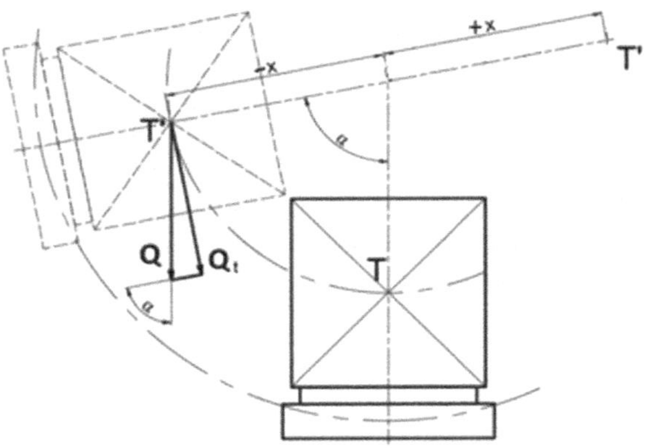

Fig. 4 Force during rotating

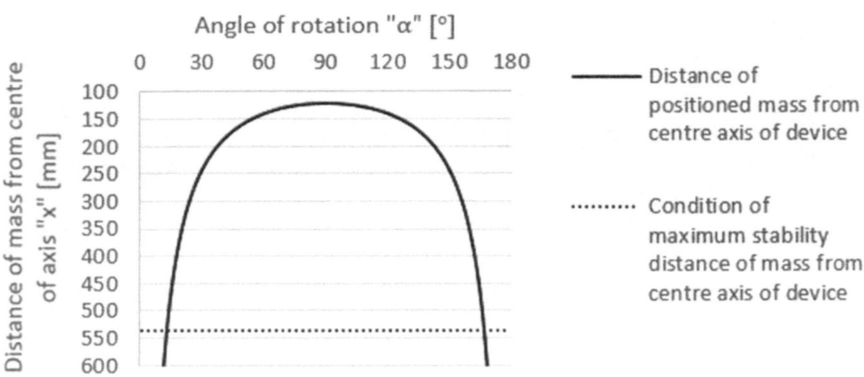

Fig. 5 Diagram of stability positioning device

4 Complex FEM

The first part of FEM contains complete deformation analysis for two extreme positions of the table: firstly for default position where the table is horizontally oriented and secondly for position where the table is moved by 90° (see Fig. 6). Final amount of deformation is relatively small therefore is negligible. There is no significant impact in case of inaccuracies while production assembling.

The mesh is created according to technical recommendations [2].

The stress analysis was done in second part for both extreme positions. The points of interest are (see Fig. 7):

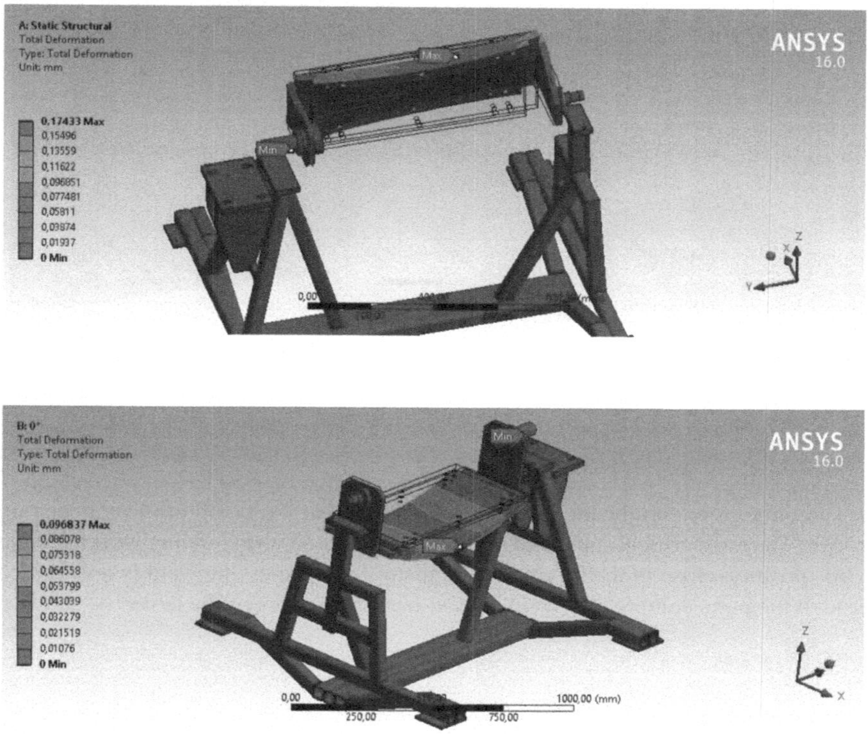

Fig. 6 Total deformation of device caused by mounted mass

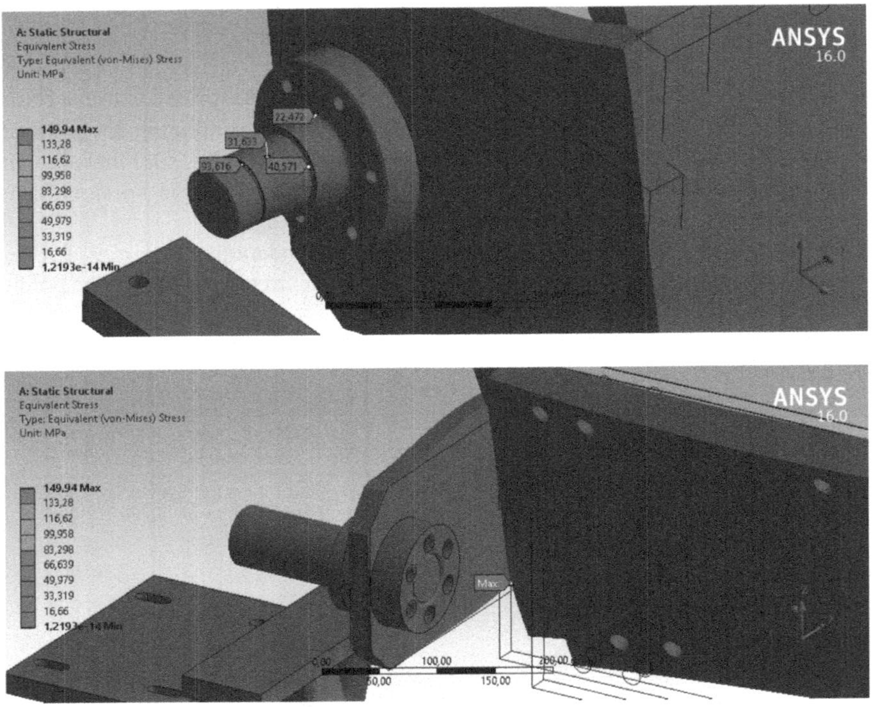

Fig. 7 Equivalent (von Mises) stress analyses of pivot and working table

- Pivots.
- Working table where welds are used.

The only more significant stress within pivots is at the position of retainer ring groove. The most critical spot at all seems to be the filled weld holding working table in 90° position close to the driving mechanism. In this spot, the weld is not welded through the parts completely. Weld is used from both sides of the table.

5 Conclusion

It can therefore be presumed that the load capacity of 500 kg is safely guaranteed for complete design. The most critical spot are the chosen ball bearings in terms of static strength. Dynamic strength is negligible in this case for very low output speeds under 10 rpm. Strength examination of bolts and keys is sufficient also with its combination while transferring torque from gearbox. Frame with housing of working table is tough enough within both extreme positions. Positioning device is currently used for 3 years and according to last information, everything works well without any crashes. In case of assembling, next series of this device is recommended using better bearings with cylindrical rollers. Angular contact bearings are unnecessary in this case because gearbox output axis is parallel to the working table axis.

References

1. Dušek F (2016) Mobilní polohovací zařízení poháněné elektromotorem. Bachelor Thesis. Vysoká škola báňská - Technická univerzita Ostrava, Ostrava
2. Cook RD (2002) Concepts and applications of finite element analysis, 4th edn. Wiley, Hoboken

Development of a New Measuring System for Verifying the Float Level Gauge

Tomáš Gajdošík, Igor Gajdáč, Ĺuboš Kučera and Jaromír Markovič

Abstract The paper deals with the possibility of increasing the accuracy of measurement of float indicators (float gauge) for level verification in high-capacity oil and hydrocarbon fuel tanks. Verification of accuracy of float level gauge is required by law. The new device will use Renishaw's metering system. The new measuring system will work with an accuracy of 10 times the current measurement system. The current measuring system uses an accurate sliding scale and the rope of the authenticated measuring system is pressed by point contact for manipulation. The new system utilizes the direct contact of the rope and the measuring wheel. The paper focuses on the overall concept of a measuring device. The paper also deals with the determination of the geometric tolerances produced by the device components to minimize the measurement error. This article was based on the solutions to common project of applied development Slovak Legal Metrology, Slovak University of Technology in Bratislava, University of Žilina, with the support of the Agency for Research and Development.

Keywords First measuring system · Float level gauge · Accuracy · Design

1 Introduction

Presented solution is based on the national legislation represented by Annex no. 68 UNMS to the regulation Nr. 210/2000 Coll., International recommendation OIML R 85: 2008 Automatic level gauges for the measurement of the level of liquid in stationary storage tanks and the standard STN EN ISO/IEC 17025: 2005 Requirements for the competence of testing and calibration laboratories.

T. Gajdošík (✉) · I. Gajdáč · Ĺ. Kučera
University of Žilina, Univerzitná 8215/1, 01026 Žilina, Slovak Republic
e-mail: tomas.gajdosik@fstroj.uniza.sk

J. Markovič
Slovak Legal Metrology n. o, Hviezdoslavova 31, 97401 Banská Bystrica, Slovak Republic

Š. Medvecký et al. (eds.), *Current Methods of Construction Design*, Lecture Notes in Mechanical Engineering, https://doi.org/10.1007/978-3-030-33146-7_5

2 Current Status

For the calibration and verification of float leveling devices, the test equipment ZOH1, which was developed on the basis of utility model no. 166-2013 [1]. This measuring system allows very effective verification of the float level meter directly at the customer, without the need for the level meter to be transported to the calibration laboratory. Figure 1 presents a schema of a device for calibrating and verifying float level gauge. The S1 is a verified device. The S2 is a steel wire 0.15 mm thickness. The device has one fixed and accurate low-resistance pulley 8 and one free accurate low-resistance pulley 3. On the pulley 3, a float S3 that is part of the level meter is suspended on the steel wire. In vessel 5, the liquid 6 has known density, and the container is placed on the weight 7. Block 1 represents the measurement system and block 2 is float drive with stepper motor and belt drive (see Fig. 1). The photograph of real equipment ZOH1 is presented in Fig. 2. By successively of wire, the level

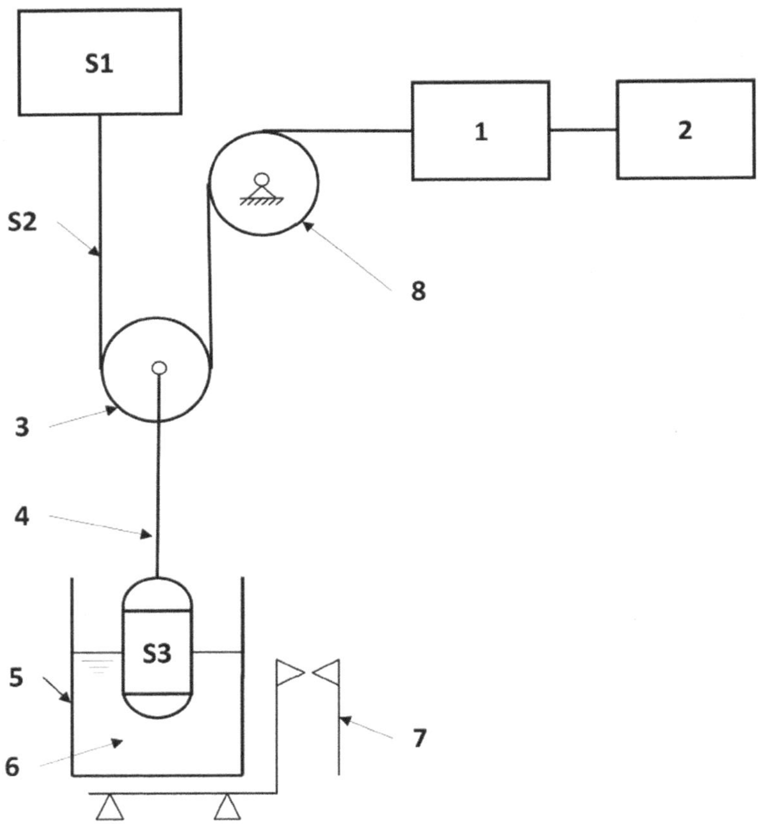

Fig. 1 Schema of the device principle—ZOH1

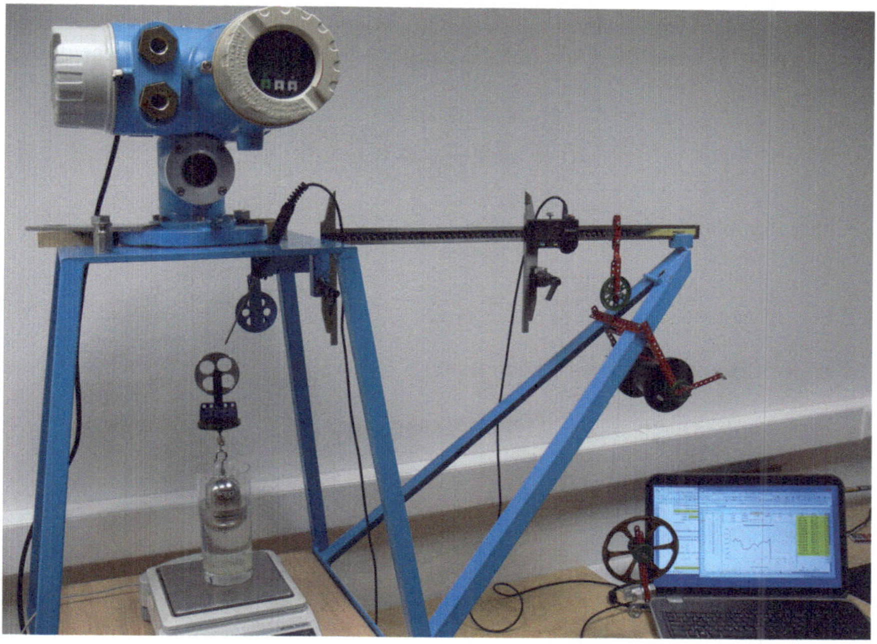

Fig. 2 Real equipment ZOH1

sensor cable S1 from point A to point B a level change is simulated and the leveling agent reacts by balancing the entire weight of the float S3 to the desired weight.

Changing the length of the S2 from the automatic level meter and subtracting the change in length to the level meter is compared with changing the length of the shift between points A and B. We need not a 35 m high liquid tube in this case. By the measurement process to change the height by 35 m, it is necessary to move the wire 106 times from point A to point B by caliper (see Fig. 3).

The caliper is calibrated, with $l = 330$ mm displacement, the deviation $\Delta l1 = \pm 0.01$ mm and total deviation $\Sigma\ \Delta l1 = 0.02$ mm.

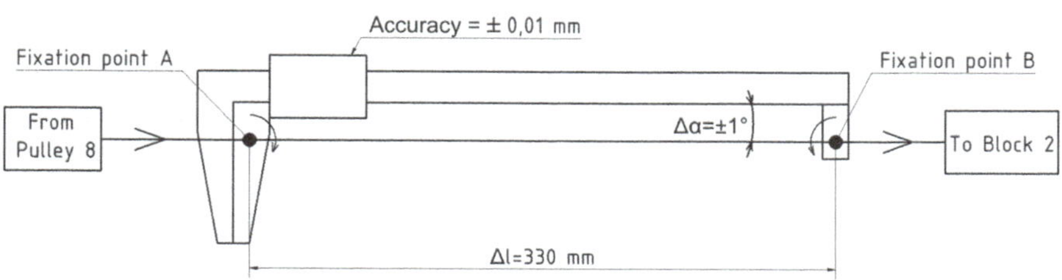

Fig. 3 Schema of block 1 of the device—ZOH1

Fixing points A and B press the steel wire around the body of the gauge, and an angular error between the wire and the caliper—$\alpha 1 = 0.1736°$ may occur during measurement.

$$\alpha l = \text{arctg}\frac{1}{330} = 0.1736° \tag{1}$$

Changing length due to angular error is Δl in mm.

$$\Delta l2 = \frac{l}{\cos \alpha 1} = \frac{330}{\cos 0.1736} = 330.0015 \tag{2}$$

The error is symmetric, so we can write

$$\Delta l2\text{max} = 330.0015 \tag{3}$$

$$\Delta l2\text{min} = 329.9985 \tag{4}$$

$$\sum \Delta l2 = \Delta l2\text{max} - \Delta l2\text{min} = 330.0015 - 329.9985 = 0.003 \tag{5}$$

Total expected deviation

$$\sum \Delta l = \sum \Delta l1 + \sum \Delta l2 = 0.023 \text{ mm} \tag{6}$$

Deviation at 106 times (length = 35 m)

$$\Delta l_{35} = 106 \times \sum \Delta l = 106 \times 0.023 = 2.438 \text{ mm} \tag{7}$$

The Schema of block 2 with test equipment (ZOH1) is presented in Fig. 4. Figure 5 presents the comparison of the measurement error in relation to the height level.

3 Design of a New Measurement System

The new metering system will use a part of the ZOH1. To increase accuracy, block 1 will use the Renishaw's VIONiC measuring system with a precision measuring angle of better than 0.1 arc s. The Renishaw's measuring ring will be part of a precision pulley with an internal diameter of 105 mm and a circularity tolerance of 0.002 mm and a wire shear angle of 350°. Block 2 will be formed by a pulley on which the level meter wire will be wound using a stepper motor and belt drive. In this case, the measuring system can be automated and the float switch in the water tank can be programmed. No manual operation will be required to perform the measurement as in the case of a caliper. Block 1 with VIONiC measuring system is presented in

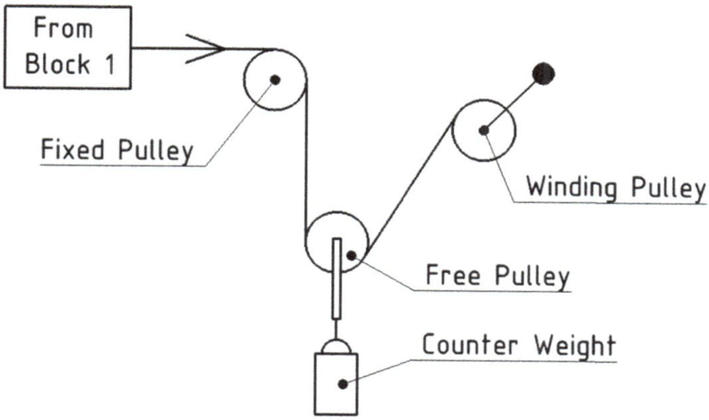

Fig. 4 Schema of block 2 of the device—ZOH1

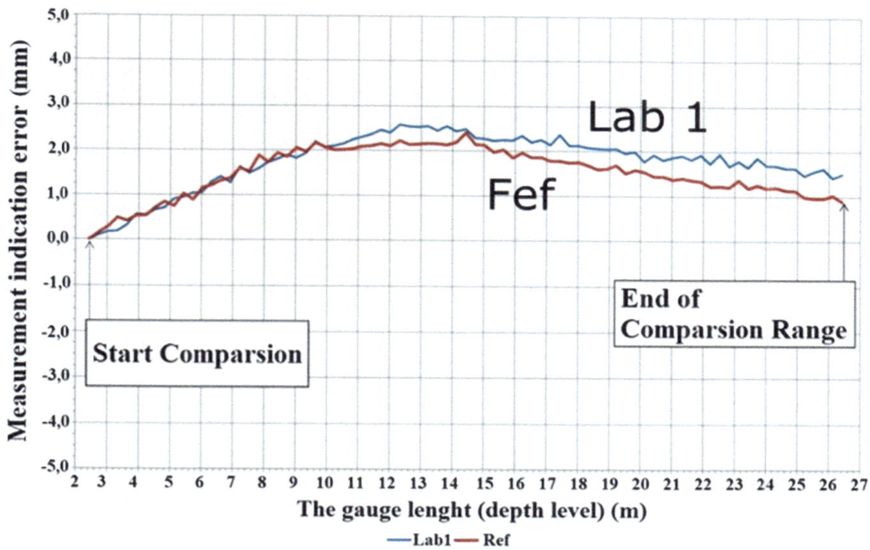

Fig. 5 Comparison of the measurement error in relation to the height level, where ref—reference measuring device, Lab1—device ZOH1

Figs. 6 and 7. Block 2—float drive with stepper motor and belt drive, is presented in Fig. 8 [3].

When we using a pulley with nominal diameter, $l1_{\text{circ}} = \pi \times 105 = 329.867$ mm.

The length per revolution with a circular tolerance of 0.002 mm:

$$l1_{\text{circ}} = \pi \times 105.002 = 329.8735 \text{ mm} \tag{8}$$

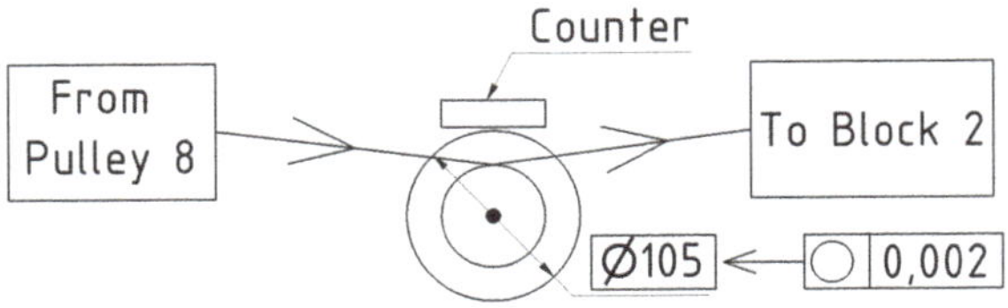

Fig. 6 Block 1 using a VIONiC measurement system

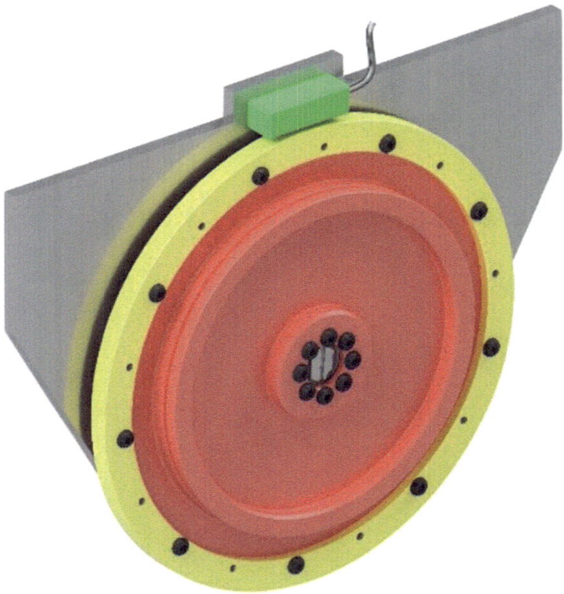

Fig. 7 Block 1 using a VIONiC measurement system—model of pulley with VIONiC measuring system [2, 4]

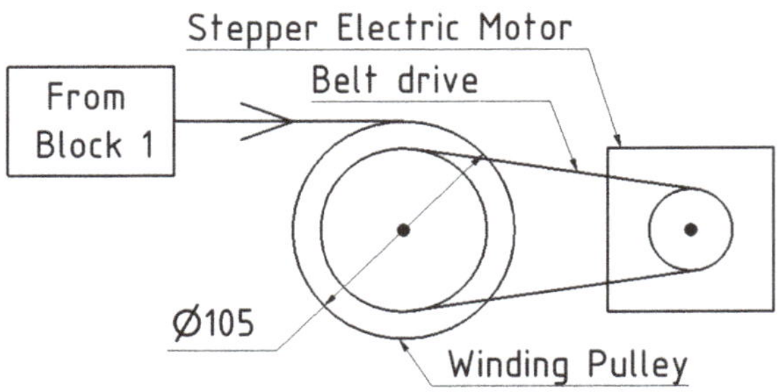

Fig. 8 Block 2—float drive with stepper motor and belt drive

The deviation per revolution:

$$\Delta l_{\text{circ}} = l1_{\text{circ}} - l_{\text{circ}} = 329.8738 - 329.867 = 0.0065 \text{ mm} \tag{9}$$

Deviation at 106 times (length = 35 m)

$$\Delta l_{\text{circ35}} = 106 \times \sum \Delta l_{\text{circ}} = 106 \times 0.0065 = 0.689 \text{ mm} \tag{10}$$

4 Conclusion

According to the calculation of the maximum deviation, we can state that by using the rotary sensor, the measuring accuracy will be 3.5 times better. It is also possible to automate and accelerate the measurement process. In addition, the device dimensions may be smaller.

Acknowledgements This article was created with the support of project APVV-15-0164—innovative technologies for calibrating measurement equipments.

References

1. Kučera Ľ (2016) Utility model no. 166–2013. ÚPV, SK
2. Šteininger J, Hrček S, Krchňavý B (2017) The design of universal loading device for grinding machines. Procedia Eng 192:869–874
3. Gramblička S, Kohár R, Madaj R (2017) Construction design automatically adjustable mechanism for crane forks. In: ICMD 2017, Proceedings of the 58th international conference of machine design departments. Czech University of Life Sciences Prague, pp 100–103
4. Renishaw Homepage. http://www.renishaw.cz/cs/1030.aspx. Last accessed 2018/5/21

Experimental Methods and Devices for the Identification of the Properties of Pneumatic Dual Mass Flywheel

Robert Grega and Peter Kaššay

Abstract The torsional vibrations generated by combustion engines installed in vehicles have an undesirable effect on the life and reliability of individual components. Vibration and excited noise reduce the comfort of the operator and crew. The subject of torsional vibration reduction is therefore highly current. One of the options for reducing torsional vibrations is the current use of dual mass flywheels that are included between the combustion engine and the vehicle's gearbox. An alternative solution to the reduction of torsional vibration is the use of a pneumatic dual mass flywheel. The pneumatic dual mass flywheel is characterized by the fact that it is possible to change its properties during the drive operation. For the purpose of investigating the use of dual mass flywheel in combustion engines, it was necessary to create experimental devices and methodology to identify their properties. The experimental device is possible to identify the operational, strength and expansion properties of pneumatic dual mass flywheels.

Keywords Pneumatic dual mass flywheel · Properties identification · Experimental methods · Devices

1 Introduction

Modern propulsion devices, new combustion engine designs and an effort to reduce emissions and weight of the whole system increase the efficiency of vehicle drives but also make those drives much more vibration—sensitive. Therefore, the problem of reducing these vibrations and their undesirable effects becomes more and more important when designing a vehicle drive [1, 2].

The vehicle propulsion system is a relatively complex dynamic system that must meet the demands placed on it in all situations—driving modes that may occur. The vehicle propulsion is still based on piston machine.

Nowadays, piston machines are the most widely used type of rotary machines [3]. Both, the driving and the driven machines are widely used [4].

R. Grega (✉) · P. Kaššay
Technical University of Košice, Letná 9, 04200 Košice, Slovak Republic
e-mail: robert.grega@tuke.sk

Š. Medvecký et al. (eds.), *Current Methods of Construction Design*, Lecture Notes in Mechanical Engineering, https://doi.org/10.1007/978-3-030-33146-7_6

In the field of driving piston engines, the dominant group is that of the combustion engines [5].

The combustion engines are characterized by varying force over the piston, which, by transformation in the crank mechanism, leads to a torque with variable rotation. Thus, the variable torque is a source of increased stress, vibration and noise, affecting other parts of the drive. Reducing these undesirable vibrations is the subject of long-term effort aimed at introducing various technical solutions for connecting piston machines to other parts of the drive [6, 7].

The basic solution is to use parts that minimize vibrations in the piston machine drive. Various torsion dampers are used for this purpose [8–10].

In the combustion engine area, the use of dual mass flywheels is ever-increasing. Consequently, the pressure on the development of dual mass flywheels is constantly rising, namely expectations of their ability to respond to the change in dynamic ratios.

The reduction of torsional vibration and its impact constitute a significant attribute in modern vehicle design. The use of dual mass flywheels is not related to passenger cars only, where the flywheel, nonetheless, has been put to significant use so far. Dual mass flywheels can nowadays be found in the truck and agricultural vehicle propulsion systems, which place new demands on these vibration eliminators [11].

Many authors focus on DMF innovations and their simulations. Mahl and Sawodny [12] focus on the development of dual mass flywheels using simulations. Pfabe proposed a driven flywheel [13]. Peghaar designed Electrical Dual Mass Flywheel [14]. Grega designed a pneumatic dual mass flywheel for low emission marine engine applications [2].

The common property of dual mass flywheels is that they must be capable of transmitting large torques. However, in order to minimize torsional vibration, their stiffness value must be low, which is achieved through a large angle of twist. In some cases, the angle of twist is measured up to 60° [15].

2 Define Problems

New design solutions in the area of dual mass flywheels, especially in developments of the pneumatic dual mass flywheels, offer a wider range of possibilities of torsional vibration tuning in vehicle propulsion systems. Each design of a pneumatic dual mass flywheel must ensure that the torsional stiffness of PDMF is low, covering a wide span of the angle of twist. Low torsional stiffness then makes the application in a torsionally vibrating mechanical system possible, where its own frequencies can be shifted to areas of very low speed. Thus, dangerous resonances at different operating speeds are avoided. In combustion engines, those resonances always follow the so-called downspeeding trend [16].

The basic design of the pneumatic dual mass flywheel was published in [17]. The basic feature of the pneumatic dual mass flywheels is that there is a pneumatic bag between the flywheel's primary and secondary part, in which the air pressure can be varied smoothly in course of the dual mass flywheel rotation. This smooth change in pressure is manifested by a continuous change in the pneumatic dual mass flywheel properties. It is precisely the change in properties, dependent on the change in the air pressure in the bag, that need to be experimentally identified. Thus, the pneumatic dual mass flywheel is characterized by a large angle of twist of up to 60° and a change in the pneumatic bag air pressure ranging from 0 to 700 kPa. In order to identify the properties of the newly developed pneumatic dual mass flywheel, it is therefore necessary to modify the methods of measurement and to design new measuring devices. The aim of the paper is thus to present the modified measuring methods and equipment that needed to be developed, designed, implemented and tested for the sake of further research in the field of pneumatic dual mass flywheels.

3 Theory and Methods

Pneumatic dual mass flywheels can be characterized as parts of machines used for flexible coupling of the driving and the driven part of the drive. Basic properties of all such devices need to be identified. Basic properties of such flexible machine parts can be divided into 3 groups, namely strength, operational and expansion properties [18, 19].

Strength properties. They are characterized by their ability to transmit different types of torque and maximum working speed without any damage.
Expansion properties. They describe how and of what magnitude the axial, radial and angular inaccuracies arising from production and assembly can be eliminated.
The operational properties. They describe the ability to harmonize the mechanical system and damp torsional vibrations.

In order to identify the above groups of properties of flexible machine parts, experimental tests need to be performed. In our case, it is necessary to carry out experimental tests for the newly developed pneumatic dual mass flywheel. Out of the above-mentioned three groups of properties, the operational property is most difficult to identify. The term "operational properties" of pneumatic dual mass flywheels denotes their static and dynamic performance.

To identify the static operational properties, it is also possible to use measurement methods and devices where the pneumatic dual mass flywheel does not need to rotate and does not need to be impacted by variable force, that is, it may not even vibrate [20, 21].

The static operational characteristics of the pneumatic dual mass flywheel include the following:

- Static characteristic (the static load torque Mk_{stat} dependent on the pneumatic dual mass flywheel φ angle of twist at a very slow change in torque).
- Static torsional stiffness of the pneumatic dual mass flywheel k_{stat} (this is a calculated value; it is a derivation of the static curve equation according to the angle of twist).

Dynamic operational characteristics of the pneumatic dual mass flywheel to be identified are:

- Dynamic torsional stiffness k_{dyn} of the pneumatic dual mass flywheel.
- Damping of the pneumatic dual mass flywheel, which can be expressed in terms of an equivalent cushioning coefficient b_e, or proportional damping Ψ.

Dynamic measurement methods and experimental devices are used to identify dynamic operational properties. In dynamic measurement, it is not necessary for the pneumatic dual mass flywheel to rotate. It is sufficient to simulate the dynamic load component. Simulation of the dynamic load component requires application of a variable load exciter. The variable loader can be based on the principle of kinematic excitation or on the principle of dynamic excitation. In view of our prior experience, we chose to use a kinetic exciter to identify the pneumatic dual mass flywheel properties [22]. The diagram of the device for identifying dynamic operational properties is shown in Fig. 1.

The measuring device for identification of the pneumatic dual mass flywheel properties should consist of the parts is defined in the diagram in Fig. 1. In particular, those include the exciting part (1), the immobilizing part (2), the pickup arm (3). Torque sensors for testing the pneumatic dual mass flywheels can be placed variously between those basic parts.

Using the measuring device of Fig. 1, which can be configured in a variety of ways, we are able to apply both static and dynamic measurement methods to identify the static and dynamic operational properties of the pneumatic dual mass flywheel.

Using the parts (2) and (3) of the measuring device, between which the pneumatic dual mass flywheel will be placed, measurements can be made to identify the static operational properties of the pneumatic dual mass flywheel. This configuration of the measuring device can also be used in application of the free vibrating method.

Measurement by means of free vibration is one of the simplest methods for establishing dynamic torsional stiffness and damping. To obtain a log with oscillation

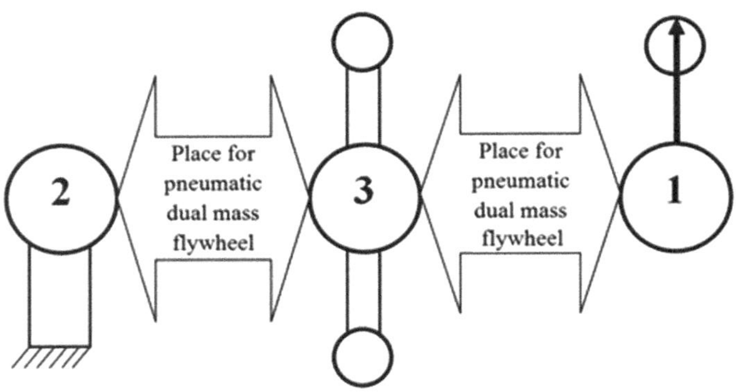

Fig. 1 Diagram of the measuring device for identification of the pneumatic dual mass flywheel properties

course, it is necessary to record the movement of part (3) of the measuring device using the acceleration sensor and the measuring apparatus. This method makes it possible to identify the properties of pneumatic dual mass flywheels even at different bias magnitudes.

The use of parts (1) and (3) of the measuring device according to Fig. 1 is suitable for detecting the pneumatic dual mass flywheel dynamic properties. The vibration pickup method of measurement is particularly suited for detecting dynamic torsional stiffness at different bias and different loading amplitudes. By combining the parts (1) and (2), it is possible to observe in particular the damping and the dynamic torsional stiffness at different frequencies and at different exciting amplitudes.

4 Results and Discussions

For the purpose of identifying the properties of the newly developed pneumatic dual mass flywheel, a measuring device was built in our laboratory, making it possible to run static and dynamic measurement tests. Verification and measuring device tests were run on dual mass flywheel designed for four-cylinder diesel combustion engines with a nominal torque 350 Nm. Figure 2 shows the measuring device configured to run static tests.

In establishing static properties, five repetitions of jolt-like load application onto the pneumatic dual mass flywheel and its subsequent release must be done. The static characteristic is then defined as a load-applying property of the fifth hysteresis loop. The rendition of the hysteresis loop describing the static property is shown in composite Fig. 4.

As already mentioned above, dynamic measurement, methods are required to identify dynamic operational properties. Our measuring device facilitates three dynamic measurement methods. Figure 3 shows the configuration of the measuring device for the application of the measuring method by means of vibration pickup.

With this method, vibration pickup amplitudes, as well as the magnitude of torque, are recorded synchronously. These data then make it possible to draw the hysteresis loop. By analyzing the hysteresis loop, we are able to establish proportional damping and dynamic torsional stiffness of the dual mass flywheel. Again, we made a record of measuring the vibration pickup in the form of a hysteresis loop. The measurements were made with different biases of the dual mass flywheel. Hysteresis loops from measurements obtained by the vibration pickup method are shown in the composite (see Fig. 4).

Referring to composite Fig. 4, hysteresis loops of vibration pickup can be made for different dual mass flywheel frequencies and different biases.

Fig. 2 Measuring device for identification of the pneumatic dual mass flywheel static properties: (1—dual mass flywheel, 2—immobilizing element, 3—swing arm, 4—measurement of the angle of twist, 5—torque sensor, 6—display unit)

Fig. 3 Measuring device facilitating the method of vibration pickup: (1—immobilizing element, 2—torque sensor, 3—angle of twist sensor, 4—dual mass flywheel, 5—exciting element)

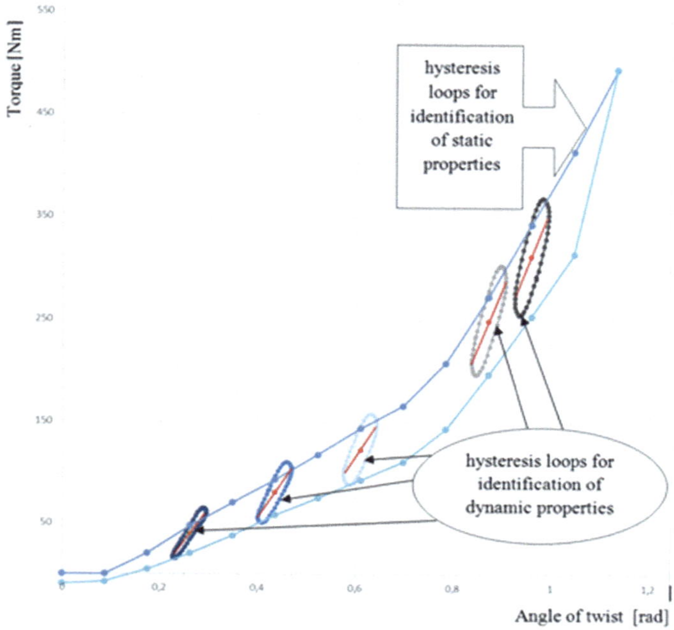

Fig. 4 Composite hysteresis loops for identification of static properties, dynamic torsional stiffness and proportional damping

5 Conclusions

Our aim was to produce and test measuring equipment and measurement methods to identify the properties of the newly developed pneumatic dual mass flywheel. We designed and built a testing device facilitating measurement methods of identification of the static and dynamic operational properties of pneumatic dual mass flywheels. Test measurements were run under our laboratory conditions. Subject to our testing measurements was a dual mass flywheel used in four-cylinder diesel engines. From the verification test results of the testing equipment, it follows that the measuring device available at our workplace, including the measuring and evaluation apparatus and the measuring methods, is suitable for identifying the properties of the pneumatic dual mass flywheels. The measuring device will be an asset of further progress in the field of research and development of pneumatic dual mass flywheel new designs.

Acknowledgements This paper was written in the framework of Grant Project VEGA 1/0473/17 "Research and development of technology for homogeneous charge self-ignition using compression in order to increase engine efficiency and to reduce vehicle emissions."

References

1. Jakubovičová L, Sága M, Kopas P, Handrik M, Vaško M (2010) Some notes on analysis of bending and torsion combined loading. Mach Dyn Res 34(3):97–105
2. Grega R, Krajňák J, Žuľová L, Fedorko G, Molnár V (2017) Failure analysis of driveshaft of truck body caused by vibrations. Eng Fail Anal 79:208–215 (2017)
3. Grega R, Homišin J, Puškár M, Kuľka J (2015) The chances for reduction of vibrations in mechanical system with low-emission ships combustion engines. Int J Marit Eng: Trans R Inst Naval Archit 157(A4):235–240
4. Žuľová L, Grega R, Krajňák J, Fedorko G, Molnár V (2017) Optimization of noisiness of mechanical system by using a pneumatic tuner during a failure of piston machine. Eng Fail Anal 79:845–851
5. Czech P, Wojnar G, Burdzik R, Konieczny L, Warczek L (2014) Application of the discrete wavelet transform and probabilistic neural networks in IC engine fault diagnostics. J Vibroeng 16:1619–1639
6. Łazarz B, Wojnar G, Czech P (2011) Early fault detection of toothed gear in exploitation conditions. Maint Reliab 1:68–77
7. Kuľka J, Mantič M, Kopas M, Faltinová E (2017) Necessity of wire rope replacement in crane lifting equipment after change of crane operational parameters. Adv Sci Technol Res J 11:226–230
8. Homišin J, Kaššay P, Puškár M, Grega R (2016) Continuous tuning of ship propulsion system by means of pneumatic tuner of torsional oscillation. Int J Marit Eng, Trans R Inst Naval Archit 158(A3):231–238
9. Homišin J, Čopan P, Urbanský M (2013) Experimental determination of characteristic properties of selected types of flexible shaft couplings. Sci J Silesian Univ Technol Ser Transp 81:51–57
10. Krajnak J, Homišin J, Grega R (2011) Effect of helium on mechanical properties of flexible pneumatic coupling. Sci J Silesian Univ Technol Ser Transp 73:63–69
11. Grega R (2014) The chance for application of the pneumatic dual mass flywheel in drive combustion engine. In: Ševčík L, Lepšík P, Petru M, Mašín I, Martonka R (eds) Modern methods of construction design, proceedings of ICMD 2013. Springer, Heidelberg, pp 43–47 (2014)
12. Mahl T, Sawodny O (2010) Modelling of an automotive dual mass flywheel. In: 5th IFAC symposium on mechatronic systems, vol 43, issue 18. Elsevier, pp 517–523 (2010)
13. Pfabe M, Woernle Ch (2016) Reducing torsional vibrations by means of a kinematically driven flywheel. Theory Exp Mech Mach Theory 102:217–228
14. Peghaar J, Lohmann B (2013) The electrical dual mass flywheel—an efficient active damping system. In: 7th IFAC symposium on advances in automotive control the international federation of automatic control, vol. 46, issue 21. Elsevier, pp 483–488 (2013)
15. Faust H (2014) Powertrain systems of the future. In: The 10th Schaeffler symposium engine, transmission and damper systems for downspeeding, downsizing, and cylinder deactivation (2014)
16. Homišin J (2016) Characteristics of pneumatic tuners of torsional oscillation as a result of patent activity. Acta Mech et Autom 10(4):316–323
17. Grega R, Krajnak J (2012) The pneumatic dual mass flywheel. Sci J Silesian Univ Technol Ser Transp 76:19–24
18. Homišin J, Urbanský M (2011) Results of measurement of transitional effects at air pressure changes in pneumatic coupling during mechanical system operation. Sci J Silesian Univ Technol Ser Transp 72:41–50
19. Urbanský M, Homišin J, Čopan P (2013) Examination of mechanical system response to gaseous media pressure changes in the pneumatic coupling. Sci J Silesian Univ Technol Ser Transp 81:143–149

20. Urbanský M (2014) Theoretic and experimental determination of the flow resistance coefficient at gaseous medium flow into and out of the pneumatic coupling. Sci J Silesian Univ Technol Ser Transp 85:119–125
21. Urbanský M (2017) Harmonic analysis of torsional vibration force excitation. Sci J Silesian Univ Technol Ser Transp 97:181–187
22. Baran P, Grega R (2015) Comparison of dynamic properties of dual mass flywheel. Diagnostyka 16(1):29–33

Innovation of Device for Production of Core Yarn with Nanofibers

Martin Chára

Abstract This work is focused on innovation of the method of application of nanofibers on yarn. A candle filter is produced from this fiber body. The goal is to increase the productivity of nanofiber deposition and to remove unwanted effects on existing device. The procedure is based on innovative methods of product innovation. At the beginning of the work, a patent survey is carried out and the QFD matrix is compiled. Several concepts are proposed by morphological matrix and the analytic hierarchy process is used for selection of concept for designing. The selected concept is designed to meet the target parameters. An experiment was performed to verify the balloon parameters and the proposed concept is divided into three sections. The designed concept is compared at the end of the thesis with the current solution. A microscope with a magnification of one hundred is used to compare the nanofibres layer applied on a yarn.

Keywords Nanofibers · Yarn · Candle filter · Innovation · Electrospinning

1 Introduction

At present, the development of nanomaterials is a trend around the world. Nanomaterials find application in various fields of human activity and one of these areas is filtration of liquid and gaseous media. For filtration, it is advisable to use nanomaterials in the form of nanofibers [1, 2], which can be applied on the substrate or used separately. Filtration is important, for example, in medicine for the creation of a health-conscious sterile environment and in many industries. Another trend that concerns almost everyone is the increasing pollution of the environment and the effort to eliminate it. A large number of people, especially in the states of Africa, have no access to clean and clean water, and a simple filtration of untreated water would help to improve their lives. Thanks to the use of nanofibres in the filter, some types of bacteria and allergen particles can be trapped to reduce water contamination. The ability to filter smaller particles can be used, for example, in the removal of yeasts and

M. Chára (✉)
Technical University of Liberec, Studentská 1402/2, 46117 Liberec 1, Czech Republic
e-mail: klamosak@gmail.com

Š. Medvecký et al. (eds.), *Current Methods of Construction Design*, Lecture Notes in Mechanical Engineering, https://doi.org/10.1007/978-3-030-33146-7_7

undesirable microorganisms from beer, and it is not necessary to use pasteurization in which the taste of beer is degraded and the process is more energy-intensive than filtration. The advantage of filters with nanofibers is their higher efficiency and lower pressure drop, which is achieved by a smaller filling of the fabric and thus increased porosity.

2 Experimental Work

In this work, an innovation of the experimental device for the production of filter candles will be carried out. Specifically, it is the production of electrostatic spinning within the framework of the project called "Coiled Coated Filter Cores of Composite Nanofibrous Yarns," which takes place at the Institute for Nanomaterials, Advanced Technologies and Innovations in Liberec (CxI). This work will mainly deal with the innovation of the device principle, i.e., the technological process, in which nanofibers are applied to the core to produce a composite nanofibrous yarn. The same type of yarn and the same filter core will be used for the upgraded device as in the previous solution.

The aim of device innovation will be above all to increase the productivity of nanofibers adding. Productivity is one of the key indicators from the point of view of industrial engineering and the improvement of this indicator is important for subsequent applications in a commercial environment. Another goal of the work will be to reduce the complexity of the current system. In this case, the reduction of this important parameter will be important in terms of maintenance (see Table 1). The device will be designed so that the composite nanofiber yarn is wound directly onto the core of the filter after drying. If the yarn was wound on the spool after the process, from which it would then be rewound to the core of the filter, wasting would be a waste of manipulation and a prolongation of the running time. For the correct

Table 1 Basic target parameters

Parameters	Values
Filter length (mm)	250
Filter inner diameter (m)	27
Filter outer diameter (mm)	45
Yarn fineness (tex)	80
Number of operators	2
Percentage of coating of yarn surface by nanofibers (%)	50
The speed of yarn passage through the device (m/min)	100
Regulation of tightening force during winding	Electronic, accuracy 0.1 N

filtering capability of the coil filter, it will be important to maintain a low twist of yarn passing through the device and the force it will tighten on the core of the filter.

3 Patent Research

3.1 Purpose of Research

The aim of the patent research is to explore technical solutions that address the issues addressed. This makes it unnecessary to rely solely on creativity on drafting concepts, but already validated solutions can be used. When researching patents, it is important to assess whether it is possible to adapt the patented solution for our case. For this purpose, the web site of the Industrial Property Office and Google Patents was used, and the search was focused on how nanofibers were created.

Search keywords: Produce nanofiber, Polymeric nanofibers, Nanofibers.

3.2 CZ 306428 B6

In view of the objective of this work, the most interesting patent No. 306428 [3] describes a new method for the production of nanofibers (see Fig. 1) and their application to the yarn. This is an AC production that is combined with the effect of "ballooning" the yarn to ensure the effective deposition of the resulting nanofibers. In the so-called ballooning yarn rotates at a larger radius than it would rotate when rotating around its own axis, so it is not necessary to use very high spinning speeds.

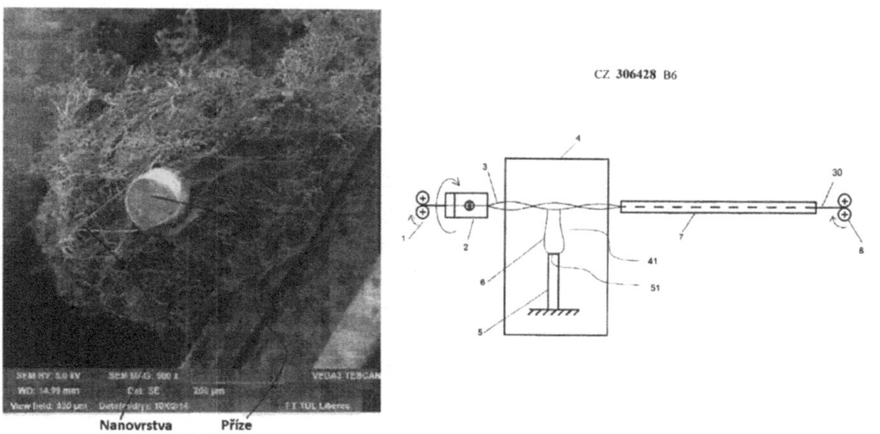

Fig. 1 Patent scheme and the layer of nanofibres [3]

The nanoparticle is uniformly applied in the form of a spiral and has a high consistency with the yarn. This method is incorporated into the following morphological matrix to be used for drafting concepts.

4 Concepts Generation

This chapter consists of draft concepts that should meet the set goals. First, a QFD matrix is developed to meet customer requirements, and a morphological matrix is used to support creativity, resulting in various concepts that are subsequently developed as 3D models. These concepts are then evaluated from various points of view and the AHP method is used to evaluate the appropriate option.

4.1 QFD

The QFD method is originated in Japan in the 1970s and is a structured process in which Voice of Customer is processed. This method ensures that customer needs will not only affect product design but also design production and service [1].

When constructing a device, it will be important to address the characteristics that have the greatest significance resulting from the QFD. The most important parameter is the use of a closed spinning space to meet the requirement for process stability and simplify cleaning. Furthermore, it is necessary to provide automatic control of yarn tension during winding and to use as many standardized parts as possible in order to make the machine simple to modify and repair. The built-up area should be as small as possible so that the facility can be operated in smaller laboratories and the target value is 10 m^2.

4.2 Morphological Matrix

The morphological matrix (see Fig. 2) serves for a clear and illustrative design of various designs of products and equipment. On the left part of the matrix, the individual functions that need to be provided are listed in the rows. There are various principles, kinematic schemes or arrangements to ensure the function. Further, various shapes and structures of the products or parts of the device may be developed. Concepts are created by linking individual cells (in each row to one cell) of the created matrix. Connections are represented by numbers 1–4, which are draft numbers [1].

As a principle of creating nanofibers, the morphological matrix includes only the production of electrical voltage, since it is the most mature solution. Both the DC voltage used in industry and the newly developed AC voltage technology are listed here. Following are the different types of yarn movement in the spinning space

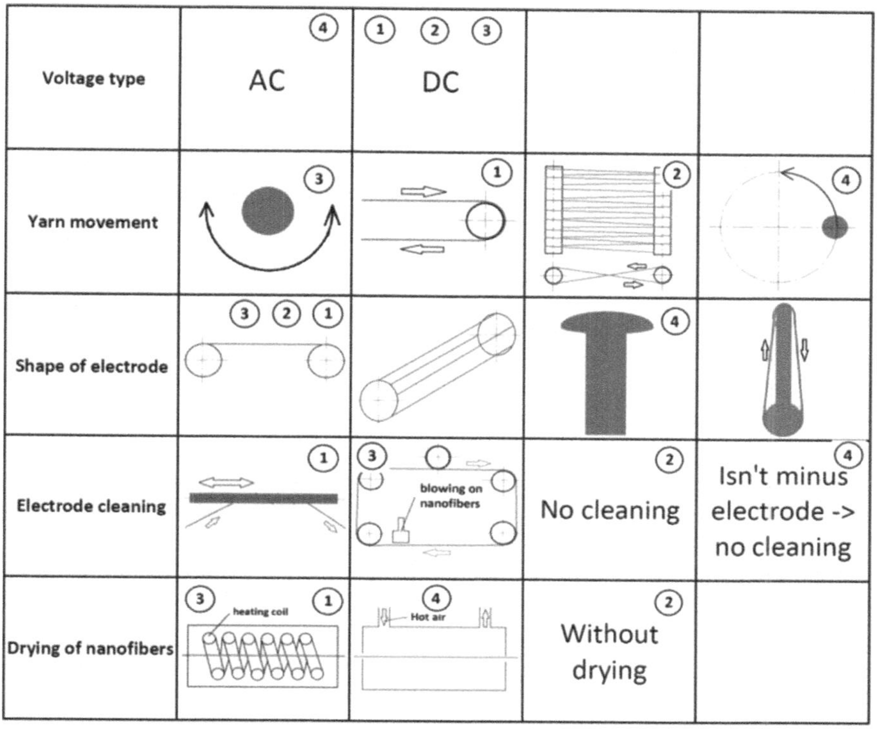

Fig. 2 Morphological matrix

and the shapes of the electrodes on which the thin film of the liquid polymer is. In addition, there are suggestions for electrode cleaning and possible ways to accelerate evaporation of solvent from liquid fibers by drying.

Four concepts have been created from this matrix, of which concept number 2 is the least radical and is rather the evolution of an existing solution. This variant was developed for a project running at the Institute for Nanomaterials, Advanced Technologies and Innovation, and the device will be made for experimental filter production. Conversely, concept number 4 is essentially different in all respects, rather than the current solution, due to the chosen voltage.

4.3 Selected Concept

The AHP method was used to select the design concept, in which the concepts were evaluated by five different weight criteria. The most appropriate concept based on AHP (see Fig. 3) was the alternating voltage variant, which was further elaborated.

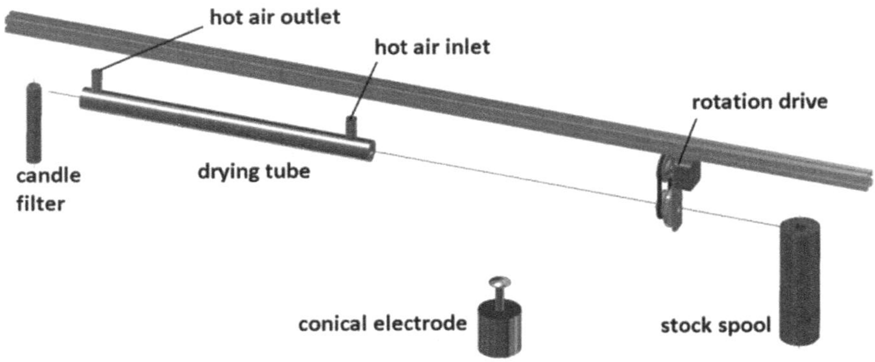

Fig. 3 Selected concept by AHP

In the selected design, the yarn is fed to a rotating mechanism that is driven over a belt drive by an electric motor. In the hollow pulley mounted on the bearing, there is an insert with holes. The holes are drilled at different distances from the axis of rotation, and by choosing the hole through which the yarn passes, we can change the thresholds of the balloon effect. Thanks to this effect, we can use a lower speed of the electric motor. The nanofiber siding is formed on the conical electrode and the liquid polymer is conveyed on its surface by an internal opening extending in the axis of the electrode. Nanofibres are applied in the form of a spiral to the yarn and, after application, passes through a drying tube. The hot air is fired at one end at the end of the drying pipe, leaving the evaporator at the other end with the evaporated solvent. Subsequently, the resulting composite yarn is wound onto the core of the filter [4].

5 Designing of the Selected Concept

5.1 Verifying Ballooning by Experiment

To verify the balloon effect, a simple experiment (see Fig. 4) was designed to find out what parameters the proposed design nodes would have to meet [4]. These are parameters influencing the shape of the balloon and thus the success of the application of nanofibres. The yarn was selected for the experiment with a maximum value of 200 tex for which the device is designed. This value has been selected to validate the intended maximum preload values. For thinner yarns, the required prestress will decrease due to less centrifugal force (Fig. 5).

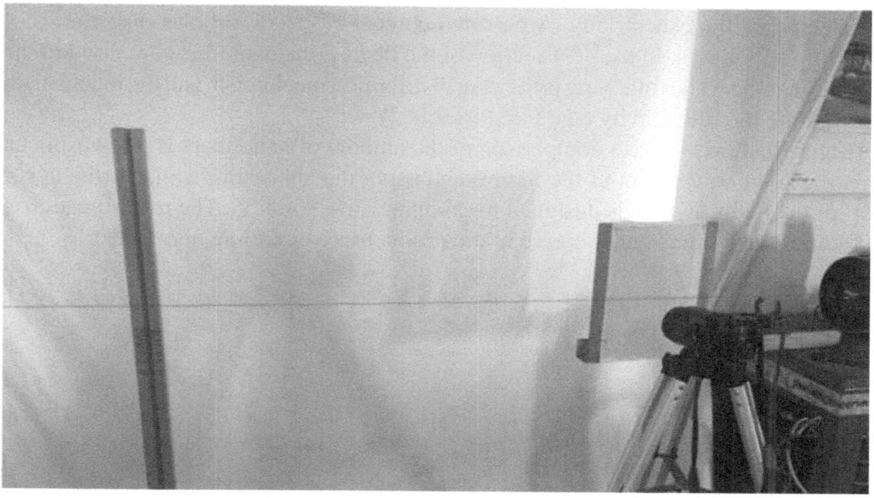

Fig. 4 Setup of experiment

Fig. 5 Picture of balloon

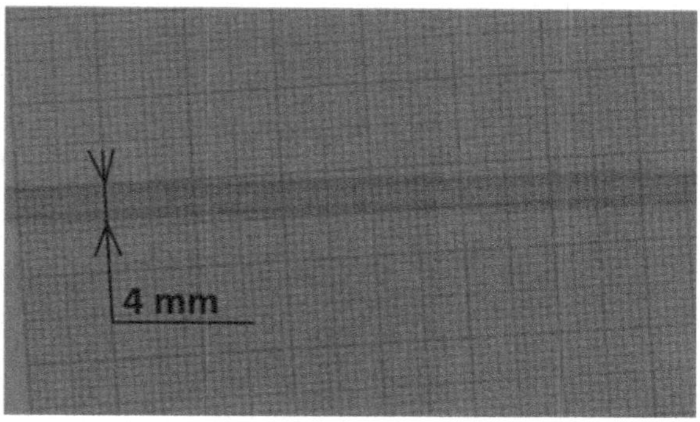

After each preparation, the two devices were started and ballooning images were taken during the stabilization. For a given speed, the balloon diameter should therefore be between 3.18 and 6.36 mm. The following figure shows that the yarn creates a balloon within a given range, and nanofibers should be applied without tearing the siding. The experiment performed provided information on the required parameters of the individual construction nodes.

5.2 Designed Device

The individual design nodes were placed into separate sections. In the first part, the yarn from the spool is unwound and the desired preload of the yarn and its rotation is selected by the eccentricity. A patented conical electrode follows the process of applying nanofibers. In the next section of the device, the nanofibers are dried in the drying chamber. The following is the yarn distribution mechanism and the mechanism for rotating the filter cavity (see Fig. 6; Table 2).

Figure 7 shows a direct comparison of the amount of nanofibers applied using an existing solution (bottom of the figure) and using the alternating voltage (the upper part of the figure) which is designed for an innovative solution. The magnification is 100 times, and there is an increase in the productivity of adding nanofibers.

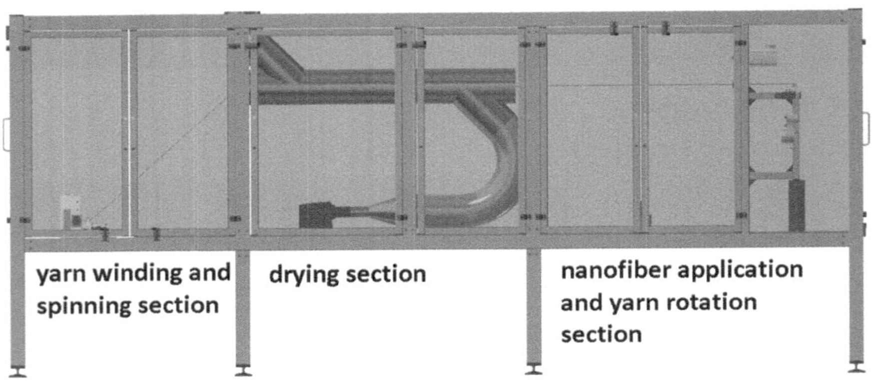

Fig. 6 Designed device

Table 2 Target and achieved values

Parameters	Values	
Filter length (mm)	250	200–400
Filter inner diameter	27 m	27 mm
Filter outer diameter (mm)	45	33–80
Yarn fineness (tex)	80	30–200
Number of operators	2	2
Percentage of coating of yarn surface by nanofibers (%)	50	70
The speed of yarn passage through the device (m/min)	100	100
Regulation of tightening force during winding	Electronic, accuracy 0.1 N	Electronic, accuracy 0.001 N

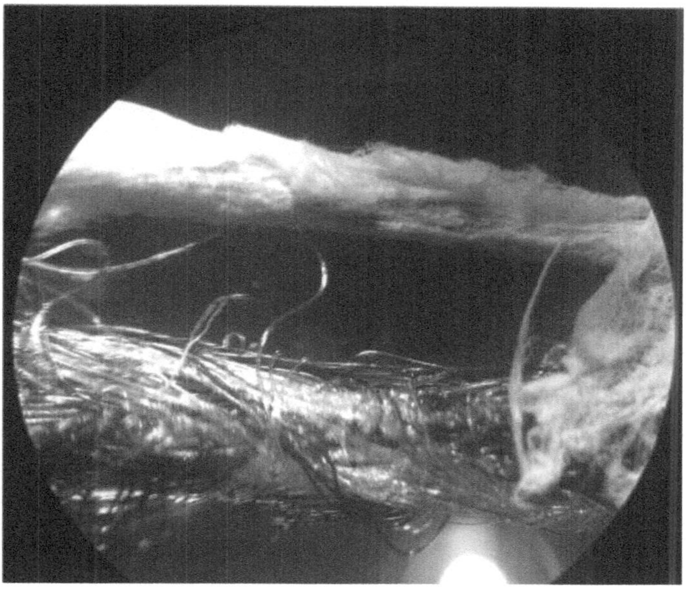

Fig. 7 Comparison of nanofibers layers

Acknowledgements This publication was written at the Technical University of Liberec as part of the project "Innovation of products and equipment in engineering practice" with the support of the Specific University Research Grant, as provided by the Ministry of Education, Youth and Sports of the Czech Republic, in the year 2018.

References

1. Mašín I (2012) Inovační inženýrství - Plánování a návrh inovovaného výrobku, 1st edn. Technical University of Liberec, Liberec
2. CZ 304137 B6. http://isdv.upv.cz/doc/FullFiles/Patents/FullDocuments/304/304137.pdf. Last accessed 2017/11/21
3. CZ 306428 B6. http://isdv.upv.cz/doc/FullFiles/Patents/FullDocuments/306/306428.pdf. Last accessed 2017/11/21
4. Lepšík P, Vodička J (2012) Plánování a řízení společných projektů, 1st edn. Technical University of Liberec, Liberec

Bearing Test Rig: Setting up and Lesson Learned

Jakub Chmelar, Pavel Mossoczy and Vojtěch Dynybyl

Abstract The paper presents recently built experimental rig for testing of rolling element bearings. Test rig comprises patented support of tested bearing that enables simulation of wide spectrum of operating conditions including external dynamic load. Forced recirculation lubrication system can maintain the temperature of oil in the bath and control the flow of supplied lubricant by a membrane pump. Sensors attached to the tested bearing provide information about inner and outer ring temperature and vibration signal. The device is additionally equipped by bearing monitoring system based on induced surface acoustic wave transmission characteristics that are sensitive to lubrication state and will be subject of prepared research. The chapter starts with a system-level overview that allows reader to get familiar with device before a detailed description of key components such as tested bearing housing, loading mechanism or control system. The experience obtained during the assembly and initial launch of rig in the form of lesson learned closes the paper.

Keywords Bearing · Lesson learnt · Lubrication · Slip rings · Temperature · Data acquisition · Lamb waves

1 Introduction

Developed test system allows experimental testing of rolling element bearings under simulated operating conditions. It serves as universal experimental device that can be equipped with wide range of sensors, to enable application-oriented research in area of bearings or applied tribology. The setup described in the current chapter is intended for testing of roller bearings, axially non-located on outer ring such as type

J. Chmelar (✉) · P. Mossoczy · V. Dynybyl
Czech Technical University in Prague, Technicka 4, 16607 Praha 6, Czech Republic
e-mail: Jakub.Chmelar@fs.cvut.cz

P. Mossoczy
e-mail: Pavel.Mossoczy@fs.cvut.cz

V. Dynybyl
e-mail: Vojtech.Dynybyl@fs.cvut.cz

Š. Medvecký et al. (eds.), *Current Methods of Construction Design*, Lecture Notes in Mechanical Engineering, https://doi.org/10.1007/978-3-030-33146-7_8

N306. It is prepared to simulate the dynamic loads in axial and radial directions in addition to radial static (pre)load. Sensors attached to the system allow detection of bearing temperatures on both rings; sense the vibration excited by tested bearing and using monitoring technology based on surface acoustic waves, the lubrication state of tested bearing is observed.

The device consists of four connected subsystems:

- Test rig.
- Control system and power system.
- Instrumentation.
- Data acquisition system (DAQ).

1.1 Test Rig

It is a cornerstone of the test system. As depicted in Fig. 1, it consists of central shaft (1) rigidly supported by preloaded pair of oversized taper roller bearings (2), housing that accommodates tested bearing (3), its supporting portal frame (4), linear hydraulic cylinder for external radial load (5) and a complete supporting frame (10). Important

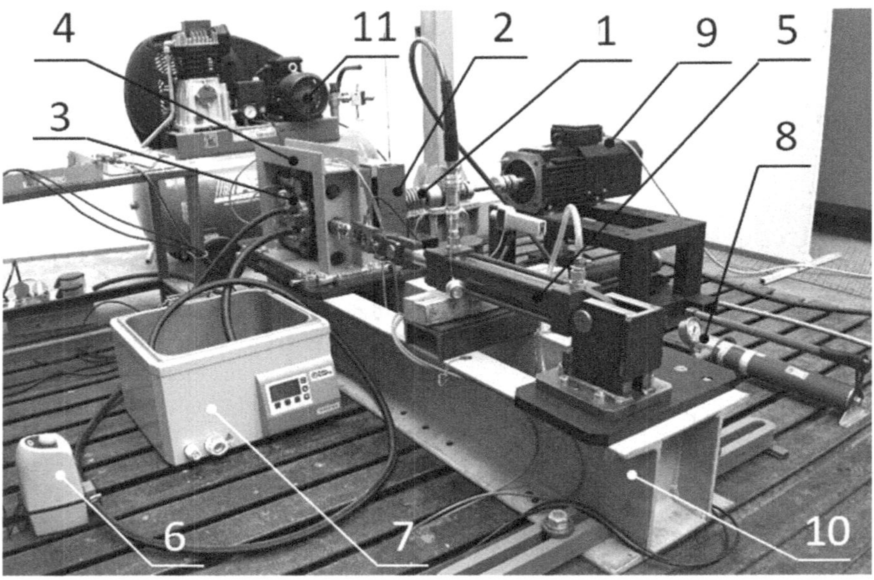

Fig. 1 Test rig: (1—central shaft; 2—support bearings housing; 3—tested bearing housing; 4—portal frame; 5—radial load cylinder; 6—membrane oil pump; 7—oil bath; 8—manual pump; 9—AC motor; 10—base frame; 11—air compressor)

part of test rig is a dedicated lubrication system for continuous oil recirculation (6, 7) and pneumatic compressor for compressed air-driven vibrators for simulation of dynamic load.

1.2 Control and Power System

It consists of a 2.2 kW AC (9) motor and a Variable Frequency Driver (VFD). The VFD drives the motor by an open-loop vector control method, allowing operation within required range of speeds, e.g., 500–3000 rpm. The motor design is optimized for operation in servo regime and includes an encoder (in current configuration not connected); separately driven fan that is especially important to improve cooling in low-speed operations, when the heat removal must be promoted to protect motor; and a thermistor (PTC) installed in the motor winding to check the temperature.

1.3 Instrumentation

It covers all sensors used on the test rig. Namely, it is a resistive temperature detector (RTD) built in the shaft to measure temperature of inner ring of tested bearing. Strain gauge-based load cell provides feedback of applied external radial load and a couple of interdigital piezoelectric transducers for excitation and sensing of surface acoustic waves (SAW), that are part of Bearing Monitoring System—BeMoS®, detection of lubrication state.

1.4 DAQ System

It aggregates and records all measured data from instruments and provides its visual output and preliminary real-time analysis to control the experiment during the course. The DAQ system consists of components by National Instruments, Ltd. (NI) and a LabVIEW data processing and visual programming environment.

2 Methods and Materials

The section provides a close overview of the most important components added with photographs from assembly and notes.

2.1 Key Components

Housing for Sample Bearing As referred earlier in the text, the construction of sample bearing housing allows testing of axially non-located roller bearings. Three rods attached to the housing, in detail depicted in Fig. 2 perform auxiliary support function, limiting the displacement of housing in axial direction. Rods are located in bushings of portal frame with clearance fit (clearance 0.05–0.1 mm), so tested bearing itself determines actual position of housing which mitigates the risk of its additional loading caused by, e.g., misalignment. The system enables clamping of tested bearing and its simultaneous loading by static radial and dynamic axial force when in the same time mitigating risk of additional undefined load received patent protection [1].

Fine-tuning of clearance between rods and bushings was conducted by extension of bushings pilot diameter during final assembly. The clearance has to allow spontaneous alignment of the housing when radial load is applied on the bearing. In ideal case, all rods shall be unloaded when no dynamic load is applied. The alignment was checked by run-out gauge attached to the front face of the central shaft and with stylus pointed to the front face of bearing housing as viewed in Fig. 2 (right). Acceptable value of maximal axial run-out over a complete revolution was 0.05 mm that was achieved.

Loading Mechanism Tested bearing operates under simulated radial load that is induced by the loading mechanism as viewed in Fig. 1. It consists of a hydraulic cylinder (5) radially attached to the housing of tested bearing, a manual hydraulic pump (8) that allows hydraulic pressure control and a load cell that provides a feedback of actual applied radial load force. Maximal working pressure of the system is

Fig. 2 Detail of sample bearing housing with guiding rods and connected hoses of lubrication system in the front. Radially are attached wires of BeMoS® sensors (left). A tool for check of misalignment attached to the front face of central shaft with stylus pointed to the bearing housing (right)

set to 100 bar that induces a force of magnitude 10 kN or 2 GPa of maximal Hertz contact stress on inner race of tested bearing. Manual pump was preferred over other considered option such as hydraulic power unit because it allows same functionality with significantly less demands on control, equipment and power. The pump works in closed hydraulic circuit with load cylinder. When pressurized, the check valve holds the pressure and force, respectively. To depressurize the circuit and release the applied load, the check valve is manually opened and the oil returns back to the built-in reservoir in the manual pump.

Lubrication System Lubrication system is one of the most important parts of the test rig. It allows control of lubricant's temperature and a supplied oil flow rate. Test rig is equipped by a closed-loop forced recirculation lubrication system, where oil bath and membrane oil pump are main components. The capacity of bath is 10 L, and it possesses of lubricant temperature control functionality up to 100 °C with precision of 0.1 °C. A membrane pump allows accurate flow rate adjustment in range of 0.3–1.3 L/min and can work with oil viscosities up to 140 cSt. Special attention was paid to the return pipe sizing during design phase, to allow free flow of lubricant from bearing housing back to the reservoir and prevent sample bearing over filling by lubricant.

In Fig. 3, there is depicted a section of housing with tested bearing. Inlet fitting was custom made with extended inside section (2), to ensure that lubricant flows directly to the bearing. The outlet (3) was designed to allow free flow of lubricant up to viscosity of 140 cSt. The chamber with sample bearing (1) is additionally equipped with a barrier (4) that holds minimal lubricant level to keep consistent lubrication state during the run.

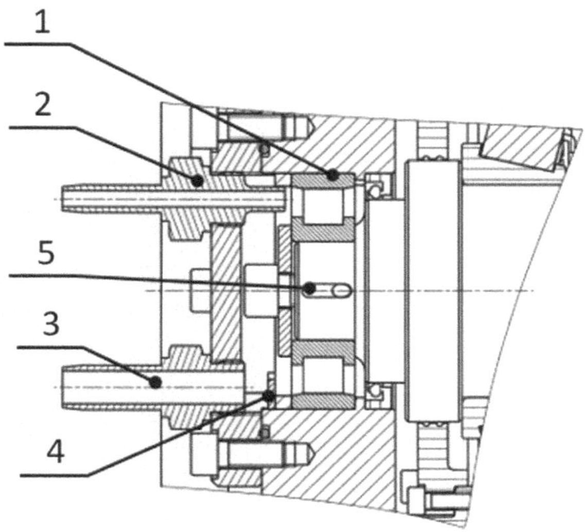

Fig. 3 Detail view: (1—tested bearing; 2—lubricant inlet; 3—lubricant outlet; 4—lubricant barrier; 5—groove for RDT)

2.2 Bearing Sensors

Temperature Sensors Temperature is a very reliable indicator of bearing's condition during operation. Test rig is equipped with sensors for monitoring both, the inner ring and outer ring temperatures. SAW sensors attached directly to the outer ring of sample bearing, described later in the text, are equipped with its own temperature sensors, so it provides information about outer ring temperature. A dedicated RTD sensor, type Pt100, integrated with the small groove under bearing seat as depicted in Fig. 5, enables monitoring of inner ring temperature. The sensor is in 4-wire configuration connected per Fig. 6, allowing full compensation of resistances due to lead. Wires are wired through a bore in the center of the shaft to the rear side of central shaft, where they are connected to slip rings and later to DAQ (see also Fig. 4).

RTD integration procedure involved the isolation of RTD's wires by a Pattex Repair Epoxy to ensure their separation and isolation on the side of probe and also to fix it in the place. It was followed by potting the sensor by high-temperature cement OmegaBond 600 in the groove, to finally seal the groove and promote thermal conduction from bearing to RTD. Repeated check of proper RTD connection was performed by measuring an isolation resistance of RTD wires against ground during the installation of sensor. After application of high-temperature cement, the isolation resistance rapidly dropped to 20 kΩ. An investigation exhibited that the cement was diluted by a small portion of water which evaporates during its solidification, but shortly after application, it caused the isolation resistance drop. When the measurement was repeated after two days, the resistance was back, at allowable value >50 MΩ. The cement was selected in this case, because after solidification it

Fig. 4 Slip rings and brushes installed at rear side of central shaft to transmit signal from inner ring RTD to DAQ system

is a hard matter that allows even repetitive change of tested bearing without making damage to it.

Lubrication State Sensors The test rig is equipped by pair of piezoelectric interdigital transducers that are used to excite and receive surface acoustic waves (SAW)—a Lambda waves or waves in transition regimes from a Lamb to Rayleigh [2]. Transmission characteristics of the waves, the decay and a group propagation velocity, possess of sensitivity to the presence of lubricant film in the race-element contact and its stiffness, so it makes them a promising bearing lubrication state monitoring method.

Sensors are the part of online lubrication monitoring system BeMoS®, produced by a BestSens AG company, which beside sensors, contains a unit that controls the excitation of SAW and runs the monitoring application, which performs online data analysis. User can observe the data in real time by a web application that runs in a control unit or can request their record to the flash drive.

The monitoring technology uses a high-frequency periodic excitation of SAW to the outer ring of bearing by the transmitter, and it is reading by sensor placed offset from emitter. There exist two basic modes of Lamb waves that could be excited by the transducer—the antisymmetric A0 and symmetric S0 [3]. Since the thickness of ring is of same magnitude as the wavelength, they interact with both surfaces during propagation as depicted in Fig. 7. Any disturbance (e.g., pass of element, presence of elastohydrodynamic lubrication (EHL) film or direct contact) on inner side of outer ring influences the propagation on the other side and measured it could be assigned with state of lubrication inside bearing. Connection example is viewed in Fig. 8. The interaction of propagation wave causes, e.g., the leakage of a portion of transmitted acoustic energy into lubricant, the transition of excited wave mode into other—usually a higher one [2], partial wave reflection because of high stiffness of contact [4], or change of group propagation velocity in the contact because of Lamb

Fig. 5 Integration of RTD sensor in the central shaft before potting

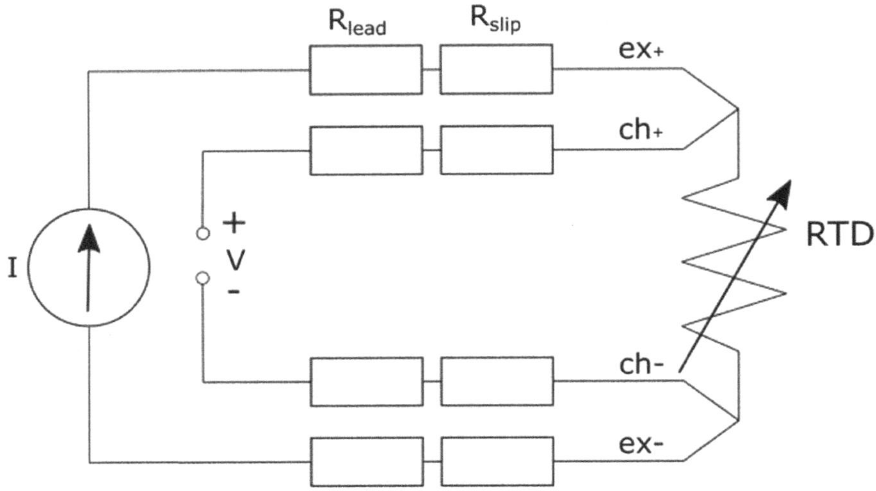

Fig. 6 Connection of RTD sensor to the DAQ system through slip rings. R_{lead} is 0.25 Ω/m and slip rings resistance R_{slip} is 0.5 Ω. Excitation current is 1 mA

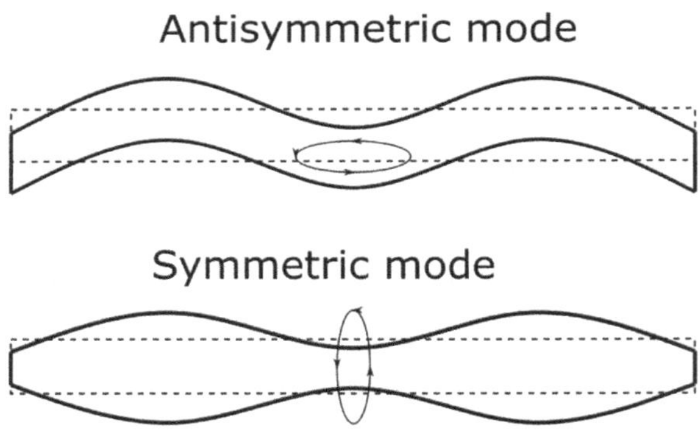

Fig. 7 Lamb wave modes. Antisymmetric mode-A0 and symmetric mode S0. The ellipses show a direction material particles motion. Modified from [3]

wave transition into other type [5]. The technology of SAW is well known and already widely explored and used for nondestructive testing (NDT), and the application in area of tribology is on the other side quite new.

DAQ Data Acquisition system is based on NI hardware (I/O device NI PCI 6221M; terminal block NI SCC-68 and modules: SCC-SG04 for strain gauge and SCC-RTD01 for RTD). The application for data record and real-time visualization was created in LabVIEW graphical programming environment. The app front panel displays following information: shaft speed and percentage of motor load that are obtained from VFD; applied radial force and inner ring temperature. All data are recorded with sampling frequency of 1 kHz, saved in text file for further detailed analysis. To promote the speed of data processing, the DAQ system uses a direct

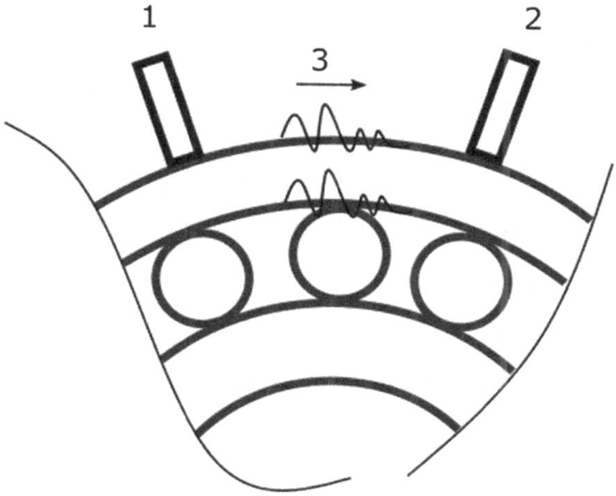

Fig. 8 Application of BeMoS® sensors to bearing. 1—transducer; 2—receiver; 3—propagating wave group. With attenuation when passed contact. Modified from [2]

memory access (DMA) that bypasses the host PC CPU, when data are transferred from first-in first-out (FIFO) buffer on DAQ device to host PC memory.

3 Lesson Learned

Whole process of test system development helped us to obtain valuable experience. Some of them are shared in this section and could be applied in various fields.

3.1 Preparation and Design and Management

It turned out to be a good move to begin the test system design with initial modeling phase of experiment that helped us drive decisions and sizing of experiment. Using of mathematical models of bearing and a lubrication layer, a range of tested bearing loads, its size and oil viscosities were balanced to enable simulation of wide range of conditions at a reasonable cost. Components that were during design phase left to be manufactured ad hoc caused during assembly majority of delays on the project schedule. So the level of virtual model preparation is very close connected with production time, not only in industrial environment but also in an agile laboratory workshop conditions. Always obtain a quotation for manufacturing of components from multiple sources. Offers may vary even by 100% and it is better having a backup company.

3.2 Assembly

It is more convenient to use adjustable connections when two or more components shall be precisely aligned instead of relying on accurate manufacturing and tolerances. Use spacer shims if there is a need for compensation of manufacturing inaccuracies. Establish a global datum plane on the device that are used not only for manufacturing purpose, but could be referenced during assembly and operation for, e.g., installing instruments. It really makes the life easier.

3.3 Sensors and Data Acquisition

Some bonding materials with isolation properties could contain conductive compounds that evaporate during solidification. Count with it and do keep calm, if, e.g., isolation resistance is not as high as it should be right after application. Electrical grounding of components is important. Always make sure that all components and sensors are properly grounded, to avoid problems during measurement.

To accelerate work with data during analysis, split the recorded data into 10 min intervals. It is easier to look up for specific events in the data when some logbook is held side-by-side the recorded measurements with reference to recorded intervals.

4 Conclusion

A description and experience from the development and construction of universal bearing test system were shared in this chapter. The test system will be used in the research of bearing lubrication layer detection by means of surface acoustic waves. Patented system is used for holding of tested bearing housing that enables simultaneous independent static radial load and dynamic axial load application. Test system is equipped by a forced recirculation lubrication system. DAQ system is based on hardware from National Instruments. The experiment control is by app made in LabVIEW environment.

Acknowledgements The project was supported by a CTU in Prague, grant: SGS18/127/OHK2/2T/12.

References

1. Chmelař J, Dynybyl V (2017) Zařízení pro testování provozních parametrů kluzných i valivých ložisek. CTU in Prague, Czech Republic. Patent No 307 380. Date: 16/5/2017
2. Lindner G, Brücker C, Schmitt M (2011) Online bearing lubricant sensing by mode conversion of surface acoustic waves. In: SENSOR + TEST conferences 2011; Proceedings SENSOR 2011, pp 45–52
3. Worden K (2001) Rayleigh and Lamb waves-basic principles. Strain 37(4):167–172
4. Drinkwater BW, Castaings M, Hosten B (2003) The interaction of Lamb Waves with solid-solid interfaces. AIP Conf Proc 657(1):1064–1071
5. Balvantín AJ (2018) Study of guided wave propagation on a plate between two solid bodies with imperfect contact conditions. Ultrasonics 83:137–145

Reverse Engineering of the Gearbox Case

Lukáš Jančar, Jiří Začal and Zdeněk Folta

Abstract The article deals with the reverse engineering procedure of the differential gear case from the racing BMW M6 for the GT3 series. The article contains a procedure for repairing a gear case with damaged functional surfaces after a bearing failure. During the drive, the bearing cage has probably been destroyed, causing the rolling elements to loosen and spill into the gearbox. The first part of the article deals with the creation of 3D data of damaged case and functional areas using 3D scanning by a portable handy scanner. The second part describes how to create CAD data for the machining of these surfaces and a description of the software used. In the end, accuracy checks are shown by comparing the generated CAD data with data obtained from 3D scanning. The article also outlines the repair procedure by re-welding the damaged surfaces and obtaining the zero areas needed for subsequent CNC machining.

Keywords Reverse engineering · 3D scanning · Gearbox case · Damage

1 Introduction

As part of the additive manufacturing and prototyping center ProtoLab [1], dealing with 3D printing and 3D scanning, we have been approached with the requirement to perform reverse engineering and design a follow-up repair of the gearbox for the racing BMW M6 for GT3 series. During the driving through the vehicle, the bearing cage has probably been destroyed, thereby releasing the rolling elements and spilling them into the gearbox. As a result, parts of the case made of magnesium alloy were damaged. Furthermore, due to bearing seizure, the functional surfaces for bearings housing and their locking were damaged. Due to the high cost of the new cabinet, the client decided to try to repair the damaged cabinet.

As a suitable repair method, 3D scanning of damaged areas including the surrounding parts of the case and subsequent virtual reconstruction of the surface was

L. Jančar (✉) · J. Začal · Z. Folta
Technical University of Ostrava, 17, Listopadu 15/2172, 70833 Ostrava, Poruba, Czech Republic
e-mail: lukas.jancar@vsb.cz

Š. Medvecký et al. (eds.), *Current Methods of Construction Design*, Lecture Notes in Mechanical Engineering, https://doi.org/10.1007/978-3-030-33146-7_9

chosen. This data was then used as input for CNC machining. Prior to this, the damaged areas were TIG welded [2]. A comparison of the accuracy of the scanned data with the CAD model was also performed.

2 Methods and Materials

2.1 The Case Preparation

The case was delivered in a disassembled state and without an oil filling. Before working, the cabinet had to be thoroughly cleaned. The rest of seals, grease, and other impurities have been removed. Subsequently, it was degreased with isopropyl alcohol.

Thanks to the good optical properties of magnesium alloy (both machined surfaces and also non-machined with a matte black finish), the cabinet did not need to be matted. The surface had a good reflection in the natural state, so there was no absorption of the laser beams or the opposite phenomenon, that is, excessive reflection and blinding of the camera. Otherwise, the surface will be dimmed with a special anti-reflective spray that will provide optimum reflectivity for the scanner, but will not change the external dimension of the part in a sensible way.

The scanned areas were then plastered with reference points (see Fig. 1). These are adhesive sticks with reflective centering and a matte periphery that serves to orient the scanner in space. It sticks at a distance of 20–100 mm apart, both on the scanned object and on the underlay, and the scanner should "see" at least 3 of them all the time.

2.2 3D Scanning

The 3D scan itself was made with the Creaform HandySCAN 700 contactless laser scanner with a high accuracy of up to 0.03 mm [3]. Scanning was utilized in the form of seven laser crosses, while the corners and holes were scanned using a single laser line. The ideal shutter time setting for the camera was 4.13 ms. Due to the high precision required of the resulting data, it was necessary to scan with the highest possible resolution of the scanner, which is 0.05 mm. This results in work with large volumes of data and the need for powerful hardware, but it is possible to precisely scan sharp edges and corners without undesirable distortion, as shown in Fig. 2. The resolution indicates the size of the sides of the triangles of the polygon network being scanned.

The creation of the scan was performed in the VX elements software (see Fig. 3). The program renders scanned data in real time and allows you to work with it when scanning. In this way, the scanned space was clipped to work with less volume of

Fig. 1 Gearbox case with damaged bearing housings. Prepared to be scanned

data, and alone points, noise, and gaps in the network were removed [4]. Due to the need to scan only the parts of the case, the housing areas and their surroundings, only one scan was needed, and the case did not need to be rotated and the scans later composed into one image. The program also captures the areas beneath the reflective targets so that it creates a continuous network where it is only necessary to guard against the absence of cracks and large areas with missing points.

The output of the scan is the cloud of points, respectively, polygonal network in STL format (see Fig. 4), which is further processed in the CAD program. In this case, it was a network with almost 5.5 million elements, which was caused by scanning in the highest possible quality.

2.3 Creating CAD Data

The STL network model created by scanning was further processed in Geomagic Design X software. This software allows you to work effectively with polygon networks and use them as a reference. It is therefore possible to extract boundary curves, cross sections, or single points from networks and use them for volume modeling.

In the program, the negative shape of cylindrical surfaces was modeled. As a basic reference, a dividing plane between the gearbox and its cover was chosen. The plane

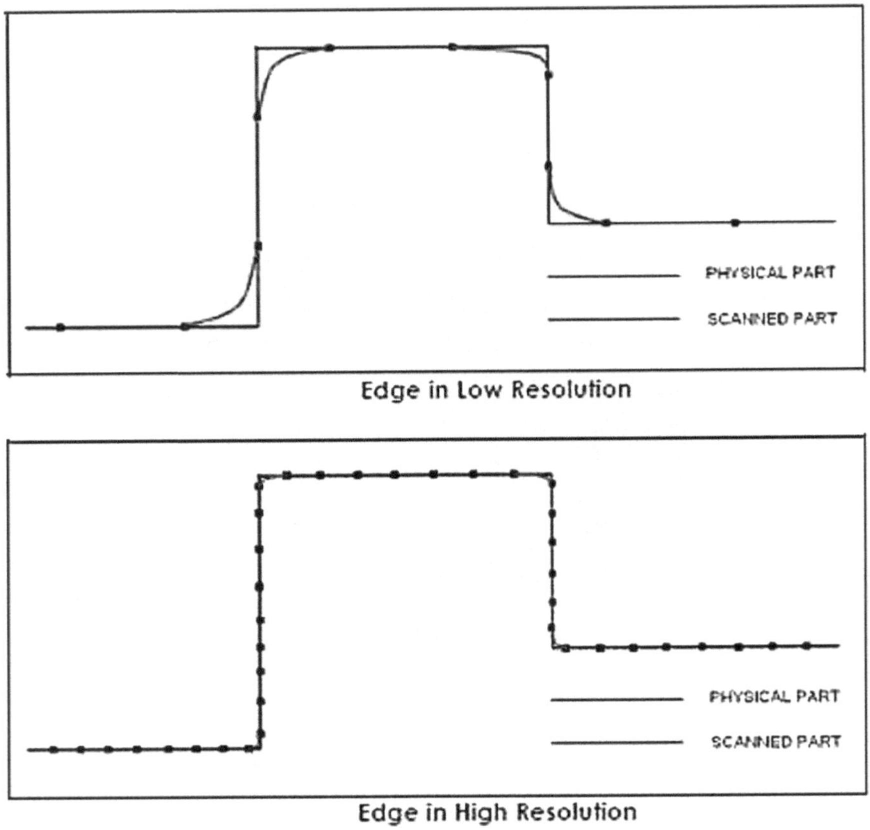

Fig. 2 Effect of the resolution setting

was extracted from the scan data using a program, ensuring that the plane is touching as many elements as possible from the upper surfaces. The second reference was the axis. This must lie in the dividing plane and be at the same time the axis of rotation of all cylindrical surfaces. It was obtained by two cross sections perpendicular to the dividing plane and made at the bearing place. In the cross section, the boundary curve of the network was always extracted and was replaced by a circle whose center lay on the perpendicular projection of the dividing plane into the plane of the cut [5]. The reference planes and vectors are shown in Fig. 5.

The second reference plane perpendicular to the dividing plane was then intersected by the axis obtained. With it, a cross section was performed through polygonal network again. The curves obtained were converted to a polygon formed by mutually perpendicular segments, and the polygon was rotated around the reference axis. The result of this rotation was the negative shape of the machined cylindrical surfaces. Some important dimensions (those for bearings and shaft seals) were then adjusted

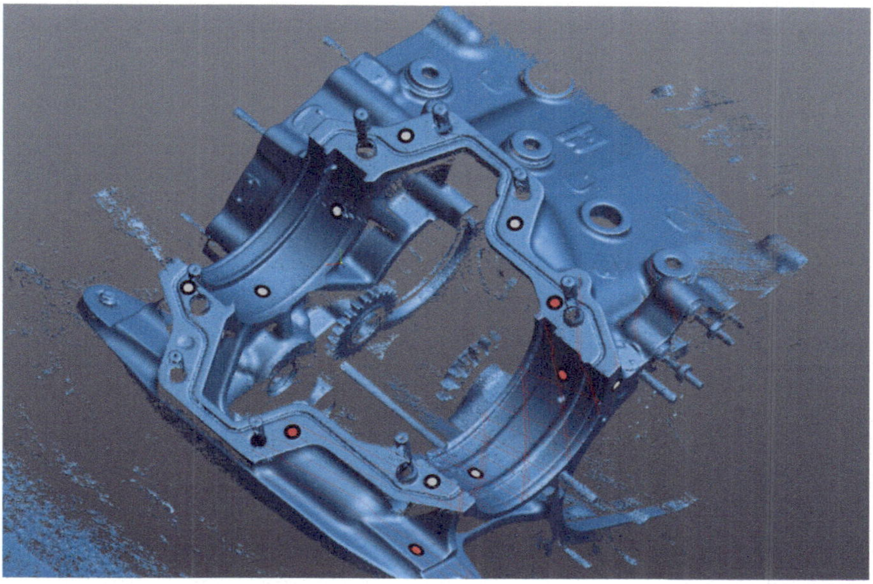

Fig. 3 Gearbox case being scanned in VX elements software

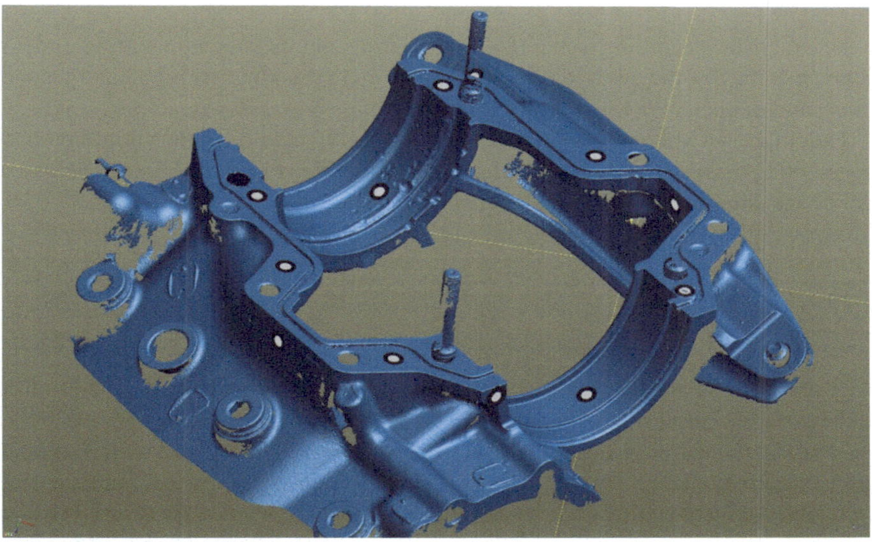

Fig. 4 Final STL model of the gearbox case, obtained by 3D scanning

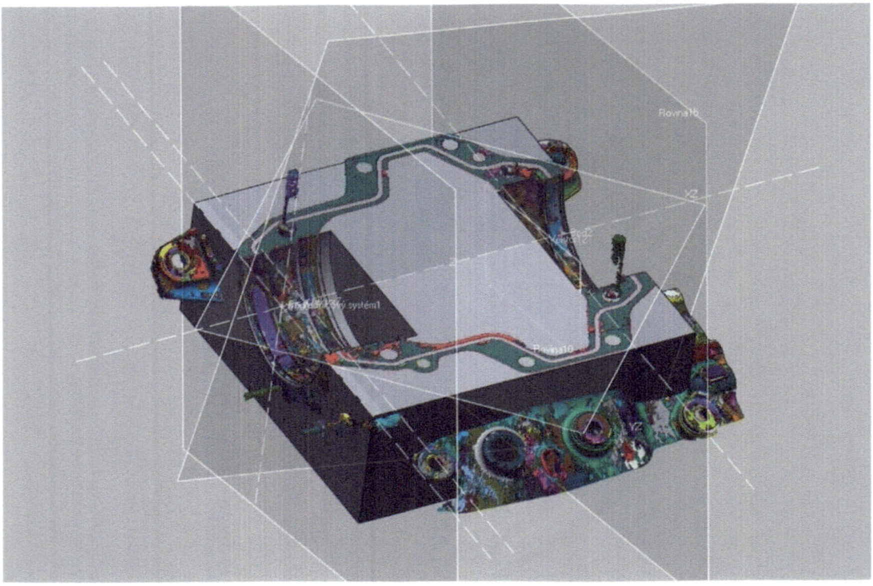

Fig. 5 Reference planes and vectors

to the exact size according to the mounted bearing and its known mounting dimensions. Specifically, they were diameters Φ 130 mm for bearings (first ball and second roller) and diameters Φ120 mm for shaft seals.

The rotational volume thus created was then subtracted from the simplified box representing the rest of the case. The resulting model is in Fig. 6. Only rotary recesses will be machined.

2.4 Accuracy Checking

After finishing the model, a final accuracy check was performed. Geomagic Control X program was used for it. The program is able to compare the deviations of the CAD model compared to the scanned network (see Fig. 7). The maximum deviation of the created model was set to ±0.1 mm. After program control, it was found that all areas surveyed were within this range. The deviations of the machined surfaces that were examined are then within ±0.05 mm. The model was found to be sufficiently accurate and suitable as a data input for the machining center.

Based on the knowledge of used bearings, locking rings, and shaft seals, tolerances of machined diameters have been modified. Specifically, the diameter Φ130G7 for bearings and slot width variations according to the manufacturer's specifications.

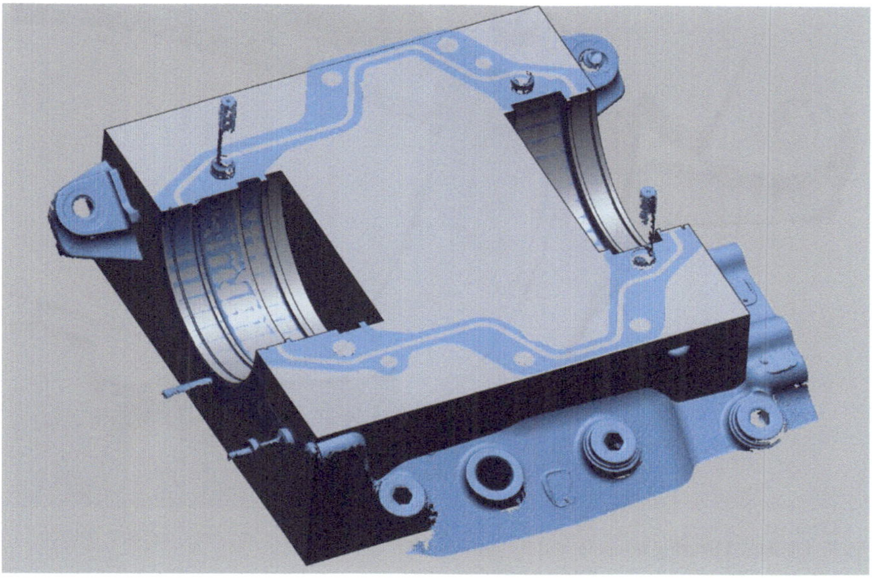

Fig. 6 Final CAD model as input for CNC machining

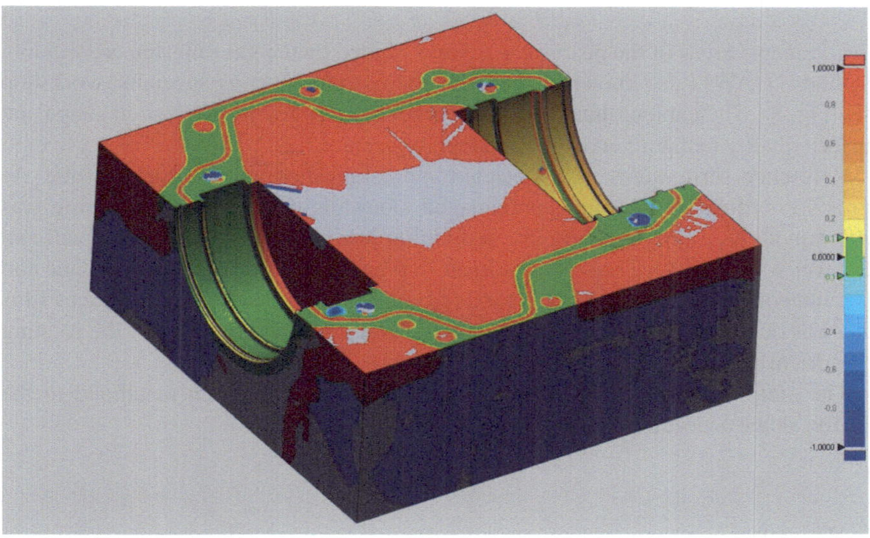

Fig. 7 Accuracy control of the CAD data

Fig. 8 Damaged bearing housing and its surrounding

2.5 Repair Procedure

The damaged areas of the gearbox were first welded by the gas tungsten arc welding method (GTAW) (also known as the TIG method) in a welding specialist workshop (see Fig. 8). The cabinet thus prepared has been machined to the desired shape and surface quality using a 5-axis CNC machining center.

Reference zero values were detected and established using a touch probe. As zero X-coordinate, one of the undamaged sidewalls of the bearing housing was determined. As a zero Y-coordinate, a dividing plane between the cabinet itself and its cover was used. As a zero Z-coordinate, a plane perpendicular to the Y plane and simultaneously passing through the axis of rotation of the cylindrical surfaces (this axis was detected by touch probe at 3, respectively 6 different points of the rotating cylindrical surfaces) was chosen.

The clamped cabinet with the set reference values was then machined to the desired shape according to the pre-created CAD data.

3 Results

The reconstruction of damaged gearbox surfaces has been carried out as described above. These areas were then professionally welded and processed on the basis of the created CAD data to its original shape. The successfully repaired case was re-mounted to the vehicle and continued to be used in races.

4 Discussion

Thanks to the use of high-precision 3D scanning, it was possible to obtain good documentation for repairing the gearbox even without knowing the original manufacturing documentation. High-quality scanning has been key to getting input data with the necessary accuracy. The data thus created could be used for surface manufacturing thanks to the use of a modern CNC machining center with high precision and accuracy. Thanks to 3D scanning, it was also possible to precisely specify zero planes for the machining center.

5 Conclusions

The article describes the reverse engineering of the racing differential gearbox case for the BMW M6 for the GT3 series. The article contains a procedure for obtaining a geometry of a case with damaged functional surfaces after a bearing failure, a procedure for the reconstruction of surfaces, and a way of repairing the case.

The main contribution of the article is the practical verification of the functionality of such a repair method and the identification of the ideal reverse engineering procedure as such. Precisely performed 3D scanning has been found to be a good way to obtain accurate input for CAD data creation and also allows you to compare the accuracy of this data with the scan data. During the virtual reconstruction of surfaces, it has been created a reliable and repeatable way to convert scanned data into an editable CAD model.

The repair method described in the article is functional and feasible using modern scanning and machining technologies. The repaired cabinet has been refitted to the vehicle and is further used.

References

1. ProtoLab Homepage. https://protolab.cz/. Last accessed 2018/06/13
2. Milroy MJ, Weir DJ, Bradley C et al (1996) Reverse engineering employing a 3D laser scanner: a case study. Int J Adv Manuf Technol 12(111):112–114
3. Creaform Homepage. https://www.creaform3d.com/en/metrology-solutions/portable-3d-scanner-handyscan-3d#section-1. Last accessed 2018/06/13
4. Zhaohui G, Bopaya B (2017) Review of reverse engineering systems—current state of the art. Virtual Phys Prototyping 12(2):161–172
5. Jovanović V (2011) Reverse engineering as a product design tool. In: Proceedings of the XV international scientific conference on industrial systems (IS'11). IS, Novi Sad, pp 66–70

Optimization of Driving Comfort of Golf Cart

Radka Jírová, Lubomír Pešík and Roman Svoboda

Abstract The paper is focused on the minimization of vibration transmitted from electric golf cartwheels to seat and seated person. This vibration appears during golf cart driving on uneven terrain of the golf court. The paper also focuses on improving the driving comfort of transported persons on an electric golf cart. This problem can be solved according to (Svoboda et al. in Bag Development s.r.o. Golfový vozík s bateriovým elektrickým pohonem a sedákem [1]) using the elastic and damping link between cart frame and seat tightly connected to battery of electric power. The designed solution of an additional vibration isolation device comes from measurement in real conditions of kinematic values of the vibration on seat and from experimental laboratory measurement of dynamic parameters of the system. Based on acquired data, the mathematical model of the system was made and dynamic parameters were optimized. This mathematical model of designed solution describes golf cart as system of masses connected with elastic and damping links. The result of mathematical description of designed solution is the reference step response of golf cart seat.

Keywords Golf cart seat · Minimization of vibration · Suspension seat

1 Introduction

Suspension seats are used in a wide range of vehicles, especially for the purpose of protecting person's health exposed to whole-body vibration during travel and also for comfortability increase. Driving on uneven terrain of golf court leads to high-intensity vibration of wheels which is transmitted to frame, seat, and seated person.

R. Jírová (✉) · L. Pešík
Technical University of Liberec, Studentská 1402/2, 46117 Liberec, Czech Republic
e-mail: radka.jirova@tul.cz

L. Pešík
e-mail: lubomir.pesik@tul.cz

R. Svoboda
BAG DEVELOPMENT s. r. o, Michalovice 20, 29301 Mladá Boleslav, Czech Republic

Š. Medvecký et al. (eds.), *Current Methods of Construction Design*, Lecture Notes in Mechanical Engineering, https://doi.org/10.1007/978-3-030-33146-7_10

Conventional battery-powered electric golf carts do not have suspension seats, for the purpose of transmitted vibration reduction to seat and seated person the elastic and damping link between chassis and frame of golf cart is being used. These links reduce transmission of vibration only partially, so the drive in golf cart can get uncomfortable.

2 Minimization of Transmitted Vibration

The principle of minimization of vibration transmitted from golf cartwheels to the seat and seated person is subject to patent application. Vibration transmission can be minimized according to [1] by using elastic and dumping link placed between golf cart frame and seat tightly connected to battery of electric power. Battery of electric power tightly connected to seat increases inertial effect of mass, which allows reducing vibration on the seat (Fig. 1).

According to [1], vibration can be effectively minimized by using elastic and dumping link placed between golf cart frame and batteries of electric power and between batteries of electric power and seat. Connecting battery of electric power to the seat using elastic and dumping link allows effective adjustment considering vibration minimization (Fig. 2).

In both cases, it is possible to add guiding mechanism to elastic and dumping link, which contains a linear guiding for inertia mass of battery of electric power and a

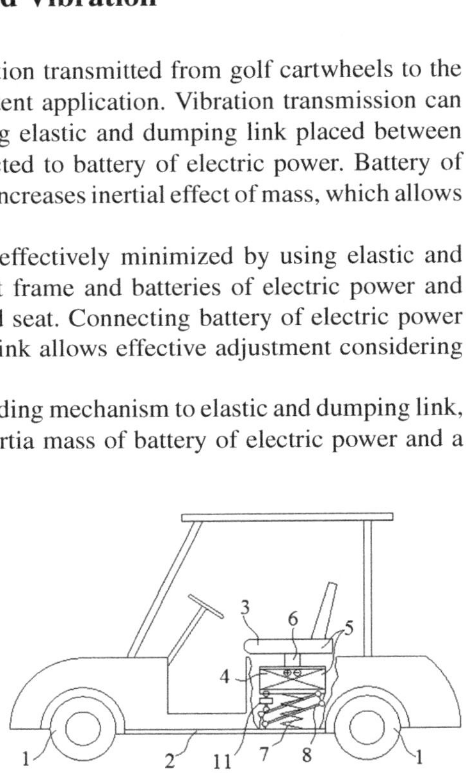

Fig. 1 First principle of minimization of vibration

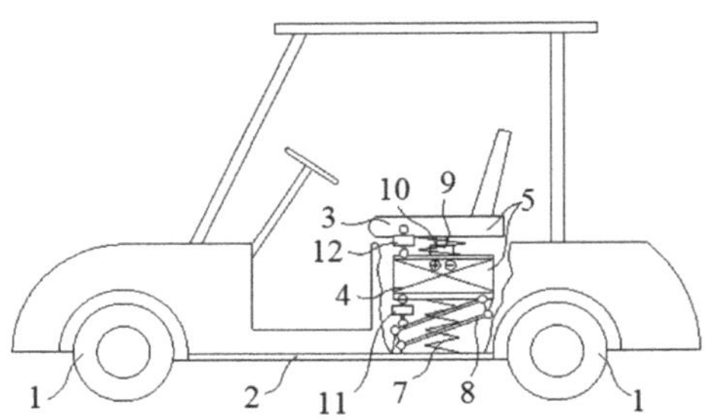

Fig. 2 Second principle of minimization of vibration

seat or for battery of electric power only. This guiding mechanism releases friction that can be conveniently used for dumping oscillating movement of seat and battery.

3 Design of Suspension Seat

Vibration transmission is minimized using elastic and dumping link placed between golf cart frame and seat tightly connected with battery of electric power. To these links, a guiding mechanism is added, which ensures desired movement of the seat against the frame and simultaneously allows optimal transformation of elastic and dumping link [2].

The design solution is displayed in Figs. 3 and 4.

The design solution preferably has two pneumatic springs that are connected to the golf cart seat and the electric drive batteries by a lever mechanism [3]. The lever mechanism is attached to the golf cart with the spherical roller bearings and allows the seat lift to be ±4 cm. A compressor is used to regulate the pressure in the pneumatic springs to allow for optimal pressure adjustment and hence the stiffness of the pneumatic spring for a driver's given weight and possibly the coach of the golf

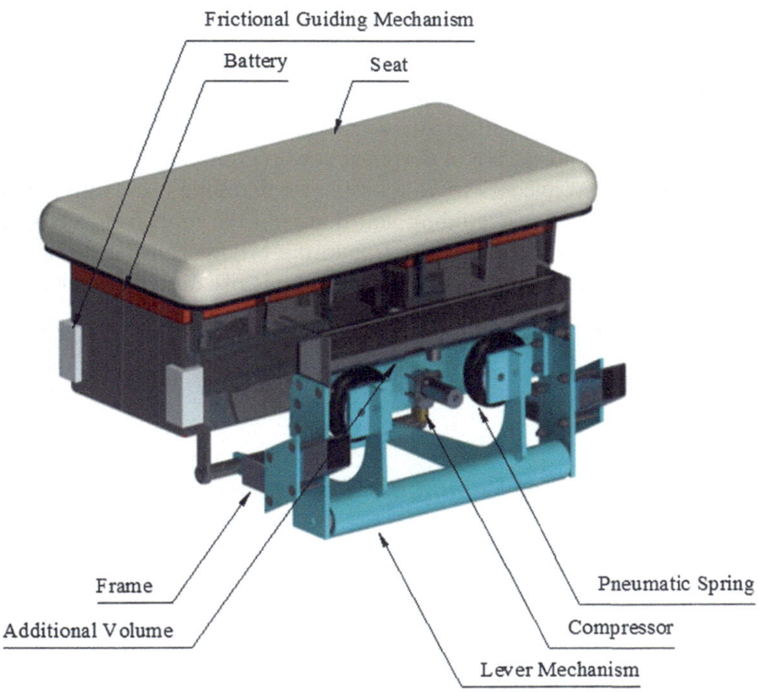

Fig. 3 Design of suspension seat—back view

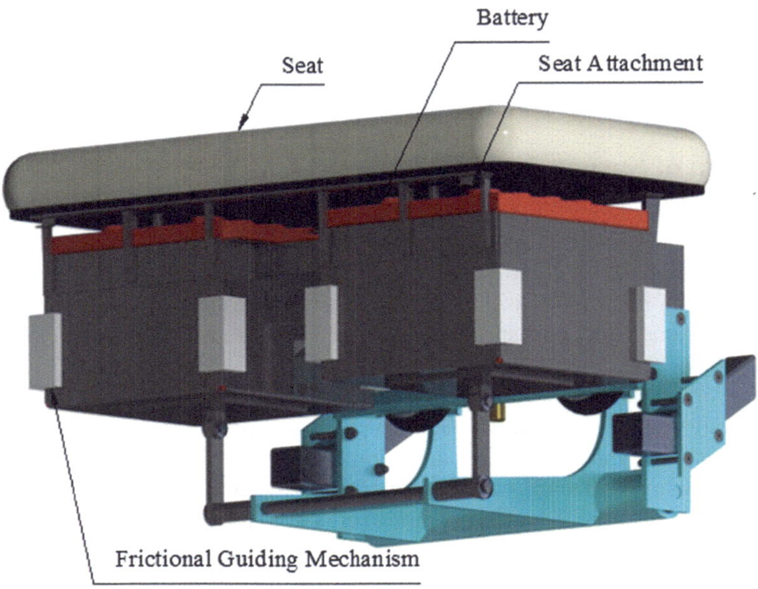

Fig. 4 Design of suspension seat—front view

cart. The compressor controls are located under the electric golf cart drive controls [4]. The pneumatic springs are supplemented with additional volume to achieve lower intrinsic frequency of the system and thus more efficient vibration isolation.

Batteries of electric power are in designed construction led linearly using linear guiding mechanism that releases friction usable for dumping oscillating movement of seat and battery. For linear guiding and dumping the movement of battery of electric power tightly connected to the frame, the original plastic design of golf cart is conveniently used. Original structural elements of golf cart are also used for connection of battery of electric power to the seat.

Batteries of electric power are connected to the lever mechanism using two rotational links.

3.1 Transformation of Pneumatic Spring Stiffness

Transformation of pneumatic spring stiffness is used for optimal adjustment of vibration isolation system. It is based on using mechanical transfer. Mechanical transfer transforms spring stiffness with transfer quadrate coefficient. Inference of stiffness transformation is evident from Fig. 5.

Equilibrium moment of lever mechanism is

$$F_{\mathrm{S}}l = F_{\mathrm{k}}b \tag{1}$$

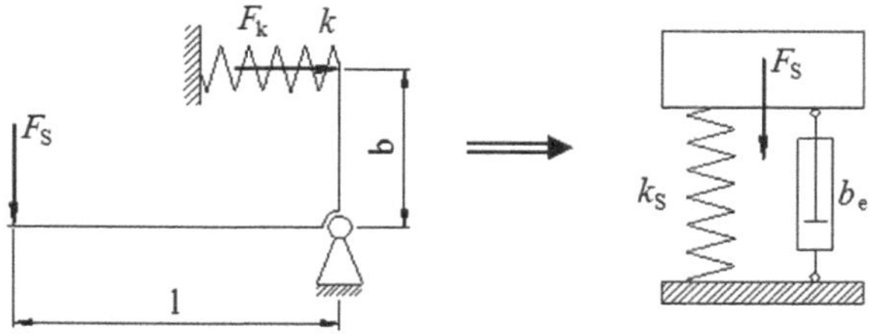

Fig. 5 Transformation of pneumatic spring stiffness

where F_S is force from batteries of electric power, seat, and seated person; F_k force from pneumatic spring; l lever length of lever mechanism; and b lever high of lever mechanism.

Lever deformation ration and length lever ratio are equal

$$\frac{\Delta l}{\Delta b} = \frac{l}{b} \tag{2}$$

and stiffness k_S at location of loading force F_S is

$$k_S = \frac{F_S}{\Delta l} = \frac{F_k b}{l} \frac{b}{l \Delta b} = \frac{F_k}{\Delta b} \left(\frac{b}{l} \right)^2 = k i^2 \tag{3}$$

where k is pneumatic spring stiffness and i transfer coefficient.

4 Effects of Vibration on Comfort

Evaluation of effect of vibration on comfort of seated person during the golf cart drive was made in accordance with [5]. Result of evaluation of effect of vibration is frequency-weighted r.m.s. acceleration of measured acceleration as a function of time during the golf cart drive and comparison this acceleration value with table in [5]. Frequency-weighted r.m.s. acceleration for the one-third octave band is shown in Fig. 6.

Comparison of frequency-weighted r.m.s. acceleration values of original state and new design solution shows that, due to the suspension of seat tightly connected to battery of electric power, frequency-weighted r.m.s. acceleration decreased almost by 57%.

5 Conclusion

The paper deals with current issue of vibration isolation of golf cart seat. The solution uses invention idea of increase inertia of tightly connected mass to seat. This allows significantly to reduce vibration transmitted from golf cartwheels to driver and passenger.

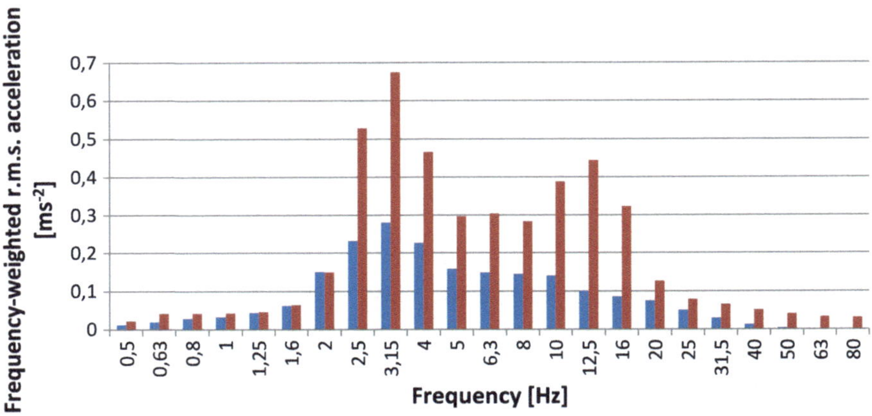

Fig. 6 Frequency-weighted r.m.s. acceleration for the one-third octave band

Solution described in the article can be used in mobile appliances and vehicles in cases when its construction allows connecting the seat to another object that has a large weight (e.g., batteries, pumps, drive units).

Acknowledgements This publication was written at the Technical University of Liberec as part of the project "Innovation of products and equipment in engineering practice" with the support of the Specific University Research Grant, as provided by the Ministry of Education, Youth and Sports of the Czech Republic in the year 2018.

References

1. Svoboda R, Jírová R, Pešík L (2018) Bag Development s.r.o. Golfový vozík s bateriovým elektrickým pohonem a sedákem. Česká Republika. Přihláška vynálezu PV 2018-112
2. Dresig H, Holyweißig F (2008) Maschinendynamik, 8th edn. Springer, Berlin
3. Pešík L, Skarolek A (2011) Tuning of vibration isolation differential pneumatic system by means of trottle valve. In: Transactions of the Universities of Košice: research reports from the Universities of Košice, vol 2. Technická univerzita v Košiciach, Košice, pp 191–196
4. Pešík L, Skarolek A (2012) Heat transfer effects on vibration isolation differential pneumatic system. In: Transactions of the Universities of Košice: research reports from the Universities of Košice, vol. 3. Technická univerzita v Košiciach, Košice, pp 109–114
5. ČSN ISO 2631-1 (1999) Vibrace a rázy – Hodnocení expozice člověka celkovým vibracím – Část 1: Všeobecné požadavky. Český normalizační institute, Praha

Creation of a Computational 2D Model of Harmonic Gearbox

Maroš Majchrák, Róbert Kohár, Michal Lukáč and Rudolf Skyba

Abstract A harmonic drive gear is, in fact, a differential gear with front-wheel drive in which the gear is achieved by elastic deformation of the flexible spline. The spline harmonic drive gear is typical for its unique structure called "strain cam wave generator". The principle of functioning of these gear systems is based on the mechanism of rotationally elliptical deformation of the flexible spline using the elliptical wave generator and subsequent tooth generating. When dealing with these types of gear systems, we come across the term "backslash", which is a surplus clearance between the generating teeth. When designing a harmonic drive gear for robots, it is important that the backslash is as low as possible. This chapter focuses on the preparation of the finite element model of a harmonic drive gear. It provides readers with a complete overview of the functions of these special, really precise and stiff gear systems and their properties. Harmonic drive gearing systems have been widely used in the area of precise control. Precise reduction and positioning have been mainly used in robotics. Its high rate of precision and weight has been used in the aviation industry. The system is also used in the engineering industry for transmission of high torque, conveyor and turbine drives and in many other machine applications.

Keywords Harmonic gearbox · Strain cam wave generator · Elastic deformation · Backslash

1 Introduction to Creating a Computational 2D Model

1.1 Introduction to the Topic

A harmonic drive gear is composed of three main parts (see Fig. 1). While generating the harmonic drive gear, flexible deformation of three reaction components occurs. After having been pressed into the gearbox bearing, wave generator deforms

M. Majchrák (✉) · R. Kohár · M. Lukáč · R. Skyba
University of Žilina, Univerzitná 8215/1, 01026 Žilina, Slovak Republic
e-mail: Maros.Majchrak@fstroj.uniza.sk

Š. Medvecký et al. (eds.), *Current Methods of Construction Design*, Lecture Notes in Mechanical Engineering, https://doi.org/10.1007/978-3-030-33146-7_11

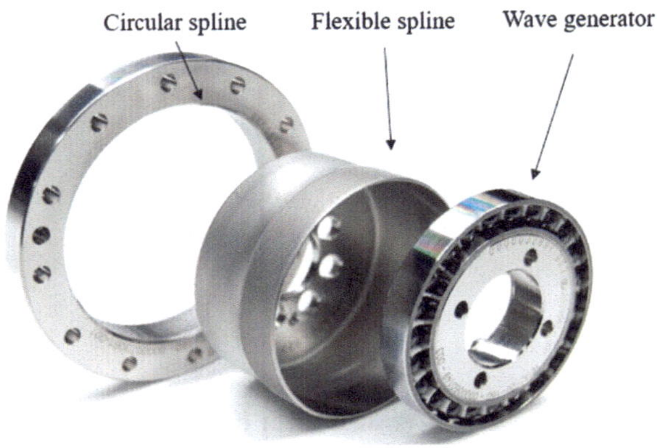

Fig. 1 Main members of the harmonic drive gear [1]

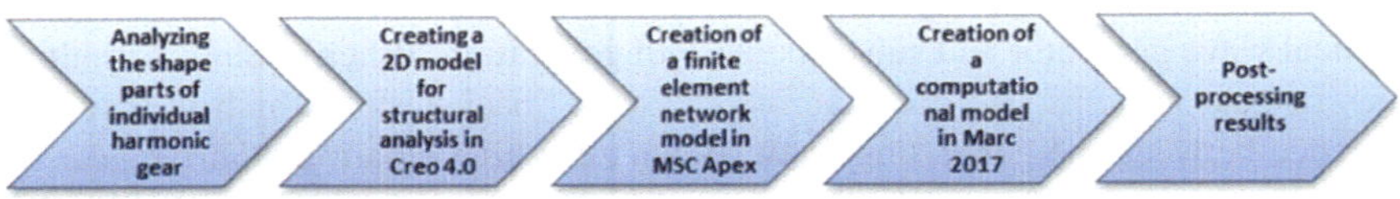

Fig. 2 Procedure of creating a computational 2D model

an internal rolling-element bearing with its elliptical shape. Through the balls, the flexible deformation is transferred through these components to the external bearing ring, deforming a flexible spline [1].

When the wave generator rotates, a complex movement occurs, a flexible spline in the circular spline, generating the basic movement of the harmonic drive gear [2].

This task is very difficult to solve from the analytic point of view; therefore, this article is centred on creating a computational finite element 2D model of a harmonic drive gear. Before creating the computational model, it is necessary to perform a detailed analysis of the geometry as well as material properties of individual harmonic gear components. Several action members are deformed simultaneously. The 2D model was created in a non-deformed shape. The procedure for creating the model is shown in Fig. 2.

2 The Description of Creating a Computational 2D Model

2.1 Analysing the Shape Parts of Individual Harmonic Gears

The analysis of the geometry shapes of individual harmonic gear parts was performed in MATLAB. Several wave generator shapes were generated based on the equation of the classical ellipse and the special equation of the ellipse with two coefficients

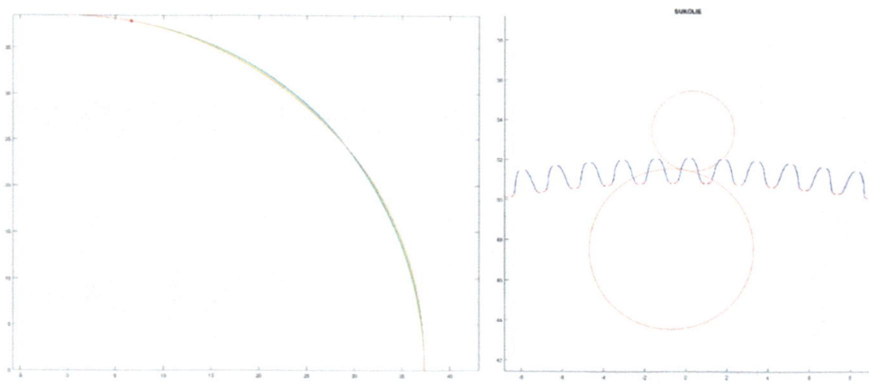

Fig. 3 Geometry shapes of the wave generator and the circular spline in MATLAB

A and B that describe the elliptical shape. In MATLAB, there were also respective shapes of teeth generated for cycloidal tooth profile of a flexible spline and a circular spline, for 200 and 202 teeth (see Fig. 3), and wave generator shapes. The geometry was saved individually in a set of points with x and y coordinates in the TXT format. The geometry and the shape of the flexible rolling bearing are indicated by the manufacturer in their catalogue [2, 3].

2.2 Creating a 2D Model for the Structural Analysis in Creo 4.0

For future analyses, a 3D model of a harmonic drive gear was created. As the deformation of the individual drive gear elements is too complex, the elements were also modelled in a non-deformed shape. The geometry of the tooth shapes was imported from MATLAB. The 3D and 2D models of the harmonic gear are shown in Fig. 4. The curves were then imported into the MSC Apex pre-processor, where the FEM network was created.

2.3 Creation of a Finite Element Network Model in MSC Apex Grizzly

After having loaded the geometry into the MSC Apex pre-processor, the body geometry was modified to make it possible to generate a mapped network with a sufficient element size. In Fig. 5, there are individual bodies decomposed to smaller units. The components of a circular spline, wave generator, the external and internal bearing ring were decomposed to four equal parts due to a better 2D network shape of the

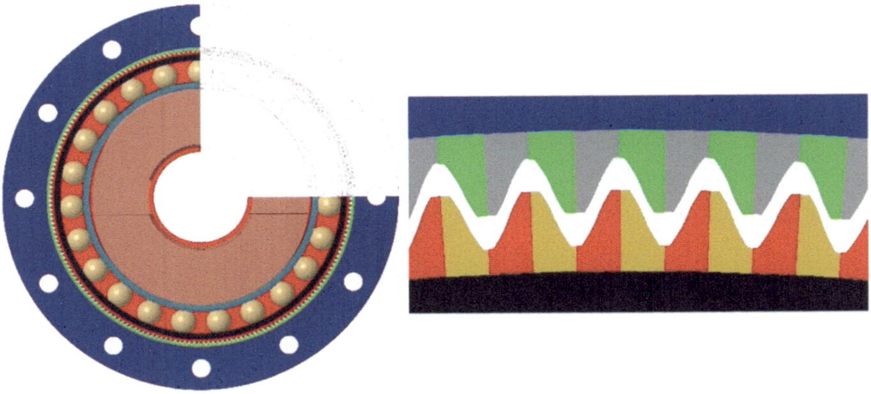

Fig. 4 Creating 2D and 3D model in Creo 4.0

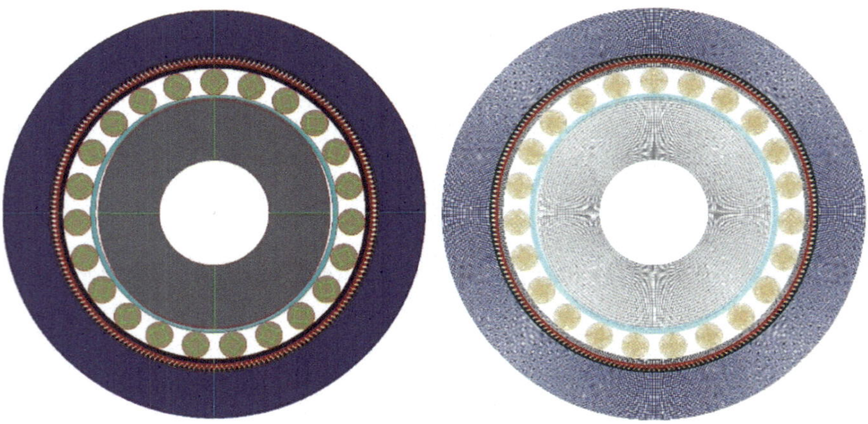

Fig. 5 Decomposition of the bodies and the 2D network

finite elements. The balls were divided inside by a square with the same number of nodes on each edge to reach a better result under the surface. These components were networked by a quadrilateral surface mesh with an element of 1 mm and teeth of 0.1 mm [4].

In Fig. 6, there is a detail of the 2D network. A total number of the elements are 117,850; a total number of the nodes are 130,115. The model was subsequently exported in BDF format.

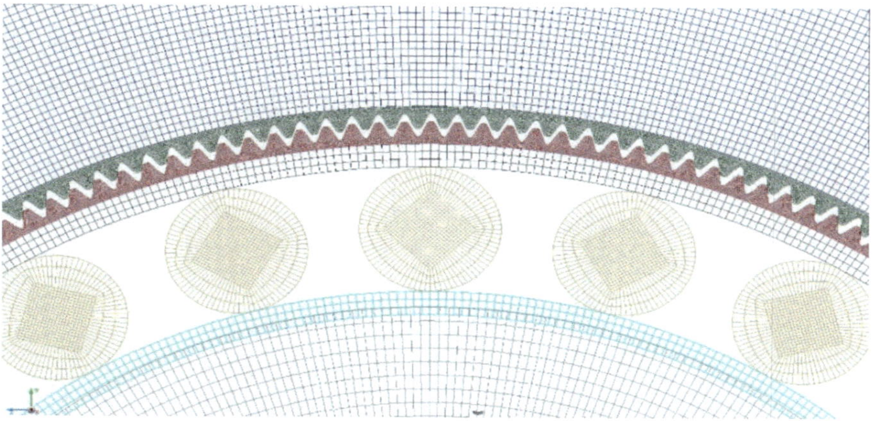

Fig. 6 A detailed image of the 2D network

2.4 Creation of a Computational Model in MSC Marc 2012

The 2D network model of the finite elements was imported to MSC Marc Mentat 2017.1. First, it was necessary to allocate material properties to each node of the 2D network of finite elements and to identify harmonic gear components as a plane stress role (see Fig. 7).

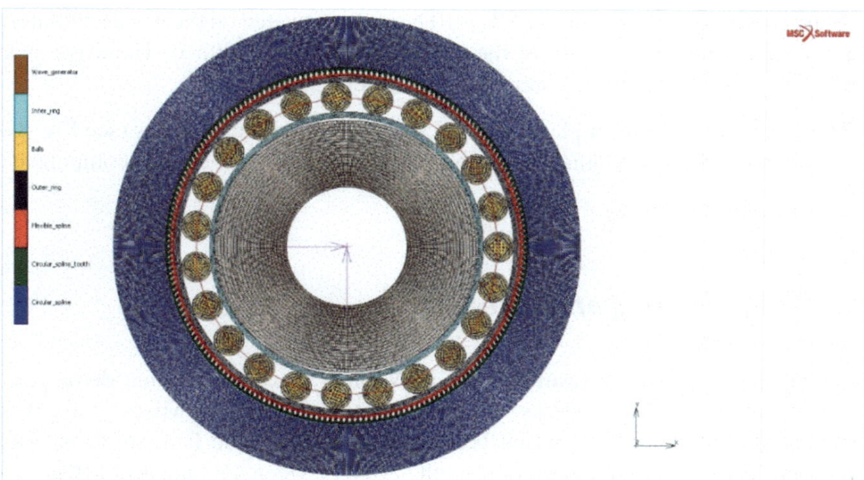

Fig. 7 Defining the components of a harmonic drive gear

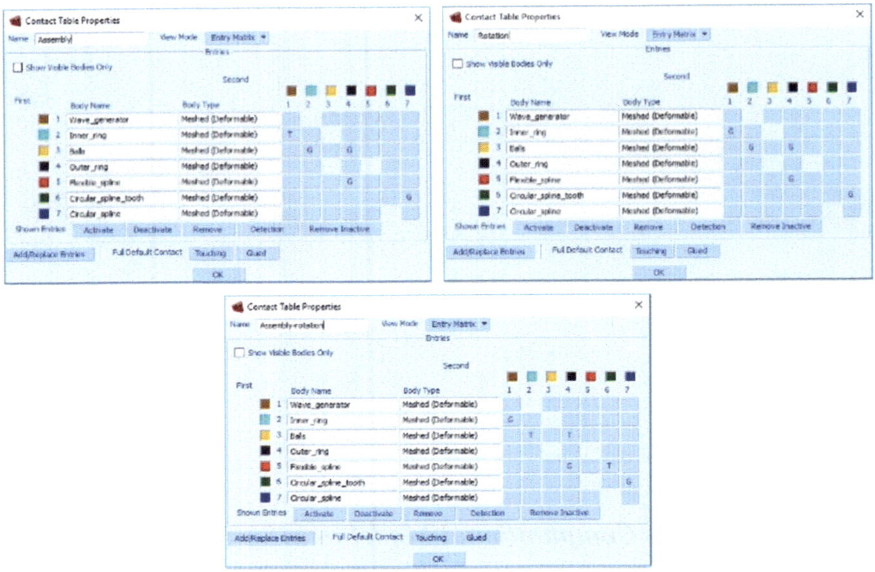

Fig. 8 Defining contacts

The analysis of the harmonic drive gear is composed of two basic parts:

- At the first stage, the internal bearing ring was pressed to the wave generator, ensuring a flexible deformation of the flexible spline, as well as providing the gear of the teeth.
- At the second stage, load torque $Mk = 10$ Nm was activated on the wave generator; at the same time, a 720° turn of the wave generator was defined. The drive gear was firmly fixed around the circumference of the circular spline.

There were three contact plates defined for respective contact states (see Fig. 8). In the pictures, there are "glue" and "touching" contacts between the components [5].

2.5 *Post-processing of the Results*

This chapter describes the results of the FEM analysis of a harmonic drive gear. Figure 9 shows von Mises stress. The maximum value of von Mises stress was created at the point of the external bearing ring touching the ball; its value was 1460 MPa. The stress value within the teeth varied between 450 and 750 MPa.

Figure 10 shows an overall stress state according to von Mises in the entire harmonic gear.

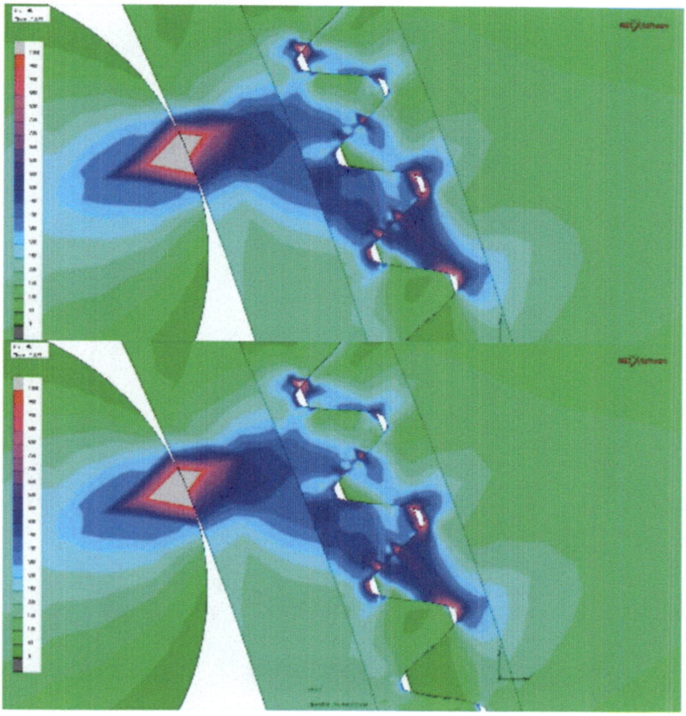

Fig. 9 Von Mises stress in touch points

3 Conclusion

The simulation of the 2D model of the harmonic drive gear was performed in 2D plane stress. Due to contact dependencies, the force is rather small. Then, after switching the contact state to generating with friction, the force settled at about F = 65 N. This is a force needed for deforming the flexible components; overcoming rolling resistance and rolling friction when gearing the drive gear (see Fig. 11). When this gearbox is rolling, then vibrations arise and this is closely related to the dead centre of these transmissions [6, 7].

The objective of this chapter was to create a functional FEM model of a harmonic gear, which then served for performing the stress analysis. It is to become a base for other analyses to examine the influence of the teeth shape on the backslash of the gear and kinematic error [8], which is the main criterion for using these gears in precise applications.

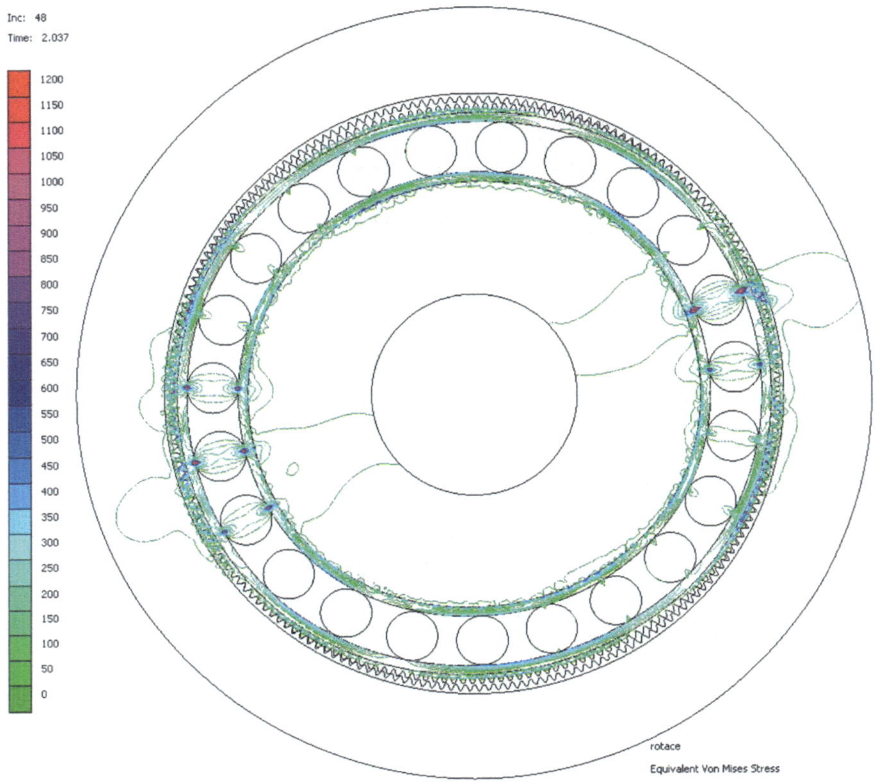

Fig. 10 Von Mises stress—entire harmonic gear

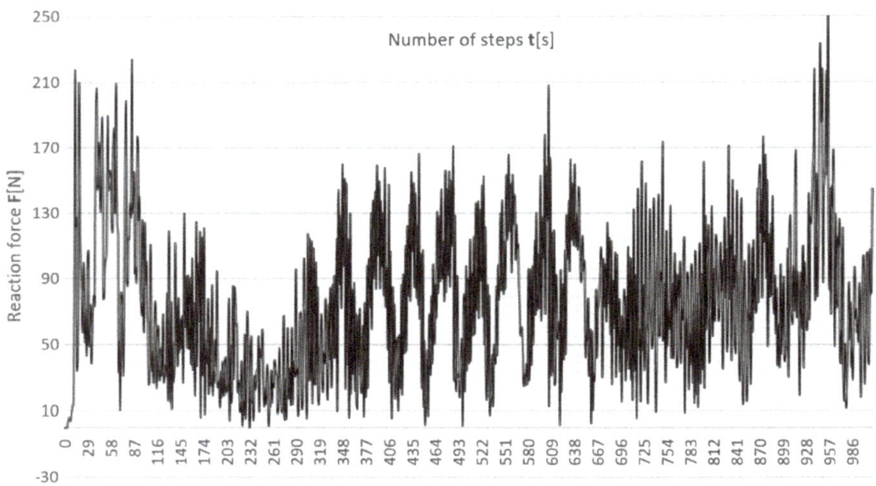

Fig. 11 Course of reaction force F in relation to time *t*

Acknowledgements Cultural and Educational Grant Agency MŠVVaŠ supported this study, contract no. 040ŽU-4/2016.

References

1. Garami B (22016) Dynamický model harmonické převodovky. Master thesis. Brno University of Technology, Faculty of Mechanical Engineering, Brno
2. Musser CW (1960) United States Patent Office 2, 932, 986 strain wave gear-species in which only one of the gears is input
3. General Catalogue Harmonic Drive AG, 10.2016, User Manual. www.harmonicdrive.net. Last accessed 2009/05/05
4. Lukáč M, Brumerčík F, Krzywonos L, Krzysiak Z (2017) Transmission system power flow model. Commun Sci Lett Univ Žilina 19(2):27–31
5. Tomasikova M, Gajdosik T, Lukac M, Brumercik F (2017) Simulation of planetary gearbox. Commun Sci Lett Univ Žilina 19(2A):48–53
6. Jedlinski L, Caban J, Krzywonos L, Wierzbicki S, Brumercik F (2015) Application of vibration signal in the diagnosis of IC engine valve clearance. J VibroEng 17(1):175–187
7. Kučera Ľ, Gajdošík T (2014) The vibrodiagnostics of gear. In: Ševčík L, Lepšík P, Petru M, Mašín I, Martonka R (eds) Modern methods of construction design, proceeding of ICMD 2013 LNME. Springer, Heidelber, pp 113–118
8. Ghorbel FH, Gandhi PS, Alpeter F (2001) On the kinematic error in harmonic drive gear. J Mech Des 123(1):90–97

Trend of Increasing Dynamicity in the Design of Flood Protection Systems

Ivan Mašín

Abstract Current innovation and mechanical engineering use, among other things, so-called trends of evolution of engineering systems (TESE). They describe natural transitions of the engineering system from one state to another and are generally valid for all engineering disciplines. In the area of flood protection systems innovation, one of the most significant trends is the trend of increasing dynamicity. This trend lies in the fact that during the development of the engineering system (components), the flexibility, dynamism, or adaptability increase. The development of engineering systems in the direction of this trend proceeds from an initial rigid structure with unchanging parameters to a more flexible and variable structure with more degrees of freedom, with adjustable parameters and a working mode adapted to changes in the external environment. The article deals with the fulfillment of this trend in the field of flood protection systems or flood barriers and presents examples from patent documents.

Keywords Design · Trend · Flood protection

1 Trends of Engineering Systems Evolution

Standard products or technology forecasting is a process that is based on the use of appropriate techniques and methods [1–4]. Data input for these methods is usually a subjective feeling and intuition, which of these methods makes a real mix of science and art. The output of these methods is the identification of the directions to be followed by the concept and design of the innovated product. It is clear that the methods of technology forecasting should more include objective rules of the development of technical systems.

Because different technical systems overcome the "same types" of problems, their general development over time is "typical". These typical trends of engineering systems evolution have been identified on the basis of a broad analysis of patent databases and historical trends in technology development [5–7]. Trends occupy a

I. Mašín (✉)
Technical University of Liberec, Studentská 2, 46117 Liberec, Czech Republic
e-mail: ivan.masin@tul.cz

Š. Medvecký et al. (eds.), *Current Methods of Construction Design*, Lecture Notes in Mechanical Engineering, https://doi.org/10.1007/978-3-030-33146-7_12

special place in the innovation science and engineering field, as they offer a view of the technical system from a variety of time perspectives—from the past, through the present to the far future. They have a great potential for problem-solving because they describe what happened in the past to successful technology and because they are leading the developer to what is likely to happen in the future for these technical systems. Trends are of a statistical nature and cannot be understood as being filled in the last letter.

Formally, trends or patterns of engineering systems evolution are systemized guides and a description of the "winning" transitions from one development phase to the next that allowed the technical solutions to occupy and maintain a leading position in the market. Knowledge of trends and patterns can also be used in the so-called evolutionary analysis, which aims to obtain information on general directions for the improvement of an innovative object or to formulate the correct tasks for the transition from one development phase to the other in the direction of the subline of evolution.

According to Lyubomirskiy and Litvin [6], trends of engineering systems evolution are eleven and together they form a hierarchical system. From the two main trends (the S-curve development and the trend of increasing ideality) are derived the other trends, respectively one trend may be the implementation of another trend (see Fig. 1).

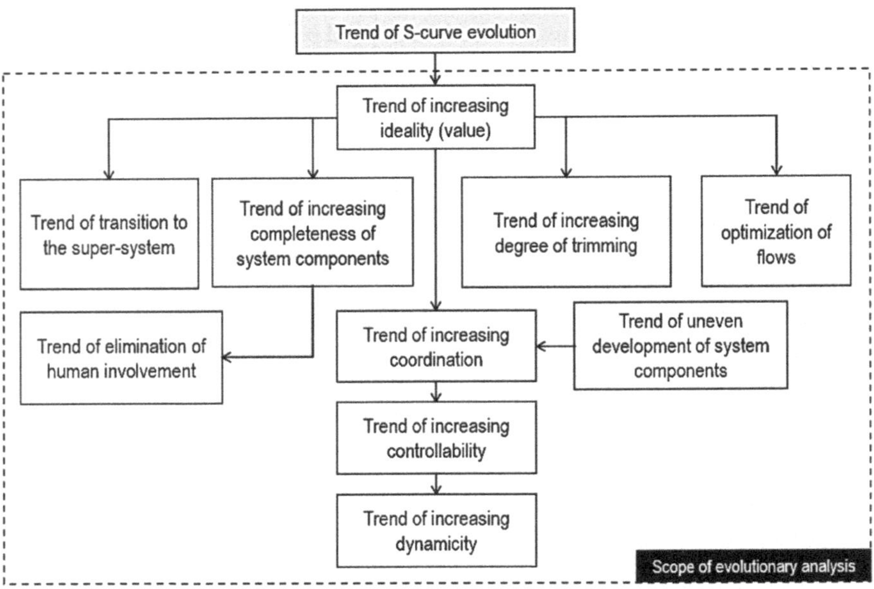

Fig. 1 Hierarchical system of trends of engineering systems evolution. Adapted from [6]

Fig. 2 Trend of increasing dynamicity of engineering system (monolithic—one hinge—several hinges—flexible—liquid—gas—field)

2 Trend of Increasing Dynamicity

Technical systems and their design evolve to become more flexible with greater degree of freedom. The trend of increasing dynamicity lies in the fact that during the development of the engineering system (components), the flexibility, mobility, variability, or adaptability increase. The development of engineering systems in the direction of this trend proceeds from an initial rigid structure with unchanging parameters to a more flexible and variable structure with more degrees of freedom, with adjustable parameters, and a working mode adapted to changes in the external environment. For example, engineering system that has rigid components is poorly adapted to operating conditions. For this reason, the evolution of rigid components becomes more flexible and dynamic. First, a single joint, then multiple joints, is inserted into the solid monolith. Moving from joints to flexible fabrics, the system gains flexibility and adaptability. Fluids or gases are even more adaptable. Finally, the system is converted to some of the physical fields that have the best controllability (see Fig. 2).

In general, increasing dynamicity lies in change of character of links among system components and change of components parameters. Increased dynamicity makes an engineering system adaptable to changes in operation conditions. To follow this trend, designers have to go through this way:

1. to design a monolithic system that is movable in one direction;
2. to increase the degrees of freedom of components;
3. to design flexible joints or hinges;
4. to design liquid-, gas-, or field-based components or hinges.

3 Following Trend of Increasing Dynamicity in the Flood Protection Systems Design

It is advisable to increase the dynamics of both individual components and entire systems when designing flood protection systems. Evidence that this trend of evolution of technical systems has been fulfilled over time can be demonstrated by examples shown in Figs. 3 and 4. As examples, mobile flood protection systems without substructure and flood protection panels for openings were selected.

Figures 3 and 4 illustrate the trend in increasing the dynamics of flood protection systems, which has been boosted especially since the first decade of this century,

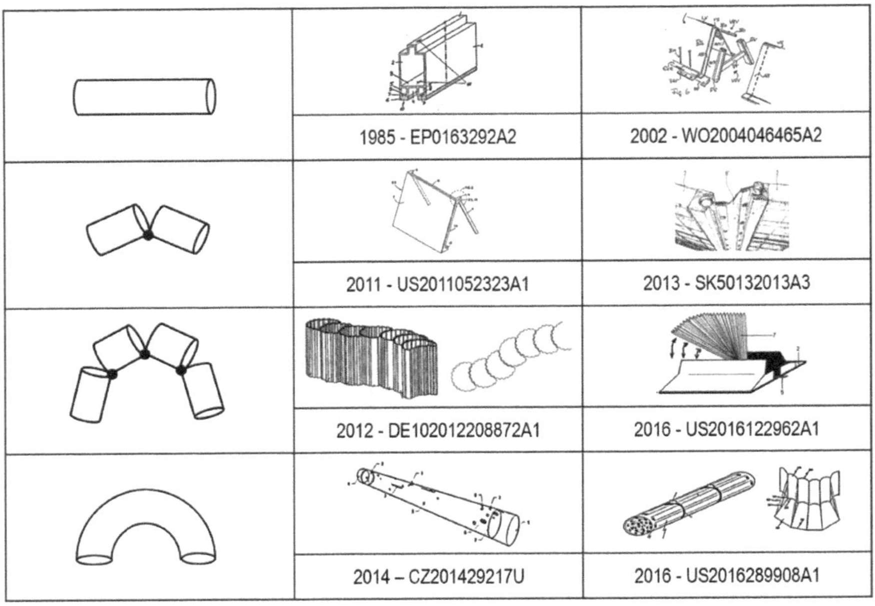

Fig. 3 Increasing the dynamicity of mobile flood protection systems without substructure

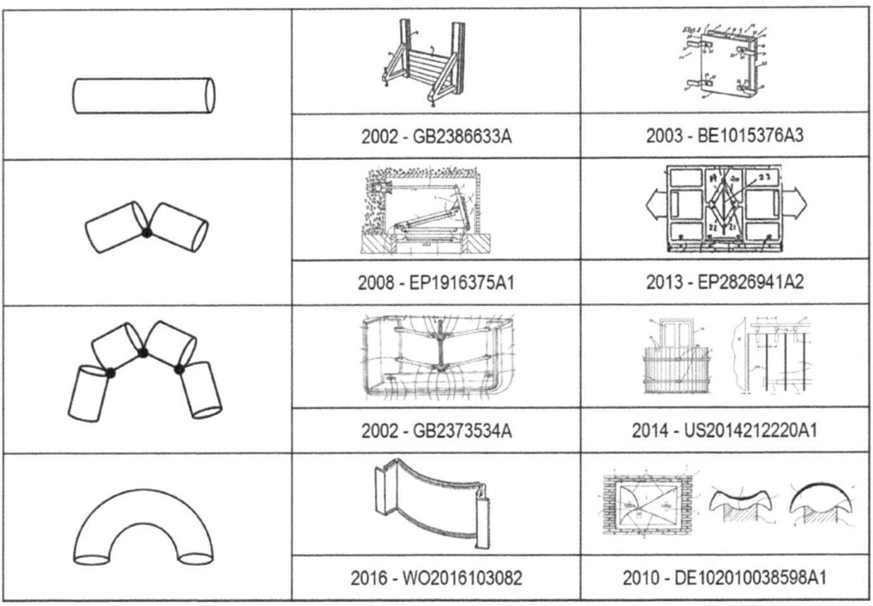

Fig. 4 Increasing the dynamicity of flood protection panels for openings

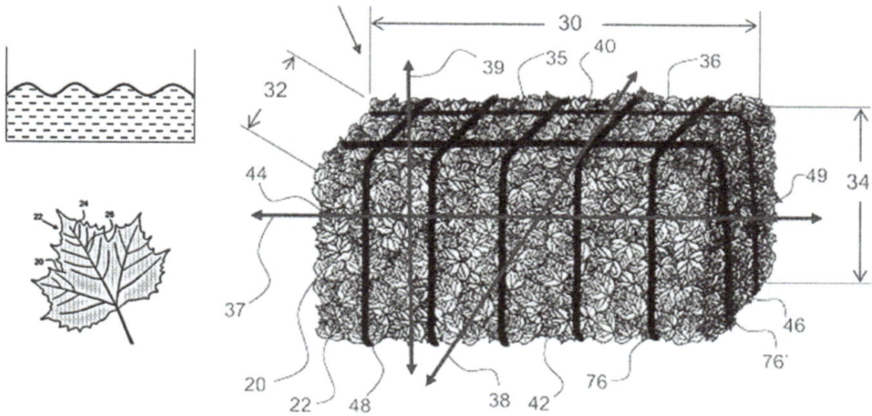

Fig. 5 Flood protection barrier in the form of high-density bale from tree leave material [8]

when climate change and related flood events began to emerge. It is clear that flood systems are still developing to the level of flexibility provided by increasing degrees of freedom by means of joints or inflatable structures.

The transition to the phase when "stop-water" function is partly performed by liquid is illustrated in Fig. 5, which shows an extraordinary technical solution using biological material. High-density bales for forming a flood barrier are made by the compaction of tree leave material. The bales are made substantially from tree leave material that retains at least a portion of the vein structure. The vein structure improves the structural integrity of the bale and produces a micro-barriers in capillaries.

It can be seen from Figs. 3 and 4 that, along with the trend of increasing dynamics, the increasing degree of trimming is also met in the case of flood protection systems design. This trend is that the number of components decreases during the development and innovation of the technical systems without compromising the functionality of the system (Fig. 6).

Maintaining functionality is ensured by following three trimming rules focused on redistributing useful functions of eliminated components to the remaining components or by transferring these useful functions to super-system elements:

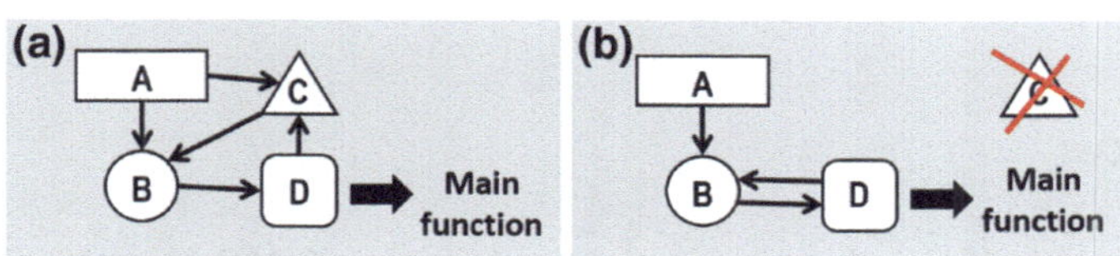

Fig. 6 Trimming implies removal of one or more components from original design of technical system (**a** original technical system, **b** technical system after trimming)

- trimming rule No. 1: Function carrier (subject) can be trimmed if we remove the object of its useful function;
- trimming rule No. 2: Function carrier (subject) can be trimmed if the object of function performs the useful function itself;
- trimming rule No. 3: Function carrier (subject) can be trimmed if another component performs its useful function.

Examples of engineering systems evolution in the direction of other trends are shown in [9].

4 Conclusion

Trends or patterns of engineering systems evolution are systemized guides and a description of the "winning" transitions from one development phase to the next that allowed the technical solutions to occupy and maintain a leading position in the market. Trends of engineering system evolution (TESE) can be a powerful tool for engineering system innovation and forecasting. The goal of TESE is to provide objective and analytical tools for problem identification of engineering systems, to assist in forecasting the future evolution of flood protection system, and to recommend solutions for innovation of engineering systems based on their evolutionary stage. Applying evolutionary analysis to the engineering systems guarantees a supply of novel ideas, trimming recommendations, and conceptual directions.

In conclusion, it is necessary to draw attention to other important trends, which also significantly influence the technical and design solution of the flood protection systems. One of the strongest and most frequently applied trends is the trend of transition to the super system. This trend reflects the fact that in the course of its evolution, the systems exhaust their internal resources and subsequently the original system integrates with other (even alternative or competing) systems and continues its development in the so-called super system (freely understood the super system as a "surround" of the system—for example, pavement or canalization). The trend of increasing coordination is reflected in the evolution of technical systems by gradually coordinating the "behavior" of the system components and consequently coordinating the "behavior" of the super system.

References

1. Kahn KB (2006) New product forecasting. An applied approach. M.E. Sharp, New York
2. Phaal R, Farrukh C, Probert D (2010) Roadmapping for strategy and innovation: aligning technology and markets in a dynamic world. Institute of Manufacturing—University of Cambridge, Cambridge
3. Miles I, Saritas O, Sokolov A (2016) Foresight for science, technology and innovation. Springer, Berlin

4. Daim TU, Pizzaro M, Talla R (eds) (2014) Planning and roadmapping technological innovations. Springer, Berlin
5. Altshuller GS (1984) Creativity as an exact science. Gordon and Breach Publishing, New York
6. Lyubomirskiy A, Litvin S (2003) Trends of engineering systems evolution—guide. Gen 3 Partners, Boston
7. Shpakovsky N (2016) Tree of technology evolution. Ways to new business opportunities. Create Space, Moscow
8. Nicholas V (2017) Highly compacted leaf bales for use as erosion control berm and methods of using same. Patent Application US2017233964A1
9. Masin I, Petru M (2018) Trends of engineering systems evolution and agricultural technology. In: Hussmann S (ed) Automation in agriculture securing food supplies for future generations. IntechOpen, London

The Gearing Temperature Shock Oscillation

Martin Mazač and Miroslav Malý

Abstract The article describes the temperature oscillation caused by shock mechanical parameters changing. The special effort on the heating curve parameters is put. The shock changes of the mechanical parameters (change of floating power) cause the extremely fast temperatures changes ongoing on the teeth flanks. The gear oil can be damaged during these processes. The lubrication parameters are changing and mechanical failures can be caused too. The automobile MQ100 gearbox for the measurements was used. The unique equipment for the temperature measurement was applied. The special stand for the operation of the gearbox was used. The gearbox operation condition was similar as in passenger car operation, without other heat sources (as a combustion engine) only. This short article describes the automobile gearbox final drive pinion teeth temperature measurements during the shock power float parameters changing. The aim of the article is a description of the tooth heating curve parameters. The goal is the maximal temperature gradient computing.

Keywords Gearing · Temperature · Power oscillation

1 Introduction

The mechanical transformation of energy by gears is one of the most widely used methods of the power parameters changing. The gearbox and gears design are special multidisciplinary process. The right function of gearbox and gears by the many physical parameters is depended. The parameters are usually interdependent. For the right gear function, the optimal operational parameters are necessary. The operational temperature is one of the main parameters and affects the parameters and lifetime of a gear.

M. Mazač (✉) · M. Malý
Technical University of Liberec, Studentska 1402/2, 46117 Liberec, Czech Republic
e-mail: martin.mazac@tul.cz

M. Malý
e-mail: miroslav.maly@tul.cz

Š. Medvecký et al. (eds.), *Current Methods of Construction Design*, Lecture Notes in Mechanical Engineering, https://doi.org/10.1007/978-3-030-33146-7_13

1.1 Final Drive Pinion Temperature Measurements

The part of the power flow is dissipated to other energy kinds (heath, chemical processes, electric … [1]). Most of the dissipated energy by the friction processes to the heath is transformed. The health increases the temperature of the parts. The temperature can be extremely high locally and the damages can be caused. The temperature measurement can be a source of very useful information for the gearbox and gears design.

The most of energy by the tooth flank friction is dissipated; tooth flank is the most important place of the heat generation. The tooth flanks friction coefficient f is very difficult to specify exactly. The part of heat by the tooth mass, wheel, and shaft is conducted. The other part of heat by the air and oil to the gearbox case is transferred. The heat during the all rotation direction is generated (left or right tooth flank "sliding" still). The temperature inside of the tooth mass can reflect the tooth surface temperature.

The next research on the manually shifted gearbox final drive pinion was aimed. The final drive gear as a heat generation and strength loading critical place was chosen.

2 Measurements and Results

The MQ100 manual shifted gearbox produced by Skoda Auto a. s. for the measurement was chosen. The MQ100 is a typical two-shaft gearbox for little cars as a Skoda Citigo. The MQ100 contains five gears with reverse. The MQ100 in [2] is described deeply.

The gearbox on a special stand for loading was mounted. The loading by the combustion engine and the electric dynamometer was realized. The stand describes for example [2] deeply.

The final drive pinion tooth body as a temperature measurement place was chosen. The temperatures were measured inside of the teeth by the miniature thermistor (NTC 2K7MCD1—telemeter electronic). The measurement place was in the half of the wheel wide on the pitch circle diameter. The temperature measurement place in Fig. 1 is depicted. The wheel modification and thermal sensors mounting describe for example [2]. The special equipment for the temperature measurement was developed.

The temperature inside of teeth by the equipment can be measured. The temperature values to the static parts of equipment by the infrared radiation are transferred. The equipment in the [3] is described.

All the assembly of the stand with the temperature measurement system in Fig. 2 is depicted.

The pinion tooth temperature during the first gear/input 3000 RPM operation was observed. The output shaft load was 1011 N m (75% of the maximal output torque, maximal input torque is 95 N m, $i_c = 14.18$).

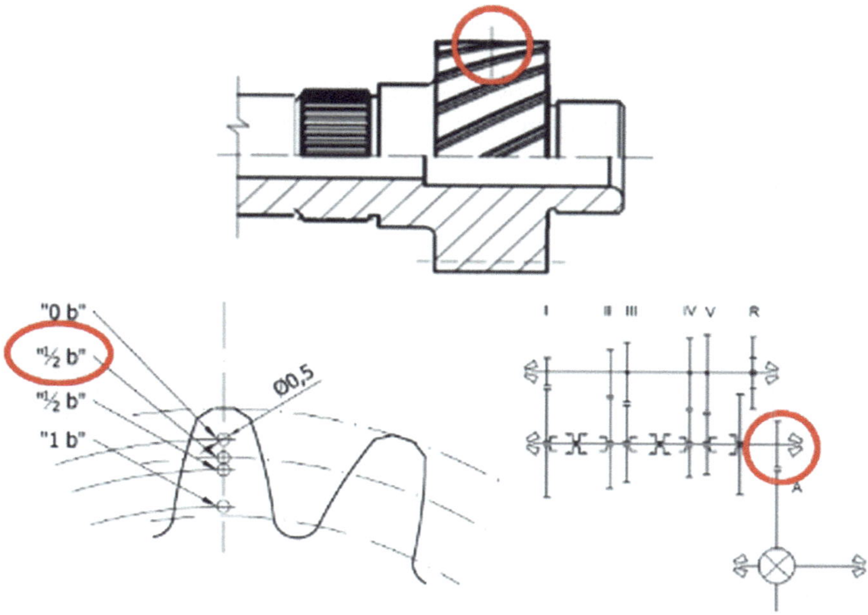

Fig. 1 The measurement place

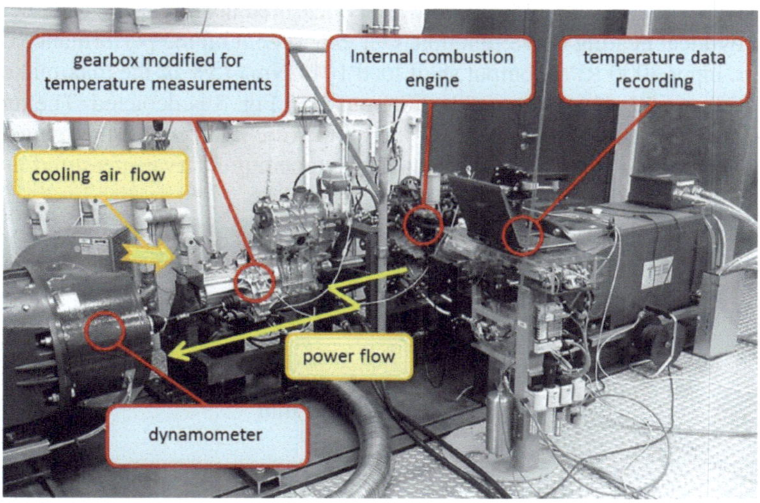

Fig. 2 The stand for the temperature measurement

2.1 The Heating Curves and Their Parameters

The heating curves are described by the standardized equations. The main parameters are defined. The temperature differences by the exponential function (1) are described. The cooling with the Eq. (2) is possible to describe.

$$\Delta\vartheta = \Delta\text{max} \cdot \left(1 - e^{-\frac{t}{\tau}}\right) \tag{1}$$

$$\Delta\vartheta = \Delta\text{max} \cdot e^{-\frac{t}{\tau}} \tag{2}$$

The parameters τ heating (cooling) time parameter are main the most important for the heating (cooling) processes. The τ parameters from the temperatures are to calculate. The (t) parameter signify the time when the temperature arise to 63.2% of the maximal temperature. The heating (cooling) process after the time $t = 3 \cdot \tau$ is finished.

2.2 Real Measured Heating Curves

The heating curves of final drive pinion teeth were measured during many different power levels and operational conditions. The measured heating curves can be interpolated by ideal heating curve function. Our experiment on the performance during first gear, input 3000 RPM, output shaft load 1011 Nm (75% of the maximal output torque) was aimed. The measured heating curve in Fig. 3 is depicted. The temperature measurements in other places of the tooth were realized; all the temperature curves are similar to some displacement. The temperature curves measured in other places of the tooth verifying the realized measurements.

The τ parameter is possible to compute from the heating curve. All input parameters and the steps of τ computing are in easy Eqs. (3), (4) and (5).

$$\vartheta_{\text{min}} = 46\ ^\circ\text{C}; \quad \vartheta_{\text{max}} = 56\ ^\circ\text{C}; \Rightarrow \Delta\vartheta = 10\ ^\circ\text{C} \tag{3}$$

$$\vartheta_\tau = 63.2 \cdot \frac{\Delta\vartheta}{100} + \tau_{\text{min}} \tag{4}$$

$$\vartheta_\tau = 52.32\ ^\circ\text{C} \Rightarrow \tau = 335\ \text{s} = 5\text{min}\ 35\ \text{s} \tag{5}$$

The first part of heating process will be approximated by a linear function (6).

$$y = kx + q \tag{6}$$

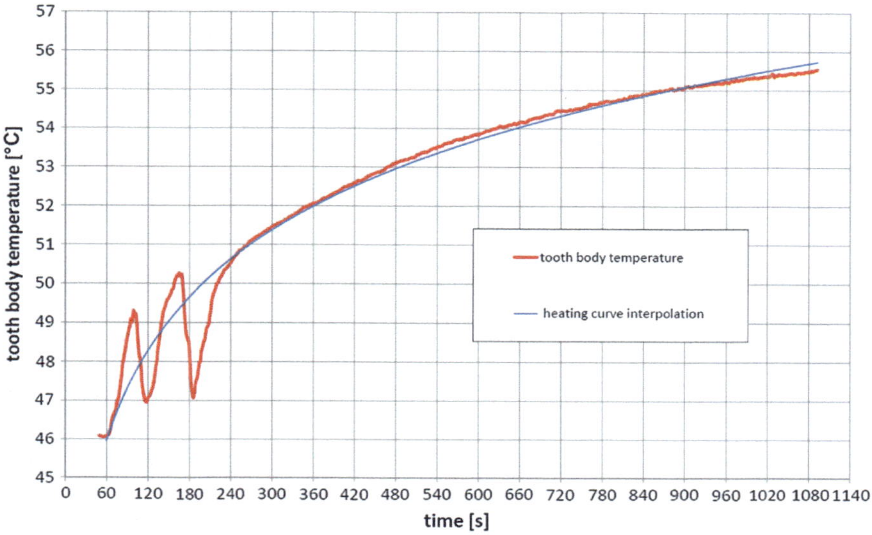

Fig. 3 Heating (cooling) process

The parameter *k* of linear function can be used for the heating rapidity demonstration. The maximal *k* slope of a linear function can be defined as a $\frac{d\vartheta}{d\tau}$ or $\frac{\Delta\vartheta}{\Delta\tau}$. The *k* parameter of the measured and approximated heating curve is 0.03.

The massive power parameters oscillation during the measurement has occurred. The temperature depends closely on the mechanical power parameters. The power oscillation caused fast temperature differences, which are on the graph, depict clearly. A similar process for the k parameter computing for the temperature oscillation curves was applied. All the temperature oscillation process to the five shorter parts was split. These five parts show three heating and two cooling processes. These short parts were approximated by the linear function and the k parameter was determined.

In Fig. 4, the graph of the temperature oscillation with the approximation is presented. The linear equation parameters are there too.

2.3 The Temperature Oscillation Parameters Comparison

The *k* parameter is the main parameter, which is possible to compare. All the *k* parameters are in Table 1.

The stable and the first heating *k* parameter are possible to compare, the beginning of the heating process is the same. The *k* parameter during the oscillating power flow heating (0.094) is more than three times higher than during stable power flow parameters (0.03). The *k* parameter of the first heating process during oscillation (oscillation_1) with the heating (oscillation_2 and oscillation_3) is possible; the parameters

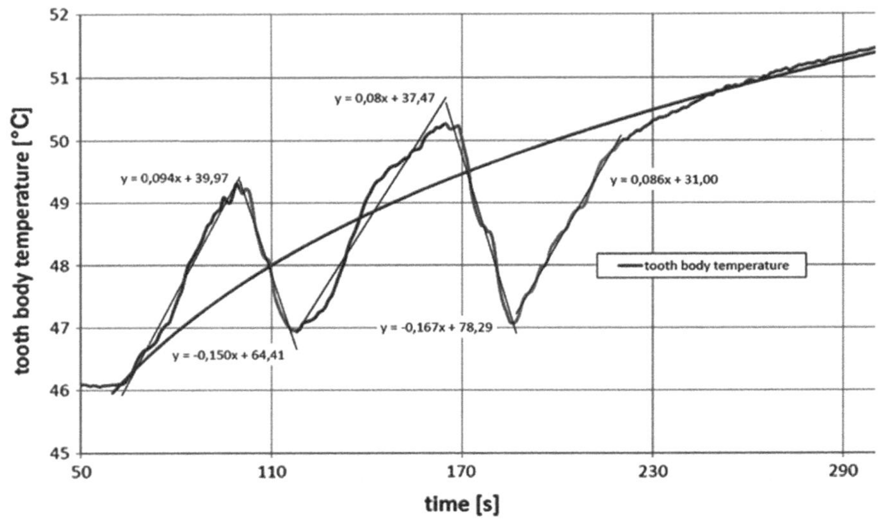

Fig. 4 Heating (cooling) process—temperature oscillation

Table 1 The *k* parameter summary table

Power flow	Kind of heat transfer	*k* parameter
stable	Heating	0.03
oscillation_1	Heating	0.094
oscillation_2	Heating	0.08
oscillation_3	Heating	0.086
oscillation_4	Cooling	−0.15
oscillation_5	Cooling	−0.167

are similar (0.94; 0.08; 0.086). The parameters of cooling during "oscillation_4 and 5" are similar too (−0.15; −0.167).

3 Conclusion and Discussion

This short article deals with temperature measurement realized inside of the final drive pinion tooth. All of the shown temperature data were measured during the mostly real operation of the gearbox. The easy mathematical operation for the temperature curves approximation was used.

The parameters of the approximated curves are included. The main *k* parameters in the table were summarized. The *k* parameter during oscillating power flow operation is more than three times higher than during stable power flow performance. The nonstationary power flow is absolutely common during the real operation of a car.

The heat is generated during both the performance of gearbox—forward, backward, drive, braking with engine use.

The oscillation process causes are not known but it may be caused by drive parts rigidity, inertia, or backlashes. The power flow direction can be one of the temperature oscillation causes too. The discussion about this topic can be large and the next research will be necessary.

Acknowledgements This work is partially supported by the Ministry of Education, Youth and Sports of the Czech Republic and the European Union in the frames of the project "Modular platform for autonomous chassis of specialized electric vehicles for freight and equipment transportation," Reg. No. CZ.02.1.01/0.0/0.0/16_025/0007293.

References

1. Štěpina V, Veselý V (1985) Maziva v tribologii, 1st edn. Veda, Bratislava
2. Mazač M, Malý M (2016) The MQ100 gearbox final drive pinion gearing temperatures. In: Kubec V, Kubec Krónerová E, Trapp Cajthamlová Š (eds) Book of proceeding, 57th international conference of machine design departments (ICMD 2016). University of West Bohemia, Pilsen, pp 447–452
3. Mazač M, Herajn P, Svoboda M (2015) The gear tooth flank temperature measurements system. Trans Electr Eng 4(1):27–30

The Device for Measuring the Stiffness of the Bearings

Václav Mosler, Jan Pavlík, Jiří Havlík and Ivo Křístek

Abstract This article deals with the design of the equipment for bearing stiffness determination of the shaft. The other part of the thesis is concerned with an experiment of the equipment. The knowledge of the bearing stiffness is important for finite element analysis (FEA) to help determine the correct type and size of the gear modification in the car gearbox. At present, this stiffness is not included in FEA software. The stiffness of the bearings is investigated statically and dynamically with rotational shaft. This is one of the requirements on the device. The first part of the article is focused on the design of the device for the bearing stiffness determination. The second part elaborates on the measurement methodology and the evaluation of the result from the measurements. The result of the measurement is not the rigidity of the bearings, but the displacement of the shafts and the tilting of the inner ring. In order to determine the bearing stiffness of these results, it is necessary to know the theoretical displacement of the shaft point in which it is measured. Shaft bearings are replaced with ideal joints. This substitution is done in FEM calculation of the shaft. The final stiffness is determined from the measured data.

Keywords Taper roller bearings · Stiffness · FEM · Modification

1 Introduction

During development of modern vehicle gearbox (see Fig. 1), there is an emphasis on low noise and requisite durability. The main causes of this noise are the operating deviations of the gearing (deformations of shafts, gears, gearbox, bearing, etc.) [1, 2]. These deviations can be reduced with correct modifications [3, 4]. The sizes of the

V. Mosler (✉) · J. Pavlík · J. Havlík · I. Křístek
Technical University of Ostrava, 17. Listopadu 15/2172, 70833 Ostrava, Poruba, Czech Republic
e-mail: vaclav.mosler@vsb.cz

J. Pavlík
e-mail: jan.pavlik.st@vsb.cz

J. Havlík
e-mail: jiri.havlik@vsb.cz

Š. Medvecký et al. (eds.), *Current Methods of Construction Design*, Lecture Notes in Mechanical Engineering, https://doi.org/10.1007/978-3-030-33146-7_14

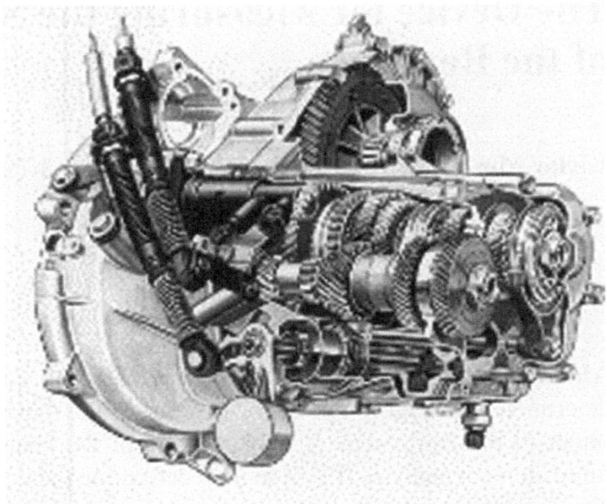

Fig. 1 Modern vehicle gearbox [4]

modifications are determined by the experiment on the actual gearbox. These procedures are very accurate but it takes a lot of time. After any modification on gearbox, experiment must be repeated. This increases costs and prolongs development time of new gearbox. An alternative method is a determination of modification with FEM method [5, 6]. This method is less accurate than experimental measurements on the gearbox. This inaccuracy causes especially parts with nonlinear properties and clearance between individual parts of the gearbox, because FEM programs are not able to count with this. Set the stiffness directly is only one way how to do it. Our goal is to explore the behavior of individual nonlinear parts and parts with the clearance on the experimental stages. The obtained results will be applied in fem analysis; it should cause approximation results of FEM analysis and real deformations on the gearbox.

2 Methods and Materials

The requirements of the device:

- Measure the shaft and bearing movements in two perpendicular radial directions.
- Measure the axial displacements of the inner bearing rings.
- Allow various prestressing forces of bearings to be introduced.
- Simulated bearing loads by different gear forces.

2.1 The Beginning of the Tooth Mesh

The stiffness is the ratio of the force and deformation that this force causes. According to this definition, the most robust way of stiffness determination is to place bearings on a rigid shaft and load it with transverse force. Deformation of the bearing is measured as a displacement of the inner ring. The stiffness obtained by this procedure corresponds to the radial bearing stiffness. The stiffness of the bearing is usually stated catalogs and the only determination would be pointless. The aim of the experiment is to obtain a bearing stiffness relative to the inner ring of the bearing. This stiffness affects the load distribution along the side of the gear tooth and its size should affect the design of the longitudinal tooth modification. This stiffness would have a significant effect on gearboxes where barrier bearings are used, so the experiment will be performed for a shaft mounted on a tapered roller bearing. Figure 2 shows a load diagram and prestress of the bearings.

An indirect method of measuring the stiffness of bearings in motion was chosen for the experiment. The disk, that will be mounted on the needle bearing shaft, will be loaded with triple bolts. These bolts will simulate gear forces. By using this substitution, it can simulate arbitrary transfer of frontal teeth, either with spur or helical gear.

Couples of sensors will be positioned along the length of the shaft to measure the transverse displacement of the shaft in two perpendicular planes. From these, displacements are possible to determine the inclination of the shaft. On the inner surfaces of the inner rings of the bearings, three sensors at a 120° pitch will be placed. With these, sensors will be measured the inclination of the bearing ring. It exists an assumption that the inclination size may change with the drive speed due to bearing tolerances and eventual manufacturing inaccuracy of the bearing cone; the drive speed will be controlled from 900 to 1000 rpm. This speed zone represents the most usual speed of the engine in traffic. Tapered roller bearings are used with prestressing. Different prestress values will be set for the same load. By varying, the prestress values will be determined the influence on the bearing stiffness. The design of the device for the bearing stiffness determination is shown in Fig. 3.

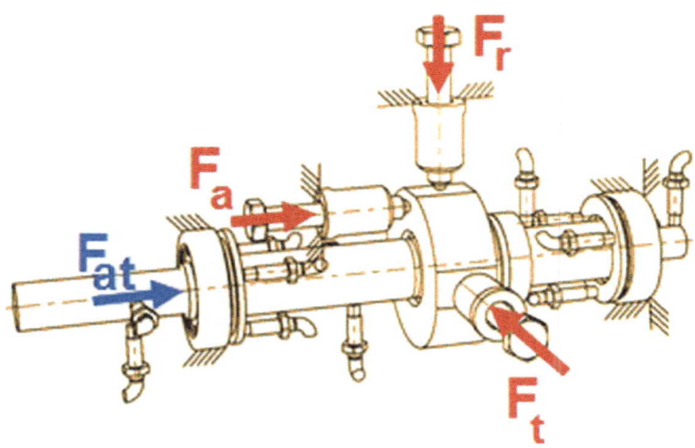

Fig. 2 Load diagram and prestress of the bearings

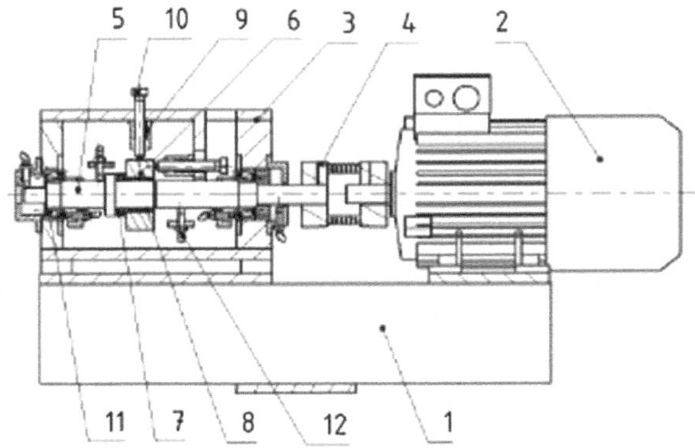

Fig. 3 The design of the device for the bearing stiffness determination, legend: 1—frame, 2—electromotor, 3—case device, 4—metal bellows couplings, 5—shaft, 6—disk, 7—axil bearing, 8—needle bearing, 9—sensor force, 10—screw, 11—tapered roller bearings, 12—sensor displacements

2.2 Evaluation

For evaluation, it will be necessary to create a FEM model of the shaft with the same forces as during the experiment. The result of FEM calculation will be deformations measured at the same points as in the experiment. The total deformation from the experiment should consist of a shaft deformation and bearing deformations (see Fig. 4 and Eq. 1). The shaft displacement should correspond to displacement from the FEM model. The difference between the calculated displacement and the displacement measured experimentally is obtained from the radial deformation of the bearing. The value is necessary to determine the radial stiffness of the bearing. This stiffness will be determined for different prestress in the bearing and for different tilting and shaft speeds.

$$Y_{\exp} = Y_A + Y_B \tag{1}$$

Another possibility of evaluation is to measure the tilt of the shaft and the tilt of the inner ring of the bearing. The result will be evaluated using FEM model. The distribution of the measured tilts will be divided into two parts $\delta_{\exp} = \delta_{\mathrm{s}} = \delta_{\mathrm{B}}$. This state will correspond to the distribution of the experiment when the tilt of the shaft is less than the clearance between the cones and the bearing tracks of the bearings.

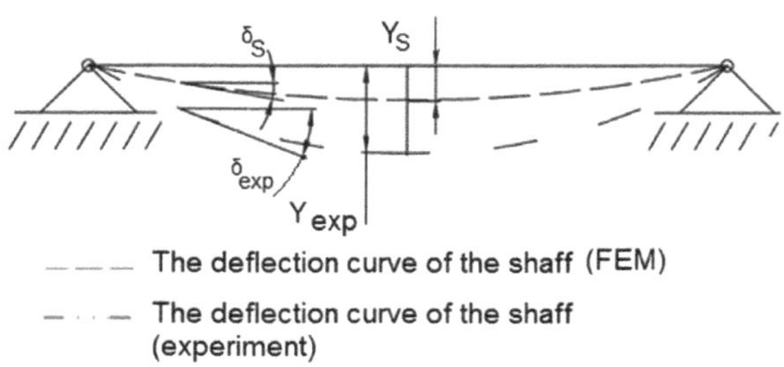

Fig. 4 The deflection curve of the shaft

In this state, the experimentally measured tilt value of the shaft will be controlled only by the stiffness of the shaft. The size of this tilt will also serve as a comparison of the experiment result and FEM calculation.

The second situation in this evaluation will correspond $\delta_{\exp} \neq \delta_{\mathrm{B}}$. Bearing clearance will be already cleared down and the experimental inclination will depend on the stiffness of the shaft and the stiffness of the bearings. Therefore, the experimentally determined tilt angle of the shaft will correspond to Eq. (2). The shaft tilt angle δ_{S} will be determined from finite element analysis.

$$\delta_{\exp} = \delta_A + \delta_B \tag{2}$$

The bearings and the shaft in the gearbox are connected in series in terms of stiffness. Therefore, the total bending stiffness C_{sexp} determined experimentally is equal to 3. The shaft stiffness $C_{\delta \mathrm{S}}$ will be determined by FEM calculation and the $C_{\delta \mathrm{B}}$ stiffness will be the stiffness of the bearing.

$$\frac{1}{C_{\delta \exp}} = \frac{1}{C_{\delta \mathrm{S}}} + \frac{1}{C_{\delta \mathrm{B}}} \tag{3}$$

Thus, the only unknown in Eq. 3 will be the bearing stiffness $C_{\delta \mathrm{B}}$. The determination of this stiffness is the main subject of the experiment. It can be determined from Eq. (3). The bending stiffness of the tapered bearing is determined from Eq. (4).

$$C_{\delta \mathrm{B}} = \frac{C_{\delta \mathrm{S}} - C_{\delta \exp}}{C_{\delta \exp} \cdot C_{\delta \mathrm{S}}} \tag{4}$$

3 Discussion

This article has proposed a device for determining the stiffness of bearings. This mechanism is able to measure only deformation of the bearings. To evaluate the stiffness of the bearings from this deformation, a methodology was proposed. By using this methodology, it can be determined radial bearing stiffness and stiffness the tilting of the inner ring tilting of. The functionality of the proposed equipment and the evaluation of methodology will be verified until the after experiment is realization.

4 Conclusion

This experiment should appoint to speed up the design of the correct value and type of gear modification by using FEM software. The stiffness values are applied into FEM model. After research on inclination of gears on needle bearings, this is another step toward to improve the FEM model of the automotive gearbox. The research is mainly done to reduce noise and vibration of automotive gearboxes, which are among the most significant sources of noise and vibration in passenger cars. This experiment is in the initial study phase. The real functionality of the proposed equipment and the methodology of the experiment will be verified after the realization of the device.

References

1. Moravec V (2001) Konstrukce strojů a zařízení II.: Čelní ozubená kola: teorie, výpočet, konstrukce, výroba, kontrola, 1st edn. Montanex, Ostrava
2. Schulze T (2007) Getriebeberechnung nach aktuellen wissenschaftlichen Erkenntnissen - Von der Lastverteilungsrechnung bis zur Analyse des Systemverhaltens, In: DMK 2007, Dresdner Maschinenelemente Kolloquium, DMK, Dresdner Maschinenelemente Kolloquium, 3. TUD press Verlag der Wissenschaften, Dresden, pp 95–12
3. Muller J (2013) How to inspect a gearbox. Gear Technol 30(6):82–86
4. Moravec V, Deil Z, Němček M, Folta Z, Havlík J (2009) Čelní ozubená kola v převodovkách automobilů, 1st edn. MONTANEX, Ostrava
5. Bizarre L, Nonato F, Cavalca KL (2018) Formulation of five degrees of freedom ball bearing model accounting for the nonlinear stiffness and damping of elastohydrodynamic point contacts. Mech Mach Theory 124:179–196
6. Tong VC, Hong SW (2017) The effect of angular misalignment on the stiffness characteristics of tapered roller bearings. Proc Inst Mech Eng Part C J Mech Eng Sci 231(4):712–727

Comments on ISO 6336-2

Milos Nemcek

Abstract For loading capacity calculation of gears, a number of standards are used. In our country, the use of the standard ČSN 01 4686 is slowly coming to an end. Unfortunately, this standard does not cover the currently designed gears with non-standard profile. Consequently, designers go to other standards—DIN 3990 or ISO 6336. ISO 6336 was essentially taken as a replacement for the ending standard ČSN 01 4686. This paper deals with this standard, specifically with its second part. "Calculation of surface durability (pitting)" is the title of the second part of ISO 6336—"Calculation of load capacity of spur and helical gears". The standard was issued in 2006. This issue is still valid, supplemented in 2008 by Technical Corrigendum 1. It is a basic standard and at the same time a guideline for a very important calculation of the contact stress on the flanks of the teeth of gears. This article points to some details in the standard. It also shows an overview of typing errors in this part of the standard. It also states their correct wording.

Keywords ISO 6336 · Part 3 · Typos · Errors

1 Introduction

This latest edition of this part of the standard [1] is comparable to the previous edition (1996—31 pages, 2006—33 pages). Similarly, to the third part also from this part, both the simplified computation method C and the method D have been completely eliminated. The new standard is more sophisticated and takes better account of the influence of some factors. Additionally, Annex A is added to the standard to determine the exact start of involute.

M. Nemcek (✉)
Technical University of Ostrava, 17. Listopadu 2172/15, 70800 Ostrava, Poruba, Czech Republic
e-mail: milos.nemcek@vsb.cz

Š. Medvecký et al. (eds.), *Current Methods of Construction Design*, Lecture Notes in Mechanical Engineering, https://doi.org/10.1007/978-3-030-33146-7_15

2 Typos and Errors

Page 7, Equation (7)—replace … σ_{HP} ref $Z_N = \sigma_{HP}$ ref (… with … $\sigma_{HP\ ref}\ Z_N = \sigma_{HP\ ref}$ (…

Page 8, Eq. (12)—replace … σ_{HP} ref $Z_N =$ … with … $\sigma_{HP\ ref}\ Z_N =$ …

Page 9, below the figure—replace… Y helix angle… with… X helix angle …

Page 9, below the figure—replace … X zone factor … with … Y zone factor …

Page 10, "Figure 3"—corrected dimension p_{bt} to the point B, (see Fig. 1 here).

Page 12, Eq. (23)—replace z with Z

Page 14, for roll angle is used ξ, although in the review in the 1. Part is ς.

Page 18, Table 2—remove two table cells, add the upper index ^{b} to the circles (see Fig. 2).

Page 20, Equation (37)—calculation of Z_L is not accurate for extreme values of υ.

Pages 20 and 21, for conversion it is advisable to use $\upsilon_{40} = 1.82411 \cdot \upsilon_{50\text{-}9}.3480569$

Page 27, Equation (57)—replace denominator 680 with 6800

Current Figure 11, in norm ISO 6336-2, (see Fig. 3)—to correct

Corrected Figure 11, (see Fig. 4)

Annex A—inappropriate symbols for the angles are not in accordance with habits

Equations A.5, A.6, A.10, A.13, A.14 replace … ϕ … with … φ …

Page 31, Figure A.2—replace … ρ_{fP} sinφ/cosβ … with … ρ_{fP} cosφ/cosβ …

Page 32, replace …υ … with …φ … (total 7 times)

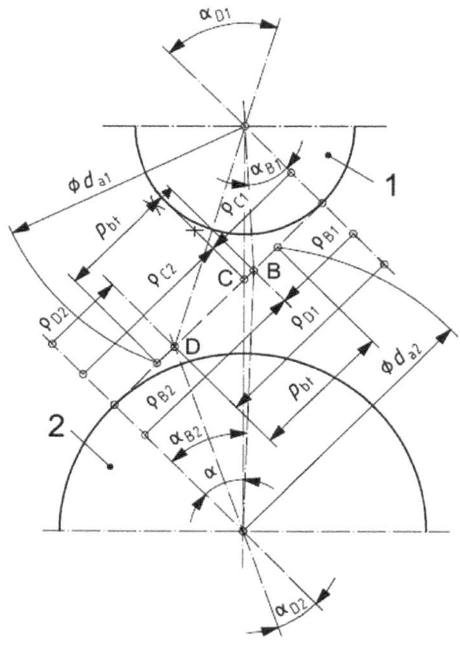

Fig. 1 "Figure 3" from ISO 6336-2 [2]

Material [a]	Number of load cycles	Life factor, Z_{NT}
St, V, GGG (perl., bai.), GTS (perl.), Eh, IF; only when a certain degree of pitting is permissible	$N_L \leqslant 6 \times 10^5$, static	1,6
	$N_L = 10^7$	1,3
	$N_L = 10^9$	1,0
	$N_L = 10^{10}$	0,85 up to 1,0 [b]
St, V, GGG (perl., bai.), GTS (perl.), Eh, IF	$N_L \leqslant 10^5$, static	1,6
	$N_L = 5 \times 10^7$	1,0
	~~$N_L = 10^9$~~	~~1,0~~
	$N_L = 10^{10}$	0,85 up to 1,0 ○
GG, GGG (ferr.), NT (nitr.), NV (nitr.)	$N_L \leqslant 10^5$, static	1,3
	$N_L = 2 \times 10^6$	1,0
	$N_L = 10^{10}$	0,85 up to 1,0 ○
NV (nitrocar.)	$N_L \leqslant 10^5$, static	1,1
	$N_L = 2 \times 10^6$	1,0
	$N_L = 10^{10}$	0,85 up to 1,0 ○

[a] See ISO 6336-1:2006, Table 2 for explanation of abbreviations used.

[b] The lower value of Z_{NT} may be used for critical service, where pitting must be minimal. Values between 0,85 and 1,0 may be used for general purpose gearing. With optimum lubrication, material, manufacturing and experience 1,0 may be used.

Fig. 2 Table 2 from ISO 6336-2

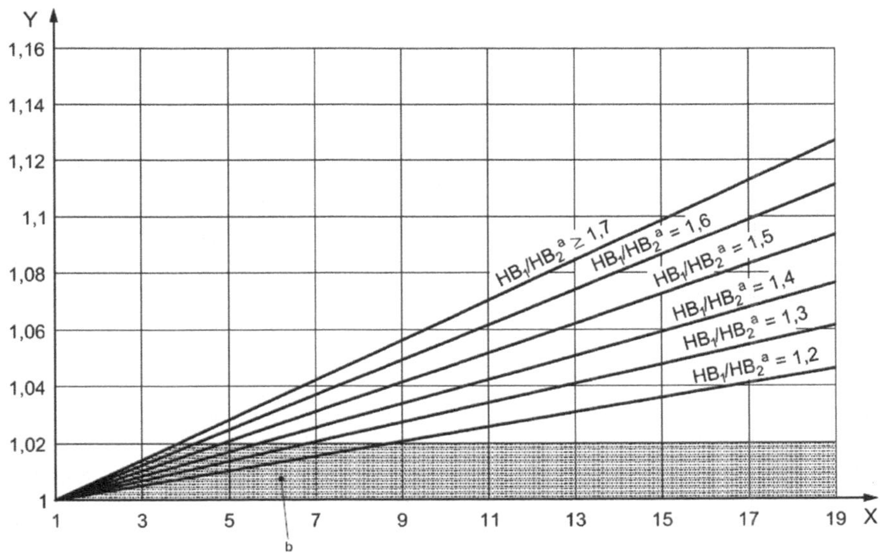

Fig. 3 Figure 11 from ISO 6336-2

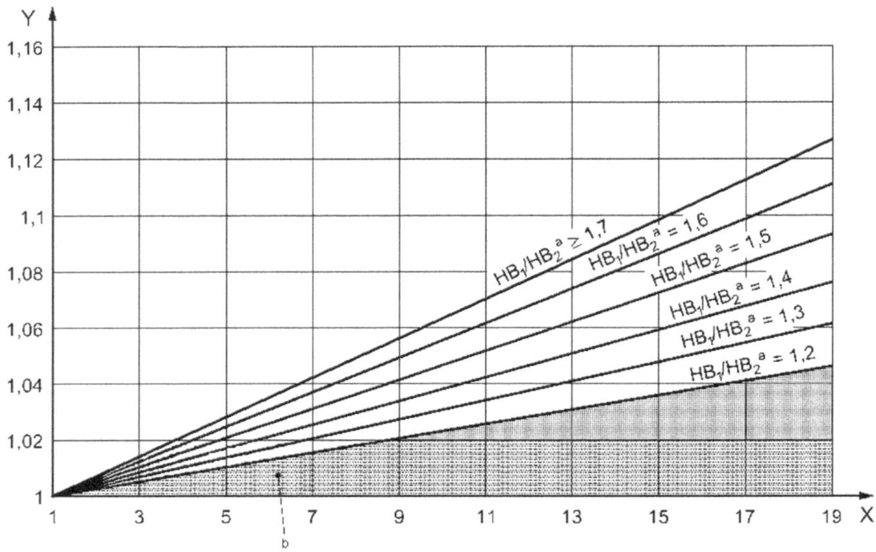

Fig. 4 Corrected Fig. 11 from the ISO 6336-2

3 Conclusion

The latest release of the second part of ISO 6336 is progress against the previous one. The calculation procedures for most of the factors are revised and refined. The scope was retained because the C and D computational methods were omitted. It is a pity that all bugs were not captured in the Technical Corrigendum 1. Nevertheless, this standard is a great asset. However, it would be appropriate to consider revising Annex A, which is unclear.

References

1. ISO 6336—3:2006(E) (2006) Calculation of load capacity of spur and helical gears—Part 2: Calculation of surface durability (pitting)
2. Linke H (2010) Stirnradvezahnung. Carl Hanser Verlag München

Contact Patterns During a Mesh of a Helical Gear Pair

Jan Pavlík and Václav Mosler

Abstract The article deals with the issue of real tooth contact on the helical gearing in the car gearbox. Theoretical images of the tooth contact in the available literature do not show all information about tooth contact looks and how are spreads on the real tooth flank surface. This article shows in the detail, how looks the contact line and the footprint (contact pattern) caused during tooth meshing on the real gearing in the car gearbox. It also shows, where the contact of the tooth begins and in what direction is expands on the tooth flank. This all is created for the comparing on different levels of load, because on the tooth are created micro-geometry modifications, which causes little bit different shapes of footprints than on unmodified gearing. These real footprints represent a better imagination for understanding how tooth meshing of helical gearing looks and how is changing the size of footprints due to increasing loading.

Keywords Helical gearing · Footprints · Involute gearing · Pattern · Load · Tooth meshing · Automotive · Modification · Noise

1 Introduction

Monitoring of the quality gear mesh is increasingly important during the design of gearboxes because automobile producers are pushed to reduce polluting emissions and noise emissions by very strict regulations. This means that the total weight of gearbox is decreasing, but this factor is counterproductive to decreasing of the noise. The gear cases have thinner walls, where cannot absorb the noise which is create from the rotating gear wheels. Also shafts are thinner; this factor causes their bigger deflection. This both factors (thinner walls of gear cases and thinner shafts) causes that overall stiffness of gearbox is lower (has higher deformations) and the gear mesh is not so ideal. This is the reason, why ideal geometry of the teeth is produced with

J. Pavlík (✉) · V. Mosler
Technical University of Ostrava, 17. Listopadu 15/2172, 70833 Ostrava, Poruba, Czech Republic
e-mail: jan.pavlik.st@vsb.cz

V. Mosler
e-mail: vaclav.mosler@vsb.cz

Š. Medvecký et al. (eds.), *Current Methods of Construction Design*, Lecture Notes in Mechanical Engineering, https://doi.org/10.1007/978-3-030-33146-7_16

modifications. Modifications reduce unwanted deformations and ensure good contact of the tooth during their mesh. Just correct tooth contact has the biggest influence on the generating acoustic emissions [1, 2]. For better understanding, how looks real footprints and define the direction of contact line on the tooth flank in the car gearbox with helical gearing, were created following pictures which are describes in the detail.

2 Methods and Materials

Engine with gearbox is mounted on the testing device (see Fig. 1) by exactly same way as in the car, but creating of the real footprints (contact patterns) on the testing device runs at very low speeds. The reason is that very small segments of rotation are required for capturing of the beginning of the tooth contact. To show tooth contact, a special marking color is applied on the tooth flank. Methodology in which way is color applied is described in [3]. There are another ways (methodology), how is possible create the footprints on the tooth flanks (e.g., [4]), but for this case of study was not so good.

Methodology of creating contact line of the tooth mesh is based on precision. It is important to apply the marking compound with brush on tooth flanks in very thin layer. With these steps are guaranteed sufficient accuracy of result footprints.

In order to capture the beginning of the footprint is necessary make segment of rotation to the point when teeth, which has applied color on the flank, goes to the contact with opposite teeth. At this point, it is necessary to stop the rotation of meshed gears. To create the picture of beginning of the footprint is need to turn

Fig. 1 Testing device with real gearbox [5]

the drive pinion in opposite direction back. Other chance how to make the photo of beginning of footprint does not exist.

To create the pictures, how contact line is spread on tooth flank, is need to make segment of forward rotation little bit bigger than previous. Then is need to turn the drive pinion backward and make the photo. These two steps are repeated as long as the tooth which has applied marking compound on his flank goes completely out from the mesh.

3 Results

This chapter is interested about better imagination for understanding where the beginning of the tooth mesh is, how tooth meshing of helical gearing looks, and how is changing the size of footprints due to increasing loading. All problems are showed on the real gearing in the gearbox.

3.1 The Beginning of the Tooth Mesh

During design of gearing is very difficult to correctly determine all geometric tooth modifications. For verify correct tooth geometry are used real footprints (contact patterns). The footprint, which has fluent beginning of the tooth mesh, has positive influence on the lower noise level [6]. Therefore, it is important to know, where beginning of the tooth mesh on the helical gearing is. On the below picture (see Fig. 2) are beginnings of the tooth mesh marked with arrow on the drive pinion and driven wheel.

The beginning of the tooth mesh on the drive pinion is from the bottom right corner. This condition applies only if the drive pinion is right-handed. If the pinion would be left-handed, the beginning of the tooth mesh would be from the bottom left corner. On the driven wheel is situation opposite to the drive pinion. The beginning of the tooth mesh is from the top right corner. This condition again applies only if the drive pinion is left-handed. For the right-handed, wheel will be begun on the top left corner.

3.2 Spread of the Meshing Line

The spreading shape of the meshing line on the tooth flank is on the helical gearing totally different then on the spur gearing. While on the spur gearing is the meshing line spreads parallelly. On the helical gearing is the meshing line spreads under the angle (in Fig. 3 is marked by black arrow). At the same time is on the picture marked with white arrows the direction of the spreading the contact line. On the drive pinion

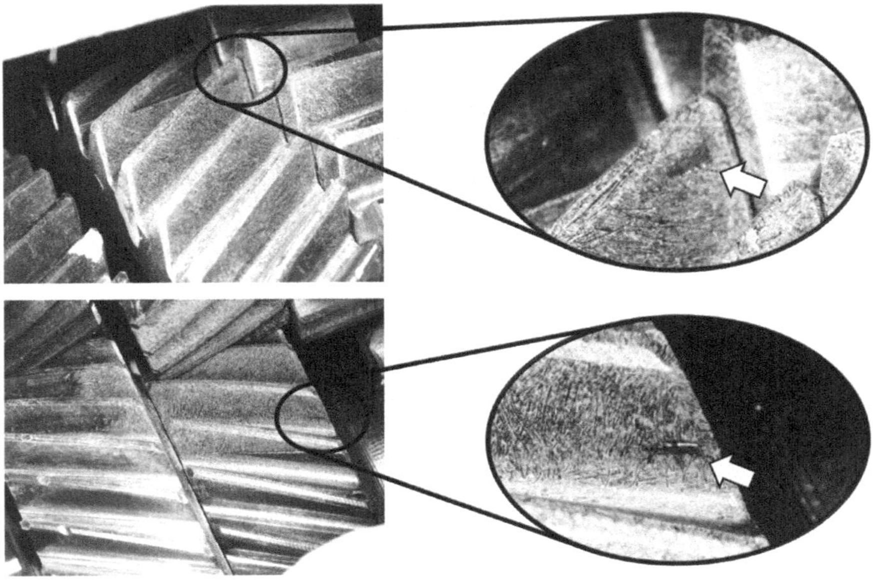

Fig. 2 Beginning of the tooth contact on the drive pinion (bottom picture) and driven wheel (top picture)

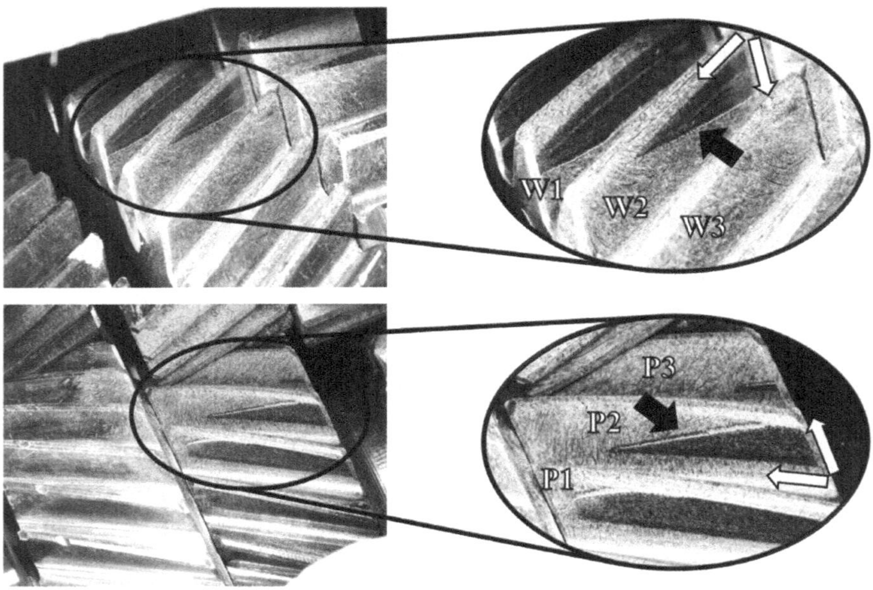

Fig. 3 Spreading of the meshing line on the drive pinion (bottom picture) and driven wheel (top picture)

is the contact line spread from the root to top of the tooth and from the right to left side (it is for right-handed pinion). On the driven wheel is the contact line spread from the top to root of the tooth and from the right to left side (it is for left-handed wheel).

During the rotation are alternate different numbers of teeth which depends on coefficient of total ratio (1).

$$\varepsilon_{\gamma} = \varepsilon_{\alpha} + \varepsilon_{\beta} \tag{1}$$

From the picture (see Fig. 3) is visible, that two and three pairs of teeth are alternating during the rotation. The teeth from drive pinion P1 is in the contact with teeth from driven wheel W1 and at the same time is in the contact P2 with W2 and P3 with W3. However, this condition will apply only for short time because teeth P1 and W1 are almost out from the mesh. Before, than pair of teeth P4 with W4 will be in the contact, the loading will be divided only between two pairs of teeth P2 with W2 and P3 with W3.

3.3 *Final Sizes and Shapes of the Footprints*

The final sizes and shapes of the footprints mainly depend on the load level, deformations (tooth, shaft, gear case, etc.) and geometric modification on the tooth. On the picture (see Fig. 4) are shown footprints on the drive pinion and driven wheel. Footprints were created on the levels 0, 30, and 120 N m, and its final sizes are

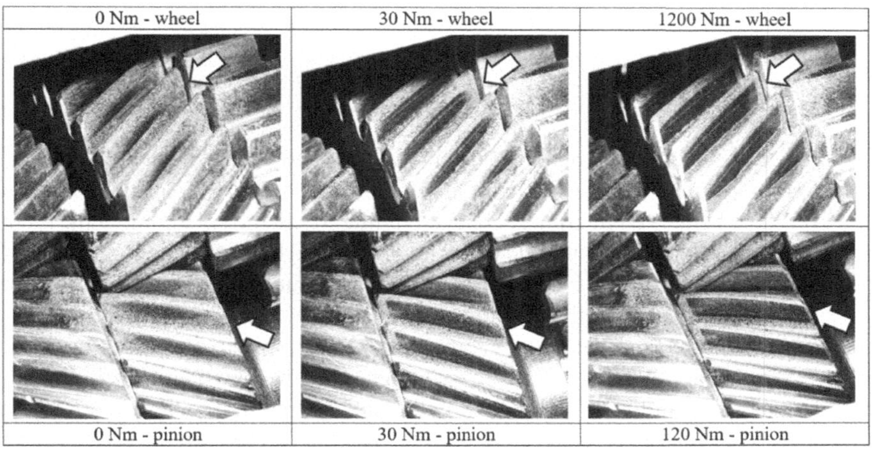

Fig. 4 Final sizes and shapes of the footprints during different load levels (drive pinion on the bottom pictures, driven wheel on the top pictures)

marked by white arrow. It is easy to see, how size and shape of footprints is changing and also, how changing ε_α and ε_β is.

With size and shape of the footprints goes hand in hand the noise level. The noise level does not affect only on the overall size of footprints but a lot of another factors they effect it [7]. This issue is very extensive and in large part still unexplored.

4 Discussion

Showed examples with footprints given a clear idea about how looks tooth mesh on the helical gearing. Theoretically, at the given load, the both footprints (on the drive pinion and the driven wheel) should have the same sizes and shapes. However, this condition ceases to apply, when specific slips on the drive pinion and the driven wheel have not the same values (if are not balanced specific slips). This difference caused that the ratio of the tracks between the roots of drive pinion is not same as on the top of driven wheel. Therefore, the showed footprints on the pinion and wheel have not the same sizes and shapes at the given load. The beginning of the tooth mesh and spreading shape of the meshing line on the tooth flank depends on the contribution if pinion (wheel) is right-handed or left-handed. Basically, if the pinion is drive, then every time the beginning of the tooth mesh is on the root, and right or left side depends on the inclination of the teeth. This condition is similar to the driven wheel. The beginning of the tooth mesh is on the top and right or left side depends on the inclination of the teeth. The overall sizes and shapes of the footprints mainly depends on the load level, deformations (tooth, shaft, gear case, etc.), and geometric modification on the tooth.

5 Conclusions

The aim of this study was approaching the issue of tooth mesh on helical gearing. Modern technology can create virtual models with simulation of tooth contact. However, for a better imagination for understanding where the beginning of the tooth mesh is, how tooth meshing of helical gearing looks, and how is changing the size of footprints due to increasing loading, is creation this issues on the real gearing. The first subchapter is deal with problematic of the beginning of the tooth mesh, which is very important for correct design of geometry modifications on the tooth. The fluent of the beginning tooth mash has positive influence on the low noise level. Spreading of contact line on the tooth flank is on the helical gearing totally different with comparing to contact line on the spur gearing. The contact line on the spur gearing is spread by parallelly, on the helical gearing is under the angle. This problematic is processed in the second subchapter. This subchapter also mentions the issue of the total contact ratio. Because during creation the contact line is also to see, how many teeth are in the mesh. The last subchapter shows, how is changed the size and shape of the footprint when is loading increasing. This experiment was performed on the gearing which has modified tooth. Therefore, footprints expand under higher load on all the directions. Showed pictures on the real gearing in this study give good background to study main issues on helical gearing.

References

1. Moravec V, Deil Z, Němček M, Folta Z, Havlík J (2009) Čelní ozubená kola v převodovkách automobilů, 1st edn. MONTANEX, Ostrava
2. Linke H (2010) Stirnradverzahnung: Berechnung - Werkstoffe – Fertigung, 2nd edn. Hanser, München
3. Pavlík J (2016) Measurement of gears footprints. VŠB—TU, Ostrava
4. Muller J (2013) How to inspect a gearbox. Gear Technol 30(6):82–86
5. Pavlík J, Folta Z (2016) Using footprints for check gearing. In: ICMD 2016: book of proceedings of the 57th international conference of machine design departments: Orea Resort Horizont, Železná Ruda, Czech Republic, September 7–9, 2016. Vydavatelství Západočeské univerzity v Plzni, Plzeň, pp 101–104
6. Moravec V (2001) Konstrukce strojů a zařízení II.: čelní ozubená kola: teorie, výpočet, konstrukce, výroba, kontrola. 1st edn. MONTANEX, Ostrava
7. Pavlík J, Začal J, Prokop J, Folta Z (2017) The growth of contact area of the tooth in depending of increasing the loading. In: Herák D (ed) Proceedings of the 58th international conference of machine design departments—ICMD 2017. Czech University of Life Sciences Prague, Prague, pp 280–283

Explanation of New Specifications Used in Geometric Dimensioning Model

Karel Petr

Abstract This article is based on rules used in ISO GPS standards. During few last years are applied to drawing documentation environment various modifiers, specifications, and symbols from 3D annotated model. There CAD model describes nominal shape of the part. Specifications are then defined by annotations directly in the 3D model [1, 2]. Specifications from 3D annotations are applied in 2D drawing documentation now. The significance of these specifications (modifiers) from field of geometric tolerance is same in 2D and in 3D documentation. This is the reason why it is necessary to know, recognize and applied these specifications. The basis for these specifications is the latest edition of the standard ISO 1101 from 2017. Main new specifications are: intersection plane indicator; orientation plane indicator; direction feature indicator; collection plane indicator. With these specifications, disappeared dependence of placement of geometric tolerances on specific 2D view in orthographic projection (ISO E). The method of evaluation for metrology was also better specified. Used these specifications, we can better and more clearly positioned geometric tolerances on a 2D drawing or also into a parametric 3D CAD model using PMI dimensions. This article explains application of these new symbols to specific components with detailed descriptions. Apart from symbols for geometrical tolerances, are also very important symbols for datum, which are defined in ISO 5459. Furthermore, the issue of linear and angular dimensions from standards ISO 14405-1 and ISO 14405-3.

Keywords Geometric specification · Intersection plane · Orientation plane · Direction feature · Collection plane · Geometrical product specification

1 Introduction

Geometrical tolerances [3–8] shall be specified in accordance with functional requirements, but manufacturing and inspection requirements can also influence geometrical

K. Petr (✉)
Czech Technical University in Prague, Technická 4, 16607 Prague 6, Czech Republic
e-mail: karel.petr@fs.cvut.cz

Š. Medvecký et al. (eds.), *Current Methods of Construction Design*, Lecture Notes in Mechanical Engineering, https://doi.org/10.1007/978-3-030-33146-7_17

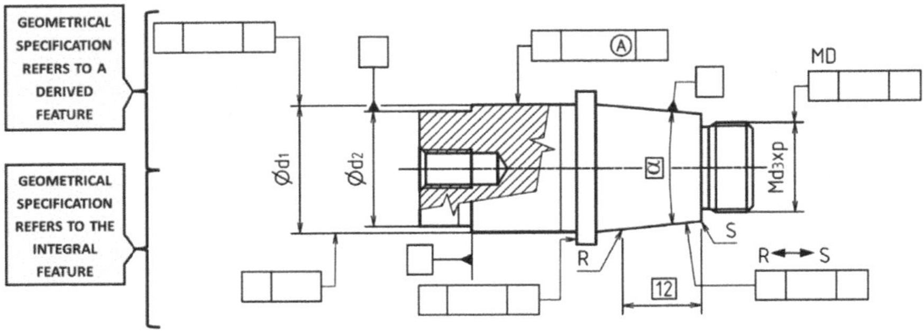

Fig. 1 Geometrical specification (ISO 1101:2017) applies to drawing [1]

tolerancing. Very important thing, which in accordance with ISO 8015, is that geometrical tolerance applied to a feature defines the tolerance zone around the reference feature within which the toleranced feature shall be contained. A feature is a specific portion of the workpiece, such as a point, a line, or a surface. These features can be integral features or derived features.

1.1 Applications of Geometrical Tolerance on Model

A geometrical specification applies to a single complete feature (see Fig. 1), unless an appropriate modifier is indicated. When the toleranced feature is not a single complete feature, see Fig. 1 (R ↔ S).

When the geometrical specification refers to the integral feature, the geometrical specification indication shall be connected to the toleranced feature by a reference line and a leader line terminating according to Fig. 1 (the bottom half of the figure). When the geometrical specification refers to a derived feature (a median point, a median line, or a median surface), it shall be indicated either by a reference line and a leader line terminated by an arrow on the extension of the dimension line of a feature of size or by modifier A in circle (see Fig. 1).

1.2 New Specifications for 2D Drawing and 3D CAD Model

Main new specifications are (see Fig. 2): intersection plane indicator; orientation plane indicator; direction feature indicator; collection plane indicator [2].

With these specifications, disappeared dependence of placement of geometric tolerances on specific 2D view in orthographic projection (ISO E). The method of evaluation for metrology it was also better specified [3]. Used these specifications,

Fig. 2 Intersection plane indicator (symbol with datum A); orientation plane indicator (symbol with datum B); direction feature indicator (symbol with datum C); collection plane indicator (symbol with datum D)

we can better and more clearly positioned geometric tolerances on a 2D drawing or also into a parametric 3D CAD model using PMI dimensions.

2 Methods and Materials

At this point, an explanation of the above symbols is given.

2.1 Intersection Plane Indicator

Intersection plane, established from an extracted feature of the workpiece, identifying a line on an extracted surface (integral or median) or a point on an extracted line. Intersection planes shall be used to identify the orientation of line requirements (straightness of a line in a plane, line profile, orientation of a line element of a feature) and "all around" specifications for lines on surfaces (line or surface profile), (see Fig. 3).

In Fig. 3 (left), any extracted line on the upper surface, as specified by the intersection plane indicator (in two direction—parallel and perpendicular to a datum A), shall be contained between two parallel straight lines 0.16 mm apart for direction parallel with datum A and 0.1 mm apart for direction parallel with datum A.

Same indication (see Fig. 3) is possible used in parametric 3D CAD model (PMI dimensions) only leader line of geometrical specifications in terminated by dot.

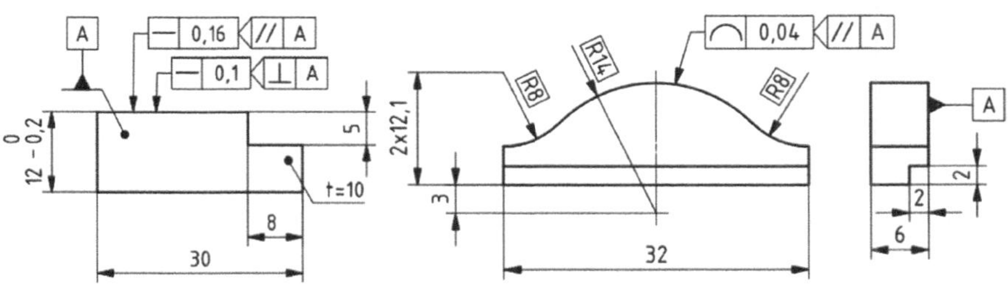

Fig. 3 Straightness specification of a line (two different direction) in a plane (left view); line profile specification on one feature—theoretical exact dimension (TED) R14 (right view)

In Fig. 3 (right), in each section, parallel to datum plane A, as specified by the intersection plane indicator, the extracted profile line (feature defined by TED R14) shall be contained between two equidistant lines enveloping circles of diameter 0.04 mm. The centers of which are situated on a line having the theoretically exact geometrical form (TEF). For indication of more than one specific feature are necessary used combination symbols "between" and UF (united feature) or CZ (combine zone) or SZ (separate zone) [4].

2.2 Direction Feature Indicator

Direction feature is an ideal feature, established from an extracted feature of the workpiece, identifying the direction of local deviations. Direction features shall be used when the toleranced feature is an integral feature and the local width of the tolerance zone is not perpendicular to the surface.

For cylindrical features, roundness applies in cross-sections perpendicular to the axis of the toleranced feature (see Fig. 4). For roundness of revolute surfaces (see Fig. 4) that are neither cylindrical nor spherical, a direction feature shall always be used to indicate the direction of the width of the tolerance zone. In Fig. 4, for both the cylindrical and conical surfaces, the extracted circumferential line, in any cross-section of the surfaces, shall be contained between two coplanar concentric circles, with a difference in radii of 0.03 mm. This is the default for the cylindrical surface and indicated by the direction feature indicator for the conical surface. The extracted (actual) cross-section line, in any cross-section perpendicular to the axis defined by datum A.

The tolerance zone defined by the specification in Fig. 5 (left), in the considered

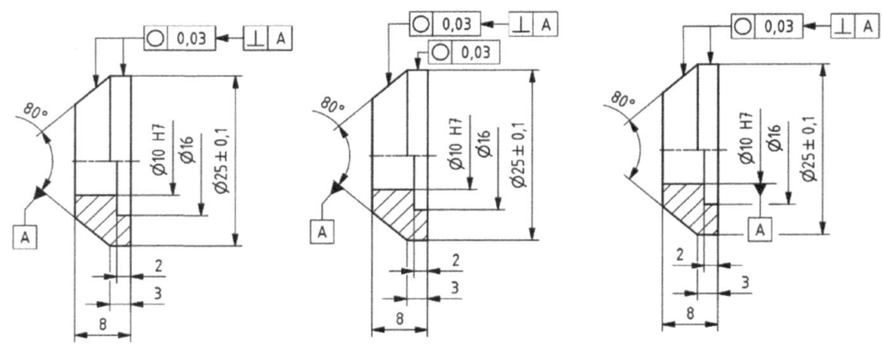

Fig. 4 Three variants of application of roundness specification with different axis. Axis defined by perpendicularity of intersection plane to datum A. Left view—datum is defined as axis of conical feature; middle view—datum is defined as axis of toleranced feature; right view—datum is defined as axis of cylindrical feature (diameter 10 H7)

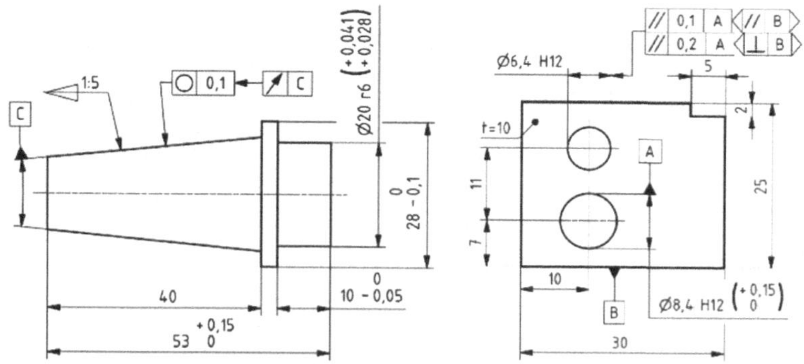

Fig. 5 Left—direction feature with symbol circular run-out. Right—orientation plane indicator for two orientations

cross-section, is limited by two circles on a conical surface a distance 0.1 mm apart along the surface. Direction feature with symbol circular run-out is used.

2.3 Orientation Plane Indicator

Orientation plane, established from an extracted feature of the workpiece, identifying the orientation of the tolerance zone. The orientation plane indicator controls both the orientation of the planes that limit the tolerance zone (directly by the datum and the symbol in the indicator) and the orientation of the width of the tolerance zone (indirectly, perpendicular to the planes), or the orientation of the axis for a cylindrical tolerance zone.

In Fig. 5 (right), the extracted median line (axis of hole diameter 6.4 H12) shall be contained between two pairs of parallel planes, which are parallel to datum axis A, and positioned 0.1 and 0.2 apart, respectively. The orientation of the planes limiting the tolerance zones is specified with respect to datum plane B by the orientation plane indicators. Deprecated practice—datum B is secondary to datum A (see ISO 1101:2014).

2.4 Collection Plane Indicator

Collection plane, established from a feature on the workpiece, defining a closed compound continuous feature. A collection plane shall be indicated when the "all around" symbol (see Fig. 6) is used to identify that a specification applies to a collection of features. A collection plane identifies a set of single features whose intersection with any plane parallel to the collection plane is a line or a point.

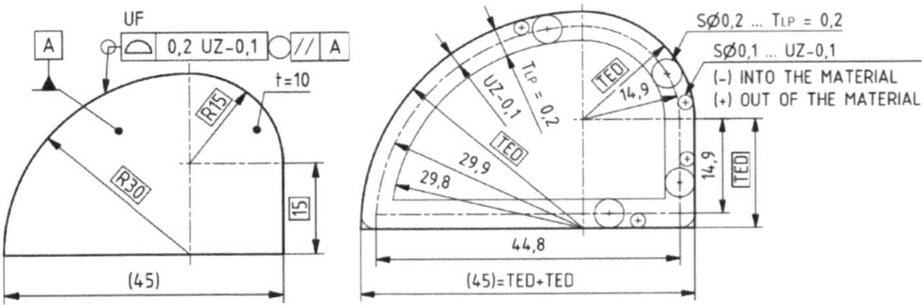

Fig. 6 Collection plane for surface profile symbol with "all around" symbol. Unequally disposed surface profile specification for a united feature (2D drawing indication—left; Tol. zone—right)

In Fig. 6, the extracted surface shall be contained between two equidistant surfaces enveloping spheres of diameter 0.2 mm, the centers of which are situated on a surface having the theoretically exact geometrical form (TEF). The UZ modifier is used to indicate that the tolerance zone is moved 0.1 into the material, see Fig. 6 (right) [1]. When using the line profile symbol, if the intersection plane and the collection plane are the same, the collection plane symbol can be omitted.

3 Results and Discussion

At present, it is necessary applied abovementioned symbols in the drawing documentation. How was say, these symbols in possible used in 2D drawing or in 3D CAD model. In the illustrations below, annotated commands are marked with a black arrow.

In Fig. 7, for "all around" symbol you can see application of collection and intersection plane indicator. First case in Fig. 7 are the intersection plane and the collection plane the same, the collection plane symbol can be omitted. For flatness tolerance is orientation plane indicator used. Common datum C-D (plane from two axes) defined orientation of flatness surface 5 × 2.

In Fig. 8, perpendicular tolerance with intersection plane indicator is used to indicate that the toleranced feature is the lines in the feature and the direction of the lines, where the toleranced feature is all lines in the cylindrical surface symmetry to datum A.

Section plane for definition of tolerance zone for roughness tolerance is perpendicular to datum, which is axis of conical feature (see Fig. 8). In standard ISO 14405-1:2016 [5] are examples for indication of direction feature indicator to linear dimension.

In Fig. 9 are shown other types of symbols which are possible to use on drawing.

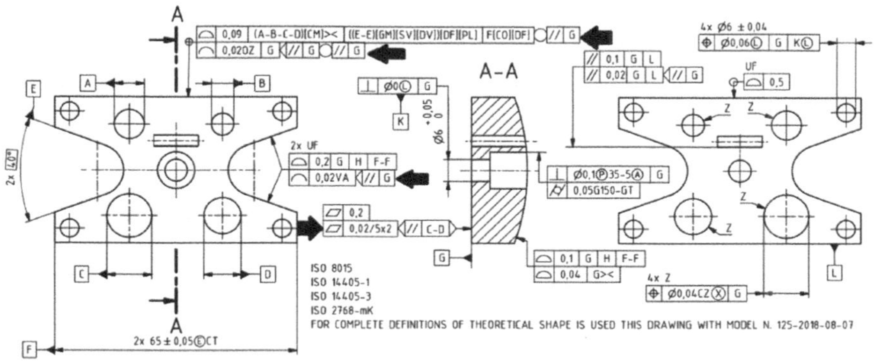

Fig. 7 Example 1—application of intersection plane indicator and collection plane indicator for line profile, orientation plane indicator for flatness. (Inspiration by ex. from prof. V. Läpple)

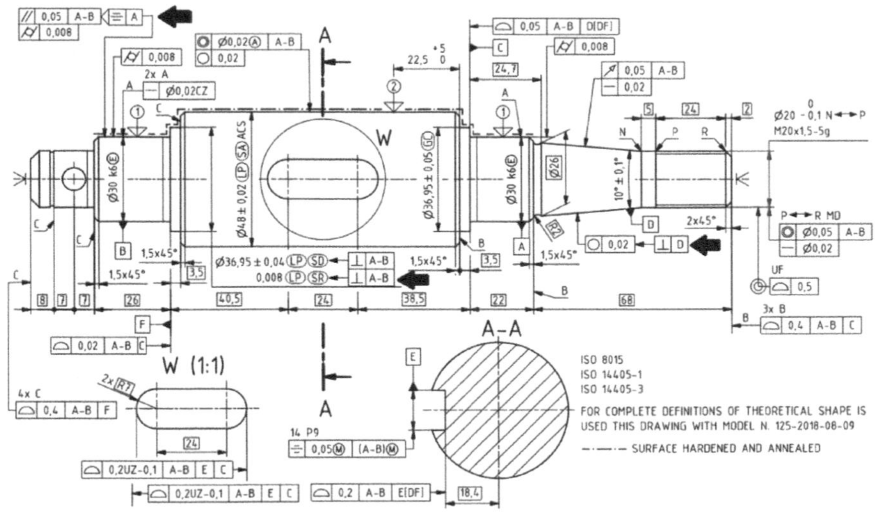

Fig. 8 Example 2—application of intersection plane indicator for parallelism symbol, direction feature indicator for roundness and for linear dimension (ISO 14405-1). (Inspiration by ex. from prof. V. Läpple)

4 Conclusions

With new specifications (orientation plane indicator; direction feature indicator; collection plane indicator), disappeared dependence of placement of geometric tolerances on specific 2D view in orthographic projection (ISO E). The method of evaluation for metrology it was also better specified.

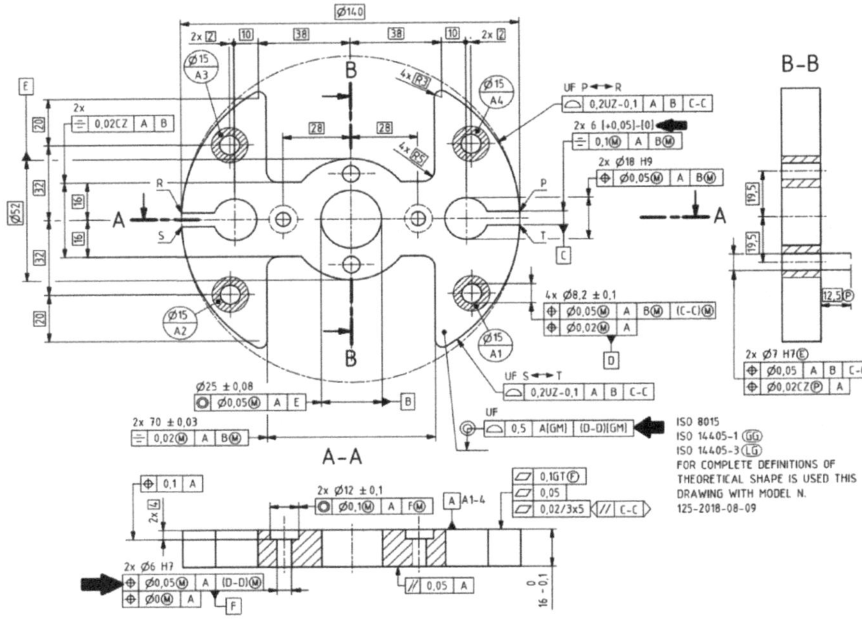

Fig. 9 Example 3—other types of symbols, which are possible to use on drawing. (Inspiration by ex. from prof. V. Läpple)

Used these specifications, we can better and more clearly positioned geometric tolerances on a 2D drawing or also into a parametric 3D CAD model. Don't forget, than apart from symbols for geometrical tolerances are also very important symbols for datum, which are defined in ISO 5459. Furthermore, is very important the issue of linear and angular dimensions according standards ISO 14405-1 and ISO 14405-3 [1].

On the drawing, are indicated many specifications (dimensions, geometrical tolerances …), but do not forget, than each specification is only for one feature or for part of feature (if we used modifier). For application of one specification on more than one feature, we must used some modifier (CZ, UF, …). Not all specifications in drawing are functional specification!

References

1. Petr K (2017) Indication of functional dimension according ISO GPS—how shall we applicate? In: Herák D (eds) Proceedings of 58th international conference of machine design departments, ICMD 2017, pp. 284–289. Czech University of Life Sciences Prague
2. Brónček J, Drbúl M, Stupavský M, Radek N (2017) The application of new rules of GPS in structural product requirement. In: Herák D (eds) Proceedings of 58th international conference of machine design departments, pp 50–55. Czech University of Life Sciences Prague

3. ISO 1101 (2017) Geometrical product specifications (GPS), geometrical tolerancing, tolerances of form, orientation, location and run-out
4. ISO 5459 (2011) Geometrical product specifications (GPS), geometrical tolerancing—datums and datum systems
5. ISO 14405-1:2016.Geometrical product specifications (GPS), Dimensional tolerancing, Part 1: Linear sizes
6. ISO 14405-3 (2016) Geometrical product specifications (GPS). Dimensional tolerancing, Part 3: Angular sizes
7. ISO 8015 (2011) Geometrical product specifications (GPS), fundamentals, concepts, principles and rules
8. ISO 1101 (2014) Geometrical product specifications(GPS), geometrical tolerancing

Creation of Calculation 3D Model of Slewing Bearing

Rudolf Skyba, Slavomír Hrček, Lukáš Smetanka and Maroš Majchrák

Abstract The article deals with the creation of 3D model of slewing bearing with the support of the 3D modeling software Creo Parametric 3.0 and with support of Apex Grizzly 16, that is especially suited for creation of a high-quality mesh on a variety of 2D or 3D models. The basic task will be, to define specific type of slewing bearing and find out the basic parameters, like ball diameter, diameter of inner and outer race, and pitch diameter of rolling tracks that are necessary for its construction. After we create suitable 3D model of slewing bearing, we need to divide the base model of ball into the several parts that will allow us to create a quality and uniform finite element mesh on the entire volume of the meshed part. After transforming the model into the program Apex Grizzly, we can create basic mesh and find optimal mesh values for each element of divided bearing separately.

Keywords Stiffness · Bearing · Model · Analysis

1 Introduction

Slewing bearings also known as slewing rings are large-sized rotary elements usually used in applications in which large rotational functional elements are involved, such as tower cranes, wind turbine generators, boring machines, etc. There are many different types of large diameter bearings, and they are divided according to the type of rolling element and its internal arrangement. There are bearings with one, two, or three rows of rolling element, and we distinguish between balls and rollers. Figure 1 points to a typical arrangement of four contact point slewing bearings and also shows the usual load system acting on bearing in axial or radial direction.

It has been studied that all angular contact ball bearings have similar features in consideration of geometry, mechanism, and structure [1, 2]. The stiffness of angular four contact point ball bearings has a very important influence on the dynamics of a rotating shaft and the machine system [3, 4]; the life and the rolling contact fatigue can be determined by full-scale bearing tests. These types of tests are released on

R. Skyba (✉) · S. Hrček · L. Smetanka · M. Majchrák
University of Žilina, Univerzitná 8215/1, 01026 Žilina, Slovak Republic
e-mail: rudolf.skyba@fstroj.uniza.sk

Š. Medvecký et al. (eds.), *Current Methods of Construction Design*, Lecture Notes in Mechanical Engineering, https://doi.org/10.1007/978-3-030-33146-7_18

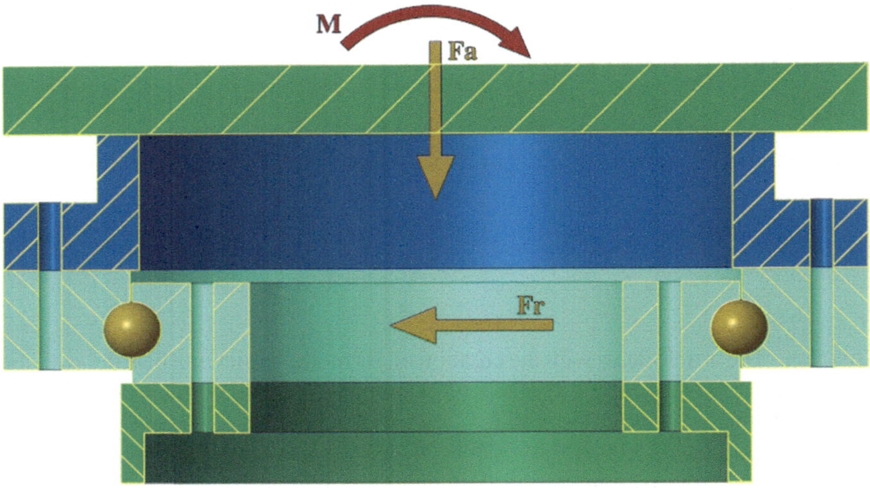

Fig. 1 Typical arrangement of four contact point slewing bearing [4]

test rigs and that is very expensive and time-consuming solution. Bearing designers would like to understand the impact of four variables like subsurface residual stress, ball material density, gradient in yield strength with depth raceway surface hardness that are thought to affect spall propagation [5, 6].

Mullick [7] researches radial stiffness of a radial and ball bearings with angular contact using the John Harris method and also the finite element method. For solving systems of nonlinear equations, we are using Newton-Raphson's method, while in contact analysis uses finite elements method. The results show that the displacement and of the bearing rings depend with high probability on the combined loads and centrifugal force. Antoine et al. [8, 9] propose two methods for determination the contact angle between inner and outer ring of the bearing depending on the preload and also speed for special cases of elastic preload. There are many methods based on the Hertz contact theory. They are based on the assumptions that the force of preload constant and does not effect on the change of contact angle or speed. In solving the system of equations, it is admitted that a certain speed, for a preload comes only to the revoke of the axial clearance, and that there is no axial deformation.

Sun et al. [10] examine relationship between contact deformation, change of stiffness, and clearance of beating. The results gained by experimental and analytical propose determining the elastic deformation of bearing elements must be taken into account radial and axial clearances in the bearing. Wei et al. [11] examined the influence of centrifugal force, gyroscopic moment, and preload on the bearing stiffness. They concluded that if the value of contact angle of raceway and ball exceed 8.9°, the value of the radial stiffness decreases with increasing speed. By the experiments have been shown that the radial stiffness decreases more than 20% if the contact angle is 40° and revs are 15,000 rpm. They also noted that the increasing of the temperature affects an increasing of preload of bearing and increase the frequency of oscillation.

The most appropriate way to determine axial or radial stiffness of four-point contact ball bearing is to create 3D model with cad software program like, for example, Creo 3.0 Parametric. In the next step, it is necessary to convert native file from 3D cad software into a suitable format that could be meshed in program such as ANSYS Workbench. After the network is created, calculations can be made. In an ideal case, experimental test will be performed to verify the calculations and confirm the hypothesis.

2 The Hertz Contact Theory

The Hertz contact theory assumes the contact of two bodies with curved surfaces, which are pushed against each other by the force Q. Each of the two bodies is characterized by a curvature in main planes (see Fig. 2) that are perpendicular to each other, and in which the maximum and minimum curvature is located. Curvature

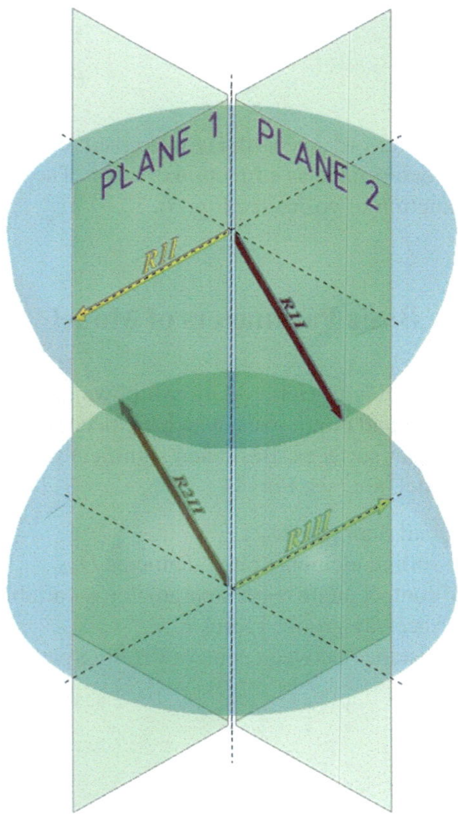

Fig. 2 Radii located in main planes

ρ is defined as the inverted value corresponding to the radius of curvature r. Curvature ρ is positive if the center of curvature is located inside the body (convex curvature $+\rho$), and negative if the center is located outside the body (concave curvature $-\rho$).

Calculation of deformation, contact area, and contact pressure according to the Hertz contact theory requires, first of all, to determine Hertz's coefficients μ, ν, $2K/\pi\mu$ and the relationship to the curvature in the contact area. These coefficients characterize the stress distribution in the contact area. In the case of rolling bearings, curvatures ρ_{1I} and ρ_{2I} as well as ρ_{1II} and ρ_{2II} are located in the same plane. The auxiliary value $\cos\tau$ is calculated from the relationship:

$$\cos\tau = \left|\frac{\rho_{1I} - \rho_{1II} + \rho_{2I} - \rho_{2II}}{\sum\rho}\right| \tag{1}$$

Sum of all radii of curvature is:

$$\sum\rho = \rho_{1I} + \rho_{1II} + \rho_{2I} + \rho_{2II} \tag{2}$$

Inverted value of radius can be expressed as following:

$$\rho_{mn} = \frac{1}{R_{mn}} \tag{3}$$

In accordance with Herts derived relations, the coefficients μ, ν, and $2K/\pi\mu$ can be determined as a function of $\cos\tau$. The values of the coefficients depending on the function of $\cos\tau$ (see Fig. 3).

3 Basic Parameters of Model

According to Hertz relationship, we can calculate basic parameters as the length of major and minor axes, which are necessary to determine the major dimensions of the contact surfaces. Basic and significant parameters, which are needed for creation a model, are (see Fig. 4):

- ball diameter D_{we},
- rolling elements pitch diameter D_{pw},
- contact angle α (this parameter is variable depending on applied load),
- raceway groove radius r_i, r_e,
- distance between center of rolling paths A.

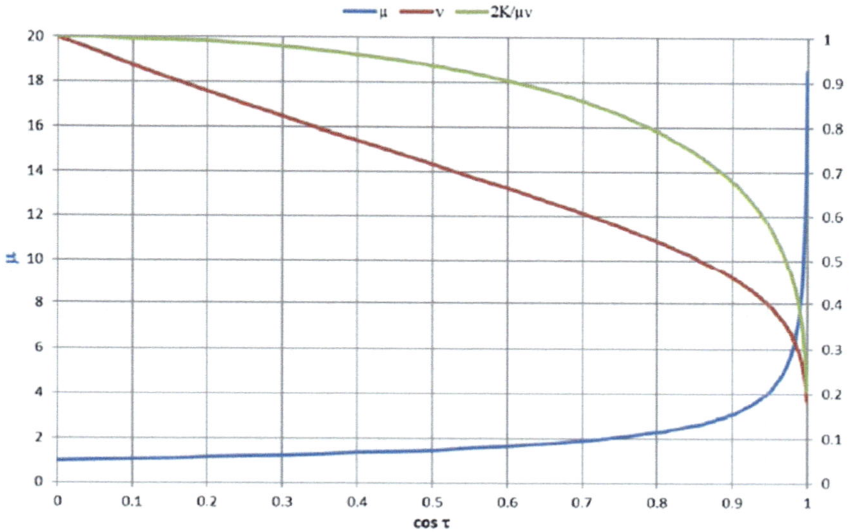

Fig. 3 Coefficients μ, ν, and $2K/\pi\mu$

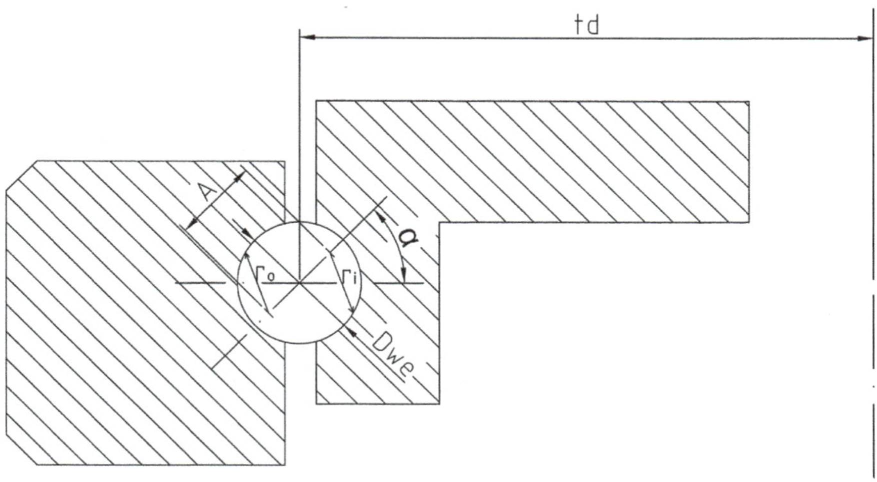

Fig. 4 Radial ball bearing under axial movement

3.1 Basic Calculations for Assembling the Bearing Model

Based on the abovementioned parameters and provided that the contact angle is 45°, we can obtain the following calculations:

Relative movement of rings in axial and in radial direction:

$$P_{eA} = A \cdot \cos_{\alpha} \tag{4}$$

$$P_{eR} = A \cdot \sin_{\alpha} \tag{5}$$

Distance between center of rolling paths A can be calculated by the following way:

$$A = B \cdot D_{we} \tag{6}$$

After we obtain parameter A, it is necessary to find the orbit diameters t_{di}, t_{do}:

$$t_{di} = D_{pw} - \frac{Dwe}{2} - \frac{\Delta r_o}{2} \tag{7}$$

$$t_{do} = D_{pw} + \frac{Dwe}{2} - \frac{\Delta r_i}{2} \tag{8}$$

After we obtain value of the radius of curvature, its sum Eq. (2) and the absolute value of the $\cos_{\tau}$ Eq. (1), we are able to calculate values of the major (9) and minor (10) semi axes by the following equations:

$$2_a = 2 \cdot \mu \cdot \sqrt[3]{\frac{1}{E} \cdot \left(1 - m^2\right) \cdot \left(\frac{3 \cdot Q}{\sum \rho}\right)} \tag{9}$$

$$2_b = 2 \cdot \nu \cdot \sqrt[3]{\frac{1}{E} \cdot \left(1 - m^2\right) \cdot \left(\frac{3 \cdot Q}{\sum \rho}\right)} \tag{10}$$

After that we need to find out the depth in which the tension work and its possible by the following equation:

$$2 \cdot z_0 = 2b \cdot 0.25 \tag{11}$$

In Fig. 5, is presented schematic representation of the model for the ball and the ball raceway contact according to [12].

4 Creation of 3D Model and Mesh

After we have obtained necessary parameters of bearing, we could create a function model of slewing bearing we need in 3D CAD software like PTC Creo 3 (see Fig. 6).

The bearing consists of ball, inner ring, and outer ring. Each part of the bearing is needed to be divided separately so that we can make fine hexagonal mesh. The creation of the mesh on the inner and outer ring is quite simple (see Fig. 7), but the

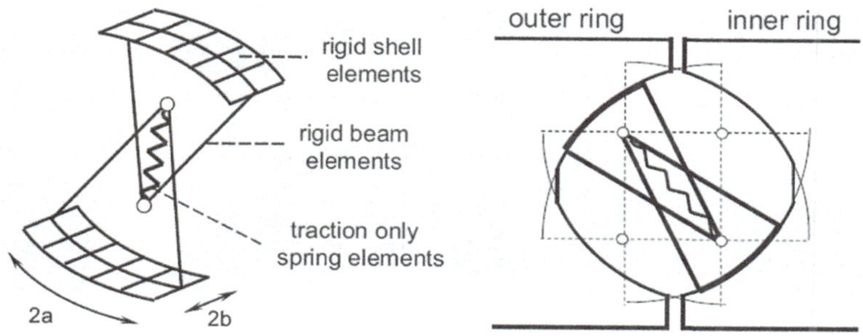

Fig. 5 Schematic representation of the bearing model

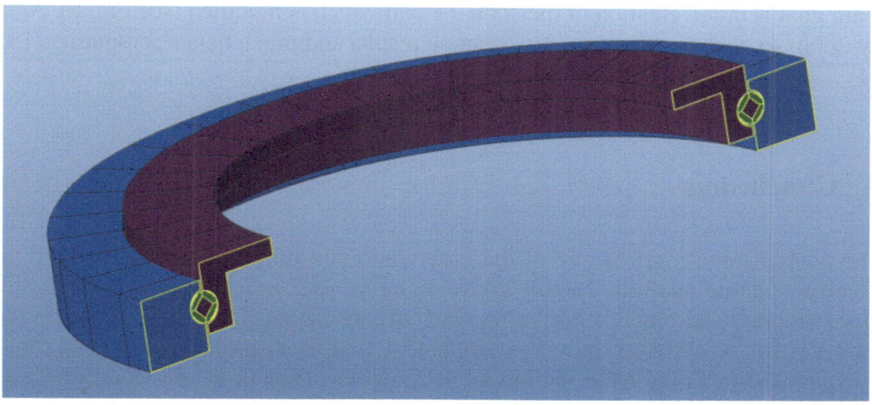

Fig. 6 3D model of slewing bearing

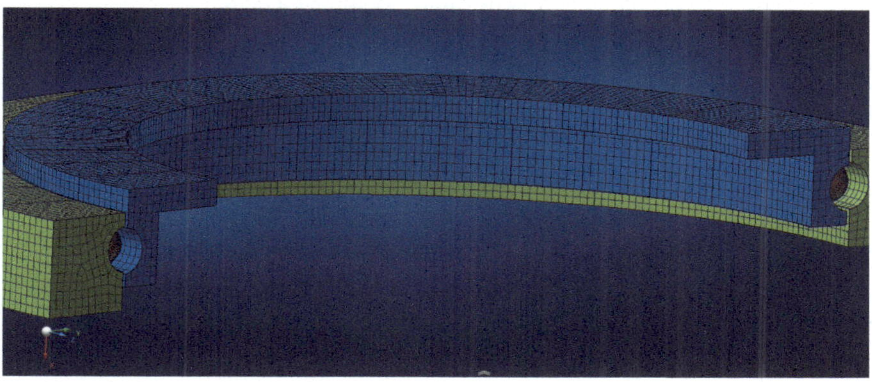

Fig. 7 3D bearing model with created mesh

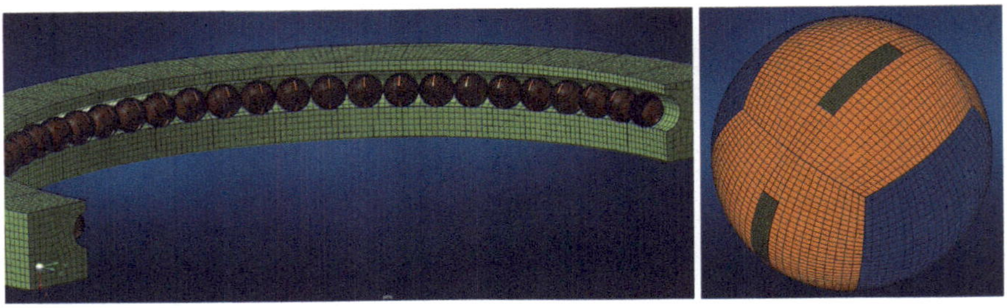

Fig. 8 Mesh created on the ball

main goal is in the creation of mesh on the ball which is modified (see Fig. 8). And by doing so, we can achieve more accurate results and much better computing [13, 14].

5 Conclusion

The quality finite mesh is one of the most important things, which is required to perform suitable analysis. Moreover, by doing so we can achieve results that are more accurate and much better computing. Before the mesh has been created, the model of ball was divided into the small pieces, and then transformation from Creo to Apex was realized. After the model in Apex was loaded, a consistent mesh on both bearing rings has been created. For creating a consistent and uniform mesh on ball, it was necessary to create a dividing-plane. Thanks to these dividing planes, it was possible to create unified mesh in whole volume of a ball.

Acknowledgements Slovak Research and Development Agency supported this study under the contract no. APVV-14-0508—development of new methods for the design of special large-size slewing rings. This study was supported by the Ministry of Education, Science, Research and Sport of the Slovak Republic under the contract VEGA 1/0595/18—optimizing the internal geometry of roller bearings with line contact in order to increase their durability and reduce their structural weight.

References

1. Liu J, Shao Y, Lin CM (2012) Vibration analysis of ball bearings with a localized defect applying piecewise response function. Mech Mach Theory 56:156–169
2. Jedlinski L, Caban J, Krzywonos L, Wierzbicki S, Brumerčik F (2015) Application of vibration signal in the diagnosis of IC engine valve clearance. J Vibro Eng 17(1):175–187
3. Lukáč M, Brumerčík F, Krzywonos L, Krzysiak Z (2017) Transmission system power flow model. Commun Sci Lett Univ Žilina 19(2):27–31

4. Kanga Y, Huang C, Lin CM, Shen P, Changa Y (2006) Stiffness determination of angular-contact ball bearings by using neural network. Tribol Int 39:461–469
5. Nathan AB, Nagaraj K, Forster N, Svendsen V (2013) Critical stresses and strains at the spall edge of a case hardened bearing due to ball impact. Int J Fatigue 47:268–278
6. Potočnik R, Göncz R, Flašker J, Glodež S (2010) Fatigue life of double row slewing ball bearing with irregular geometry. Procedia Eng 2:1877–1886
7. Mullick A (1990) An investigation in stiffness of rolling contact. Dissertation thesis, Chung Yuan Christian University, Department of Mechanical Engineering
8. Antoine JF, Abba G, Molinari A (2006) A new proposal for explicit angle calculation in angular contact ball bearing. J Mech Des Trans ASME 128:468–478
9. Antoine JF, Visa C, Sauvey C (2006) Approximate analytical model for hertzian elliptical contact problems. J Tribol Trans ASME 128:660–664
10. Sun MK, Kang JL, Sun KL (2002) Effect of bearing support structure on the high-speed spindle bearing compliance. Int J Mach Tools Manuf 42:365–373
11. Wei L, Jay FT, Kamman J (2003) An integrated thermomechanical—dynamic model to characterize motorized machine tool spindles during very high speed rotation. Int J Mach Tools Manuf 43:1035–1050
12. Daidie A, Chaib Z, Ghosn A (2008) 3D simplified finite elements analysis of load and contact angle in a slewing ball bearing. ASME 130(8):082601
13. Kohár R, Hrček S (2014) Dynamic analysis of rolling bearings with an elastic cage. In: Modern methods of construction design. Lecture notes in mechanical engineering. Springer International Publishing, Switzerland, pp 249–254
14. Medvecký Š, Kohár R, Hrček S (2012) Dynamic analysis of cage behavior in a tapered roller bearing. In: KOD 2012, Konstruisanje - Oblikovanje – Dizajn, 7th international symposium machine and industrial design in mechanical engineering, Balatonfüred, Hungary. Faculty of technical sciences, Novi Sad, pp 45–50

Design Optimization of Micro-hydro Power Plant

Pavel Srb and Michal Petrů

Abstract This article deals with the design optimization of the micro-hydropower plant. This mini-power source is designed as an additional power source for small recreational objects or remote places with access to a water stream of flow rate approximately 0.4 m^3/s. A paddle wheel with 3 m diameter is welded from sheet metal. The paddle wheel is designed for a rotational speed of 5 rpm. The electric generator with planetary gear is integrated into central supported tube. The original construction was very robust and heavy. Particularly due to the high weight, it was required optimizing of design. A finite element model of the whole structure of the hydropower plant was created. Mechanical stresses and deformations in main part have been analyzed. Based on these analyzes, optimization has been proposed with the result of weight reduction while maintaining of use properties and with good durability.

Keywords Optimization · Hydropower · FEM analysis

1 Introduction

Hydropower is derived from the potential and kinetic energy of flowing water. If the water head is allowed to move a mechanical component then that movement involves the conversion of the potential energy of the water into mechanical energy. Hydropower is a very clean source of energy and only uses the water which can be used for other purposes after generating energy [1]. A hydropower resource can be evaluated by the availability of the power. The extracted power P is a function of the hydraulic head H, rate of fluid flow Q, density of water ρ, the converter efficiency η, and the local acceleration due to gravity g [2]. The power is formulated as:

$$P = \eta \cdot \rho \cdot Q \cdot g \cdot H \tag{1}$$

P. Srb (✉) · M. Petrů
Technical University of Liberec, Studentská 2, 46117 Liberec 1, Czech Republic
e-mail: pavel.srb@tul.cz

M. Petrů
e-mail: michal.petru@tul.cz

Š. Medvecký et al. (eds.), *Current Methods of Construction Design*, Lecture Notes in Mechanical Engineering, https://doi.org/10.1007/978-3-030-33146-7_19

Hydropower started with the wooden waterwheel. Many various types of waterwheel had appeared in many parts of Europe and Asia for more than 2000 years, mostly for milling grain. During the industrial revolution, waterwheel technology had been developed to efficiencies 70%. The development of the industry combined with the need to develop smaller and higher speed devices to generate electricity, led to the development of modern turbines. At the end of nineteenth century, many mills were replacing their waterwheel with turbines. The biggest boom of hydropower was the first half of the twentieth century, before oil took over as the dominant force in energy provision [3].

Small hydro is in most cases run-of-river; in other words, any dam or barrage is quite small, usually just a weir, and generally little or no water is stored. Therefore, run-of-river installations do not have the effect on the local environment as large hydro. Small hydropower has several degrees of size. A maximum of 10 MW is the most widely accepted value of small hydro, devices below 2 MW are mini-hydro, and below 500 kW micro-hydro [2, 3].

Hydropower installations generate about 17% (2700 TWh/yr) of the world's supply of electricity which is 97% of all renewable power generation. Globally, hydropower potential is estimated at 40,000 TWh/yr of which 14,400 TWh/yr is technically feasible for development, and 8000 TWh/yr is economically feasible [4].

Three forms of waterwheel are commonly defined along with simple explanations of their working principles. These are the undershot, overshot, and middleshot, which are shown in Fig. 1. The classification depends on the water entry mechanism as follow, according to [4]:

- Undershot: The water enters beneath the wheel, which is predominantly driven by impulse, resulting from a difference between the lower peripheral velocity of the wheel and the faster velocity of the water.
- Overshot: The water enters from above the wheel, which rotates in the opposite direction to undershot and middleshot waterwheels. These are predominantly driven by the water's potential, resulting from gravity.
- Middleshot: The water enters around the same height as the axle. These require a curved section of channel bed called a breast or shroud, and extract energy from

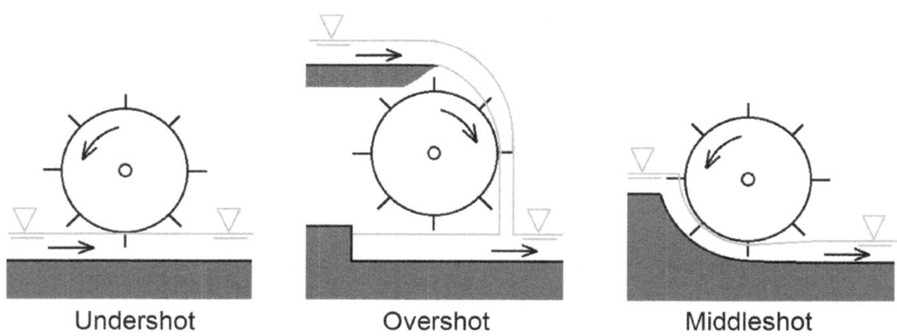

Fig. 1 Waterwheel classification

Fig. 2 Poncelet waterwheel [4]

the water through impulse and potential, lowering the water through a vertical distance under influence of gravity.

- The design of the traditional undershot waterwheel was improved by Jean-Victor Poncelet who replaced the straight blades on ordinary wheel with curved or cylindrical blades [5]. His design used curved blades positioned so the water met the blade flat to its edge instead of the side. This eliminated the bounce that robbed power from typical design. Poncelet estimated that practical wheels would reach efficiency between 70 and 85% [2]. Figure 2 shows Poncelet waterwheel.

2 Methods and Materials

The optimized micro-hydropower plant is based on a Poncelet waterwheel with a diameter of 3 m. The electric generator of power 2 kW with planetary gear with ration 100:1 is integrated into central supported tube. Figure 3 shows CAD model of optimized micro-hydropower. Optimization will be performed using finite element method FEM [6, 7].

2.1 Preparation of FEM Model

The preparation of the analysis for the technical assessment of the design of the hydroelectric power plant was based on the supplied CAD data. Model data included the geometry of the critical exposed locations (shaft, blades, and bearings) that will be most loaded during operation. Standard steel materials (steel 11523 non-alloy steel of the standard quality with a minimum yield strength of 355 MPa, fatigue limit 160 MPa) was used for such structures. From the CAD data of the complete

Fig. 3 Micro-hydropower

hydroelectric drive assembly, all the construction nodes and parts that directly affect the drive operation have been used for computational model analysis transmit forces. This geometry was imported into the preprocessor of the FEM software ABAQUS [8]. Subsequently, initial and boundary conditions were imported into the model. The shaft ends are fixed 50 mm from the end. The bearings have been replaced by rings. The inner surface of the rings was tied connected to the shaft. A non-friction contact was used between the outer surface of the rings and the inner surface of the hub. The welded and screwed parts were connected by a tie constrain. Flexible couplings have been replaced by coupling bonding the surfaces through the control point. Initial and boundary conditions are shown in Fig. 4.

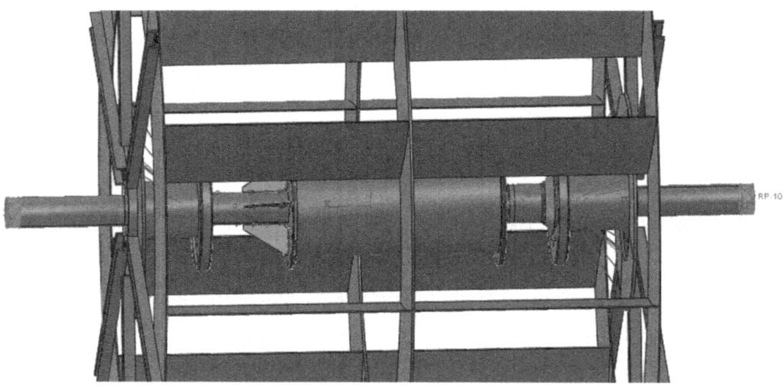

Fig. 4 Initial and boundary conditions

2.2 *FEM Model*

A finite element net was created, consisting of 114,965 elements and 206,877 nodes. The material model was chosen as the elastic corresponding steel 11,523, the Young's elastic modulus $E = 210{,}000$ MPa, the Poisson number $\mu = 0.3$, the specific weight $\rho = 7800\ \text{kg.m}^{-3}$. Simulation was divided into two steps. In the first step, gravity was applied $g = 9.81\ \text{m.s}^{-2}$. In the second step, pressure was applied to one pair of blades of value corresponding to a torque of 5500 Nm.

3 Results

The simulation results are shown in the following figures. Figure 5 shows stress distribution. Highest value of stress 441 MPa was found in key shaft groove at singular point. This value is not valid and is not included in Fig. 5. Key shaft will be recalculated. Except singular point, the highest value of stress was 72 MPa near fixed ends of shaft. This value is significantly lower than yield stress. Figure 6 shows displacement distribution, highest displacement was 21 mm, which is relatively low in relation to the size of the structure.

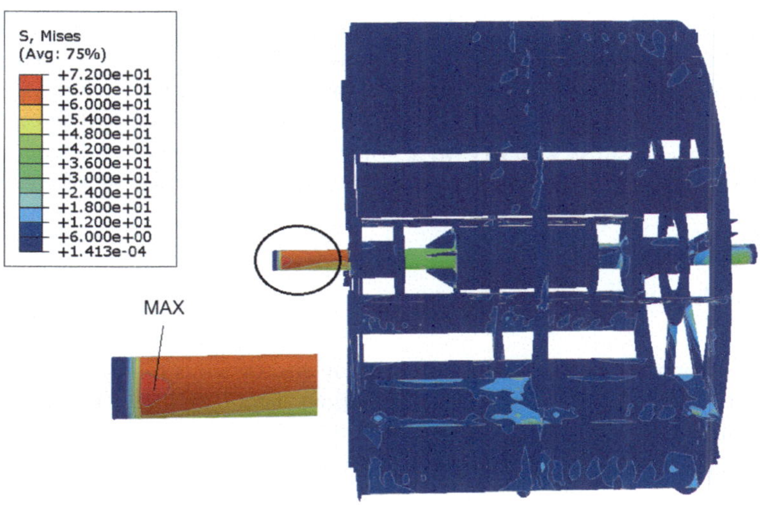

Fig. 5 Analysis results—stress distribution

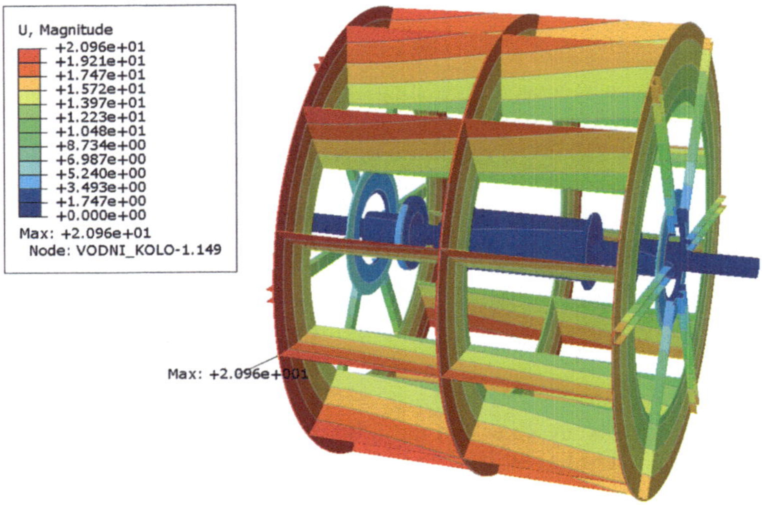

Fig. 6 Analysis results—displacement distribution

4 Discussion

Simulations of the selected parts of the hydroelectric power plant have determined that the individual parts are dimensioned with adequate safety for the material S355J0. The higher value of stress was found in the shaft key groove where at the singular point the stress reached 441.3 MPa. Stress in shaft key was recalculated and the resulting value of stress was 139 MPa, which is higher value than recommended limit of 120 MPa. It is recommended to use two shaft keys rotated of 120°.

5 Conclusions

FEM simulation of the micro-hydropower plant was carried out. Based on technical judgment and FEM simulations, critical locations were identified. Design changes have been recommended to ensure trouble-free operation of the hydropower for a specified period of 5 years.

Acknowledgements The results of this project No. LO1201 were obtained through the financial support of the Ministry of Education, Youth and Sports, Czech Republic in the framework of the targeted support of the "National Programme for Sustainability I."

References

1. Mohd M, Hakim A (2004) Basic design aspects of micro hydro power plant and its potential development in Malaysia, In: National power and energy conference (PECon) 2004 Proceedings, pp 220–223. IEEE, Malaysia
2. Helmizar H (2016) Turbine wheel—a hydropower converter for head differences between 2.5 and 5 m. Doctoral Thesis, University of Southampton, Malaysia
3. Paish O (2002) Small hydro power: technology and current status. Renew Sustain Energy Rev 6(6):537–556
4. Senior JA (2009) Hydrostatic pressure converters for the exploitation of very low head hydropower potential. Doctoral Thesis, University of Southampton, Malaysia
5. Gilgili M, Ozbek A (2016) An overview of micro-hydropower technologies and design characteristics of waterwheel systems. Çukurova Üniversitesi Mühendislik Mimarlık Fakültesi Dergisi, Haziran 31(1):117–134
6. Srb P, Petrů M (2018) Numerical simulation of reinforced polyurethane foam. In: Lufinka A, Sivcak L, Petrikova I (eds) 56th conference on experimental stress analysis, pp 381–386. Czech Society for Mechanics, Czech Republic
7. Kulhavý P, Srb P, Petrů M (2015) Numerical and experimental analysis of the real load arising in the cushion of the car seat. Manuf Technol 15(6):999–1005
8. Hibbitt, Karlsson, Sorensen (2013) ABAQUS/CAE 6.13 User's manual, Dassault Systemes

The Loading of the Crew Protection Structure of the ROPS According to Different World Standards

Ladislav Ševčík and Ha Van Nguyen

Abstract In contribution are compare two standards for testing the roll-over protective structures (ROPS) frame. The ROPS can be tested by different standards. Mostly used standard is the European Union (EU) regulation for protective frames of agricultural and forestry machines No. 1322/2014 for ROPS "Tractor" or ČSN EN ISO 3471:2008 Earth-moving machinery-laboratory tests and performance requirements for roll-over protective structures. The aim of the study was to investigate the protective structure of the cab of a truck MERCEDES 4 × 4, to cater for both European standards. Furthermore, the calculations were compared with Australian standard AS 2294.2-1997 Earth-moving machinery-protective structures. Also the laboratory tests and performance requirements for roll-over protective structures was made. Suggest a change ROPS a proposal to verify calculations deformation, stress, and energy loss when objects falling on the car cabin. The car is built into mining operations and large excavators an object may fall on a passing truck. Based on the calculations, the design of the cabin protection equipment was optimized by different standards.

Keywords ROPS · Lorry · Tractor · FEM calculation · Testing standards

1 Introduction

The aim of the study was to investigate the protective structure of the cab of a truck MERCEDES 4 × 4 to cater for both EU regulation for protective frames of agricultural and forestry machines No. 1322/2014 for ROPS "Tractor" [1] and ČSN EN ISO 3471:2008 Earth-moving machinery-laboratory tests and performance requirements for roll-over protective structures [2]. The aim was to design ROPS to meet both regulations. CAD proposal to verify by fine element method (FEM) calculations of deformation, stress and energy absorption. The car is built into mining operations and large excavators an object may fall on a passing truck. Design the proposed ROPS protection solution was tested in the hydrodynamic laboratory.

L. Ševčík (✉) · H. Van Nguyen
Technical University of Liberec, Studentská 2, 46117 Liberec, Czech Republic
e-mail: ladislav.sevcik@tul.cz

Š. Medvecký et al. (eds.), *Current Methods of Construction Design*, Lecture Notes in Mechanical Engineering, https://doi.org/10.1007/978-3-030-33146-7_20

2 Materials and Methods

2.1 The Values Specified for the Design and Calculations ROPS by Different Standards

The input data and documents geometry were shaped contours cabin, which was carried out according to the design of the safety of prolonged ROPS to the frame of the lorry as can be seen in Fig. 1 MERCEDES 4 × 4 ACTROS.

Design of ROPS was made in program PTC CREO. Furthermore, the maximum values were given by static forces onerous safety ROPS acting in a lateral direction, longitudinal, and vertical direction and in the lateral effects of kinetic energy (see, Table 1), common for all tests and calculations.

The main dimensions of safety frame, the size of the forces, and their sphere of activity are the same for all admissible solutions of the protective frame. Location of protective beams is given by norm. Maximum static compressive force loads that load ROPS in the vertical direction and the minimum amount of kinetic energy absorbed during longitudinal and lateral loading, determined in accordance with the

Fig. 1 MERCEDES-BENZ ACTROS 2040, 4 × 4 tractor

Table 1 Vehicle weights

Type of chassis	4 × 4
Vehicle standby weight (kg)	8500
The maximum technical weight of the vehicle (kg)	20,500
Reference mass of the vehicle m = maximum technical mass/1.75 (kg)	11,714

The equation for calculating the forces and energy (force and energy equations), line 7 rigid frame machines—only for ROPS (rigid frame dumpers—ROPS only option) and the International Standard ISO 3471:2008

Table 2 Load variants by "EU regulation for ROPS 1322/2014" ROPS "Tractor"

Type of chassis		4 × 4
1	Min. energy absorbed at longitudinal load E (IL1) = 1.4 M (J)	16,400
2	Min. energy absorbed at lateral load *E* (IS) = 1.75 M (J)	20,500
3	Strength of vertical (pressure) loading *F* = 20 M (N)	234,286
4	Strength of vertical (front) loading in front *F* = 20 M (N)	234,286
5	Min. energy at longitudinal load in the opposite direction *E* (IL2) = 0.35 M (J)	4100

Table 3 The value of forces acting in different directions of the coordinate system by International Standard ISO 3471:2008

Direction	x (lateral)	y (longitudinal)	z (vertical)
Force[a]	$F_2 \equiv F_x =$ 105,778 N	$F_3 \equiv F_y = -$ 84,630 N	$F_1 \equiv F_z = -$ 235,320 N
Energy[a]	$U_2 \equiv U_x =$ 18,839 J	–	–

[a]Values of all forces and energies in the direction of the lateral were calculated weight of the machine $M = 12{,}000$ kg according to Australian standard AS 2294.2-1997

"EU Regulation for ROPS 1322/2014" ROPS "Tractor" lon MERCEDES-Benz, type ACTROS series, 4 × 4 chassis (see Tables 1 and 2).

Noting that the reference mass of the vehicle *M* is the mass chosen by the manufacturer for calculating the input power and the compressive forces to be used in the tests. It must not be less than the weight without weights and must be sufficient to ensure that the weight ratio does not exceed 1.75 times.

The orientation of the model in a rectangular coordinate system of safety frame:

- x-axis is the direction of the lateral (lateral) forces,
- y-axis is the direction of the longitudinal force (against cavalry),
- z-axis is the direction vertical forces.

In Table 3, is displayed the value of forces acting in different directions of the coordinate system.

2.2 Location of Force Distributors

The length of the dispensing element for the longitudinal and transverse loading of the frame can be at least 250 mm and maximum 700 mm by multiples of 50 mm, height is always 150 mm.

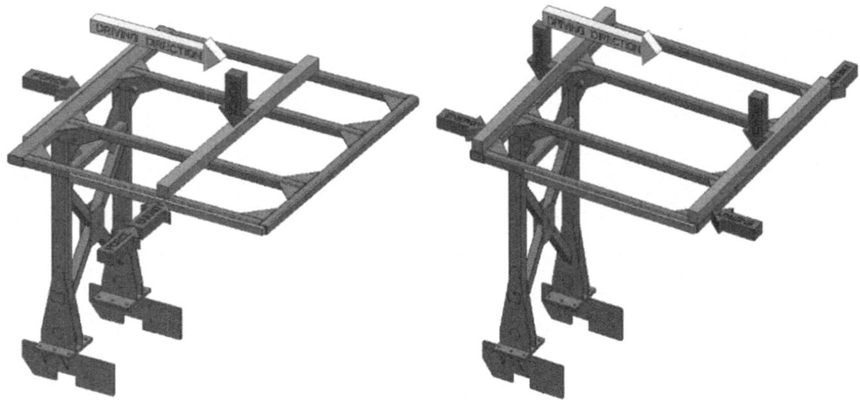

Fig. 2 Load comparison according to ISO 3471 (left) and EU regulation for ROPS 1322/2014 (right)

The positions of loads according to EU regulations are shown in Fig. 2.

For longitudinal rear loading (in the direction of travel), the force is at a distance of 1/6 of the frame width W from the edge, closer to the operator's seat (driver), in the solved task of 400 mm.

With regard to the optimized top frame design, the length of the trailer for the longitudinal load from the rear E (IL1) is 600 mm.

For longitudinal front loading, the force is at a distance of 1/6 of the frame width W from the edge on the opposite side of the top of the frame, from the driver's seat (in the 400 mm task). The length of the E2 (IL2) longitudinal load handler is 700 mm.

For lateral load E (IS), the force field is determined by the position of the protected area, precisely by the position of the operator's (driver's) operator seat reference point (SIP). In the solved task, the side loading is only 119 mm from the front edge of the frame, and therefore, the distributor has the shortest possible length, i.e., 250 [3].

Vertical loading (pressure test) is carried out on the rear and front of the upper frame by force $F = 20$ M by means of a pressure beam (not part of the ROPS frame) which is pressed against the frame by two hydraulic cylinders connected by Cardan joints.

2.3 Material of Structural Elements

For the main welded beams of the vertical frame (*s*) and the supporting upper part of the frame welded material was added. With the high strength steel S690QL, which has the strength of 770–940 MPa, the yield strength of 690–790 MPa is mostly used. In the case of wind voltages, it is possible to use materials of higher strength, such as high strength steel S890QL with the strength of 940–1 100 MPa, yield strength 890–944 MPa, or high strength steel S960QL, strength 980–1150 MPa, yield strength 960 MPa. For other parts of the structure, non-alloy structural steel S355J2+N is used, the strength of 470 MPa, yield strength of 355 MPa.

The design of the protective frame allows its adjustment in two positions. At work, movement position the frame is secured by two pins whose movement is derived from the linear hydraulic motors.

3 Results

Assuming perfect penetration welds, the parts connected to the entire construction simplified so that the load is transferred only sections of the workpieces structure, i.e., it is not considered material located outside the weld these sections. This simplification is to be done, and it can be assumed that significantly affect the results of calculations. FEM calculations are performed for the material constants: modulus of elasticity $E = 200$ GPa and Poisson's ratio $\mu = 0.3$.

Strains and stresses were calculated for lonely forces amid distributors load (distributors) located:

- For lateral loads in a direction (+*x*) (lateral load) on the side beam of the frame, the distributor has a length of 66.6% of the distance (*L*) between the outer front and rear lateral of the upper rectangular frame ROPS and is positioned according to a distance of 1/3 L from the rear lateral of the upper rectangular frame ROPS.
- For longitudinal load in the direction of travel (*y*) (longitudinal load) in the middle of the rear frame beam width, the distributor is max. 80% of the distance (*W*) between the outer left and right sides of the upper rectangular frame ROPS. For the vertical load in the axial direction (*z*) (vertical load) in the middle of the vertical load beam (not part of the frame ROPS), located midway between the front and rear outer side of the upper rectangular frame ROPS.

On color, maps are shown the calculated values of deformation (displacement) and equivalent (reduced) stress by HMH. Results in the *z*-direction are shown in Fig. 3. The original design compliant with the ISO 3471 standard did not comply with the calculation under the EU regulation for ROPS 1322/2014. Therefore, a completely new design had to be proposed in Fig. 3, right.

Evaluation of energy absorption will be considered according to the mechanical energy expended on its deformation in bend—mechanical energy is determined by the following Equation [4]:

$$U = \int_{x_1}^{x_2} F(x)\mathrm{d}x, \qquad (1)$$

in the preceding formula represents:

U mechanical energy [J],

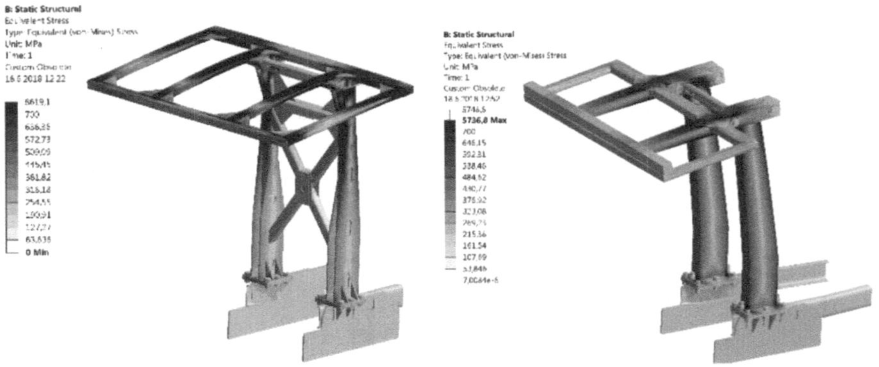

Fig. 3 Sample of the result of maximal stress calculation according to ISO 3471 (left) and EU regulation for ROPS 1322/2014 (right)

Fig. 4 Testing real sample of ROPS in laboratory

$F(x)$ function of the force-deformation of the ROPS [N],
X deformation of the test [m]

The energy calculation was performed on elastoplastic theory. Designed designs were tested in the laboratory (see Fig. 4).

4 Conclusions

The load under EU regulation for ROPS 1322/2014 has much more challenging for the design of ROPS. Design conforming to ISO 3471 does not comply with EU regulation for ROPS 1322/2014. It had to be completely overdesigned.

Acknowledgements This publication was written at the Technical University of Liberec as part of the project "Innovation of products and equipment in engineering practice" with the support of the specific University Research grant 117, as provided by the Ministry of Education, Youth and Sports of the Czech Republic in the year 2018.

References

1. Regulation (EU) No 167/2013 of the European Parlament and of the council, of 5 February 2013, on the approval and market surveillance of agricultural and forestry vehicles
2. ISO 3471 (2008) Earth-moving machinery—roll-over protective structures—laboratory tests and performance requirements, 4th edn., Switzerland
3. Stockton AD, O'Neill DH, Hampson PJ (2002) Methods for optimizing the effectiveness of roll-over protective systems. Silsoe Research Institute, UK
4. Mangado J, Arana JI, Jaren C, Arazuri S, Arnal P (2007) Design calculations on roll-over protective structures for agricultural tractors. Biosys Eng 96(2):181–191

Design and Manufacturing of Multipurpose Steering Wheel

Dana Šišmišová, Erik Mikuláš, Stanislav Zeman and Ĺuboš Kučera

Abstract This paper approaches the design, modeling, and manufacturing process of the multipurpose steering wheel for a formula-style race car. The Multipurpose steering wheel is a device for lateral control of the formula student and also for changing the formula's parameters. Steering wheel's parts are designed with an aim to comfort, usefulness, manufacturability, and cost of prototype. Subsequently, the main body modular mold design and manufacturing process is presented, taken into the account the technology of production and material with suitable attributes selection. For the steering wheel, it is utilized the technology of pre-impregnated carbon fiber fabric reinforced polymer (CFRP) cured in an autoclave. Design of layers of composite material is described in the detail, followed by the material's bend test. Custom internal parts made by 3D printing technology with a purpose of mounting electrical components and mechanical control components are demonstrated. The final product is part of formula student electric racing car SGT-FE18.

Keywords Carbon fiber · Steering wheel · Pre-preg

1 Introduction

The paper deals with the design of formula student racing car steering wheel, its ergonomics and the technology of composite materials production. The multipurpose steering wheel is a device for lateral control of the formula student and also for changing its parameters. It is made using modern methods, whether mechanical or electrical. It must meet the criteria prescribed by the formula student rules. The steering wheel is an essential component that the driver interacts with vehicle. Therefore, it is necessary to adapt its anatomic shape, dimensions, and layout to

D. Šišmišová (✉) · E. Mikuláš · S. Zeman
Slovak University of Technology in Bratislava,
Nam. Slobody 17, 81231 Bratislava 1, Slovak Republic
e-mail: dana.sismisova@stuba.sk

Ĺ. Kučera
University of Žilina, Univerzitná 8215/1, 01026 Žilina, Slovak Republic
e-mail: lubos.kucera@fstroj.uniza.sk

Š. Medvecký et al. (eds.), *Current Methods of Construction Design*, Lecture Notes in Mechanical Engineering, https://doi.org/10.1007/978-3-030-33146-7_21

avoid any negative influence on driving performance. The SGT-FS18 is based on an older version inspired by formula 1.

The formula student, similarly to the F1 rules, has requirements on the design and construction of the vehicle and individual components. The following rules serve as a baseline for the design of the steering wheel:

- The steering wheel must be attached to the column with a quick disconnect. The driver must be able to operate the quick disconnect while in the normal driving position with gloves on.
- The steering wheel must have a continuous perimeter that is near circular or near oval. The outer perimeter profile may have some straight sections, but no concave sections.
- All parts of the vehicle which may become electrically conductive (e.g., completely coated metal parts, carbon fiber parts, etc.) which are within 100 mm of any TS or GLVS component, must have a resistance below 5 Ω to GLVS ground [1].

2 Modeling and Prototyping

The steering wheel was designed according to the ergonomics rules taken in account the workspace where steering wheel is operated as shown in Fig. 1. After modeling the body of the steering wheel, the envelope was taken and suitable material was selected to minimize the cost of prototype 3D printing [2].

After review and testing the prototype, it was necessary to correct the final shape and ensure the fitting of all the elements in place. The encountered issues occurred

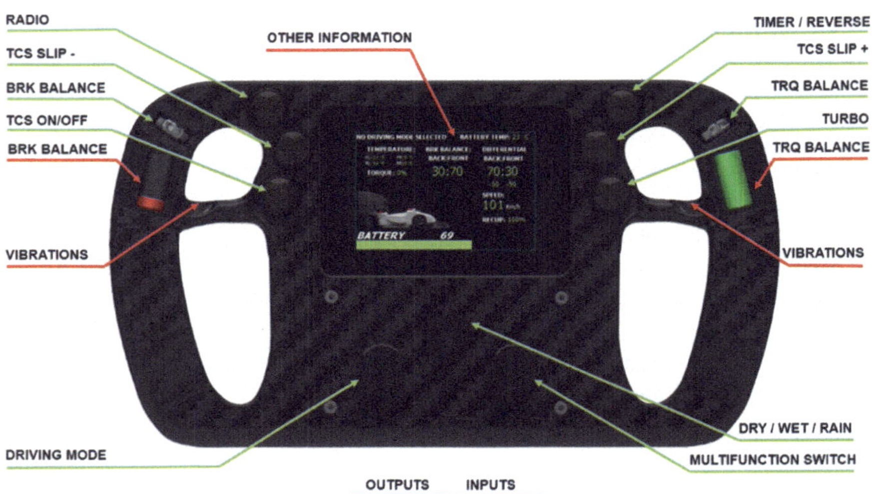

Fig. 1 Control and information elements layout according to ergonomic rules

mainly due to 3D software distortions. After testing and tapping the prototype, the correction phase followed to revise the errors made in the design. Once this phase was completed, design of the molds took place [3].

The most important part of the design was to select the splitting plane of the steering wheel, to avoid negative angles, that the product can be easily removed from the mold. It was necessary to create technical documentation and to determine tolerances for the dimensions and roughness of the surfaces. When designing a mold, it was necessary to take into account the technology of production and select material that is resistant to higher temperature, thermally stable, thermally conductive, well machinable and must not have a porous structure. Aluminum alloy ENAW 5083 was chosen as the ideal material of these properties. It is relatively soft, its surface can be fine polished to the mirror surface, which is necessary to make the composite material unbounded to mold. The hollow body shape of the steering wheel can not be easily produced, so the steering wheel was split into two separate parts.

Two-piece form of the steering wheel rear panel serves as a joint surface that was smaller in diameter than the edge. Therefore, this part had to be solved as a removable part which was easily screwed together before laminating and when the mold product was pulled out, it was decomposed as shown in Fig. 2.

The main part of the lower mold part had to be milled from one piece of the aluminum alloy blank ENAW 5083 casting made, that it did not have internal tensions which cause rolling and the mold did not twist during the autoclave process. The surface of this mold was fine machined to a final roughness Ra of 0.4 with a mirror surface as shown in Fig. 3.

The front steering wheel mold is a more complicated piece. This mold is composed of the base part and the inserts. Which are placed into the holes and the shape of the mold itself ensures their position. Based on this knowledge, a third insert has also been made to reduce mold production costs in future. In order to change the display unit, it is no longer necessary to design and manufacture the whole mold, which significantly saves costs and enables the mold modularity. The green-colored surface in Fig. 4 shows the main part of the mold, the yellow bars are marked to create a space for the OLED display and the rotary switches. Blue represents mentioned insert for display module that can be easily changed. With pink color are marked the cones for the button-locking position inside the steering wheel.

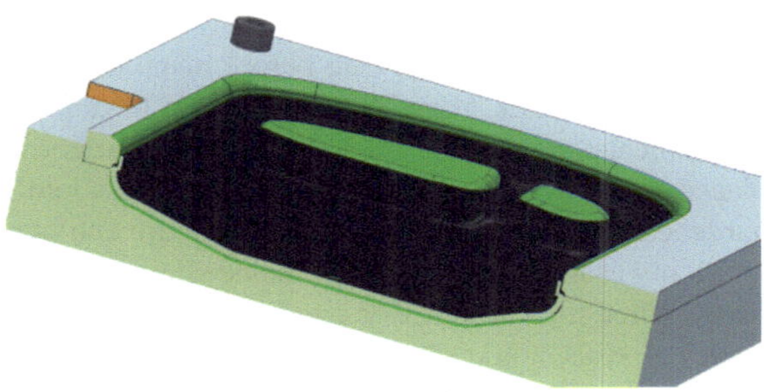

Fig. 2 Two-part mold cut, the black part of the figure is a composite part that can be pulled out of the mold after the mold has been dismantled

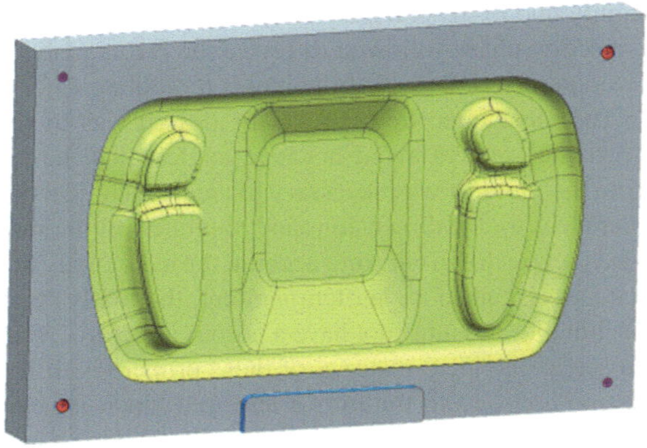

Fig. 3 The main part of the steering wheel lower part form with the polished surface marked yellow

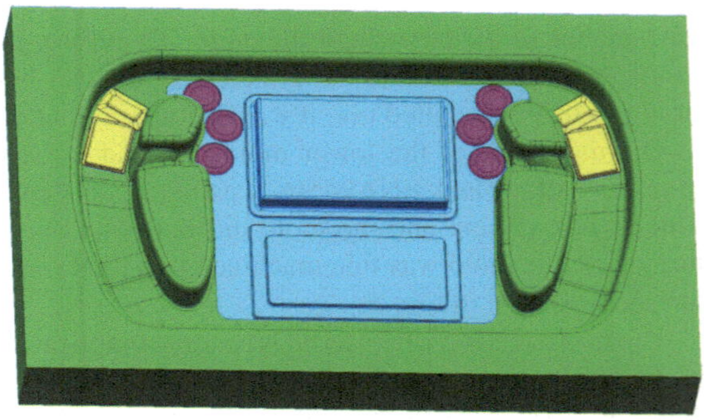

Fig. 4 The shape for laminating the front half of the steering wheel on which the blue, yellow, and pink colors are marked by the interchangeable runners

3 Mechanical Assembling

Based on the requirements, it was determined that 6 buttons, 2 rotary switches, 2 rotary encoders, and a three-position switch are necessary. Three displays were specified as output devices. A large, center display provides complete information about the vehicle and two smaller serve for the setup of the virtual inter-axle differential and the braking force distribution. This resulted in a mechanical design with the layout of buttons and switches according to the ergonomics rules. Given design was modeled by computer software before manufacturing drawings, mold milling, and polish process took place. After the molds have been manufactured, the necessary mechanical parts were cleaned, degreased, separator was applied and all inserts were placed. The carbon fiber pre-preg material was laid into the molds, prepared, and hardened in an autoclave at high temperature and pressure. After pulling out of the mold, it was necessary to cut off the needless parts, drill holes, polish, and paint the parts. When the carbon fiber part was prepared, it was necessary to place buttons, switches, seals, holders, and electronics parts inside. Finally, both half of the steering wheel was sealed and bonded together [4].

4 Composite Materials

The steering wheel design for the SGT FE-18 has to comply with the requirement of the lowest possible weight. As the steering wheel body was the largest component, the choice of the right material was the key factor. It was necessary to choose a light and solid material from which it was possible to create a product with a complicated shape. In the 1980s and 1990s carbon fiber reinforced composites began to be used racing cars production. By combining the epoxy matrix and carbon fibers, the toughest and lightest form of composite material is produced. The characteristic of the material used is on Fig. 5. Just because of its properties such as high stiffness, low weight, high ductility, or heat resistance, this form of composite material had been chosen to manufacture the steering wheel [5].

Due to the complicated shape of the steering wheel, the production technology was chosen using pre-preg composite material with high temperature and pressure curing in the autoclave. Pre-preg or "pre-impregnated" is the technology where the carbon fiber fabric is impregnated with activated epoxy resin directly in production process. An autoclave is a device in which there is an artificially increased pressure to ensure the best adhesion of the composite material to the mold and in between liars.

To comply with the formula SAE/student rules for the electrical conductivity of parts (these rules require that all parts of a vehicle that are less than 100 mm from the low-voltage or high-voltage system and become electrically conductive must have a resistance less than 5 Ω to the ground of the low voltage system), the copper mesh had to be implemented in the composite material in order to reduce the electrical resistance [1, 6].

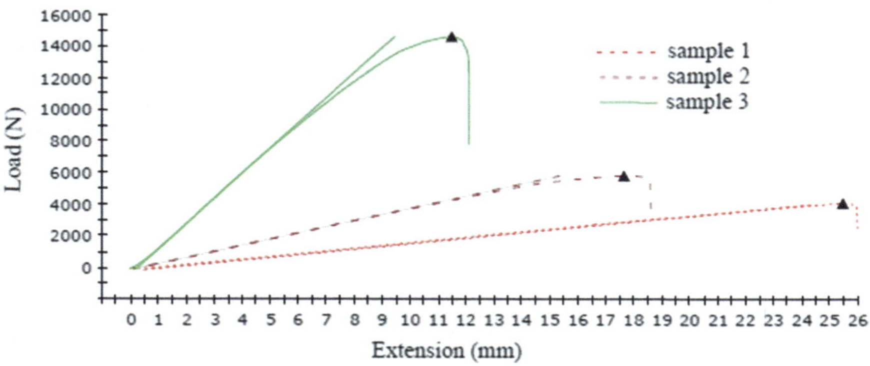

Fig. 5 Diagram of bend test for used carbon fiber composite

4.1 Design of Layers of Composite Material

The layers design had to take in account the anisotropic properties of the material, which are compensated by the rotation of the carbon layers relative to each other. This ensures at least partially similar properties of the composite in different directions.

Because of the low steering wheel load, the front part is composed from 3 layers of carbon fiber and one layer of copper mesh. The first layer serves as a "visually nice" surface and aims to provide the material with a good visual effect. Due to the complexity of the steering wheel, the first layer is visual grade carbon fiber twill with a weight parameter of 220 g/m^2. It is created from one piece, in order to achieve the best visual effect. This layer is very thin and enables to copy complex shapes. This is followed by the second layer with a copper grid and finally two support layers of twill carbon fiber with the weight 600 g/m^2. These layers are rotated 45° with each other and are shown in Fig. 6. The steering wheel handles are reinforced with an additional layer weighing 220 g/m^2, and the edge is reinforced with a layer of 800 g/m^2 for easier removal from the mold.

The steering forces are transferred through the rear of the steering wheel those must be rigid enough because the flexibility may lead to interfere with the vehicle's response. Therefore, the back part consists from 5 layers of carbon fiber, and 1 layer of copper mesh. Identically to the front part the first layer is from visual grade carbon fiber twill with weight 220 g/m^2, followed by the copper mesh and 4 carrier layers of carbon fiber, and rotated 45° with each other. These layers have a weight of 600 g/m^2 and the composition of them can be seen in Fig. 7. An extra layer weighing 800 g/m^2 is added to the mold edges and the part where the steering wheel quick release

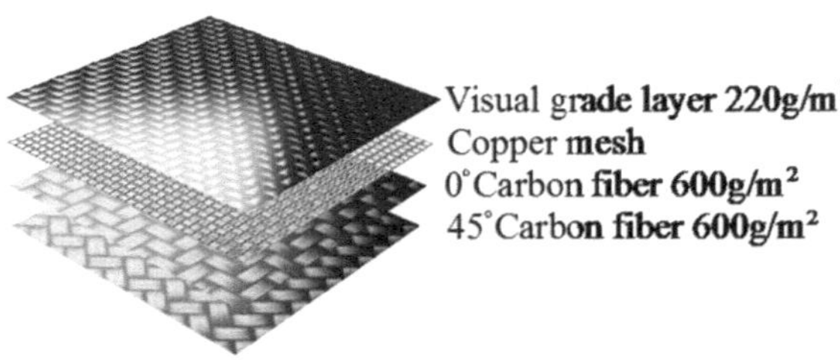

Fig. 6 The composite material layers at the front of the steering wheel

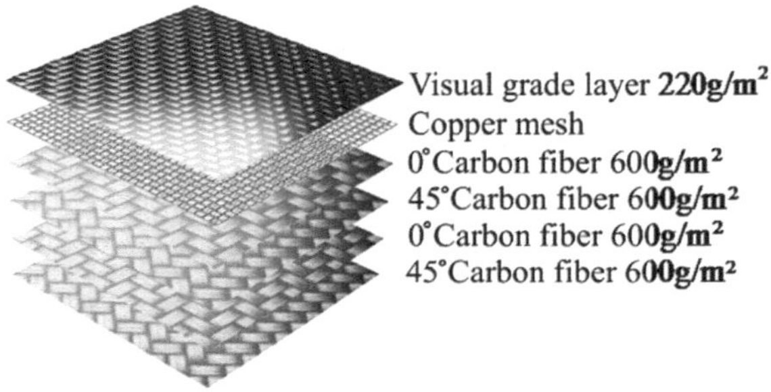

Fig. 7 The layout of the composite layers in the rear of the steering wheel

coupling is attached. On the circumference of the mold, an extra material is added in order to ease removing of the fart from the mold [1, 6].

4.2 Composite Forming and Preparing Molds Procedure

The mold must be ready before using. Its preparation consists of a thorough cleaning and applying of separator. Mold cleaning is necessary for the sealing of the separator. The mold must be cleaned several times with the chemical agent Ferkote PMC from the Loctite manufacturer until it is absolutely clean, which is checked by the control strip. When the mold is clean, separating begins. It is a process whereby the pores of the material are clogged and a thin layer is formed which serves as a barrier against adhesion of the composite material to the mold. Separation is performed by the Loctite manufacturer's Ferkote 770NC. It is a liquid-shaped separator that is deposited on ten-layer paper towels. Between the applications of the individual layers, there must be a break of the separator for at least 15 min.

The separation process follows the composite placing. The prepared, cut carbon fiber layers according to the drawings will begin to be lied in the mold. It's a tedious process when it is necessary to insert one layer of carbon into the mold and push it into all corners by using the tools so that it does not move when touching it gently. It is necessary to prepare rolls of two centimeters wide strips of unidirectional carbon fiber, which are placed on all edges of the mold. These rolls ensure that carbon fibers are pushed up to the corners of the mold. After the rolls of carbon fibers have been deposited, the mold must be prepared for the first layer vacuuming.

Vacuuming after the first layer is necessary because of the careful pushing of the carbon fibers into all corners and edges in the mold and at the same time sucking air bubbles that could be found between the composite material and the mold surface. For vacuuming, it is necessary to make perforated film to be laid on the composite material in sufficient quantity to avoid bridging. The aerated layer, the so-called breather, must be placed on the perforated foil. The final procedure is to wrap the form into a vacuum bag, which is sealed with a vacuum sealing tape, after following the suck of the air out from the mold and the treatment of the foil to avoid any bridging. The mold has to stand for 30–60 min. After vacuuming, the mold is unwrapped and begins to deposit additional layers of composite. After the remaining layers of composite have been laid, it is necessary to add layer of perforated release film in order to enable the excess resin flow into the breather. After wrapping the mold in the bag, the air bag is tested by a measuring device. The composite material thus prepared is shown in Fig. 8 and is waiting until the autoclave is released, a device in which the composite materials are hardened at high temperature and pressure.

A 4-h curing cycle in autoclave was selected for the steering wheel and the temperature and pressure diagram can be seen in Fig. 9.

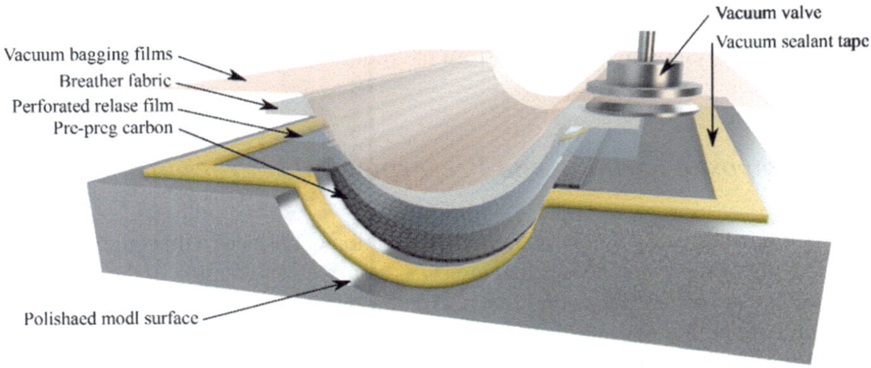

Fig. 8 Laid composite material prepared for vacuuming

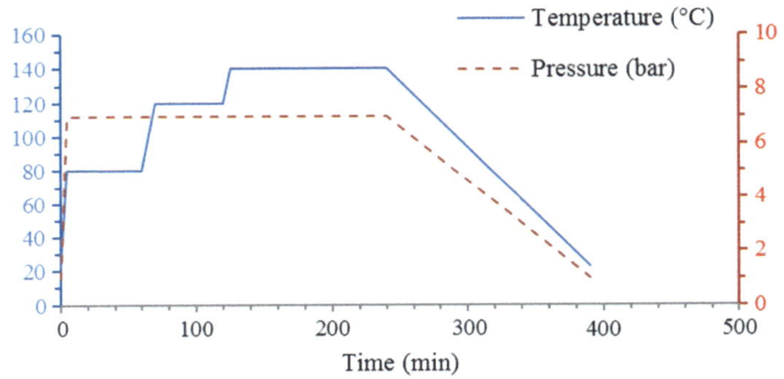

Fig. 9 Time course of composite curing

5 Conclusion

The paper presents the design and manufacturing of the control unit—steering wheel STORM for the SGT-FE18 racing car. The correct design is the essential to achieve best results in the formula student international competition. Based on a detailed analysis, a concept for the steering wheel parameters, layout of controls and display elements, number of buttons and functionality was created. After modeling, a prototype was created using 3D printing using FDM technology. Errors were checked for the prototype that had to be removed before making the final steering wheel product. The key factor was ergonomics, 3D model of molds, and the most suitable composite material selection for the steering wheel production. For given technology of steering wheel production, the molding material ENAW 5083 was selected based on consultation with Karlovensky Composites. This mold was made by CNC milling and finishing methods to achieve surface roughness Ra 0.4. The mold was assembled with composite materials using the pre-preg technology and such created the steering wheel body. At the same time as design and mold manufacturing, design of electronics was made in Altium Designer software. These designs have been realized onto PCBs and then components placed. After testing the functionality, the entire electronic system was inserted into the steering wheel, which was then bounded. Our future goal is to test this steering wheel on the SGT-FE18 racing car, which was not available at the time of the steering wheel design. Recursive neural network to identify the pilot and automatically set the vehicle parameters implementation into steering wheel is planned in the future.

Acknowledgements The authors would like to thank the Slovak University of Technology in Bratislava and the Slovak Institute of metrology and the grand agency APVV project no.: 15-0164 and 15-0295, agency VEGA grant no.: 1/0317/17, and KEGA project no.: 006STU-4/2018 and 027STU-4/2017.

References

1. 2017-18 Formula SAE® Rules (2016) SAE International. https://www.fsaeonline.com/content/2017-18%20FSAE%20Rules%209.2.16a.pdf. Last accessed 08 July 2018
2. Mihailidis A, Samaras Z, Nerantzis I, Fontaras G, Karaoglanidis G (2009) The design of a formula student race car: a case study. Proc Inst Mech Eng Part D J Autom Eng, Sage Journals 223(6):805–818
3. Palenčar R, Ďuriš S, Dovica M, Palenčár J (2015) Application of Monte Carlo method for evaluation of uncertainties of temperature measurement by SPRT. In: XXI IMEKO World Congress, Measurement in Research and Industry, Prague CZ, pp. 1194–1999. IMEKO, Hungary
4. Hirz M, Rossbacher P, Gulanová J (2017) Future trends in CAD-from the perspective of automotive industry. Comput-Aid Des Appl 14(6):734–741
5. Pulc V, Hrnčiar P, Gondár E (2008) Material science. STU Publishing, Bratislava
6. Mohit S, Shanglin G, Edith M, Himani S, Leong YW, Jayashree B (2014) Carbon fiber surfaces and composite interphases. Elsevier-Paris, Composites Sci Technol 102:35–50

Sealing Technology for Vacuum Applications Working by Increased Temperatures

Michal Tropp, Michal Lukáč, Milan Benko, František Brumerčík, Zbigniew Krzysiak and Aleksander Nieoczym

Abstract This article is focused to design of vacuum chamber for deep drawing machine working by high temperatures. The high temperatures are demanding because of use of molybdenum sheet, which should be drawn by temperatures approx. 300 °C. The thermal analysis is important for the correct design of the deep drawing machine for two reasons: The first is the correct temperature of the molded material—molybdenum sheet and the second is the temperature around the seals of vacuum chamber. The simulation forms the basis for selecting the appropriate type of seals. The result of the simulation is a thermal field of working components: the punch, the die, the blank holder, the working plate, and the working table. The second important analysis result is the temperatures of components, which are in contact with seals in particular places.

Keywords Molybdenum · Vacuum · Sealing · Deep drawing

1 Introduction

Thermal analysis is important for the correct design of the deep drawing machine for two reasons: The first is the correct temperature of the molded material—molybdenum sheet and the second is the temperature around the seals of vacuum chamber. The simulation thus forms the basis for selecting the appropriate type of seals [1].

For the proper process of deep molybdenum sheet drawing, a higher temperature of working material is required due to poor drawability of molybdenum at temperatures around 20 °C. During the process, the working material must be tempered to a temperature in the range of 200–300 °C [2].

M. Tropp · M. Lukáč · M. Benko · F. Brumerčík (✉)
University of Žilina, Univerzitná 8215/1, 01026 Žilina, Slovak Republic
e-mail: frantisek.brumercik@fstroj.uniza.sk

Z. Krzysiak
University of Life Sciences in Lublin, 28 Głęboka Street, 20612 Lublin, Poland

A. Nieoczym
Lublin University of Technology, 36 Nadbystrzycka Str, 20950 Lublin, Poland

Š. Medvecký et al. (eds.), *Current Methods of Construction Design*, Lecture Notes in Mechanical Engineering, https://doi.org/10.1007/978-3-030-33146-7_22

2 Boundary Conditions of Simulation

The aim of the simulation is focused on the design of gaskets, which are placed on the contact surfaces between the components of the vacuum chamber and parts of the work table (static seals) and seals between the vacuum chamber and the working components of the device, which are in this case hydraulic linear motors (dynamic seals) [3]. The simulation also provides the answer to the question of whether it is necessary to equip the vacuum chamber with cooling. The following Fig. 1 shows the following components included in the simulation [4, 5]:

- vacuum chamber,
- hydraulic linear motors,
- working components: (die, punch, blank holder, working plate, and working table).

The following Fig. 2 shows the boundary conditions of thermal simulation. The areas in the figure, marked in blue, represent the internal surfaces of the vacuum chamber on which heat dissipation is applied [6]. On the remaining surfaces were applied both types of heat dissipation, the convention and the radiation [7]. The heater power was set at 5200 W and the duration of the simulation was 1800 s.

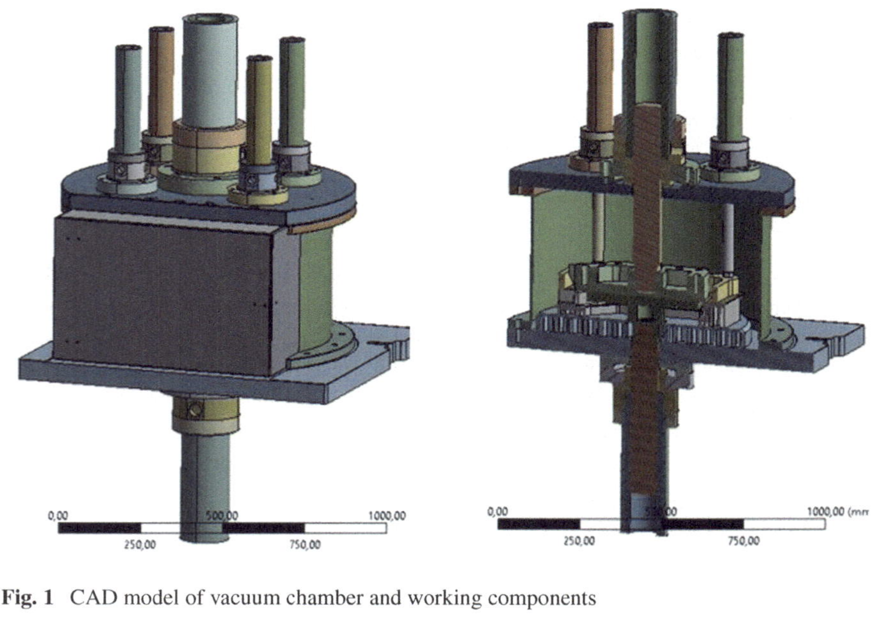

Fig. 1 CAD model of vacuum chamber and working components

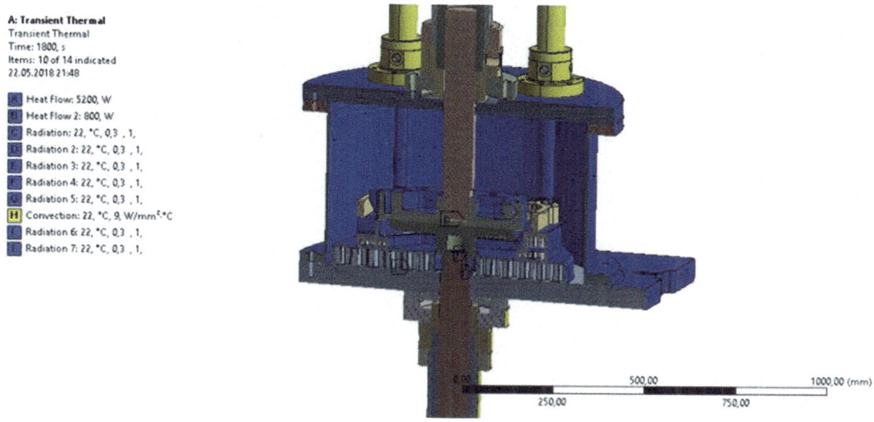

Fig. 2 Boundary conditions of the simulation

3 Results of the Simulation

Figure 3 shows the simulation results—thermal simulation of the complete machine with all the working components and the vacuum chamber [8]. It is clearly visible that the heat transfer between the working components and the walls of the vacuum chamber is so small that the chamber does not need an active cooling system. The outer temperature of the vacuum chamber is 22 °C [9, 10].

The location of the seals is shown in Fig. 4. Sealing places are marked with red arrows and letters [11]. Safety factor for determining the minimal temperature of the gaskets is set 2. The reason why the coefficient is so high is that if the gasket is damaged due to exceeding its working temperature, the vacuum will be braked and

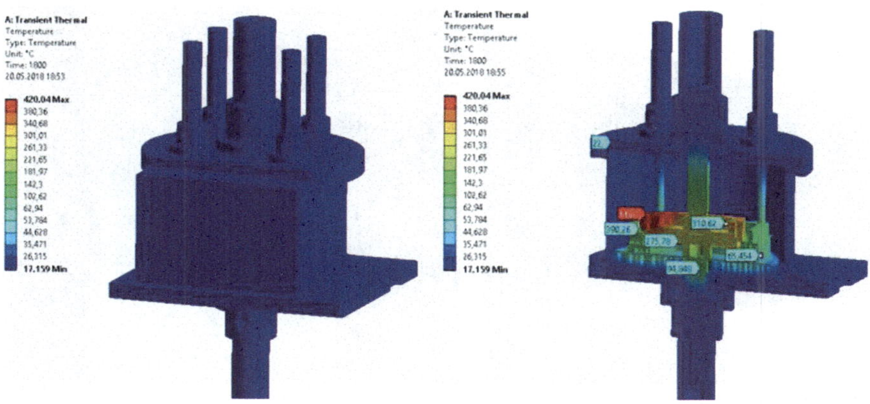

Fig. 3 Heat transfer through the working components and the vacuum chamber

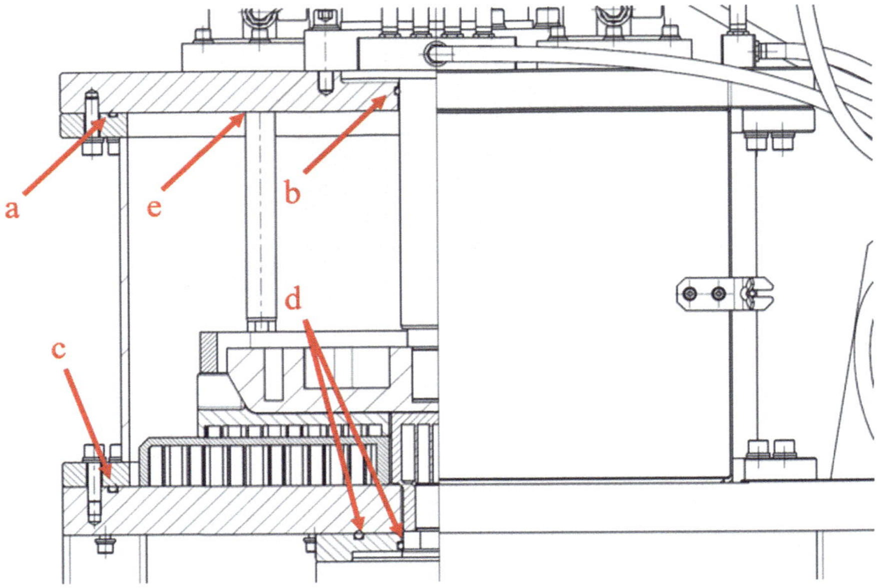

Fig. 4 Locations of vacuum seals

the molybdenum sheet will be destroyed due to the penetration of oxygen into the chamber and subsequent oxidation.

Place "a" is shown on the left-hand side in Fig. 5, and the temperature of the contact is only 22 °C. Location "b" is shown on the right side in Fig. 5, the temperature in the contact of seal and the piston is 42.44 °C.

Seal placed in "c" location is shown on the left side in Fig. 6. The temperature

Fig. 5 Temperatures in locations of contact between the seals and working components—vacuum chamber and punch

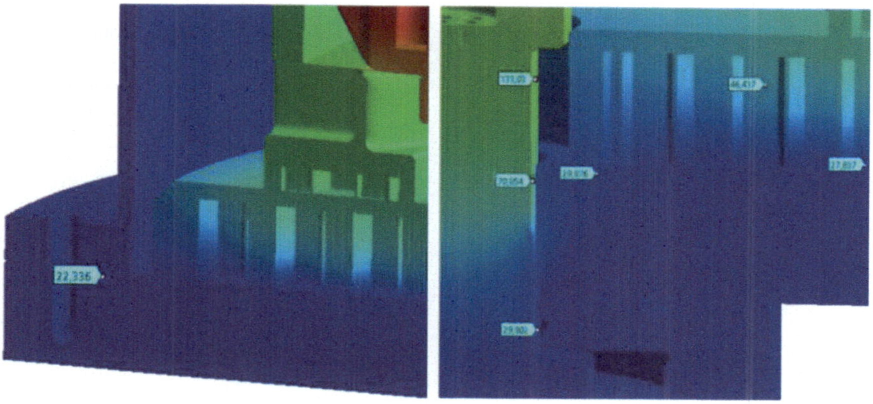

Fig. 6 Temperatures in locations of contact between the seals and working components—working table bottom piston

measured in the joint is 22.34 °C is shown in Fig. 6. The contact between the bodies is static, there is no, negligible mutual movement [12]. Of the right side are visible locations marked with "d" letter. Temperature of the contact surfaces of static sealing is 22.01 °C and temperature, which have withstand the dynamic sealing, is 70.85 °C.

Figure 7 shows the location of contact of the vacuum chamber with the linear hydraulic motor of the blank holder marked with the letter "e". The contact is dynamic, the bodies move toward each other during the work process [13]. As a seal, an O-ring with a diameter of 6 mm will be used. The temperature measured in the contact is 22.49 °C (Fig. 7; Table 1).

For proper operation of the seals is also necessary to keep the recommended compression values. The following graph shows the dependence between the O-ring

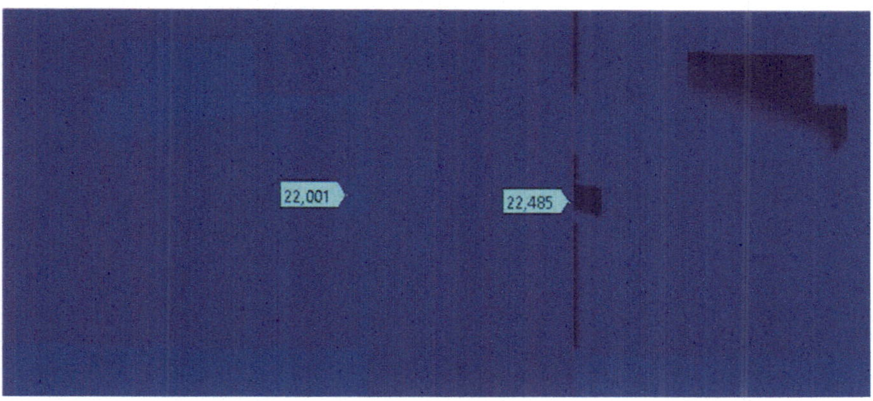

Fig. 7 Temperatures in locations of contact between the seals and working components—blank holder

Table 1 Max. temperature in sealing locations

Type of sealing	Temperature from simulation (°C)	Max. temperature (°C)
Static	22.00	
	22.34	22.34
	20.01	
Dynamic	42.44	
	70.85	70.85
	20.49	

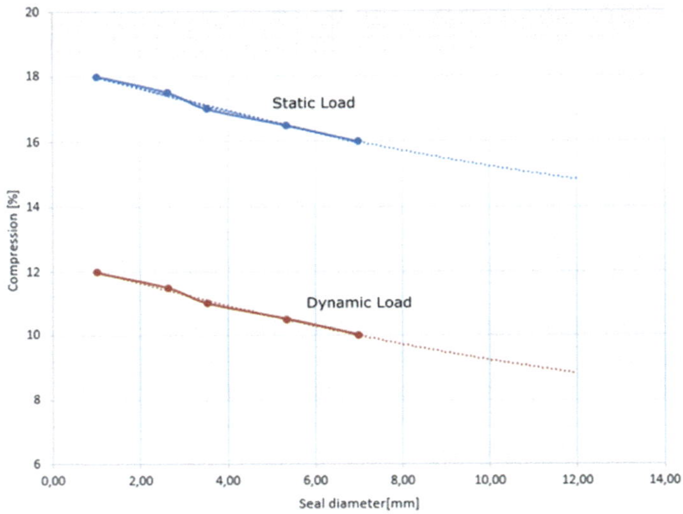

Fig. 8 Recommended compression rate of static and dynamic O-ring sealing

thickness and the recommended compression (Fig. 8). Suppliers recommend using these values for the working temperature of seals made from elastomers from 21 to 200 °C. To use static seals at temperatures above 200 °C, it is necessary to reduce compression values by 2% [14].

After all requirements have been taken into account, the necessary sealing parameters appropriate to the proposed equipment have been established (Tables 2 and 3) [15, 16].

These last two tables are the main result of the simulation, they give us the answer, which type of sealings have to be used for proper sealing of vacuum chamber of the deep drawing machine with the risk of failure as low as possible [17, 18].

Table 2 Main parameters of static seals

Static sealing		
Material	Synthetic rubber, NBR, CR (NEOPREN®)	
Working temperature	90 °C	
Demanded work temperature	44.68 °C	
Shape	O-ring	
Compression	Seal diameter (mm)	Compression (%)
	8	15.5
	10	15.2
	10	15.2

Table 3 Main parameters of dynamic seals

Static sealing		
Material	Fluoroelastomer, FKM (VITON®)	
Working temperature	150 °C	
Demanded work temperature	141.7 °C	
Shape	O-ring	
Compression	Seal diameter (mm)	Compression (%)
	8	9.7
	8	9.7
	6	10.3

Acknowledgements The research is supported by the Cultural and Educational Grant Agency of the Ministry of Education, Science, Research and Sport of the Slovak Republic under the project No. KEGA 046ŽU-4/2018.

References

1. Faturik L, Trsko L, Hrcek S, Bokuvka O (2014) Comparison of structural design in high and ultra-high cycle fatigue regions. Trans FAMENA 38(4):1–12
2. Korolev VN, Shofman LA (1967) Deep drawing of molybdenum and its alloys. Kuznechno-Shtampovochnoye Proizvodstvo (Forging and Stamping) 10:17–23
3. Glowacz A (2014) Diagnostics of DC and induction motors based on the analysis of acoustic signals. Measur Sci Rev 14(5):257–262
4. Martikan M, Brumercik F, Bastovansky R (2015) Development of mechatronic deformation system. Appl Mech Mater 803:173–178
5. Bumercik F, Brumercikova E, Bukova B (2015) Mechatronic and transport system simulation. Appl Mech Mater 803:201–206

6. Kohar R, Hrcek S (2014) Dynamic Analysis of a rolling bearing cage with respect to the elastic properties of the cage for the axial and radial load cases. Commun Scientif Lett Univ Zilina 16(3A):74–81
7. Hrcek S, Kohar R, Medvecky S (2012) Determination on the maximum roller bearing load with regards to durability thereof using FEM analysis. Commun Scientif Lett Univ Zilina 14(3):55–61
8. Zmindak M, Donic T, Jurik P (2016) Numerical simulation of rectangular deep draw containers made from thin molybdenum sheet. In: IOP conference series: materials science and engineering, 159(8)
9. Caban J, Drozdziel P, Vrabel J, Sarkan B, Marczuk A, Krzywonos L, Rybicka I (2016) The research on ageing of glycol-based brake fluids of vehicles in operation. Adv Sci Technol 10(32):9–16
10. Caban J, Marczuk A, Sarkan B, Vrabel J (2015) Studies on operational wear of glycol-based brake fluid. Przem Chem 94(10):1802–1806
11. Komsta H, Brumercikova E, Bukova B (2016) Applications of NFC technology in passenger rail transport. Transp Probl 11(3):43–53
12. Caco M, Tribula R, Scerba P, Dzubriak J, Kohar R (2016) Automatic E-frame—increase of the AGV system utility value. ProIN 17(1):34–37
13. Brumercikova E, Bukova B, Krzywonos L (2016) NFC Technology in Public Transport. Commun Scientif Lett Univ Žilina 18(2):20–25
14. Sonderegger K (2006) Vacuum sealing technology. CERN Accelerator School, Platja D'Aro, Spain, 16–24 May 2006
15. Drozdziel P, Komsta H, Krzywonos L (2013) Repair costs and the intensity of vehicle use. Transp Probl 8(3):131–138
16. Drozdziel P, Krzywonos L (2009) The estimation of the reliability of the first daily diesel engine start-up during its operation in the vehicle. Mainten Reliab 41(1):4–10
17. Kohar R, Hrcek S, Medvecky S (2012) Usage of dynamic analysis to determine force interactions between components of rolling bearings. Commun Scientif Lett Univ Zilina 14(3):62–67
18. Malik L, Hrcek S (2016) General calculation model of the gear mechanisms with step and continuously gear ratio change, University of Zilina

Contactless Measuring Device for Flexible Shaft Coupling Twist Angle

Matej Urbanský and Robert Grega

Abstract The very important current trend in machinery is the reduction of vibration and noise, mainly in terms of human health and lifetime and safety of machines. Our current research is focused on continuous tuning of torsional oscillating mechanical systems during their operation, mainly in terms of torsional vibration magnitude, using pneumatic flexible shaft couplings (pneumatic tuners) and electronic control systems. Our electronic control systems work with the manipulated variable—gaseous media pressure in pneumatic couplings and with controlled variables, which can be, for example, the magnitude of torsional or rectilinear vibration, the noise level of a mechanical system, etc. In our case, the twist angle of a pneumatic coupling is very important variable, which can be directly the controlled variable, or it can be used for the computation of controlled variable, for example, for the computation of the load torque transmitted by the pneumatic coupling. Therefore, the objective of this paper is to present a contactless device for the measurement of the twist angle of a flexible shaft coupling. The device was newly developed by us and uses a pair of optoelectronic sensors and real-time data processing.

Keywords Flexible shaft coupling · Twist angle · Contactless measuring device

1 Introduction

Nowadays, the reduction of vibration and noise in machinery is very important task, mainly in terms of human health and lifetime, safety and efficiency of machines. Therefore, many researchers keep making an effort to optimize and improve existing machines using modern computational methods, technical diagnostics and experimental data evaluation, e.g., [1–5], or to develop new prototypes, e.g., [6–9], and to test their properties, e.g., [9–14]. Likewise, working conditions in practice are

M. Urbanský · R. Grega (✉)
Technical University of Košice, Letná 9, 04200 Košice, Slovak Republic
e-mail: robert.grega@tuke.sk

M. Urbanský
e-mail: matej.urbansky@tuke.sk

Š. Medvecký et al. (eds.), *Current Methods of Construction Design*, Lecture Notes in Mechanical Engineering, https://doi.org/10.1007/978-3-030-33146-7_23

frequently controlled in terms of negative influence of vibration and noisiness on human health, e.g., [15].

At our department, we deal with optimal dynamic tuning of torsional oscillating mechanical systems (TOMS), mainly in terms of torsional vibration magnitude (but also in terms of rectilinear vibration magnitude or noise arising from torsional vibration), e.g., [4, 6, 16–19]. As means of this dynamic tuning, we use pneumatic flexible shaft couplings (pneumatic torsional vibration tuners, hereinafter "pneumatic tuners") developed at our department, e.g., [6, 7].

The torsional stiffness of given pneumatic tuners and so the natural frequencies of torsional systems can be changed by adjusting the gaseous media pressure (the most commonly air) in their pneumatic flexible elements. By suitable value of torsional stiffness k ($k_2 < k_1 < k_3$), resonances from individual harmonic components of excitation (see Fig. 1) can be moved from the operational speed (n) range (OSR) of a mechanical system and herewith the value of dynamic component M_D of the transmitted load torque can be reduced.

Our current research is focused on development of following electronic control systems for continuous tuning of torsional oscillating mechanical systems during their operation:

- Electronic system for pneumatic tuners with auto-regulation, e.g., [6, 18],
- Electronic extremal control tuning system, e.g., [6, 16, 17, 19].

Our electronic control systems work with the manipulated variable u—gaseous media pressure in pneumatic tuners (Fig. 2).

For the electronic control system No. 1, the controlled variable y is directly the mean value of pneumatic tuner twist angle. For the electronic control system No.2, the controlled variable y can be, for example, the magnitude of torsional or rectilinear

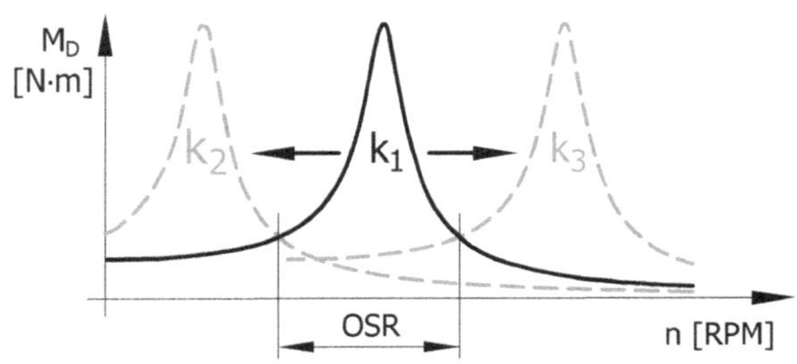

Fig. 1 Mechanical systems tuning principle

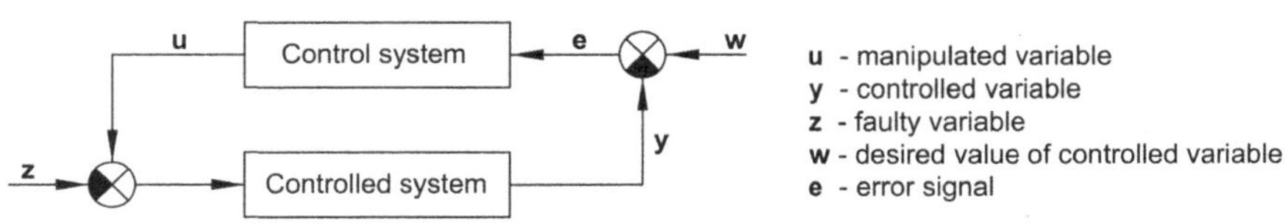

Fig. 2 Basic block scheme of control circuit

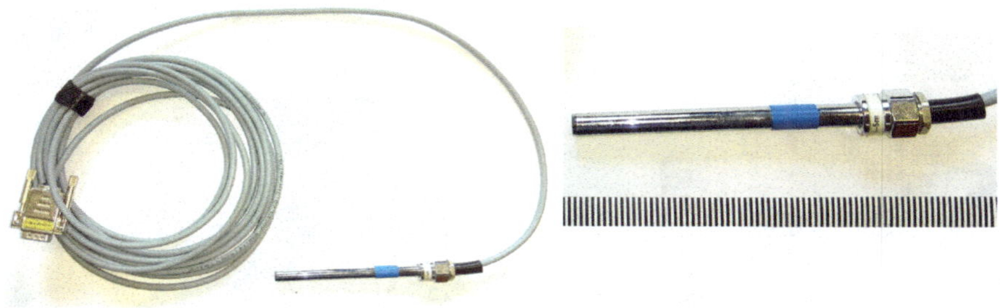

Fig. 3 Optoelectronic sensor SE-TACHO-PROBE-01

vibration, the noise level of a mechanical system, etc. The twist angle of a pneumatic tuner can be used for the computation of controlled variable, for example, for the computation of the load torque transmitted by pneumatic tuner.

Therefore, in our case, the twist angle of a pneumatic tuner is very important variable. The objective of this paper is to present a contactless device for the measurement of the twist angle of a pneumatic flexible shaft coupling, newly developed by us.

2 The Hardware of the Contactless Measuring Device

2.1 *Optoelectronic Sensors*

According to our specific requirements, we use the pair of Dewetron optoelectronic sensors of type SE-TACHO-PROBE-01 [20] (Fig. 3).

These sensors detect the reflection from reflective black-and-white moving tapes (see Fig. 3 right). The sensors react to the edges between black and white stripes. There is a distinct change of electrical output voltage in the moment when the edge is crossed. These sensors can work with maximal frequency of 10 kHz but the cleanness of the reflective tapes, cleanness of the optical parts of the sensors and the sharpness of the edges between black and white stripes must be excellent [20].

2.2 *Pneumatic Torsional Vibration Tuner*

In the following Fig. 4, we can see the mechanical system of a piston compressor drive.

We can see that the piston compressor (in this case the torsional vibration exciter) is driven by an electromotor through a pneumatic flexible shaft coupling—pneumatic torsional vibration tuner (orange-painted). The pneumatic tuner is equipped with

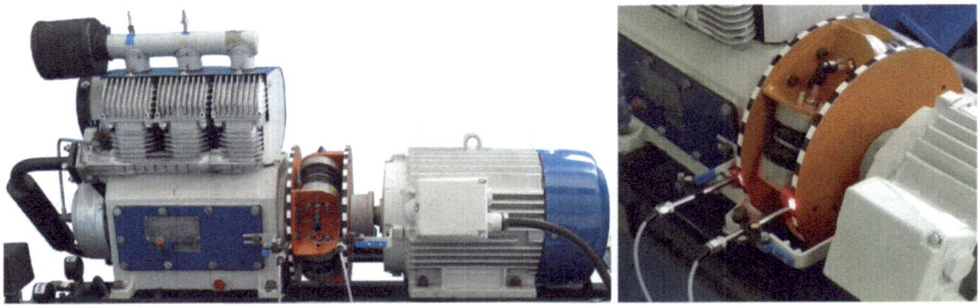

Fig. 4 Adjustment of a pneumatic tuner in order to use the optoelectronic sensors

black-and-white tapes, which are glued to the circumference of driving and driven flange of the pneumatic tuner and they are scanned by the pair of the optoelectronic sensors.

2.3 *Multifunctional Module*

The multifunctional module has following functions:

1. it serves as power supply for the optoelectronic sensors;
2. it measures the edge-crossing times for both flexible coupling flanges and in our case, only the edges at the crossings from black to white stripes are considered;
3. it communicates with personal computer (PC), the measured data is sent to PC in order to further processing in the real-time.

3 Data Measuring and Processing

A dynamic load torque transmitted by a flexible shaft coupling causes the mutual dynamic angular twisting of the driving and driven flange of the flexible coupling. The measurement is based on determining the time dependence of the twist angle of a flexible coupling.

The number of black-and-white stripe pairs for the driving and driven flange should be equal and chosen with respect to the maximal twist angle of a flexible shaft coupling and the character of the transmitted load torque. It is also important that the length of all the black and white stripes should be equal.

The edge-crossing times t_i need to be measured, where i stands for the sample order number. In our case, only the edges at the crossings from black to white stripes are considered. The times t_i are computed from the counted number of impulses from the counter of a microprocessor. One impulse represents the time 1/14745600 s.

Considering the times t_{1i} for a driving flange and the times t_{2i} for a driven flange of a flexible coupling, the time delays Δt_i can be computed using the following formula:

$$\Delta t_i = t_{2i} - t_{1i}[\mathrm{s}]. \tag{1}$$

In the next step, the total twist angle φ_c of a flexible coupling can be computed according to the following formula:

$$\varphi_{ci} = \frac{\pi \cdot n \cdot \Delta t_i}{30}[\mathrm{rad}], \tag{2}$$

where n is the immediate rotation speed of a mechanical system [min^{-1}].

From the total twist angle of a flexible coupling φ_c, its static component φ_{stat} can be computed as the mean value:

$$\varphi_{\mathrm{stat}} = \frac{\sum_{i=1}^{k} \varphi_{ci}}{k}[\mathrm{rad}], \tag{3}$$

where k is number of samples. In practice, the floating average method is used for this computation.

PC performs the computations according to Eqs. (1–3) in real-time. This way, the controlled variable for the electronic system for pneumatic tuners with auto-regulation can be computed. It is very suitable for this system, because the torsional vibration does not directly affect the control device, as in the case of the mechanical controller solution, e.g., [6, 18].

In the case of our electronic extremal control tuning system, the most suitable controlled variable seems to be the effective value RMS of dynamic component M_{dyn} of the load torque, transmitted by a pneumatic flexible coupling. It can be measured directly, for example, using torque sensors built in a mechanical system (Fig. 5).

Fig. 5 Torque sensor built in a mechanical system

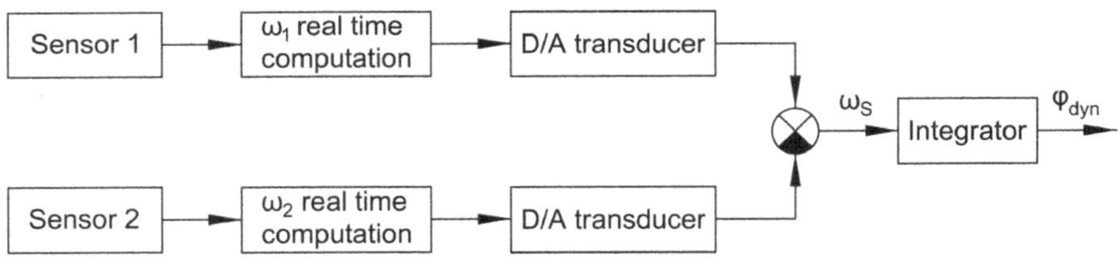

Fig. 6 Scheme for computing the dynamic twist angle of a flexible coupling φ_{dyn}

Significant disadvantages of the torque sensors which are built in mechanical systems are their size and relative high price. Their usage is therefore often unsuitable or even impossible.

In the following Fig. 6, the most suitable way to compute the dynamic twist angle of a flexible coupling φ_{dyn} and the twisting angular speed of a flexible coupling ω_S is described. The φ_{dyn} and ω_S signals are subsequently needed for computing the load torque dynamic component M_{dyn} transmitted by a flexible coupling.

In Fig. 6, we can see that ω_1 and ω_2 are the immediate angular speeds of the driving and driven flange of a flexible coupling. These angular speeds can be computed using following formula:

$$\omega_i = \frac{2 \cdot \pi}{x \cdot t_{pi}} \left[\text{rad} \cdot \text{s}^{-1}\right], \tag{4}$$

where x is the number of black-and-white stripe pairs and t_{pi} are the times between two consecutive edge crossings from black to white stripes . Subsequently, by subtracting the ω_2 from the ω_1, we get the twisting angular speed of a flexible coupling ω_S, but both the digital signals ω_1 and ω_2 must be transformed to analog signals using D/A transducers in advance. By integrating the analog ω_S signal, we get the analog φ_{dyn} signal, which is the dynamic component of total twist angle φ_c of the coupling.

Our research is focused on the characteristic static and especially dynamic properties of pneumatic flexible shaft couplings in the long term. We investigate mainly the influence of various factors like static component of load torque, loading amplitude and frequency, temperature, loading cycles amount, throttling, gaseous media type, design, etc. on the dynamic properties of pneumatic flexible shaft couplings, especially on their dynamic torsional stiffness and damping, e.g., [6, 7, 9, 11, 12, 16–19].

Therefore, for our purpose of developing the electronic extremal control tuning system, following computation can be, for example, used with sufficient accuracy for computing the load torque dynamic component M_{dyn} transmitted by a pneumatic flexible coupling:

$$M_{\text{dyn}} = \varphi_{\text{dyn}} \cdot k_{\text{dyn}} + \omega_S \cdot b \,[\text{N m}], \tag{5}$$

where k_{dyn} [N m rad^{-1}] is the dynamic torsional stiffness of a pneumatic tuner, b [N m rad^{-1} s] is the viscous damping coefficient of a pneumatic tuner.

In the Eq. (5), we can see that the dynamic component M_{dyn} of the load torque transmitted by a pneumatic coupling consists of its "flexible part" ($\varphi_{dyn} \cdot k_{dyn}$) and its "damping part" ($\omega_S \cdot b$).

Therefore, according to pneumatic tuner type and the operating conditions of the continuous tuned torsional oscillating mechanical system, we are able to take into consideration the influence of the various above-mentioned factors on the proper computation of the pneumatic tuner dynamic properties k_{dyn} or b.

4 Conclusion

Every technical solution has advantages and disadvantages. The use of the presented contactless measuring device for flexible shaft coupling twist angle measurement provides the following general advantages:

- the dynamic mass properties of a mechanical system are not influenced by additional masses, because there are not built-in sensors;
- there is no change of overall dimensions of a mechanical system itself;
- it allows the usage for any huge torsional oscillating mechanical systems with very high transmitted load torques;
- there is no friction between the sensors and a mechanical system;
- it is possible to quick replace the broken, damaged or malfunctioning sensor.

It is also necessary to mention the general disadvantages of the device:

- the device function is very sensitive to dirty the optical part of the optoelectronic sensor or the reflective black-and-white tape. This issue could be fixed using similar, for example, magnetic sensors;
- the load torque dynamic component M_{dyn} transmitted by a pneumatic flexible coupling is obtained by the indirect computation method and therefore its accuracy is lowered;
- with increasing maximal twist angle of a flexible coupling, the number of samples per revolution for the computation of the total twist angle of a flexible coupling φ_c and its static component φ_{stat} decreases and therefore the accuracy also decreases.

Both our above-mentioned electronic systems for continuous tuning of torsional oscillating mechanical systems during their operation consist of a hardware part and a software part. We aim for designing the same hardware part for both the electronic systems in order to create the universal control system. Considering the above-mentioned advantages and disadvantages, it seems to be advantageous to incorporate the contactless measuring device into this universal control system.

Acknowledgements This paper was written in the framework of Ph.D. Student's and Young Researcher' s Project: "Solution of a control system element for mechanical systems continuous tuning".

References

1. Czech P, Wojnar G, Burdzik R, Konieczny Ł, Warczek J (2014) Application of the discrete wavelet transform and probabilistic neural networks in IC engine fault diagnostics. J VibroEng 16:1619–1639
2. Sapietová A, Dekýš V (2016) Dynamic analysis of rotating machines in MSC.ADAMS. In: Vavro J, Ondrušová D (eds) MMS 2015, Elsevier Science BV, Amsterdam, pp 143–149
3. Łazarz B, Wojnar G, Madej H, Czech P (2009) Evaluation of gear power losses from experimental test data and analytical methods. Mechanika 80(6):56–63
4. Homišin J, Kaššay P (2014) Optimal tuning of torsional oscillating mechanical systems. In: Sevcik L, Lepsik P (eds) 54th ICMD 2013, LNME. Springer Int. Publishing AG, Cham, pp 63–69
5. Vojtková J (2017) Determining and verifying the geometric characteristics of helical grooves in the worm in planetary toroidal drives in a more effective way. Scientif J Silesian Univ Technol-Ser Transp 96:197–204
6. Homišin J (2002) New Types of flexible shaft couplings: development-research-application. Vienala, Košice
7. Homišin J (2016) Characteristics of pneumatic tuners of torsional oscillation as a result of patent activity. Acta Mechanica et Automatica 10(4):316–323
8. Puškár M, Fabian M, Kádárová J, Blišťan P, Kopas M (2017) Autonomous vehicle with internal combustion drive based on the homogeneous charge compression ignition technology. Int J Adv Robot Syst 14(5)
9. Kaššay P, Homišin J (2013) Determining the properties of pneumatic flexible shaft couplings with wedge flexible elements. Scient J Silesian Univ Technol-Ser Transp 81:59–67
10. Medvecká-Beňová S, Miková Ľ, Kaššay P (2015) Material properties of rubber-cord flexible element of pneumatic flexible coupling. Metalurgija 54(1):194–196
11. Krajňák J (2014) Influence of gaseous media on damping of pneumatic couplings. Scientif J Silesian Univ Technol-Ser Transp 85:83–87
12. Kaššay P (2014) Comparison of pneumatic flexible shaft coupling static load characteristics obtained experimentally and by calculation. Scientif J Silesian Univ Technol-Ser Transp 85:57–65
13. Mantič M, Kuľka J, Krajňák J, Kopas M, Schneider M (2016) Influence of selected digitization methods on final accuracy of 3D model. In: Majerník M, Daneshjo N (eds) ICESPM 2015, PMES. CRC Press, Boca Raton, pp 475–480
14. Kyslan K, Rodič M, Suchý Ľ, Ferková Ž, Ďurovský F (2017) Industrial controller-based dynamometer with dynamic emulation of mechanical loads. Electr Eng 99(4):1245–1254
15. Liptai P, Moravec M, Lumnitzer E, Gergeľová M (2017) Proposal of the sound insulating measures for a vibrational sorter and verification of the measured effectiveness. Adv Sci Technol-Res J 11(3):196–203
16. Homišin J (2014) Research and application of universal regulation system in order to master the source of mechanical systems excitation. Scientif J Silesian Univ Technol-Ser Transp 84:7–12
17. Homišin J, Kaššay P (2014) Experimental verification of the possibility using pneumatic flexible shaft couplings for the extremal control of torsional oscillating mechanical system. Diagnostyka 15(1):11–16
18. Homišin J, Kaššay P, Puškár M, Grega R, Krajňák J, Urbanský M, Moravič M (2016) Continuous tuning of ship propulsion system by means of pneumatic tuner of torsional oscillation. Int J Maritime Eng: Trans Royal Inst Naval Archit 158(A3):231–238
19. Čopan P (2014) The application of new method of torsional oscillating mechanical systems continuous tuning. Doctoral thesis, Technical University of Košice
20. DEWETRON Homepage (2018). https://www.dewetron.com/products/components-and-sensors/rpm-angle-sensors/. Last accessed 1 June 2018

Profile Design of Groove Cam with Roller Follower

Ha Van Nguyen and Ladislav Ševčík

Abstract To determine the size of the cam mechanism is an important task of the cam designer to produce the contour of the cam with high precision. There are two major factors which affect cam size, the pressure angle and the radius of curvature. Both of these involve either the base circle radius on the cam R_b when using flat-faced follower, or the prime circle radius on the cam R_p when using roller or curved followers. Therefore, the paper presents the relations between the geometric parameters and the cam angles of the groove cam mechanism are established. To calculate each contact point at any instant moment by using the method according to velocity and geometric relationship of instant velocity centers. Hence the cam profile was defined by contouring of the contact points. Moreover, a design example of constant-diameter groove cam is illustrated in this article.

Keywords Groove cam mechanism · Pressure angle · Profile design · Instant velocity center

1 Introduction

Cam mechanisms are widely used in many types of modern machines because of their excellent property for operation speed, motion accuracy, structural rigidity and low production cost. Generally, plate cam mechanisms are only one of the larger number of the cam and follower combinations and can be classified in several ways: by type of follower motion, by type of joint closure, by type of motion constraints and so on. According to the motion type, the follower may be translating or oscillating. The translating follower is simple, common to use and easy to design, the latter is known to be complicated and quite difficult to design but has less friction and higher efficiency and work more smoothly. Depend on how the cam and follower contact is maintained, cam mechanisms can be divided into force-closed and form-closed cam mechanisms. For kind of force-closed cam mechanism, in order to ensure constant contact between the cam and follower, an external force is required to be applied.

H. V. Nguyen (✉) · L. Ševčík
Technical University of Liberec, Studentská 2, 46117 Liberec, Czech Republic
e-mail: nguyenha.hut@gmail.com

Š. Medvecký et al. (eds.), *Current Methods of Construction Design*, Lecture Notes in Mechanical Engineering, https://doi.org/10.1007/978-3-030-33146-7_24

This force is usually provided by a spring of sufficient stiffness or sometimes by an air cylinder. As a result, of the force effect, the driving torque and contact stress between the cam and follower in the rise are increased; it makes the wear on the cam and follower. Whereas in the form-closed cam mechanism, contact is obtained by letting the roller follower run in a cam groove or by using a conjugate condition. So no force is required to maintain the contact between its cam and follower; therefore, the contact stress between the cam and follower will consequently be smaller and the driving torque in the rise can be reduced [1].

In recent years, with the rapid development and popularization of mechanical products in modern society, it has been such an important question on designing and producing mechanisms precisely. Besides, with the assistance of numerical controlled manufacture system being widely applied in this time, which are useful tools for the designer to develop a new type of cam profile. But, the conventional type of cam profile is usually defined in approximation by lines or curves which have a bigger error and require more node points, result in chattering of the principal axis. In fact, the design process of cam must make accurate solutions for precise cam profile.

Therefore, a novel cam model of Fig. 1 is designed for converting the cam rotation to a desired output motion of the output shaft. The mechanism is composed mainly of a rotary input camshaft, a rotary output shaft, a middle part and two balls and a frame to mount all the parts [2–4]. With this model of the cam mechanism used two steel balls to transmit rotary motion between two concentric shafts, the main function of the balls is the transmission force and motion from input camshaft into output shaft. So when cam works each ball can be easy to move up and down in each straight groove on the output shaft and middle part corresponding. Moreover, these balls must move along in the circular groove on the input camshaft.

Due to each steel ball only moves up and down in each straight groove of the middle part. Therefore, each ball plays the role of a roller follower of a cam mechanism, which means that the cam mechanism model is designed with a pair of followers for one cam but it is not a type of conjugate cam system. However, two balls always turn around in the circular groove of the input camshaft, which creates a constant-diameter between two of the centers of the balls (rollers). In this case, a constant distance is equal to the diameter of the circular groove on the input camshaft. Hence the cam mechanism is designed, taking some characteristics of a breadth cam [5].

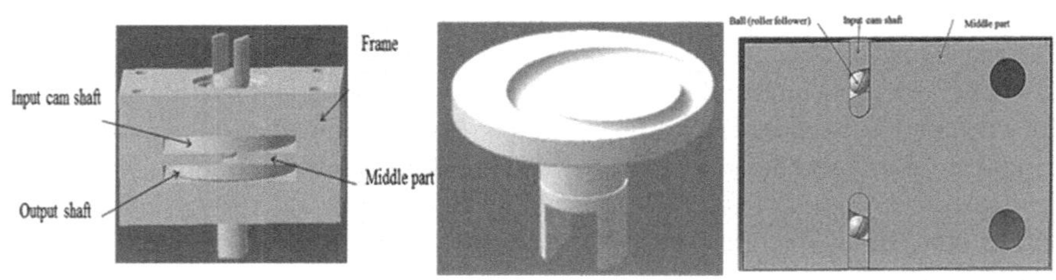

Fig. 1 From left to right: Cam mechanism model. Shape of input camshaft. Assembly of the balls in the middle part and input cam

As mentioned above, the cam mechanism was designed which takes some characteristics of a form-closed and a breath cam, so the cam mechanism gets some advantages to other cam mechanisms because of following considerations: compared with a force-closed cam mechanism, the mechanism withstands lower contact stress because it is a form-closed cam mechanism and no force is required to maintain the contact between its cam and follower; compared with conjugate cam mechanism in which a set of two cams must be used, the cam mechanism is simpler in construction because only one cam is needed. So, it occupies small space and has lower cost; compared with a constant-breadth cam mechanism of which the cam profile should be convex everywhere, the cam mechanism possesses wider adaptability to the output motion because its groove cam profile could be concave. Another point in its favor of this model is the cam mechanism used the ball for followers, which are commonly used in production machinery where their ease of replacement and availability from bearing manufacturers stock in any quantities are advantages.

In [1], it was shown that the major drawback of the groove cam is the phenomenon of crossover shock. Every time the acceleration of the follower changes sign, the inertial force also does so. This causes the follower to abruptly shift from one side of the groove cam to the other. Due to the clearance between the roller followers and the circular groove, even if the clearance is very small, there will still be an opportunity for the follower to develop some velocity in its short trip across the groove, and it will impact the other side. With this groove cam model in Fig. 1, a typically fail at the points where the acceleration reverses sign, due to many cycles of crossover shock. Note also that the single ball has to reverse its angular velocity direction every time it crosses over to the other side of the groove. This causes significant balls slip and high wear on the roller followers compared to an open. Therefore, the crossover shock can be reduced or eliminated by improving the cam surface to get high precision to control the clearance, but it makes the cam tends to be more expensive.

However, until now there are very few publications that can be found in the literature that addresses the issue of the kinematics of the groove cam. Therefore, the design of this mechanism remains to be a challenging task. Although the cam mechanism of Fig. 1 has received some attention from any industrial machine parts and mechanisms as well as automated machine production and so on.

In this paper, a method for analytically designing the groove cam contour is suggested. The method according to velocity and geometric relationships at instant velocity centers can determine the coordinates at contact points easily and correctly. The results are useful in designing to get a suitable groove cam profile for operating the mechanism.

2 Materials and Methods

2.1 Kinematic Relationships of Cam Mechanism Design

The groove cam mechanism as shown in Fig. 2 consists of a circular groove cam

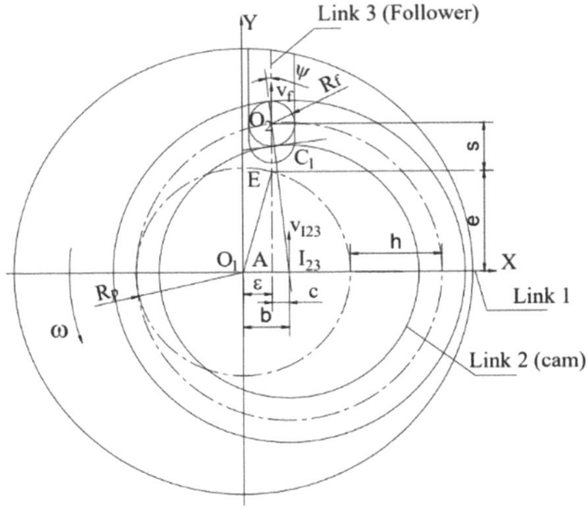

Fig. 2 Instant velocity center of groove cam with roller follower

and a ball as a follower. The cam rotates in counterclockwise (CCW) with a constant angular velocity and drives the follower to translate up and down in reciprocating motion.

Figure 2 shows the geometry of a circular groove cam and translating roller follower in an arbitrary position. This shows the general case in that the axis of the motion of the follower does not intersect the center of the cam. There is an eccentricity ε, which is the perpendicular distance between the follower's axis of motion and the center of the cam. Often this eccentricity ε will be zero, making it aligned follower.

As can be seen in Fig. 2, the axis of transmission is extended to intersect effective link 1, which is a frame (the ground link). This intersection is instant center I_{23} (labeled I_{23}), which is defined as a point, common to two rigid bodies (links) in plane motion that has the same instantaneous velocity in each body.

In other words, there is no relative velocity between these two points at that instant. Thus, one body can be considered to be in pure rotation with respect to the other about their common instant center. By definition above, the velocity in link 2 (the cam) and link 3 (the ball followers) has the same value. All points on the follower have identical velocities *Vf*, which are equal to the velocity of I_{23} in link 2. We can write an expression for the velocity at I_{23} by cam rotation:

$$V_{I_{23}} = b\omega = \dot{s}, \tag{1}$$

where ω is the constant cam angular velocity and the radius b from cam center to I_{23}, s is the instantaneous displacement of the follower from the s diagram and $\dot{s}$ is its time derivative in units of length/sec. Now applying the chain rule with ω constant:

$$V_{I_{23}} = b\omega = \dot{s} = \frac{ds}{dt}, \tag{2}$$

also, the Eq. 2 can express by

$$\frac{ds}{dt} = \frac{ds}{dt}\frac{d\theta}{d\theta} = \frac{ds}{d\theta}\frac{d\theta}{dt} = \frac{ds}{d\theta}\omega = v\omega. \tag{3}$$

Thus, the following equation is obtained:

$$b\omega = v\omega. \tag{4}$$

According to Eq. 4 we get

$$b = v. \tag{5}$$

From Eq. 5, we conclude that the distance b to the instant center I_{23} is numerically equal to the velocity of the follower v in units of length per radian.

2.2 Pressure Angle and Geometry Parameters

The pressure angle ψ is defined as the angle between the direction of motion (velocity) of the follower and the direction of the axis of transmission [1], as shown in Fig. 2. As a rule of thumb, we would like the pressure angle has a value as small as to avoid undesirable levels like as jam, follower sliding or pivot friction. So determining the pressure angle must be done in design. From Fig. 2, the pressure angle can be expressed in terms of displacement s, velocity v, eccentricity ε and the prime circle radius R_p. The construction is shown in Fig. 2. The point E is the intersection of the arc of radial R_p and the axis of motion of the follower at point E, where the length e is defined by the distance from link 1 to this intersection. For any chosen prime circle radius, it is a constant value. From triangle AO_2I_{23}:

$$c = (s + e)\tan\psi, \tag{6}$$

and

$$b = (s + e)\tan\psi + \varepsilon. \tag{7}$$

Substituting Eq. 5 into Eq. 7, we obtained

$$v = (s + e)\tan\psi + \varepsilon. \tag{8}$$

As clear from triangle AEO_1

$$e = \sqrt{R_p^2 + \varepsilon^2} \tag{9}$$

substituting Eq. 9 into Eq. 8 we get

$$\psi = \arctan \frac{v - s}{s + \sqrt{R_p^2 - s^2}} \tag{10}$$

According to Eq. 10, it clearly shows that the pressure angle depends on the displacement s, velocity v, eccentricity ε and prime circle radius R_p. Therefore, when determining cam we need to choose R_p and ε to get an acceptable maximum pressure angle. If R_p is increased, ψ will be reduced but it makes the size of the cam is bigger, also increase cost. In general practice, we assume a trial value for R_p and an initial eccentricity of zero to calculate the values of ψ for the entire cam. Then adjust R_p and repeat the calculation until an acceptable is found. The value of a pressure angle desire is to be between zero and ±30.

2.3 *Shape Design for Groove Cam*

As shown in Fig. 3, a groove cam mechanism is suggested to design, which has a pair of balls as followers. The cam rotates in counterclockwise at a constant angular velocity and drives the pair of the followers to translate up and down in reciprocating

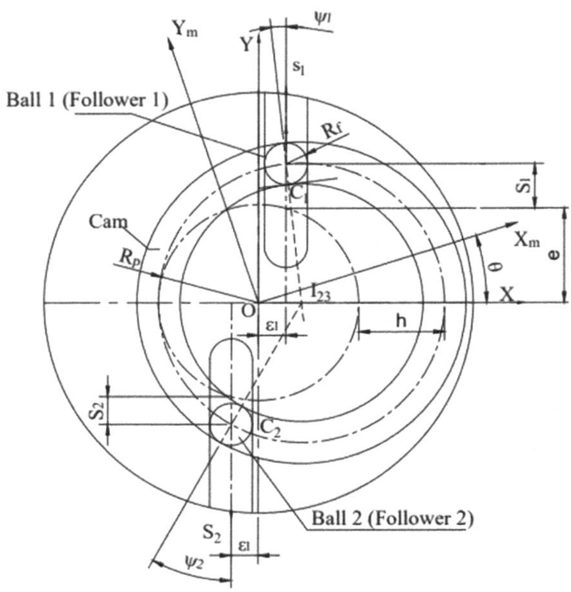

Fig. 3 Geometry of groove cam with reciprocating follower

motion. The locus of the center of the followers is called the pitch curve of the cam. In this case, both centers must overlap over one rotation of the cam.

Hence, the groove cam mechanism must satisfy the kinematic conditions: Roller followers on upper and lower sides have the same displacement, velocity and acceleration. In order to design the groove cam profile to satisfy the kinematic conditions, instant velocity center for I_{23} by two followers must be met at the same point [5]. This means that the normal lines at the contact points by the upper ball and lower ball must pass through the same point. Also, when the cam rotates one revolution, the position of the roller 1 and roller 2 with respect to the cam is opposite to the position. Therefore, the eccentricities ε_1 and ε_2 of the ball 1 and ball 2 have the same magnitude ε and must be located on both sides of the Y-axis as seen in Fig. 3.

$$\begin{cases} |\varepsilon_1| = |\varepsilon_2| = \varepsilon \\ \varepsilon_1 = -\varepsilon_1. \end{cases} \tag{11}$$

For deriving the profile equation of the groove cam, two coordinates systems are defined, as shown in Fig. 3, where ψ_1 and ψ_2 represent contact angle at the contact points C_1 and C_2, respectively. $S(X, Y)$ represents a stationary reference system and $S_m(X_m, Y_m)$ represents a mobile reference system. The reference system S_m is defined by the input camshaft rotation angle θ.

Substituting Eq. 5 into Eq. 10, we get the contact point pressure angle for the upper and lower followers are represented by the following expressions, respectively:

$$\psi_1 = \arctan \frac{v - s}{s_2 + \sqrt{R_p^2 - s^2}} \tag{12}$$

$$\psi_2 = \arctan \frac{v - s}{s_2 + \sqrt{R_p^2 - s^2}} \tag{13}$$

The coordinates $S(X_{c1}, Y_{c1})$ and $S(X_{c2}, Y_{c2})$ of the contact points between ball 1, ball 2 and groove cam respectively are represented by the following expressions:

$$\begin{cases} X_{c1} = \varepsilon + R_f \sin \psi_1 \\ Y_{c1} = e + s_1 - R_f \cos \psi_1 \end{cases}, \tag{14}$$

$$\begin{cases} X_{c1} = -\varepsilon + R_f \sin \psi_2 \\ Y_{c1} = -e - s_2 - R_f \cos \psi_2 \end{cases}, \tag{15}$$

where θ is the cam angle and h is the maximum lift of followers with $h = s(\theta)$, which is the function of the follower.

Therefore, the coordinate of contact points C_1 and C_2 of the upper and lower followers in the fixed coordinate systems S are converted into the S_m coordinate system can be expressed in terms of the displacement S_1 and S_2 of the followers

$$\begin{bmatrix} m_{Xc1} \\ m_{Yc1} \end{bmatrix} = \begin{bmatrix} \cos\theta & \sin\theta \\ -\sin\theta & \cos\theta \end{bmatrix} \begin{bmatrix} \varepsilon + R_f \sin\psi_1 \\ e + s_1 - R_f \cos\psi_1 \end{bmatrix} \quad (16)$$

$$\begin{bmatrix} m_{Xc2} \\ m_{Yc2} \end{bmatrix} = \begin{bmatrix} \cos\theta & \sin\theta \\ -\sin\theta & \cos\theta \end{bmatrix} \begin{bmatrix} -\varepsilon + R_f \sin\psi_2 \\ -e - s_2 + R_f \cos\psi_2 \end{bmatrix} \quad (17)$$

From Eqs. 16 and 17, the whole cam contours are obtained. The method proposed to design the cam contour in this paper is simple in principle; it does not include complicated calculations so that which can be suitable for applying in practice of cam design.

3 Results

An example was used to confirm the design of a groove cam mechanism with the data given in Tables 1 and 2.

A groove cam rotates in a constant speed and two balls followers move to reciprocating displacement with double-harmonic function.

The equations for double-harmonic function are given followings:

for the rise

$$s = \frac{h}{2}\left\{1 - \cos\left(\pi\frac{\theta}{\beta}\right) - \frac{1}{4}\left[1 - \cos\left(2\pi\frac{\theta}{\beta}\right)\right]\right\}, \quad (18)$$

for the fall

$$s = \frac{h}{2}\left\{\left[1 + \cos\left(\pi\frac{\theta}{\beta}\right)\right] - \frac{1}{4}\left[1 - \cos\left(2\pi\frac{\theta}{\beta}\right)\right]\right\}. \quad (19)$$

Table 1 Displacement conditions for cam

Segment	Cam angle [°]	Total angle of segment β [°]	Motion type [mm]	Function
1	0–180	180	12 rise	Double harmonic
2	180–360	180	−12 return	Double harmonic

Table 2 Design parameters for cam

Parameters	Value
Prime circle radius R_p [mm]	14
Eccentricity(ε)/mm	0
Ball radius (follower) R_f [mm]	3

The corresponding displacement, velocity and acceleration diagram of groove cam are depicted in Fig. 4.

Also, the pressure angle at the contact point is shown in Fig. 5, and the maximum

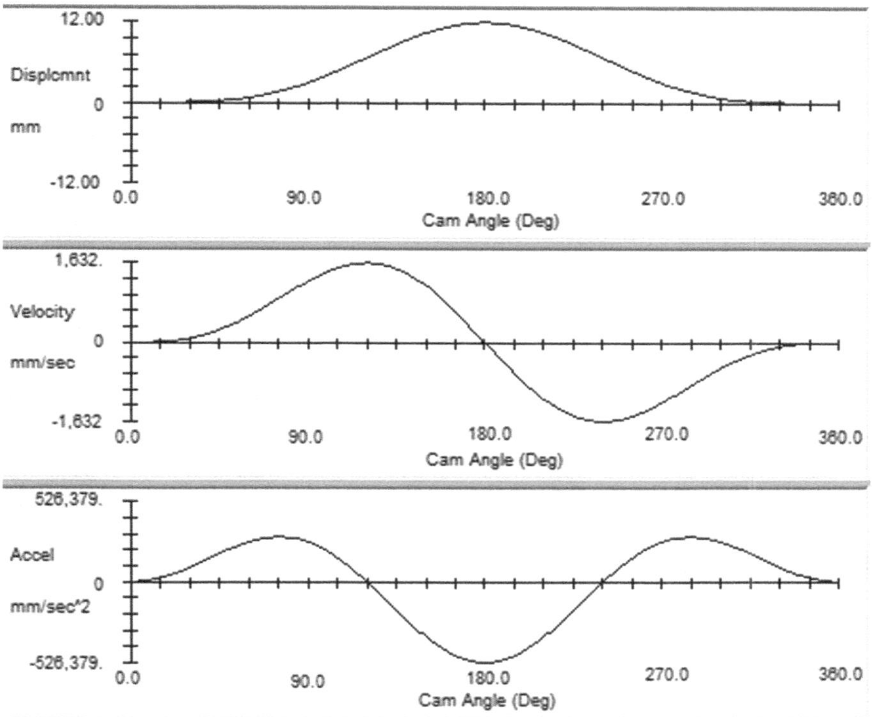

Fig. 4 Displacement, velocity and acceleration of the groove cam

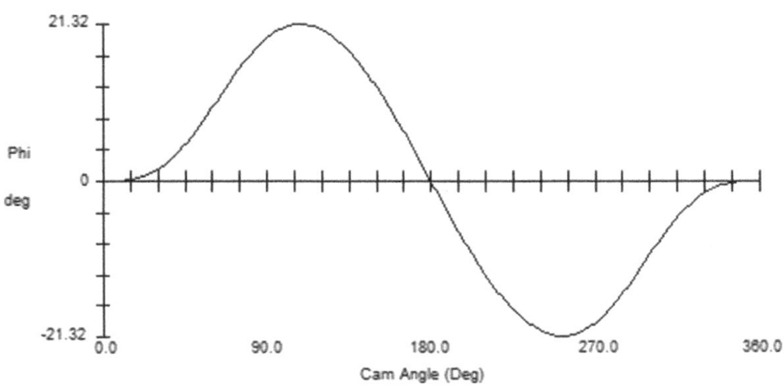

Fig. 5 Pressure angle at the contact point

value of pressure angle is 21.32°. This value is satisfactory with theory. Hence the radius of curvature of the groove cam is obtained as can be seen in Fig. 6.

The example used in this study illustrated for a special case of the groove cam, where the groove cam without eccentricity and also no dwell motion. This means that the cam will rotate with no rest. The application of the groove cam profile is given in Fig. 7. However, the 3D model of the groove cam can be obtained by using CATIA software as shown in Fig. 8.

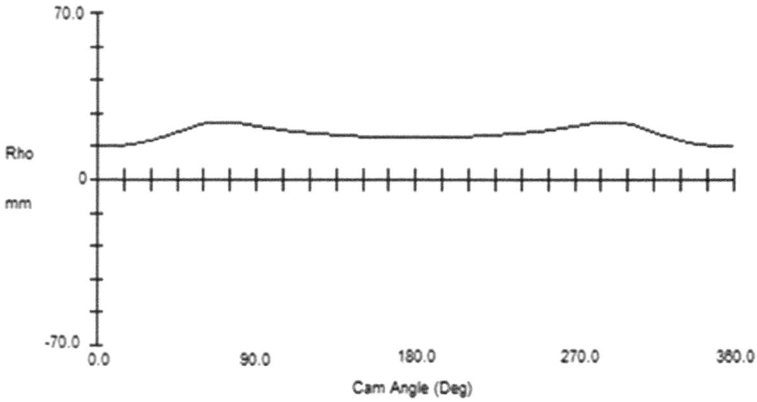

Fig. 6 Radius of curvature of groove cam

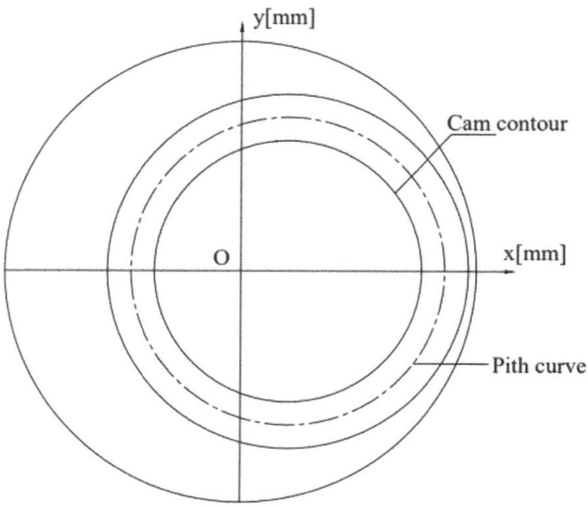

Fig. 7 2D model of groove cam profile

Fig. 8 3D model of the groove cam mechanism by CATIA

4 Conclusions

The method is based on the geometric and velocity relationship at instant velocity center is proposed. By using this method, the radius of curvature of the cam contour and the pressure angle at the contact point is easily obtained, which is significant in practical cam design to optimize a cam mechanism.

Acknowledgements This publication was written at the Technical University of Liberec as part of the project "Innovation of products and equipment in engineering practice" with the support of the specific University Research grant, as provided by the Ministry of Education, Youth and Sports of the Czech Republic in the year 2018.

References

1. Robert LN (2002) Cam design and manufacturing handbook. New York Industrial Press Inc
2. Van Nguyen H, Ševčík L (2016) Design calculation of the cam mechanism for transmission between two concentric shafts. In: 57th International conference of Machine Design Departments—ICMD 2016, University of West Bohemia, Plzeň, pp 137–140
3. Van Nguyen H, Ševčík L (2017) Apply steel balls for designing the cam mechanism. In: Proceeding of 22nd International Conference Mechanika, Kaunas University of Technology, Lithuania, pp 118–121
4. Van Nguyen H, Ševčík L (2017) The contact analysis of designing the cam mechanism by using ANSYS. In: Herák D (eds) Proceedings of 58th international conference of machine design departments—ICMD 2017, Czech University of Life Sciences Praha, Praha, pp 271–274
5. Shin JH, Li L, Yang HD, Kwon SM (2011) Kinematical system of bread the cam profile design. J Cent South Univ Technol 18:451–457

Geometric Specification of Complex Spatially-Oriented and Compliant Components II

Václav Vaněk, Martin Gorschenek and Roman Polák

Abstract The article is directly linked to the article called "Geometric Specification of Complex Spatially—Oriented and Compliant Components I". Manufacturers of complicated and flexible machine parts, particularly in the automotive industry, are finding themselves in almost insoluble problems when designing TS and preparing their technical documentation. These problems arise mainly in describing the geometrical properties of technical products in order to ensure their errorless production and control in sub-suppliers. Unfortunately, GPS standards are not flexible enough to meet these specific requirements, and from this reason manufacturers or their consortiums respond to this situation by defining their own procedures and regulations that allow them to solve these problems. However, the negative consequence of this state is that there is no general agreement and consistency in the description between the individual internal rules of the firms and this creates considerable problems for the subcontractors, who cannot properly read the technical documentation and correctly set up production and control procedures, which can lead in some cases to fatal consequences. The aim of this paper is to identify these inconsistencies and propose ways to solve the above-mentioned problems.

Keywords Reference points system · Drawing datum · Degrees of system freedom

1 Introduction

Removal of degrees of system freedom and related fixation of movement of tolerance fields of geometric tolerances, determination of exact position of datum and directions of measurement for complex parts, definition of pressure and free states of flexible parts, uncertainty of parameter specification in dimensioning and resulting uncertainty for control and measurement, and a number of other uncertainties are the impetus for introducing a certain system and establishing clarity in the specifications of individual GPS parameters.

V. Vaněk (✉) · M. Gorschenek · R. Polák
University of West Bohemia in Pilsen, Univerzitni 22, 30614 Pilsen, Czech Republic
e-mail: vanekv@gmail.com

Š. Medvecký et al. (eds.), *Current Methods of Construction Design*, Lecture Notes in Mechanical Engineering, https://doi.org/10.1007/978-3-030-33146-7_25

2 Reference Points System (RPS)

Reference points can be used to set, ensure the correct position and orientation of a component. Designing RPS points should primarily be based on an analysis of how and under what conditions the machine part works and what are its links to the surrounding technical system and follow the specific purpose of using RPS points.

According to Vaněk and Polák [1], the bases for the use of RPS are defined, their types are identified and RP designation is proposed in the technical documentation. Because it is impractical and often unfeasible to provide a full description of the RP directly at its designation, it is necessary to define the supplementary table in such a way that it is possible to record all the necessary parameters related to the RPS definition. In the technical drawing, RP is then marked as defined in [1] and supplemented by a unique identifier (ID). Figure 1 shows the RPS points on the drawing in practice. Those are RPS for control on metering machines (CMM). The marking of point P_5u_x means that measuring contact track in direction of coordinate of LCS. Other required point parameters are specified in the RPS table.

2.1 Identification of Parameters Specified in the Drawing

Identifying the specifications on the drawing by dividing the drawing into sectors by means of the coordinate network has proven to be inadequate and therefore it is necessary to assign each geometric product specification (dimension, dimensional tolerance, shape and position tolerance, surface quality, technological processing, note ...) unique code (ID). ID is also assigned to the reference point, line/curve, surface, coordinate system, etc., but also generalized design/drawing elements—plane in the drawing space, rotation axis, overall symmetry axis, local structural axis, areas and curves, and more. ID is given in so-called tears (see Fig. 2), or in rings. The

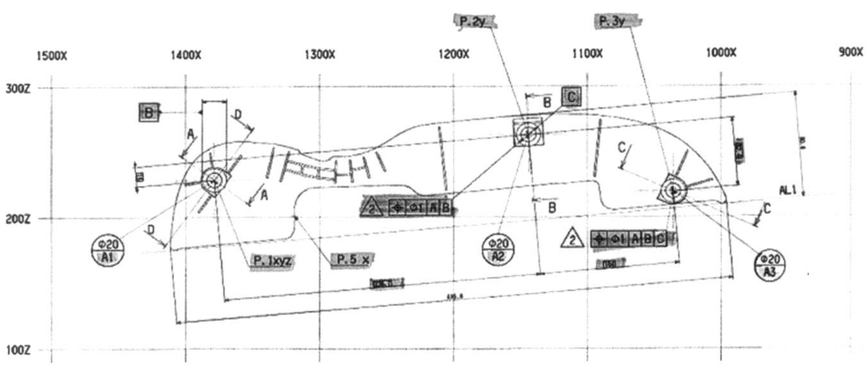

Fig. 1 RPS points on a technical drawing

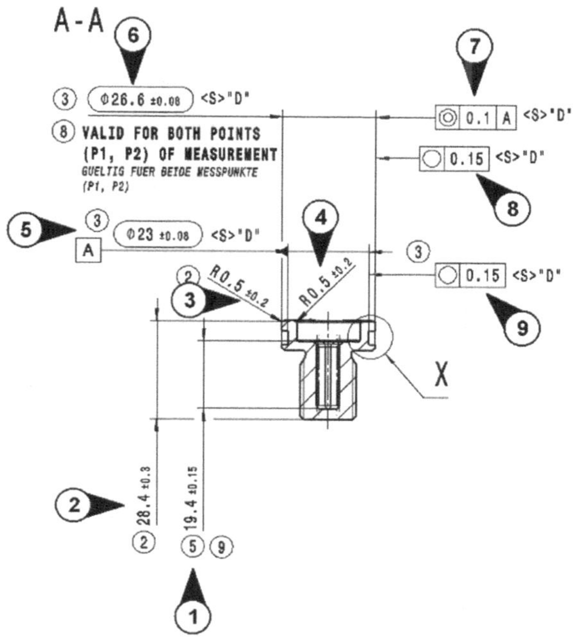

Fig. 2 Identifying specifications on a drawing

ID of the individual specifications can be then assigned an information envelope, which as a graphic symbol is capable of transmitting a variety of information in a visual way. Subsequently, additional information can be added through software products for managing technical documentation product data management (PDM).

2.2 Reference Points System Parameters Are Specified in the RPS Table

From Figs. 3, 4 and Table 1, it is possible to trace the considerable inconsistency of the RPS tables resulting from the above-mentioned problem that the GPS standards lag behind the requirements laid down in the technical documentation on the production documentation. This state of affairs is due in particular to new and often more stringent demands on the characteristics of current technical systems and to the rapid development of modern equipment to support the design and control of technical and design processes as well as of the technical systems themselves.

In Fig. 3, we can see a very simplified table for assigning LCS coordinate of the RPS points. Therefore, the definition is incomplete and the understanding given by the worthy can occur only from the context of the drawing. In this case, it is RPS for specifying reference curves for tool making. Since it is a reference element, it is the theoretically exact coordinates by which the curve reference is passed. Again, it

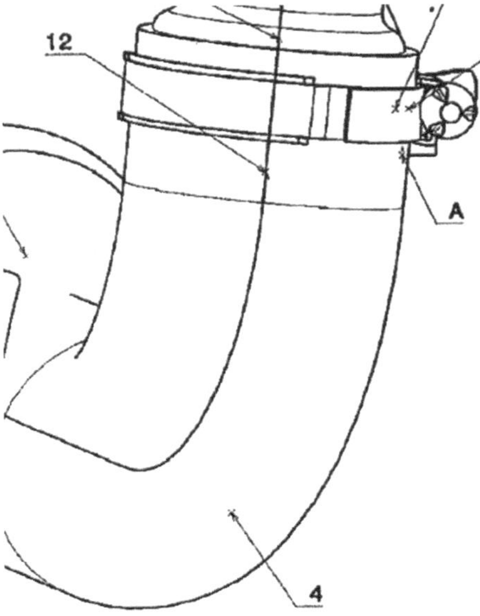

PART COORDINATE AXIS

REF.	X	Y	Z	Radius
1	-454,06	356,31	-170,79	
2	-403,51	295,22	-157,71	R75
3	-411,52	127,26	-192,62	R80
4	-358,37	-26,97	-216,56	R70
5	-347,26	-34,93	-109,78	
6	-8,74	-7,3	11,73	
7	-34,24	-28,61	45,92	R50
8	-86,81	-40,62	48,83	R50
9	-112,2	-39,92	36,45	R45
10	-118,57	-38,78	31,31	R45
11	-153,99	-38,78	17,14	
12	-350,36	-32,71	-139,53	
13	-341,87	-38,78	-57,99	
14	-124,28	-38,78	29,03	

Fig. 3 RPS for specifying reference curves (the way used in French firms—(mothers and daughters))

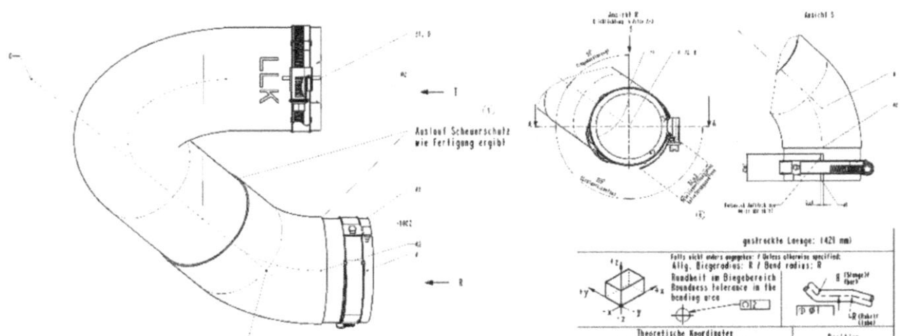

	Theoretische Koordinaten Theoretical coordinate dimensions				Position		
	Punkt Bending point	x	y	z	Strecke Distance	t	R
Schlauch	A	-668,1	459,4	-129,0	--	0	--
	B	-675,7	526,0	-129,0	AB	±8	75
	C	-876,2	680,0	0,0	BC	±8	75
	D	-876,2	498,0	0,0	CD	±8	--
Markierung	A1	-668,1	459,4	-93,9	--	--	--
	B1	-911,2	498,0	0,0	--	--	--
Scheuerschutz	A2	-671,6	488,8	-129,0	--	±8	--
	B2	-730,2	567,8	-94,0	--	±8	--

Fig. 4 RPS for specifying reference curves (mainly used in German companies)

Table 1 RPS detailed table (mainly used in German companies)

Drawing sector	Point label	GCS absolute coordinates			Contact field or contact hole
DSec	PLbl	ACx	ACy	ACz	$F\phi D(a \times b)$; $H\phi D(a \times b)$
K22	$P_1H_{x,\ y,\ z}$	1376.60	116.91	228.00	$H\varnothing 8$
L18	P_2F_z	1143.9	126.20	262.91	$F\ 5 \times 5$
F10	P_{3y}	1340.0	104.41	201.30	–
K21	P_{4x}	1240.0	114.22	250.00	–

LCS [ACx, ACy, ACz, $r_x^{GCS}, r_y^{GCS}, r_z^{GCS}$]

LCS Related coordinates			Tolerances			
x_{LCS}	y_{LCS}	z_{LCS}	ST_x	ST_y	ST_z	⊕ ϕT A_{PrRef} B_{PrRef} C_{PrRef} ⌓ T A_{PrRef} B_{PrRef} C_{PrRef}
0.00	0.00	0.00	0.00	0.00	0.00	0.00
−57.60	14.30	−32.00	–	–	–	–
0.00	−4.60	−24.60	±0.02	+0.00 −0.03	+0.05 −0.01	–
−20.00	30.60	−10.60	–	–	–	⊕ 0.01 A_{PrR} B_{PrR} C_{PrR}

is not specified how the curve was created. From the context and production of the shaping mandrel again, it is advisable that the reference centerline of the hose be composed of a line and tangent arcs whose spatial position and orientation is given by RPS points. It is not recommended to create a centerline in the CAD system using a spline curve ⇒ problematic hose production.

In Fig. 4, then we can observe the considerable confusion of the established definitions and the misunderstanding of the relationship between GPS tolerance and reference elements. From the context of the drawing, it is clear that the RPS points are reference points for creating a reference curve suitable for determining and checking the geometric specifications of the hose. Thus, there is no doubt that the coordinates of RPS points are theoretically accurate, and the reference curve is theoretically accurate and lies in a theoretically accurate position in the 3D space. Any geometric tolerance of the shape and position of this reference curve is not in place here and everything can be uniquely addressed by established procedures in GPS standards. The axis of the actual hose is, of course, not theoretically accurate, but lies within a cylindrical tolerance field defined by the position tolerance of the respective reference element or the tolerated surface shape.

Table 1 is the most sophisticated RPS table used in the technical practice. The first point is referred to as the reference point for the LCS location. The CMM then proceeds from the direction of all three coordinate axes, and all three results must conform to the specifications. Since this is a reference point, it is an element located in the theoretically accurate position and therefore the coordinate tolerances are zero. The second point is in general tolerances and has the shape of the $F\ 5 \times 5$ add-on area. From this, it can be deduced that this is not a point suitable for construction gauges, LER and/or the measuring device, but a point suitable for the construction of cubings [1]. The third and fourth points are measured points. It is different that point 3 has coordinate tolerances and point 4 has geometric tolerances. Again, it is not explicitly specified what type of point it is and what purpose it is intended for and what its use or purpose is. Also, surface tolerance and tolerance field tolerance are of no use since they are only points.

2.3 New RPS Table Proposal

We are trying to solve the problems of existing solutions by proposing a new RPS table (see Tables 2 and 3), which already respects the current needs of technical practice and respects the RPS proposal set out in [1]. The RPS definitions given in [1] are respected and projected into the table header. It should be stressed that such an unplanned RPS point will allow to solve a number of problems that the current system of standards for technical documentation cannot solve satisfactorily. It is important to note that if the introduced tolerance or geometric position of the RPS point is not a reference node but a point for measurement and control.

Function and control dimension GPS point (based on FUN—SPEC/MAN—SPEC)

Table 2 RPS table—proposal of the new table

<table>
<tr><td colspan="3">Drawing sector</td><td colspan="2">Identification code</td><td>Point label</td><td colspan="3">Measurement directions (MDir) or load on the contact direction (LCDir)</td><td>Measurement method</td><td colspan="2">Measuring contact if C-contact geometrical specifications-contact pressure force [other specifications when NC]</td></tr>
<tr><td colspan="3">DSec</td><td colspan="2">ID</td><td>PLbl
$^{MDa}_{MDb}P_n^{Rf/Ms}$</td><td>u_x</td><td>u_y</td><td>u_z</td><td>C/NC contact/non-contact</td><td>$SR\,x(F\phi D(a \times b))$
$[Other\ Spec.]$</td><td>$PFx/da[\%]$
$[Other\ Spec.]$</td></tr>
<tr><td colspan="3">GCS
Absolute coordinates</td><td colspan="4">Touch point or touch field or touch hole and constrain load</td><td colspan="3">Contact indicator</td><td colspan="2">TCS
Target coordinate system</td></tr>
<tr><td>Cx</td><td>ACy</td><td>ACz</td><td colspan="2">$SRx; (F/H\phi D(a \times b))$
$[Other\ Spec.]$</td><td colspan="2">$CLx/da[\%]$
$[Other\ Spec.]$</td><td colspan="3">$[TC/PC/FC/FIX]$</td><td colspan="2">$TCS_{Pn(XA,\ Y[A],\ Z[A],\ rx^{LCS},\ [rx]^{LCS},\ [rx]^{LCS})}$</td></tr>
<tr><td colspan="8"></td><td colspan="4">LCS [ACx, ACy, ACz, r_x^{GCS}, r_y^{GCS}, r_z^{GCS}]</td></tr>
<tr><td colspan="8">LCS/TCS
Related coordinates
[CTF—Technical and Functional Characteristic]
[CSE—Monitored Essential Characteristic]
[SC—Significant Characteristic]
[CC— Critical Characteristic]
[DmbA—Classification of Documentation and Records] <S> Critical Safety Element</td><td colspan="4">Tolerances</td></tr>
<tr><td colspan="3">$x_{LCS/TCS}CTF/CSE^{SC/CC}_{\langle S\rangle;DmbA}$</td><td colspan="2">$y_{LCS/TCS}CTF/CSE^{SC/CC}_{\langle S\rangle;DmbA}$</td><td colspan="3">$z_{LCS/TCS}CTF/CSE^{SC/CC}_{\langle S\rangle;DmbA}$</td><td>$ST_x$</td><td>$ST_y$</td><td>$ST_z$</td><td>| ⊕ | T | A_{PrR} | B_{PrR} | C_{PrR} |</td></tr>
</table>

Table 3 RPS supplementary table—proposal of the new table

Identification code	Feature type	Feature label
ID_F	FType	FLbl
1	Directional vector	$\overrightarrow{DV_n}$
	Line segment	$LS_n^{Rf/Ms}$
	Line straight	$LStr_n^{Rf/Ms}$
	Offset line	$Ofs_n^{BaseLine}$
	Arc tangent curve	$\mathrm{ArcTC}_n^{Rf/Ms}$
	Spline (NURBS)	$\mathrm{SplC}_n^{Rf/Ms}$
m	Offset curve	$Ofs_n^{BaseCurve}$

RPS—controlling P_{na}— primary NP

	P_1	P_2	–	P_n	–	P_m
Nodal Points (NP)	P_{1a}	–	–	$[P_{nb}]$	–	–
Collinear DV (note)	[✓]	–	–	[✓]	–	[✓)
NP	P_{1a}	$[P_{2end}]$	–	–	–	$[P_{mb}]$
NP	–	$[P_2]$	–	P_n	–	–
OfsP	–	$[P_2]$	–	–	–	–
NP/ $[R_T^{P_i,P_s}]$	$\left[P_1/R_T^{P_i,P_s}\right]$	$\left[P_2/R_T^{P_i,P_s}\right]$	–	$\left[P_n/R_T^{P_i,P_s}\right]$	–	$\left[P_m/\mathrm{R}_\mathrm{T}^{\mathrm{P_i,P_s}}\right]$
NP and smooth points	P_1	P_2	–	P_n	–	$[P_m]$
OfsP	–	$[P_2]$	–	–	–	–

Courses LCS/TCS			Segment length	Offset distance	Offset gaptype	Spline curvature type
$\alpha_i^{LCS/TCS}$	$\beta_i^{LCS/TCS}$	$\gamma_i^{LCS/TCS}$	SL	OfsD	OGT	SCT
$[\alpha]$	$[\beta]$	$[\gamma]$	–	–	–	–
–	–	–	–	–	–	–
$[\alpha]$	$[\beta]$	$[\gamma]$	[SL]	–	–	–
–	–	–	–	–	–	–
–	–	–	–	[OfsD]	0/1/2	–
–	–	–	–	–	–	–
–	–	–	–	–	–	0/1/2/3
–	–	–	–	[OfsD]	0/1/2	–

(continued)

Table 3 (continued)

Structural plane modifiers	Tolerances
StPM	GTol
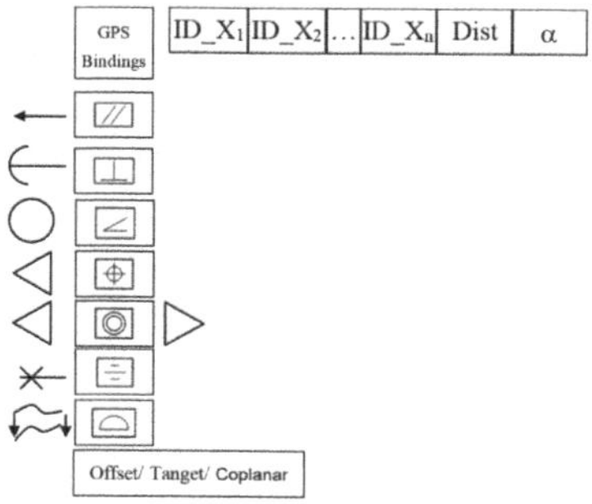	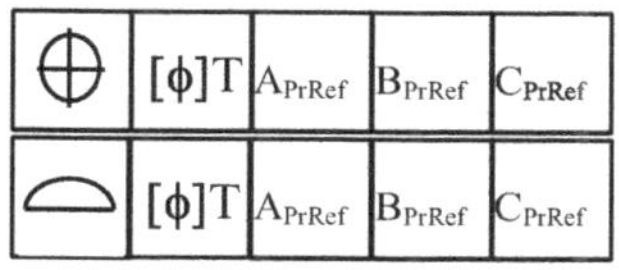

Note Assign the collinear DV to smooth out points

- CTF—Technical and Functional Characteristic of the final product, compliance with which ensures that the product meets all the requirements specified in the technical specification.
- CSE Definition of a Monitored Essential Characteristic. Based on the CTF list, related customer risks, provisional or known complexity of processes and variability, the supplier proposes a CSE list.

3 Conclusions

The proposed system allows not only to specify RPS points more precisely but also to created reference or measured structural elements. These elements can be not only basic lines, segments, curves, planes (see Table 4) but also directional vectors and planes and structural planes (see Table 4). Furthermore, Structural Plane Modifiers (see Table 4) are defined, which allow for a wide variety of structural planes. Another option is to create boundaries of existing surfaces. The boundaries in this case may have any shape and direction in 3D. Obviously, in a similar way as the line elements are defined (see Table 4), it is also possible to define flat elements. This problematics already exceed the scope of this contribution and is therefore no longer mentioned here.

Table 4 Definition structural planes using RPS elements. The meaning of images in CAD systems is given in [2]

Plane	Definition
	Creates the construction plane passing through selected FType
	The median plane of the torus ID_X
	Tangent to the surface ID_Xa across the edge ID_Xb
	Normally to the curve ID_Xa at the point ID_Xb
	Normally to axis ID_Xa through point ID_Xa
	Offset from plane ID_X
	Angle to the plane ID_Xa along the edge ID_Xb
	Tangentially to the surface ID_Xa via RPS point ID_Xb
	Center plane between two planes ID_Xa a ID_Xb
	Parallel to the plane ID_Xa through the RPS point ID_Xb
	Construction plane passing through three RPS points ID_Xa,b,c
	Tangent to the surface ID_Xa and parallel to plane ID_Xb
	Two coplanar edges ID_Xa a ID_Xb

References

1. Vaněk V, Polák R (2017) Geometric specification of complex spatially—oriented and compliant components I. ICMD 2017, pp 414–419. CULS, Prague
2. Dassault Systèmes SolidWorks Corporation, SOLIDWORKS Web Help (2017) http://help.solidworks.com/HelpProducts.aspx
3. Hosnedl S, Vaněk V, Borusíková I (2001) Design science for engineering design practice. In: Culley S et al. (eds) Proceedings of the ICED 01, vol 3, pp 363–370. IMechE, London

Engineering Analyses

Unconventional Carbon Springs

Milan Benko, Ľuboš Kučera, Tomáš Gajdošík and Michal Herda

Abstract The purpose of this contribution is to invent new spring, which will be made of carbon woven material. In this case, we examine spring properties in FEM software intended for testing of composite materials. This research is based on the need of special lightweight spring, which will have long length of compression and excellent resilience. Design of this unconventional spring we want to use as spring with damper in our project for extremely terrain lightweight vehicle. We designed two types of springs. First is made of corrugated carbon strip aligned into square and second is helical conical compression spring of rectangular wire. These springs were tested for the maximum load of 2000 N and we monitored the stress and deformation. Both springs were also tested with different angle direction of forming layers. The article wants to find new field of usage of carbon material through good elastic properties in spring sphere.

Keywords Carbon spring · FEM analysis · Compression · Resilience

1 Introduction

The usage area of carbon composite materials is day by day wider and wider. Problem with this kind of material is in their price and expensive cost of production [1]. Cost of it going to be stable for few years. But sphere of research of carbon construction should become less expensive thanks to FEM analysis computer software [2]. It is also the aim of this study to design new shapes of springs which will be made of carbon material. After design phase, the springs will be tested in FEM software with compressive force. Values of pressure stress and deformation show us the limit of proportionality and so we can find out the resilience of newly designed springs. This can help in the phase of development and design to confirm if the carbon should be satisfying material in the field of springs or not. Also, this study wants to invent new spring shapes that were never used before and see their properties. In the end, we

M. Benko (✉) · Ľ. Kučera · T. Gajdošík · M. Herda
University of Žilina, Univerzitná 8215/1, 01026 Žilina, Slovak Republic
e-mail: Milan.Benko@fstroj.uniza.sk

Š. Medvecký et al. (eds.), *Current Methods of Construction Design*, Lecture Notes in Mechanical Engineering, https://doi.org/10.1007/978-3-030-33146-7_26

will be able to design carbon spring parts, which can stand the operating load based only on simulation results.

2 Methods

For the purpose of the study was used ANSYS Workbench for FEM analysis [3]. For clarification, we used module static structural for load tests and ACP module to define direction of angle arrangement of each layer of carbon material [4, 5]. Material of the fabric used for study purpose was epoxy carbon woven prepreg. This material has different properties at different direction. For example, our chosen one has at X and Y direction Young's modulus 61,340 MPa and in third Z direction has 6900 MPa. Ultimate strength for this material is declared at 230 GPa. But we tested similar material before on real physical tests and we found out that this number is far from true. Selection of prepreg material is based on technology of manufacture for our springs. We designed shapes that should be made with this kind of technology, because prepreg means that fabric is pre-impregnated with a resin system. Resin is mostly epoxy. Prepreg is ready to lay into mold without the addition of any more resin. In order for the laminate cure, it is necessary to use combination of vacuum, pressure and heat during process of curing the carbon fibers in the furnace.

First set (see Fig. 1) of material layers is for final thickness of 1 mm and contains 5 layers of carbon woven. It means that each layer measures 0.2 mm and each layer has different angle direction. In this case, it was from bottom to top surface 0°, 90°, 0°, 90° and 0°. The fibers are arranged in consideration of the applied load. So layer of 0° angle direction is parallel to compressive force and layer of 90° angle direction is perpendicular to this force.

Second set (see Fig. 2) of material has thick of 1 mm and consists of 5 layers. From bottom to top surface, it is 45°, −45°, 45°, −45° and 45°.

Third set (see Fig. 3) of material is also thick of 1 mm and consists of 5 layers.

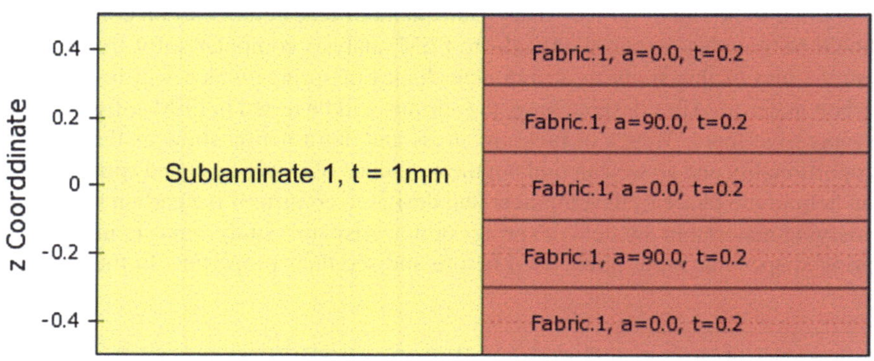

Fig. 1 Angle direction of layers for first set

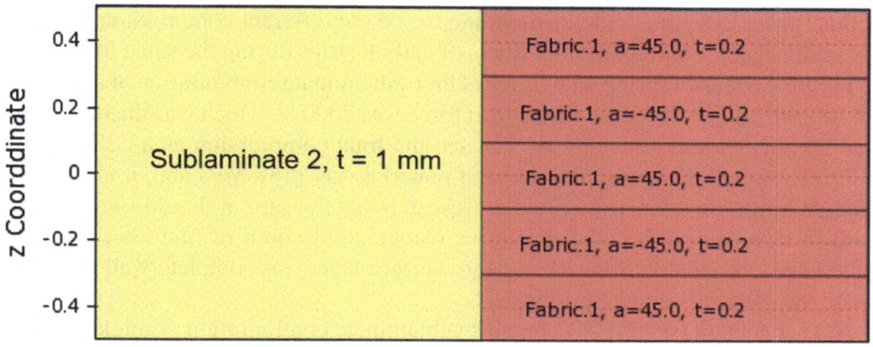

Fig. 2 Angle direction of layers for second set

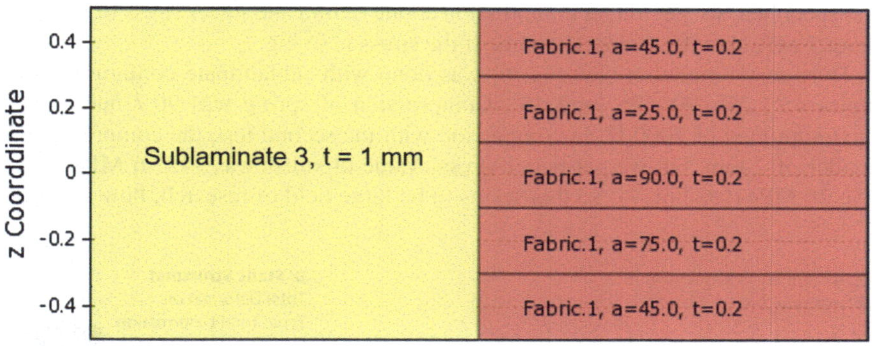

Fig. 3 Angle direction of layers for third set

From bottom to top surface, it is 45°, 25°, 90°, 75° and 45°.

3 Results

3.1 A Spring of Corrugated Strip Carbon Parts

First design of spring is made of more parts. Flexible element of spring is created by a wavy strip, which has the similar shape as corrugated sheet. That comparison is just for imagination. These strips made of carbon are inserted into circular plates with rectangular holes, where holes are made near the edge. Plates serve as a solid component of spring and define a trajectory of compression. Circuit arrangement of strips allows to put a damper in the middle of the circular plates. Damper is not shown in simulation pictures. Geometry of the spring is shown in the picture. Length of this spring is 50 cm, and we are going to see the different compression between different angular configuration of fibers of carbon strips during the same load force.

Figure 4 is shown spring with usage of first sublaminate configuration of angles 0°, 90°, 0°, 90° and 0°. It is case where load force was 2000 N, which was the maximum load for this springs tests and we can see the final compression as a deformation of 77.75 mm. At this point, the stress of material was 1189 MPa and, it is far from strength limit. We know from previous research that the carbon should stand the big loads as this, but on the other hand, we recognize the limit of first fiber cracking somewhere around 400 MPa. That means surface layer was completely all right, but inside the material some fibers crash.

Next specimen (see Fig. 5) contains sublaminate configuration of angles 45°, −45°, 45°, −45° and 45°. It is an universal arrangement of layers, and it is most used layout for carbon parts in the world. At the maximum load of 2000 N, we can see the compression of 92.92 mm and stress was 1182 MPa. In comparison with the first specimen, we get longer compression about 15 mm and lower stress of material about 7 MPa, but it's almost the same if the stress is so big.

Third test simulation (see Fig. 6) was done with sublaminate configuration of angles 45°, 25°, 90°, 75° and 45°. Compression of spring was 90.7 mm during maximum load of 2000 N. In comparison with the second test, the compression is smaller of 2 mm, but we get better stress. Value of stress was 1157.6 MPa so we save 26 MPa (see Fig. 7). So this seems to be large field of research, how to adjust

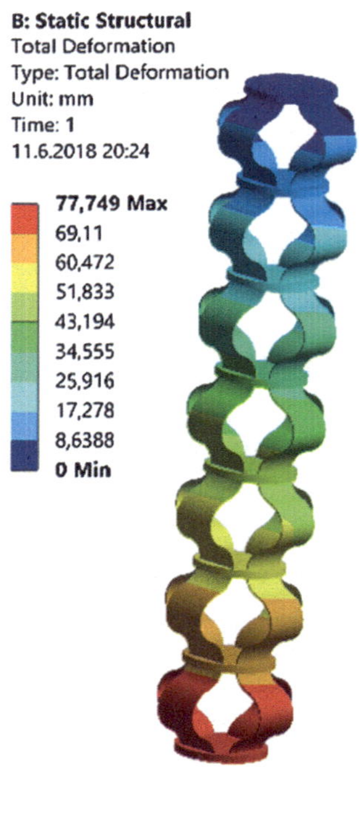

Fig. 4 Spring compresion of sublaminate 1

Fig. 5 Spring compression of sublaminate 2

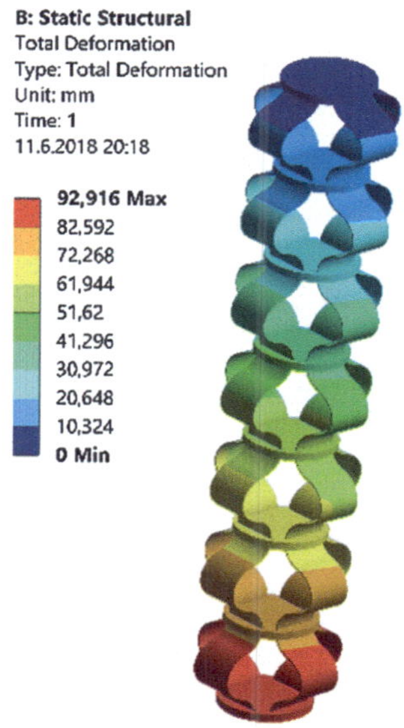

Fig. 6 Spring compresion of sublaminate 3

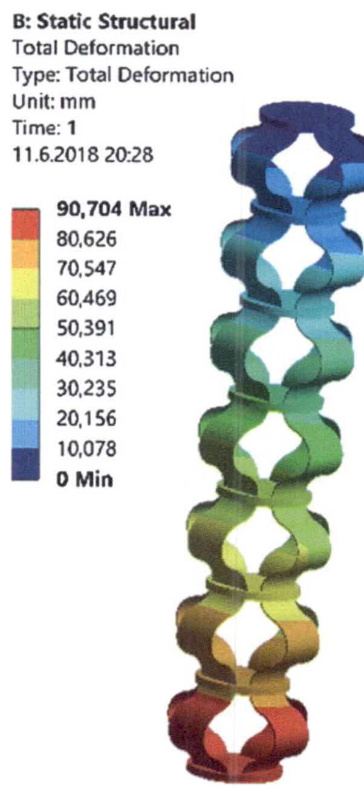

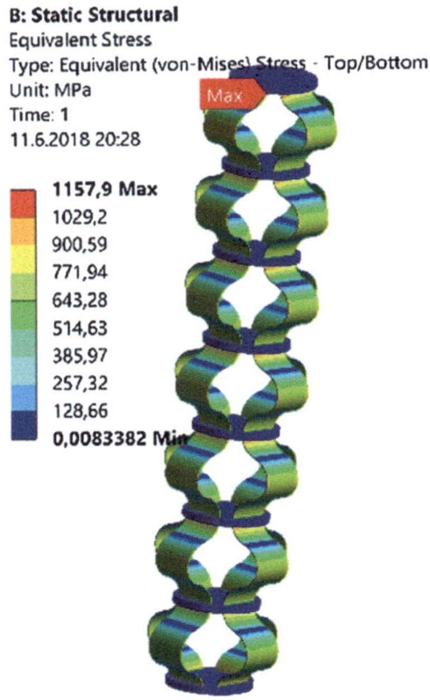

Fig. 7 Equivalent stress of sublaminate 3

the right angle configuration for specific load. As we can see if we mix the angle of fibers, we can get better condition from this material as with usual 45° angle. Size of load was chosen on base of conditions for our project of all-terrain vehicle for extreme usage.

3.2 *Helical Conical Compression Spring of Rectangular Wire*

This shape of spring is based on helical conical compression spring. At this case, the conical wire has rectangular cross section and is wounded in vertical position. Width of wounded strip is 30 mm and thickness is 1 mm as it was described before. Greater thickness should be too strong for elastic purpose of use as a spring. Thicker wall should be better, but we find out that this wall thickness can be sharp as blade and the usage will be dangerous [6].

The first test spring (see Fig. 8) has sublaminate configuration of angles 0°, 90°, 0°, 90° and 0°. It is case where load force was 2000 N, which was the maximum load for this springs tests and we can see the final compression as a deformation of 20.54 mm. At this point, the stress of material was 292 MPa. The deformation and stress increased linearly. That means during half load of force 1 000 N the deformation was 10.2 mm and stress von misses was 146 MPa. This in fact is feature of carbon material [7].

Second test simulation (see Fig. 9) was done with sublaminate configuration of

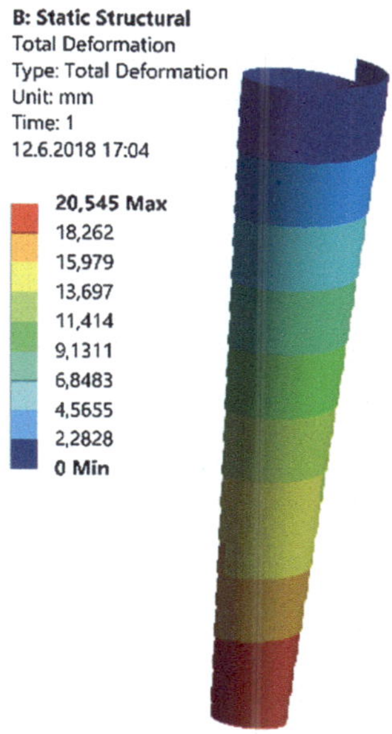

Fig. 8 Spring compresion of sublaminate 1

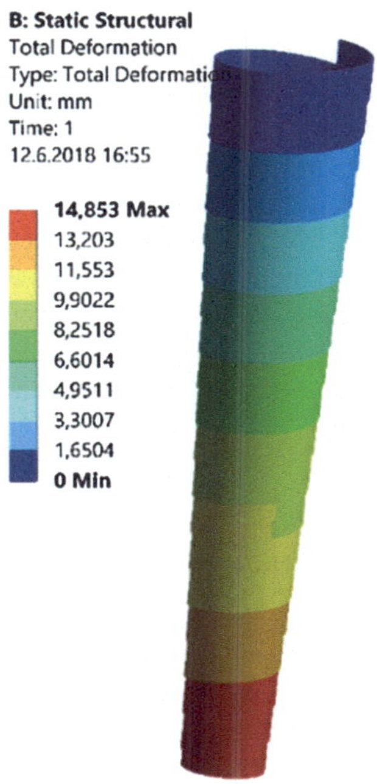

Fig. 9 Spring compression of sublaminate 2

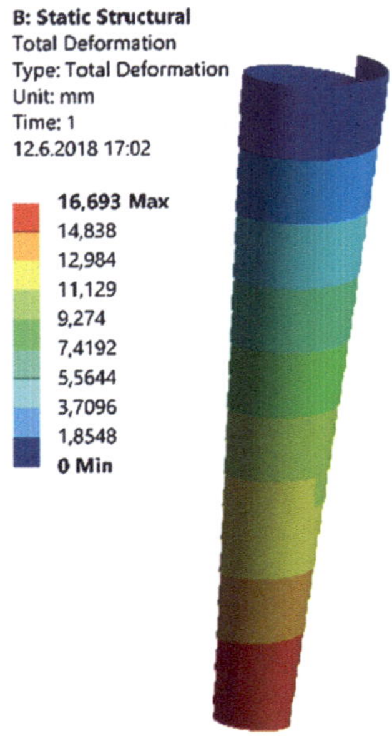

Fig. 10 Spring compresion of sublaminate 3

angles 45°, −45°, 45°, −45° and 45°. Compression of spring was 14.85 mm during maximum load of 2000 N. In comparison with the first test, the compression is smaller around 5.5 mm, but we get better stress. Value of stress was 283 MPa so we save 9 MPa to the detriment of shorter compression.

Last simulation (see Fig. 10) contains sublaminate configuration of angles 45°, 25°, 90°, 75° and 45°. At the maximum load of 2000 N, we can see the compression of 16.69 mm and stress was 309.13 MPa (see Fig. 11). In this case, we get the worst results: shortest compression and biggest stress. It is interesting because first spring made of carbon strips had with this angle arrangement the best result and on the other side for this helical conical spring is material with this kind of structure totally useless [8].

4 Conclusion

The evidence above shows that different angular configuration of carbon fibers in material should improve the material features. As we could see the first spring made of elastic carbon strips had best properties with sublaminate of 45°, 25°, 90°, 75° and 45°. This mixture was designed by us to try to find different layers with good conditions for our purposes [9]. However, what we did not expect was that our angle configuration failed during testing conical helical spring. Better properties had material with angles of 0°, 90°, 0°, 90° and 0°. It follows that every specific component should have its own structure depended on load and carbon should be

Fig. 11 Equivalent stress of sublaminate 3

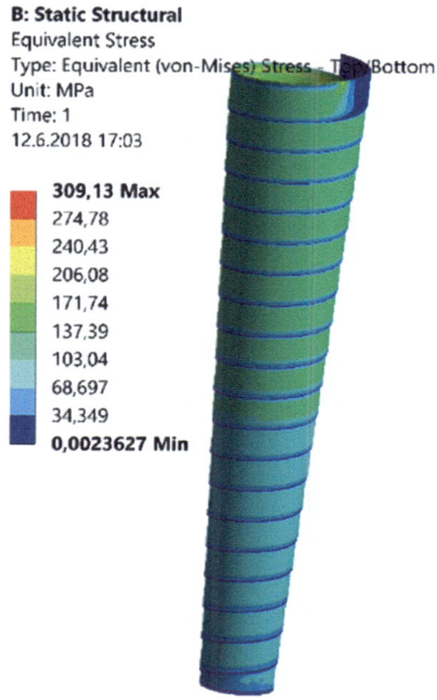

used in future in field of making springs to save weight and to get better features [10].

Acknowledgements The research is supported by the Cultural and Educational Grant Agency of the Ministry of Education, Science, Research and Sport of the Slovak Republic under the project No. KEGA 046ŽU-4/2018.

References

1. Kyriakides S, Arseculeratne R, Perry EJ, Liechti KM (1995) On the compressive failure of fiber reinforced composites. Int J Solids Struct 32(6–7):689–738
2. Mazov IN, Kuznetsov VL, Krasnikov DV, Rudina NA, Romanenko AI, Anikeeva OB, Suslyaev VI, Korovin EY, Zhuravlev VA (2011) Structure and properties of multiwall carbon nanotubes/polystyrene composites prepared via coagulation precipitation technique. J Nanotechnol 2011, Article ID 648324, pp 1–7
3. Lukáč M, Brumerčík F, Krzywonos L, Krzysiak Z (2017) Transmission system power flow. Commun Scientif Lett Univ Žilina 19(2):27–31
4. Coleman JN, Khan U, Blau JW, Gun'ko YK (2006) Small but strong: a review of the mechanical properties of carbon nanotube-polymer composites. Carbon 44:1624–1652
5. Andrews R, Berkovich A, Hower JC, Jacques D, Rantell T (2001) Fabrication of carbon multiwall nanotube/ polymer composites by shear mixing. University of Kentucky, USA
6. Jedlinski L, Caban J, Krzywonos L, Wierzbicki S, Brumerčík F (2015) Application of vibration signal in the diagnosis of IC engine valve clearance. J VibroEng 17(1):175–187

7. Medvecká I, Biňasová V, Kubinec L (2017) Planning and performance evaluation of the manufacturing organizations. Procedia Eng 192:46–51
8. Bašťovanský R, Tropp M, Lukáč M, Brumerčik F (2017) Molybdenum sheet metal test device. Commun Scientif Lett Univ Žilina 19:124–127
9. Kohár R, Hrček S (2014) Dynamic analysis of a rolling bearing cage with respect to the elastic properties of the cage for the axial and radial load cases. Commun Scientif Lett Univ Žilina 3A:74–81
10. Gramblička S, Kohár R, Stopka M (2017) Dynamic analysis of mechanical conveyor drive system. Procedia Eng 192:259–264

Impact of the Spin Cycle on the Acoustic Power Level of a Washing Machine

Marek Bisták, Štefan Medvecký, Michal Belorit and Ján Ďungel

Abstract This contribution describes the impact of the spin cycle of a washing machine on its acoustic power level and examines the adverse effects occurring during the spinning. Such effects result in an increase in the rated acoustic power level shown on the rating plate of a particular major appliance type. To enable an objective assessment of this problem, a series of measurements is required since many of those effects are accidental and occur sporadically. The rating declaration for a particular type of washing machine is based on three repeated measurements made on three sample units. The measured values are then used in the final calculation according to the relevant standard, in a manner that allows the repeatability of measurement. The wash cycle ends with the final water discharge, which is followed by a series of spin phases at different speeds and varied drum imbalances. The test code for the determination of airborne acoustic noise foresees the specification of a 50-second interval of the acoustic power maximum profile for use in the declaration calculation. If the measured acoustic power levels differ between the repeated measurements, it can reasonably be supposed that an unwanted effect is present that needs to be identified and eliminated to avoid a further increase in the total acoustic power level of the washing machine during the spin cycle.

Keywords Acoustic diagnostics · Source of noise · Declaration measurement

1 Introduction

Comparing or analyzing the acoustic performance of an automatic washing machine during operation is a very difficult assignment, which is particularly due to changes

---

M. Bisták · Š. Medvecký (✉) · M. Belorit
University of Žilina, Univerzitná 8215/1, 01026 Žilina, Slovak Republic
e-mail: stefan.medvecky@fstroj.uniza.sk

M. Bisták
e-mail: marek.bistak@fstroj.uniza.sk

J. Ďungel
CEIT Engineering Services, s.r.o, Univerzitná 8413/6, 01008 Žilina, Slovak Republic

Š. Medvecký et al. (eds.), *Current Methods of Construction Design*, Lecture Notes in Mechanical Engineering, https://doi.org/10.1007/978-3-030-33146-7_27

in the operating conditions of the appliance. During a spin cycle, the drum stops and then restarts to take up speed, which causes the redistribution of laundry in the drum. The highest speed is reached during the last spin phase at the end of the cycle and it is, therefore, necessary to minimize the imbalance of the drum during that phase. Further, it is also important to ensure that water is drained from the machine at the right time and that the spinning frequency is not identical to the intrinsic frequencies of the washing machine [1].

2 Design of Experiments for an Objective Assessment of the Effects Occurring in the Different Phases of a Spin Cycle

The measurements were performed in a certified acoustic laboratory (Fig. 1), and their aim was to assess the acoustic noise emissions generated in the different phases of a spin cycle. The acoustic noise emissions measurement is mentioned in [2, 3].

Measurements conditions [1]:

- Laboratory: Anechoic acoustic chamber
- Measuring system: PULSE system, B&K 3053-B-120
- Microphones: B&K 4190
- Calibrator: B&K 4231
- Accelerometers: B&K 4517
- Weight of laundry: 5.02 kg
- Acoustic camera: Noise Inspector
- Appliance program setting: Spin 1200 rpm.

3 Declaration Measurement

The procedure involves three repeated measurements made on three sample units. The values for the wash and spin cycles are recorded separately and, after evaluation, entered in the rating plate. The final spin cycle comprises a number of phases during which the drum reaches the spin speed, water is discharged and the drum stops. After each phase, a drum balancing system is activated whose function is to achieve an even distribution of laundry over the circumference of the drum.

During the spinning action with the maximum acoustic power level, a 50-second steady-state interval is taken and the noise values from this interval are used in the final report. The interval normally corresponds to the maximum spin speed. If this is not the case, it is an indication of a spinning action-related problem. The proper diagnosis and elimination of the problem will lead to a reduction in the overall acoustic power level.

Fig. 1 Position of the washing machine within the anechoic chamber

Figure 2 shows the acoustic power profile for the four phases of the spin cycle. The measured imbalances of laundry weight distribution in the drum vary at the different speeds between 0.28 and 0.82 kg. The maximum acoustic power profile over the 50-second interval was recorded during the second spin phase (650 rpm), while the target interval should be associated with the last spin phase at the maximum speed of 1170 rpm (Fig. 3).

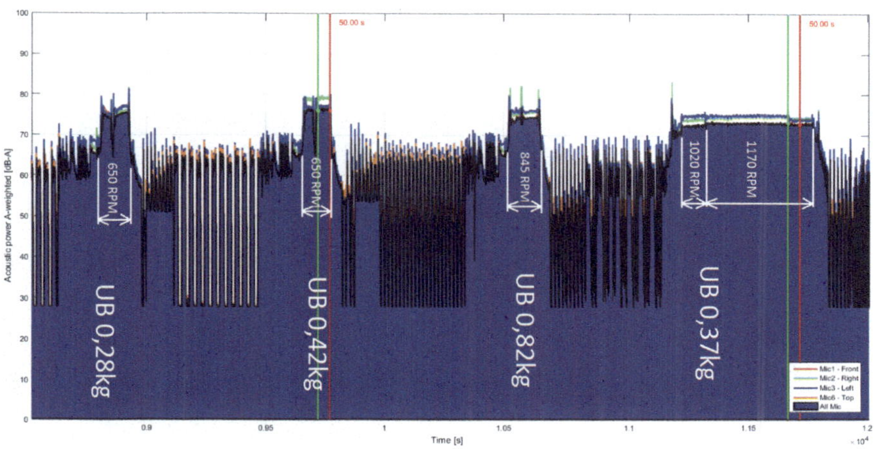

Fig. 2 Acoustic powers at different spin speeds

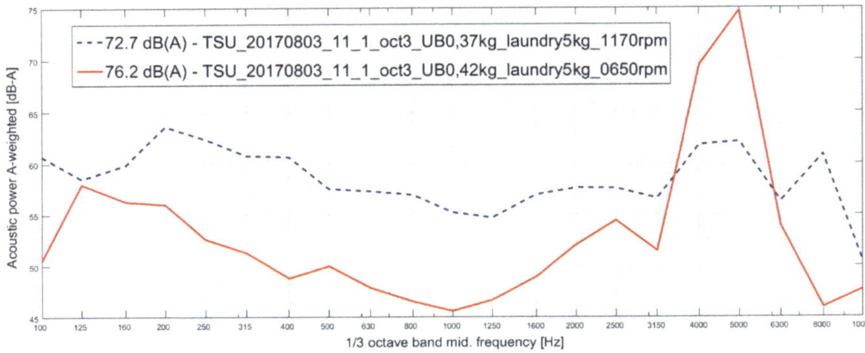

Fig. 3 Specific one-third octave FA at the second and third final spin phases

The one-third octave frequency analysis indicated that a certain proportion of the increase in total acoustic power was caused by frequencies around 4000–5000 Hz, while the variation of imbalances of the drum assembly occurring in the different spin phases was low (Fig. 3).

Placed appropriately on the washing machine drum assembly (Fig. 4), the sensor measured speeds with an error of 0.6% and laundry weight distribution imbalances with an error of ±0.01 kg.

It clearly follows from the comparative measurements (Table 1) of the washing machines 1, 2 and 3 that an increased drum imbalance generates a higher acoustic power level at the maximum spin speed. From measurements 7–9 for the washing machine 3, the drum imbalances were similar, and so were the resulting total acoustic power levels (Fig. 5).

With laundry evenly distributed over the circumference of the drum, noise during the spinning is strongly supposed to decrease as much as 1.7 dB(A), as seen between the measurements 6 and 9.

Fig. 4 Sensor placement on the drum assembly

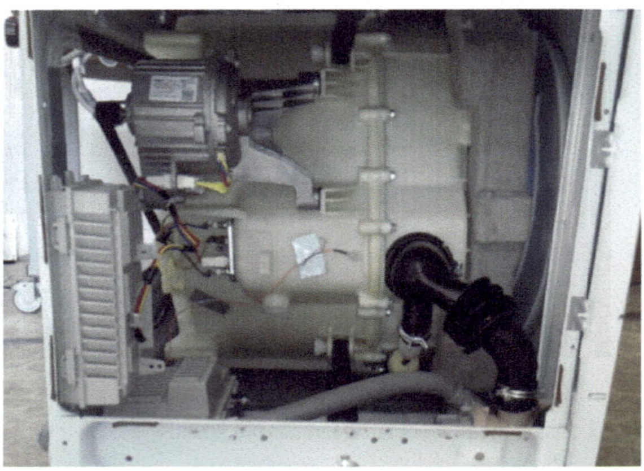

Table 1 Result of measurement comparisons

Measurement	Washing machine	Load	Weight imbalance (kg)	Drum speed (rpm)	Acoustic power level (dB(A))
1	1	Laundry	0.21	1163	73.9
2	1	Laundry	0.33	1163	73.2
3	1	Laundry	0.64	1163	74.6
4	2	Laundry	0.28	1163	74.1
5	2	Laundry	0.43	1163	74.5
6	2	Laundry	0.12	1163	73.2
7	3	Laundry	0.47	1163	74.7
8	3	Laundry	0.5	1163	74.6
9	3	Laundry	0.59	1163	74.9

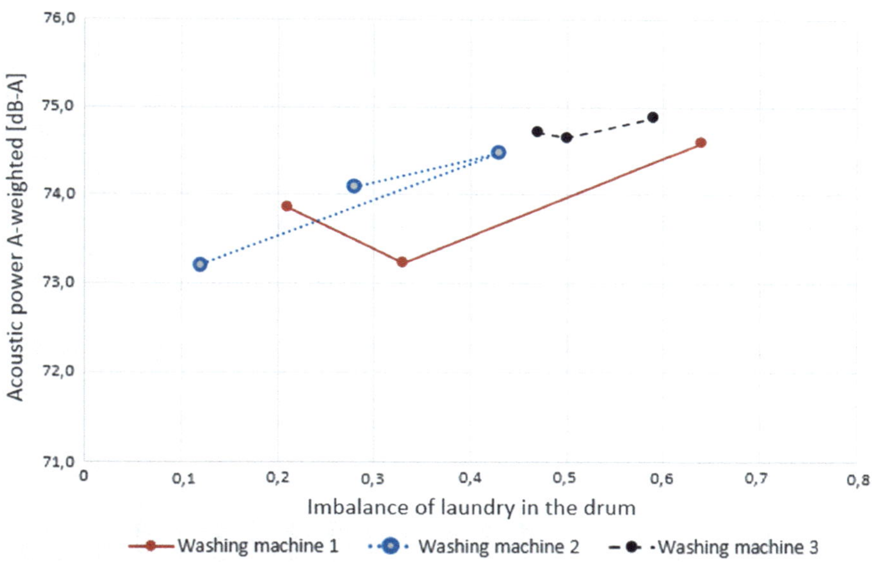

Fig. 5 Total acoustic power at the different imbalances of laundry in the drum at maximum spin speed

Measures aiming to optimize an existing product (reduction of frame stiffness, replacement of mufflers and springs, use of smaller sheets for the housing, etc.) [4] may reveal various acoustically relevant defects that were not detected at the development stage. Such defects may be driven by cost-saving goals, since with the large-series production involved, even a small saving has a substantial impact on the total annual manufacturing costs. Many product models made by the same manufacturer feature a very similar, if not even identical, structural design [5]. As a result, most washing machine manufacturers resolve any detected major imbalances occurring during the spin cycle by reducing the spin speed to avoid worsened noise performance of the products. The methods and accuracy of detection of laundry weight distribution imbalances, as the critical factor, and of the spatial distribution of laundry in the drum are very important elements in reducing the noisiness of washing machines. During washing, as laundry tumbles in the drum and the rotating speed increases, the centrifugal force causes the laundry to spread over the circumference of the drum [6–8]. The distribution of laundry in the drum is random, and this effect occurs several times during the spin cycle. After each spin phase, the drum stops and then restarts to spin with different laundry imbalances, as shown in Fig. 2.

Figure 6 indicates the one-third-octave band of the washing machine 1 at different weight distribution imbalances with laundry positioned mostly centrally in the drum at 1163 rpm. The recorded spectrum frequency shift indicates that a growing weight imbalance drives higher frequencies, tending to approach 1600 Hz. A significant frequency shift, from the 125 and 160 Hz regions to levels close to 200 and 250 Hz, was observed at the weight imbalance of 0.64 kg.

The best practice is using a special microphone set for the detection and identification and an acoustic camera for the subsequent graphical representation of the acquired acoustic data along with the object being measured. The acoustic camera record may reveal numerous faults of the appliance (such as insufficient contacts between the washing machine housing parts and assembly errors) which contribute to the noisiness of a washing machine. Figure 7 indicates that at 630 Hz, acoustic energy is transmitted through the front door where the water discharge pump is placed. As soon as all such faults are cured, it is possible to proceed to designing noise control measures; it should be noted that besides the frequency spectrum that clearly determines the noise damping material to be used, another important factor in noise control is the paths by which sound is transmitted from the appliance. This will

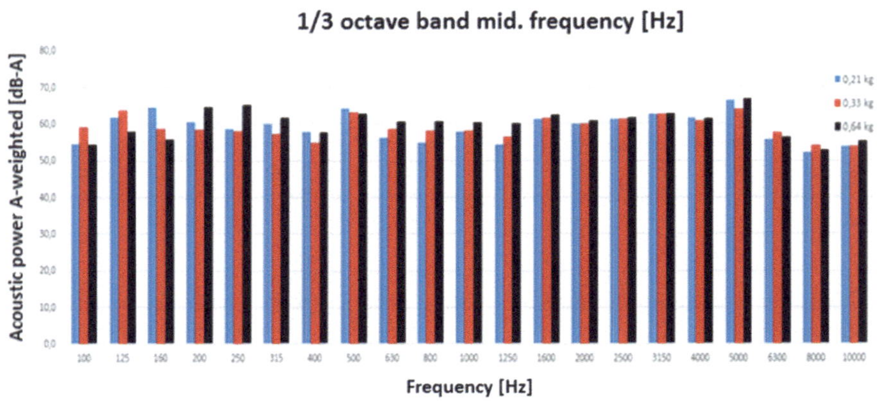

Fig. 6 "Specific" one-third octave FA and the resulting acoustic power values in dB(A) at different imbalances of the washing machine 1

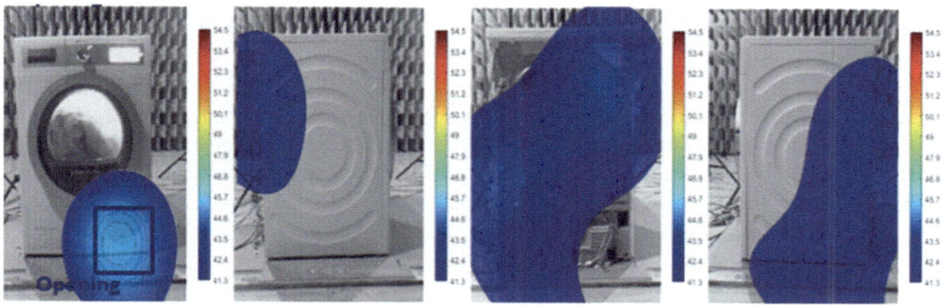

Fig. 7 Record of Measurement 1 for the washing machine 1, made by an acoustic camera at 630 Hz

prevent loss of materials for the manufacturer, in addition to allowing many effective combinations, such as for the placement of soundproofing insulation on the panel of the product.

4 Conclusion

The maximum acoustic power level observed during the spin cycle was 76.2 dB(A) at 650 rpm. During the final spin stage, at the spin speed of 1170 rpm, the acoustic power level of 72.7 dB(A) was achieved.

The analysis of the two spin cycles revealed that most of the increase in the acoustic power level was not due to laundry imbalances, but due to insufficient water discharge, causing the accumulation of a "water ring" around the drum. This causes frequency to rise to 3000 Hz, as shown in Fig. 3, which significantly contributes to the higher overall acoustic power value.

A large drum assembly imbalance was observed during the third spin phase and, therefore, a modification of the balancing system has been proposed to prevent the occurrence of laundry weight distribution imbalance exceeding 0.5 kg, which was observed during the third spin phase. If this value was achieved during the final, high-speed spin cycle, a substantial increase in noisiness could be expected.

It can reasonably be assumed that if these undesirable effects are eliminated by modifications of the program of the appliance, the standardized section of the maximum acoustic power level will correspond to the highest-speed spinning action occurring during the final spin stage.

The random distribution of laundry in the drum has a significant impact on the overall noisiness of the washing machine during the spin cycle. Measurement deviations will increase along with growing imbalances of the drum during the spinning action, which may lead to the declaration of a higher acoustic power level in the rating plate of the appliance. This may adversely affect the sales of the product. It is, therefore, advisable to make efforts to improve the accuracy of weight distribution imbalance detection before the spin cycle and ensure that before taking up speed for the SPIN cycle, the washing machine first distributes the laundry weight in the drum as evenly as possible.

Acknowledgements This article is a result of the implementation of the "Innovation Technology Knowledge Centre for Innovation of Production Systems in Industry and Services", project (ITMS code: 26220220155) supported by the Research and Development Operational Programme co-funded by the European Regional Development Fund.

References

1. STN EN 60704-1 (2010) Acoustics. Household and similar electrical appliances. Test code for the determination of airborne acoustical noise. Part 1: General requirements
2. Poljak S, Gajdáč I (2014) Acoustic Camera, Ai magazine. J Autom Ind Mech Eng Econ 7(3):86–87
3. Poljak S, Gajdáč I (2014) Acoustic camera and her utilize in the praxis. Technológ 6(3):23–26
4. Bisták M, Medvecký Š, Kohár R, Hrček S (2016) The jib crane structural design using the TRIZ method. In: ICMD 2016: book of proceedings of the 57th international conference of Machine design departments, Železná Ruda, Czech Republic, pp 9–14. University of West Bohemia in Pilsen, Pilsen
5. Lukáč M, Brumerčík F, Krzywonos L, Krzysiak Z (2017) Transmission system power flow model. Commun Scientif Lett Univ Žilina 19(2):48–53
6. Jedliński Ł, Caban J, Krzywonos L, Wierzbicki S, Brumercik F (2015) Application of vibration signal in the diagnosis of IC engine valve clearance. J VibroEng 17(1):175–187
7. Weis P, Kučera Ľ, Pecháč P, Močilán M (2017) Modal analysis of gearbox housing with applied load. Procedia Eng 192(2017):953–958
8. Kučera Ľ, Gajdošík T (2014) The vibrodiagnostics of gear. In: Ševčík L, Lepšík P, Petru M, Mašín I, Martonka R (eds) ICMD 2013, Modern methods of construction design. Lecture Notes in Mechanical Engineering. Springer, Heidelberg, pp 113–118

Sensor for In-time Identification of Deep Core Drilling Parameters

Martin Dub and Vojtěch Dynybyl

Abstract The deep core drilling has currently wide application—construction drills, geological exploration, drills for measuring pendulums in dams, etc. It is very important and useful to know the drilling process parameters precisely. The most important is the used torque, pressure and rotation speed of the drilling machine necessary for the optimal drilling process. These parameters depend on the strength of the drilled material, for example, rock, sand, clay, etc. and of course on the drilled diameter and applied pressure. In-time identification of the drilling parameters enables the optimization of the drilling process directly during the performance. Furthermore, it can serve for the real data acquisition, which could be used for later drilling planning in similar geological environment. Obtained data are also very useful for designing new drilling machines. They could be designed according to their future usage to the optimal performance. This contribution deals with the design of the new sensor for in-time identification of above-mentioned core drilling parameters. The sensor is based on the strain gauge measurement and enables online monitoring of the drilling process. The measured data are also stored and could be post-processed in longer time context to obtain also the drilling capacity of the drilled material.

Keywords Core drilling · Sensor · Measurement · Strain gauge · Monitoring · Process optimization

1 Introduction

The deep core drilling has currently wide application—construction drills, geological exploration, drills for measuring pendulums in dams, etc. [1]. It is very important and useful to know the drilling process parameters precisely. The most important is the used torque, pressure and rotation speed of the drilling machine necessary for the optimal drilling process. These parameters depend on the strength of the drilled

M. Dub (✉) · V. Dynybyl
Czech Technical University in Prague, Technická 4, 16607 Prague 6, Czech Republic
e-mail: martin.dub@fs.cvut.cz

V. Dynybyl
e-mail: Vojtech.dynybyl@fs.cvut.cz

Š. Medvecký et al. (eds.), *Current Methods of Construction Design*, Lecture Notes in Mechanical Engineering, https://doi.org/10.1007/978-3-030-33146-7_28

material, for example, rock, sand, clay, etc. and of course on the drilled diameter and applied pressure. In-time identification of the drilling parameters enables the optimization of the drilling process directly during the performance. Furthermore, it can serve for the real data acquisition, which could be used for later drilling planning in similar geological environment. Obtained data are also very useful for designing new drilling machines. They could be designed according to their future usage to the optimal performance. This contribution deals with the design of the new sensor for in-time identification of above-mentioned core drilling parameters. The sensor is based on the strain gauge measurement and enables online monitoring of the drilling process. The measured data are also stored and could be post-processed in longer time context to obtain also the drilling capacity of the drilled material.

2 Methods and Materials

2.1 Concept Design

The designed sensor should be able to identify the drilling parameters—torque, pressure and rotation speed. These quantities will be wirelessly transmitted to the device for displaying and storing, usually the laptop or tablet. The concept sketch of the sensor usage can be seen in Fig. 1.

At first, it is necessary to analyze the types of drilling machines which should be used in combination with the designed sensor. This range of machines is given by the cooperating company drilling hardware but it could be easily extended for other types of drilling machines to get the universally usable sensor. All the scoped drilling machines are used for core drilling. The drill bit is connected with the drilling machine by means of the rod system. The output of the drilling head is usually formed by the hollow shaft with inner hexagon 55 mm. Single drilling rods are connected also by the inner and outer hexagon 55 mm and they are axially locked by the pins, see Fig. 2. These drilling rods have various lengths according to the used drilling machine. Hexagonal joint is typical for core drilling and it should not be changed. That is why the new sensor is designed to be compatible with this type of connection.

As it was mentioned, the main goal is to identify the drilling parameters—torque, pressure and rotation speed. Strain gauge measurement will be used for torque and pressure identification. The deformation member has to be designed and tested. This deformation member is to be in accordance with the connection type because it has to be a part of the drilling column. It has to satisfy the limited installation dimensions including the protective case for the electronic parts of the sensor.

The deformation member is designed with an emphasis on maximum simplicity and robustness because of the demanding conditions during the drilling process. It is designed by the hollow tube with welded inner and outer hexagon (see Fig. 3) for the interconnection with the drilling head and the rest of drilling column. This solution should not influence the strength and stiffness (both torsional and axial) of the drilling column.

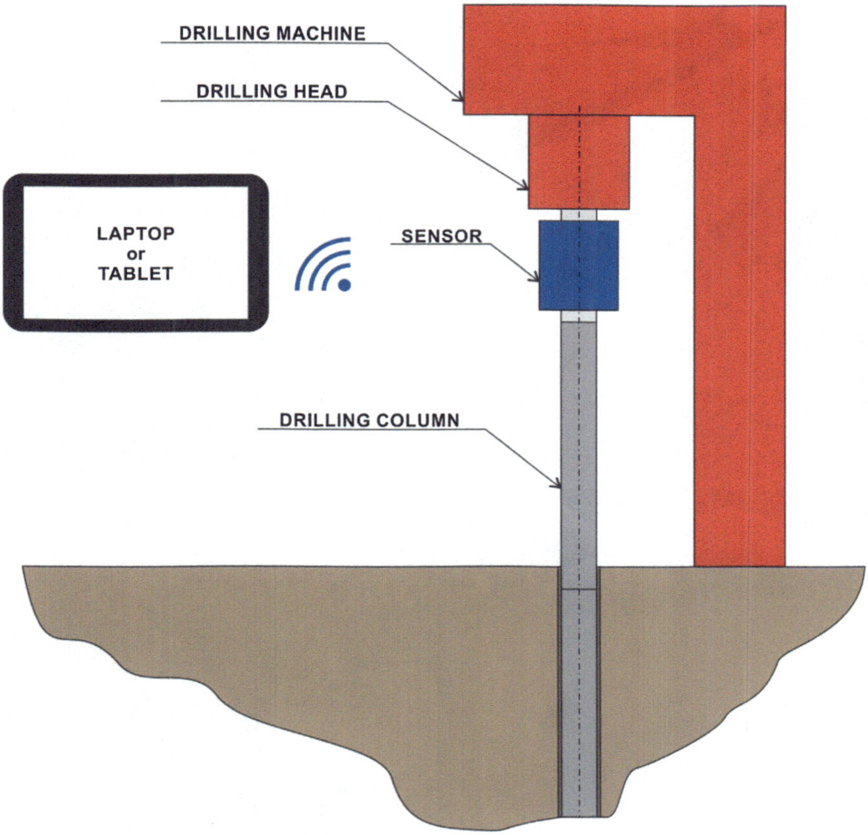

Fig. 1 Concept sketch of the designed sensor usage including the drilling machine

2.2 *Data Acquisition and Telemetry*

Measurement of the torque is realized by means of the foil strain gauges connected to the Wheatstone full bridge to get the maximum sensitivity and temperature compensation. Axial force, especially the pressure, is measured also by the strain gauges. For this purpose, the Wheatstone half-bridge is used. For measuring the rotation speed, a Hall probe is used. Measured data have to be processed, wirelessly transmitted and also stored. To fulfill these goals, special data logger is developed (further designation LPDR) in cooperation with the company CleverTech. Furthermore, the LPDR has to have compact dimensions and low consumption of electric power. LPDR can be seen in Fig. 4.

The LPDR is set to three-channel acquisition with sampling frequency 500 Hz per channel. The number of channels can be increased up to five channels with sampling frequency up to 1 kHz. The power voltage of the LPDR has to be from 3.3 to 5 V. The telemetry is realized by means of the Digi XBee 2.4 GHz transmission. The size

Fig. 2 Drilling rods—connection by means of the inner and outer hexagon

Fig. 3 Deformation member of the designed sensor

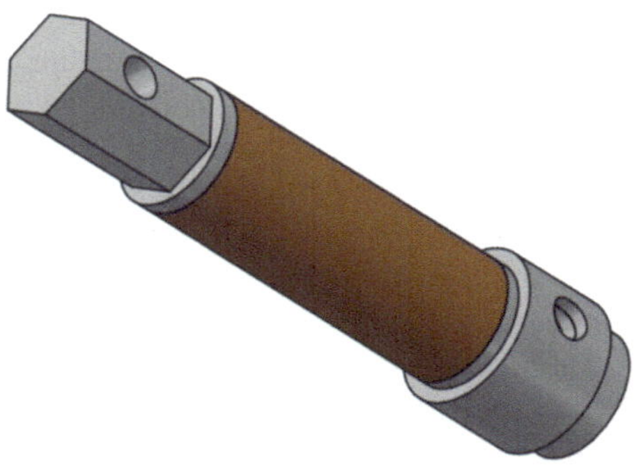

Fig. 4 Special data logger (LPDR) with XBee receiver for data acquisition, transmission and storage

of the LPDR is 70 × 55 × 25 mm with weight approximately 165 g. Measured data are processed, transmitted and also stored on the micro SD memory card. Measured quantities are transformed into the corresponding units of torque, axial force and rotation speed. These data can be then immediately and easily evaluated by the drilling machine operator, and according to this, the drilling process can be optimized.

3 Results

Before the manufacturing the deformation member the laboratory testing was carried out. The main goal is to verify new hardware LPDR and also test the sensitivity. For this purpose, the torque measurement was chosen. Test deformation member is manufactured from the part of used drilling rod. It is the hollow tube with outer diameter 76.5 mm and thickness 6 mm. Foil strain gauges HBM XY41-6/120 are installed in the full-bridge arrangement on the deformation member to measure torsion, see Fig. 5. The test deformation member is then embedded into the test rig for pure torque loading, see Fig. 6. This rig works on the lever principle.

The weights of 19.8 kg are hung up on the 1 m long arm so the torque up to the approximately 1100 Nm can be produced. Loading torque is measured by the LPDR and by the National Instruments PXI-1073 system as a verification. The raw data acquired both by the LPDR and NI PXI-1073 are recalculated into the torque according to the following formula:

$$M_k = \frac{2W_k G U}{k U_b}, \tag{1}$$

where W_k is the torsional constant, G is the shear modulus, U is the indicated voltage of the bridge, k is the k-factor of the strain gauges and U_b is the excitation voltage of the bridge.

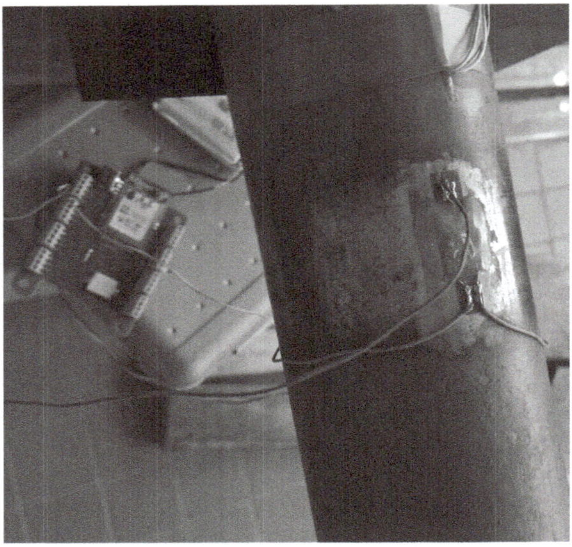

Fig. 5 Strain gauges for torsion measurement connected to the LPDR

Fig. 6 Test rig for the torque loading

4 Discussion

There is a very good agreement of the torque measured by means of LPDR and NI PXI-1073 measuring system, see Fig. 7. Only small difference, maximum 3.9%, from the analytical calculation occurred. This deviation is caused by the imperfection of the tube inner diameter, because even very small difference of the diameters will strongly

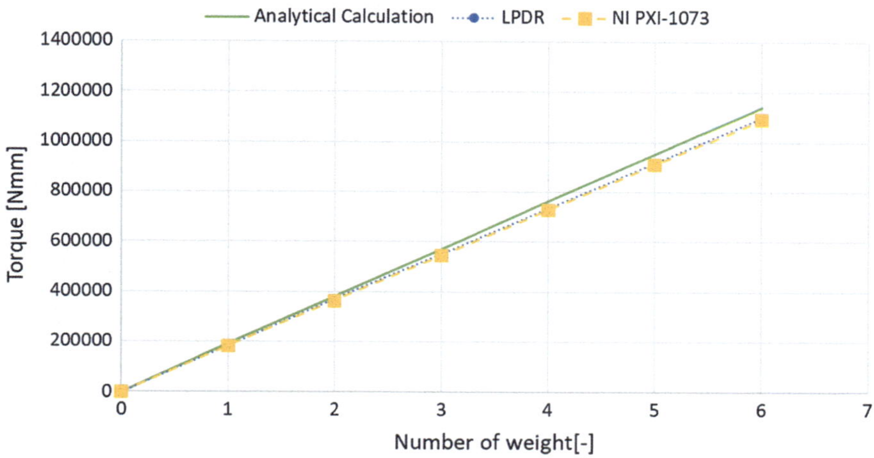

Fig. 7 Comparison of calculated torque and measured values by means of LPDR and NI PXI-1073

influence the torsional constant W_k. Test deformation member is manufactured from the old drilling rod, which has very rough and rusted inner surface. The machining of this surface would have been very complicated. New deformation member will be manufactured for the final sensor so that very small deviations are supposed.

5 Conclusions

The new concept of the sensor for in-time identification of deep core drilling parameters was introduced in this proceeding. This sensor can serve for the optimization of the deep core drilling process and also for long-time data acquisition. These data could help in designing new drilling machines in the future. The basic design and verification experiments are actually done. Further work will focus on the final deformation member design and also on the design of the protective cover. After that, the sensor will be tested with various drilling machines in various geological areas to prove its versatility.

Acknowledgements This work was supported by The Technology Agency of the Czech Republic project No: TH02010958 "Sensor for measuring of operational parameters of deep core drilling".

Reference

1. Dub M, Janda M, Mrázek J, Kavka M, Dynybyl V (2017) Core drilling machine for perfectly vertical drills. In: Herák D (ed) Proceedings of 58th International conference of machine design departments—ICMD 2017, pp 66–68. Czech University of Life Sciences Praha, Praha

The Energy Assist for the Electric Car Edison

Igor Gajdáč, Tomáš Gajdošík and Ján Šteininger

Abstract One of the biggest barriers to the development of electromobility is the fear of people, because electric car has a significantly shorter range than conventional car. In modern electric cars to reduce electricity consumption and the range, complex technologies are used as energy assist. The intelligent energy assist role is to optimize the way of driving to minimize energy consumption by means of video and audio communications with the driver. The energy assist is not only predicting the range calculated from the difference between the received and consumed energy. Energy assist controls, manages, updates, and communicates with the electric car driver. It even foresees a compromised run to the nearest charger, so the driver warns and proposes austerity measures. Energy assist performs the tasks before driving, and then throughout the drive to the charging station. The aim of our research and development is the energy assist of the electric car driver. Modern components in the Edison electric car, built at the Department of Design and Mechanical Elements, enable the collection of a wide range of data, allow for changing set parameters, monitor their impact on energy consumption and measurements on the roller dynamometer, and verify the change of parameters on the road in various driving and climatic conditions.

Keywords Electric car · Energy assist · Edison

1 Introduction

Range of electric car has long been considered a major barrier in acceptance of electric mobility due to electric car having a significantly shorter range than conventional car. The range of electric car influences not only the design of the car but also driving style and operational factors. The main design factor is battery capacity. On the other hand, battery size influence car curb weight and price. Right size of battery pack can also extend range. If electric car is using, for example, to commute to work and daily route not exceed 50 km it is wasteful using electric car, for example, with 24 kWh battery pack because for 50 km distance can be used electric car with half battery

I. Gajdáč (✉) · T. Gajdošík · J. Šteininger
University of Žilina, Univerzitná, 8215/1, 01026 Žilina, Slovak Republic
e-mail: igor.gajdac@fstroj.uniza.sk

Š. Medvecký et al. (eds.), *Current Methods of Construction Design*, Lecture Notes in Mechanical Engineering, https://doi.org/10.1007/978-3-030-33146-7_29

capacity. The electric car is then lighter and has lower electric energy consumption. Nowadays modular design is becoming new trend. The modular design allows the customer to choose the appropriate size of the battery pack. Climatic conditions have significant impact to the electric car range. The range is decreasing with extremely low or high temperatures. This problem is manufacturers trying to reduce by using various cooling or heating systems for battery packs. Manufactures of electric car sometimes use an electric pre-heat function to warm up the battery during charging. Technical condition of the car is also important especially accurate tire pressure. The range affects also using of comfort features such an air condition or heating. The driving style has significant influence to the electric car range. Although electric car and car with internal combustion engine may appear to be identical, drivers must adjust their driving habits to fit the different mode of driving required by the electric car. It is beneficial using coasting or recuperation of kinetic energy during braking. One of the possibilities of how to influence the driving style is a system which gives driver the information on how to behave in order to reduce energy consumption and extend range. Such a system is currently developing at the University of Žilina. The modern electric car has only systems that evaluate driving style. The proposed system will inform driver how to change driving style immediately.

2 Energy Consumption Monitoring in Electric Car

Energy consumption monitoring of the electric car can be divided into three groups of indicators [1].

- Design indicators—the most important negative indicator is weight which can be affected during the car design,
- Operational indicators—to some extent can influence the driver, for example, if intelligent traffic management system helps him with the choice of optimal route in heavy city traffic, then it is the impact of climatic conditions and the environment character in which the electric car moves,
- Driver impact—to the range of the car and recommendations: driving style (minimizing unnecessary braking and acceleration, which are the most critical in terms of consumption, smooth driving, suitable distance behind the car, using recuperation, and coasting), preparing car for drive (removing useless load-weight, tire pressure monitoring, care of the technical condition, and itinerary of charging stations), and use of electrical equipment of the car (wise use of air conditioning, heating, and other electrical devices) (Table 1).

Table 1 Breakdown of consumption monitoring indicators

Design indicators		Operational indicators	Driver impact
Influence on road resistors	**Design structure of modules**	– Overcrowded city traffic – The road surface – Select a rout – Climatic conditions	– Driving style – Preparing car for drive – Use of electrical equipment of the car
Rolling resistance: – Car **weight** – Tire design – Road quality *Air resistance:* – Aerodynamic shape – Driving mode *Climb resistance:* – Car **weight** *Resistance to acceleration:* – Car **weight**	*Pre-rated battery capacity:* – **Weight** – Price – Product life *Incorrect choice of components:* – Low efficient energy recuperation – Little power – Minimal compatibility		**Additional power sources to extend the car range** – Energy recuperation – Energy from photovoltaic panels – Another range extender

3 Energy Assistant Tasks from the Start of the Journey to the Charging Station

Standard onboard computers in standard-drive cars and in electric cars after counting of refueling fuel, or charging of batteries, they measure fuel consumption (l/100 km) or electric energy (kWh/100 km) The value of car range is continuously updated according to the amount of fuel in the tank or the remaining charging state of the batteries. The aim of our research and development is the EA for the electric car Edison. Modern components in the Edison electric car are capable of collecting a wide range of data, enabling them to change set the parameters, and monitor their impact on energy consumption. The data can be verified on the roller dynamometer or on the road. The role of the EA is not only to establish the value of the car range calculated from the difference between the received and consumed energy. Energy assist assists, controls, manages, updates, and communicates with the electric car driver. Predicts the risk of being discharged before drive range to charging station and warns the driver and proposes austerity measures to the nearest charger, so the driver warns and proposes energy-effective measures. EA performs the tasks before driving, and then throughout the drive to the charging station (Fig. 1).

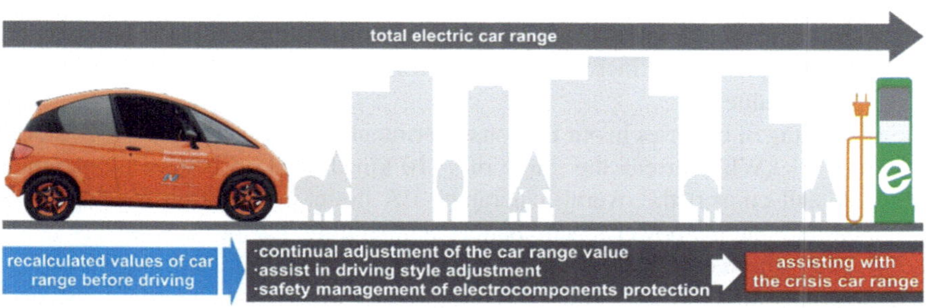

Fig. 1 Energy assistant tasks through the total car range

Fig. 2 Development of electric car Edison

The role of the intelligent energy usage assist is audiovisual communication with the driver in order to optimize the driving style and minimize energy consumption. The proposed energy usage assist system will be applied into the experimental electric car Edison. The energy usage assist will be alone system which will be receiving the data from the car control unit and the battery management system. Integrated GPS receiver will be used for detecting speed limits. Based on powertrain measured efficiency maps, system will inform driver through display how to behave.

The Edison is experimental electric car build at University of Žilina. The curb weight of the experimental car including the battery is 1048 kg. Propulsion provides a compact lightweight all-aluminum, air-cooled asynchronous electric motor AKOE with a nominal output of 16 kW and maximum power of 30 kW, with motor controller Curtis and traction LiFeYPO4 24 kWh battery pack with the battery management system and onboard charger 110–240 V/16 A. The main part of the car is a tubular steel space frame. EV Edison enables data logging of a wide range of data such as speed, acceleration, temperatures, state of charge, and information from battery management system (Fig. 2).

4 Tasks of the Energy Assist (EA) Before Driving Electric Car

4.1 Recalculation of Driving Range on Climate Conditions

When operating an electric car at low temperatures below freezing, the low temperature has an impact on the increase in energy consumption, which means shortening the car range value.

Self-heating of batteries in our Edison experimental electric car with a total stored energy of 24 kWh shortens the range of 7–10 km. Based on battery temperature data; EA will convert the overall car range. EA limits power from heat recovery and maximum engine power to safe battery temperature. EA protects under-cooled batteries from high currents.

Low temperature reduces battery performance due to increased internal resistance. On the other hand, increased resistance speeds up battery heating [2]. The Edison

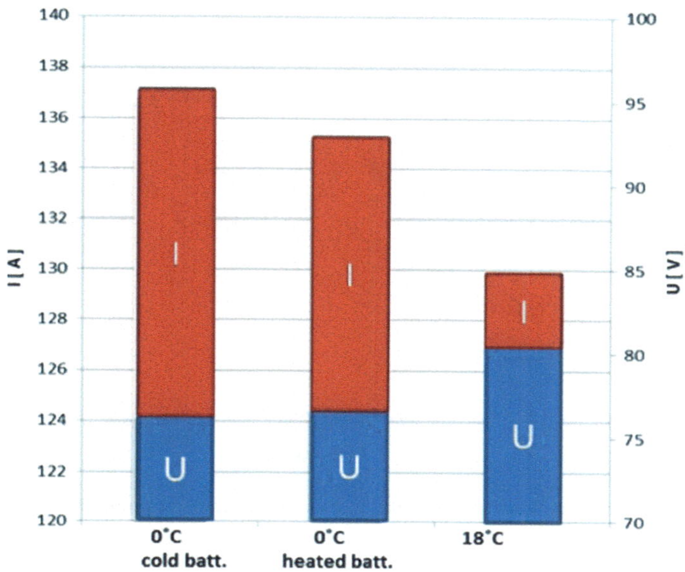

Fig. 3 Impact of different climatic conditions on traction batteries properties

electric car has a LiFeYPO4 battery with a capacity of 300 Ah and a traction voltage of 80 V. Low temperatures reduce battery performance and, in addition, the need to heat the vehicle also takes off energy, which consequently reduces downtime. The first results from the electric car test in the winter prove this (Fig. 3).

Edison was parked for several days, and the temperature did not rise above 0 °C. During the tests, the two-member crew ran the same route in the premises of the University of Žilina. The low temperature has resulted in increased energy consumption and, therefore, a reduction in the onset of travel. A comparison of consumption during test runs shows that if and when the electric car is gradually heating up, energy consumption is reduced. At first, the consumption was 18.5 kWh/100 km, at the last 14.9 kWh/100 km. For a better comparison, three cases were chosen when the car needs the same power output of 10.4 kW from its batteries. In the first case, the batteries were cold at start-up, in the second case the cells were heated because the battery was loaded for 10 km of driving. In both the first and second cases, the ambient temperature was 0 °C. In the third case, the ambient temperature was 18 °C. Figure 4 shows that at low temperatures, increased internal resistance causes a significant drop in voltage, which means an increased current gauge. This means that the electric car in the winter at the same power takes a higher current from the battery.

In order to eliminate this phenomenon, electric car manufacturers use battery heaters. Nissan Leaf has a direct-on-battery heating system. Tesla acquires heat for heating the accumulators from the heat emitted by the electric motor. This solution is more efficient because it does not use battery energy, but uses heat that would otherwise not be used. Since the electric motor has a high efficiency it takes a few

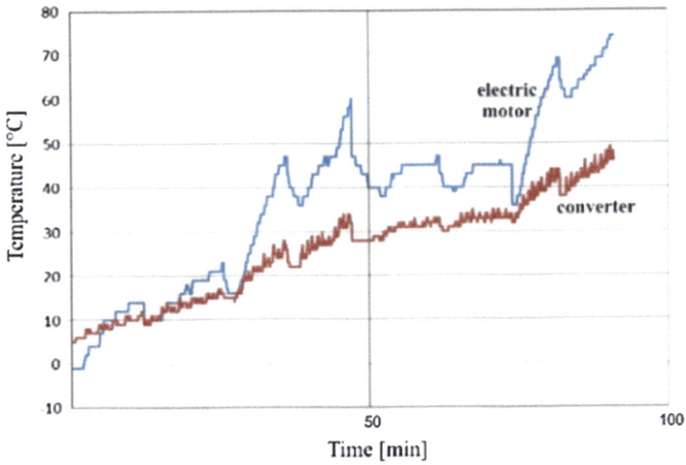

Fig. 4 Impact of different climatic conditions on traction accumulator properties

minutes to heat the batteries to the temperature when they can deliver full power. EV Edison does not have a battery heating system. We have analyzed heat recovery options for battery heating.

The picture shows the temperature of Edison's electric motor and transducer. Temperatures were recorded during 8 runs that lasted 1.5 h. During these rides, the style of defensive sport was changed, which is also evident from the fluctuations in temperature. Ambient temperature was 0 °C. The electric car has gone 13.4 km. The temperature of the electric motor reached 74 °C at the end of the run and the inverter temperature was 46 °C.

4.2 Recalculation of Driving Range on the Selected Route

The route selection and traffic indicators are determined by the number of planned stops, traffic density, and so on. The useable input value may be more accurate when intelligent traffic management data is available. EA corrects the value of the runtime based on the database of available digital maps. At the same time, the system allows you to perform your own measurements on commonly used routes.

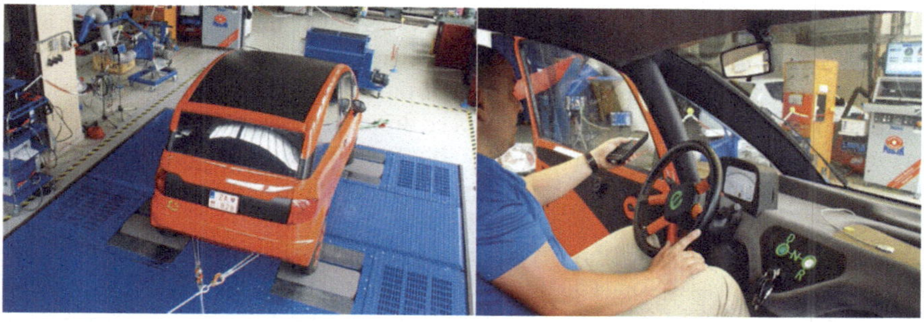

Fig. 5 Electric car Edison on the roller dynamometer MAHA

5 The Tasks of the Energy Assist (EA) While Driving

5.1 Driving Style Assistance—Coasting, Energy Recuperation, Braking

Driving style is purely a subjective indicator of the influence of the driver on the car range. From the point of view of energy intensity, it is most advantageous to slow down, for example, when entering the village or stopping at a crossing to take coasting of the vehicle. Do not use the brakes, but at the right moment to use the kinetic energy of the vehicle, its speed reduction or complete stop if the traffic situation requires it.

Driving route profile and traffic situation, vehicle weight and speed, relevant recuperation effect, and ramp curve database; this is just a few of the main input parameters for the decision process of the EA system, which communicates in the audiovisual communication with the driver using appropriate picture information, supplemented if necessary with sound. The aim is to give the driver information on how to handle the smallest electricity consumption (Fig. 5).

5.2 Driving Style Assistance—Acceleration and Powertrain Efficiency

Acceleration is another subjective indicator of driving style. It is more energy-intensive than run-up. With an appropriate combination—correcting the time needed to overcome a certain distance and mapping the overall efficiency of the electric drive; it is possible to optimize the engine start. EA, like the one used in the race, in addition to the required parameters, uses the map of the overall efficiency of the electric drive in order to decide and inform the driver whether it is in the most efficient mode when it is started—in the best efficiency mode [3], which means the least energy-intensive start.

In the first phase of the development, the system will inform the driver if he is currently in the most efficient mode during acceleration—highest efficiency mode. If we assume that the driver needs to accelerate from an intersection in the city. System will use GPS position to evaluate the end speed during acceleration to 50 km/h. Based on powertrain efficiency map the system will inform driver by graphic indication on the dashboard which smooth acceleration mode he should choose quick or slow. This will help the driver to achieve the optimal mode.

It was created analysis of three types of accelerations (Fig. 8) from 0 to 50 km/h. Traveled distance was 200 m at all three accelerations. Duration of acceleration 1 was 27 s; acceleration 2 was 22 s and acceleration 3 was 19 s. Five measurements were done for each acceleration test. The average value of electric energy consumption was 0.069 kWh for acceleration 1, 0.072 kWh for acceleration 2 and 0.072 kWh for acceleration 3 (Figs. 6, 7 and 8).

This measurement demonstrated that in the EV Edison case is preferably accelerate aggressively and then use coasting. Although the vehicle needs for faster acceleration more energy to overcome drive resistances, the EV Edison has higher overall powertrain efficiency at higher loads. This caused that the EV Edison has almost same energy consumption during slow and fast accelerations. It is more effective to use coasting then recuperation during deceleration. The disadvantage of coasting is a higher time to slow the vehicle. This disadvantage could be compensated by faster acceleration. It is not influence energy consumption due to EV Edison powertrain

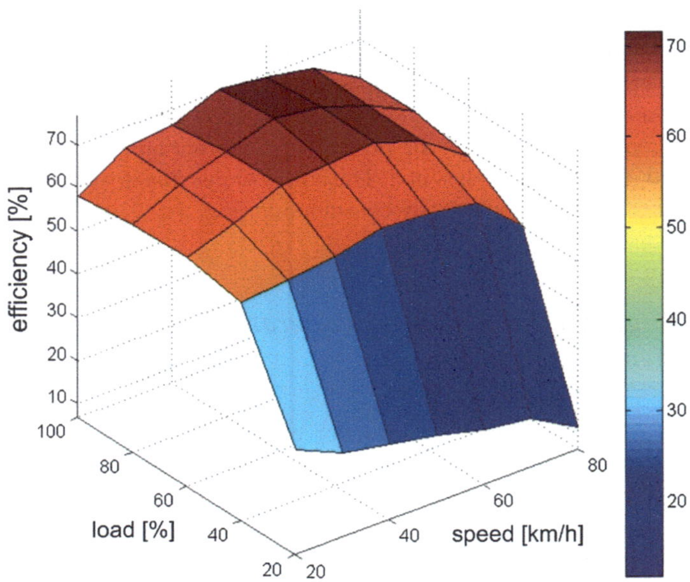

Fig. 6 Efficiency map of drive Edison

Fig. 7 Dashboard information during acceleration

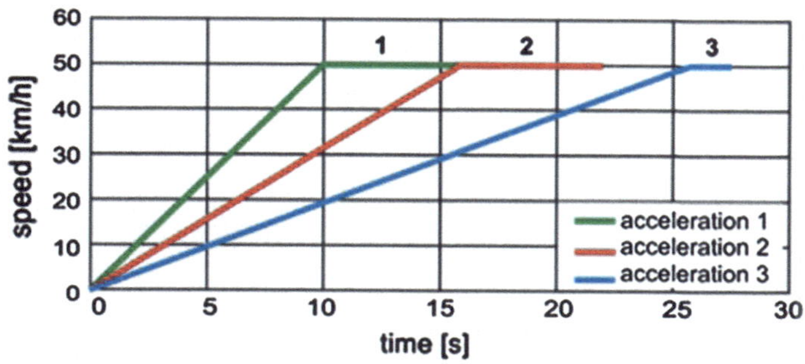

Fig. 8 Accelerations profiles

efficiency. Faster acceleration and then use coasting driving style can reduce energy consumption.

5.3 *Driving Style Assistance—Depending on the Way of Driving and Consuming Electricity*

Updating the remaining action radius requires continuous monitoring of electricity consumption. The monitoring has three basic indicators: operational (under what conditions the car is moving), the subjective influence of the driver (driving style), and the use of other appliances (vehicle lighting, heating) or the use of energy sources (energy recuperation, photovoltaic panels) (Fig. 9).

EA updates the run-up value based on the current stored energy value in traction batteries (SOC) and the monitoring of the power consumption mode. This is not the final value. If EA is working with data on the current electric car movement route, the value will be recalculated again. At present, we measure the relationship between the discharge characteristic of the traction accumulators used and the capacity drop in the areas of 80% SOC and 20% SOC, which has a direct effect on the vehicle's interior.

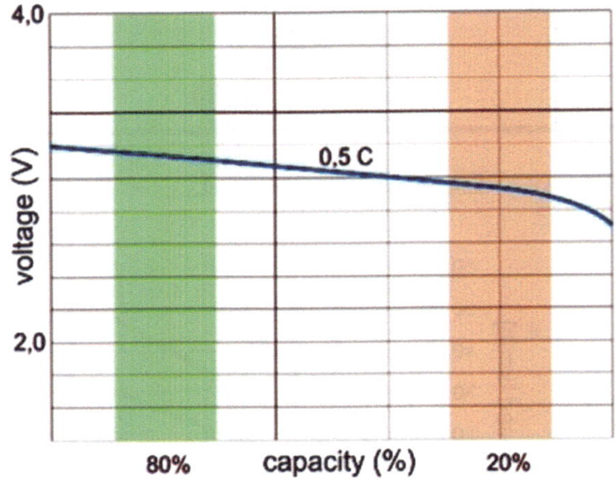

Fig. 9 Example of battery discharging characteristics in Edison electric car

6 Control and Protection of Electrocomponents

It is not directly assisted by the driver of an electric car, but it is a condition of EA's operation. To enable EA, for example, to assist the driver in adjusting the style of driving and to regulate the conversion of electrical energy to traction, it must control all the appliances in the vehicle, including the ability to independently switch off selected systems. This is also indirectly controlled by their status. The protection of electrocomponents is filled with EA, for example, at low and high temperatures of batteries, electric motors, and converter. Our goal is to get the driver through the EA communication panel not only updated driver information, driving advice, but also warnings and notifications of security interference in the system, for example, at low battery temperatures.

7 Energy Assist in Crisis Situations

The best case for understanding the assisting in a particular crisis situation is an example. AE identifies the location of the nearest charging station on our route. At the same time, based on our current route, based on the profile of our route, and the traffic situation, we compare the theoretical possible approach with the distance of the nearest charging station. Comparison is ongoing. When a critical condition occurs EA decides to assist and recommends driver savings (Fig. 10).

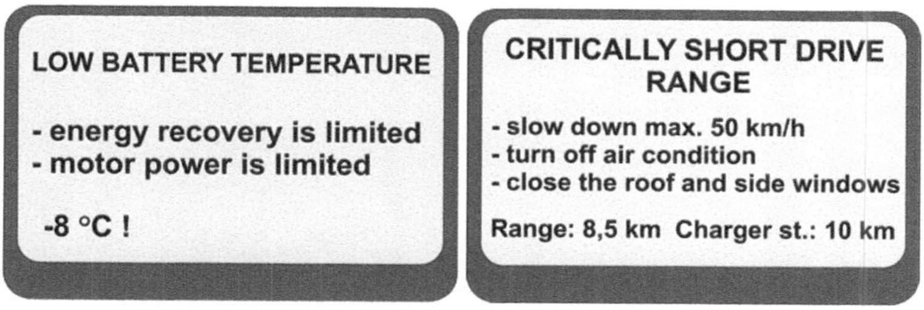

Fig. 10 Example of EA safety alert when battery temperature is low and assistance in crisis situations

Acknowledgement This publication is the result of the Project implementation: Competency Center for Knowledge technologies applied in Innovation of Production Systems in Industry and Services, ITMS: 26220220155, supported by the Research & Development Operational Programme funded by the ERDF.

References

1. Kučera L, Gajdáč I, Mruzek M (2016) Energeticý asistent vodiča elektromobilu. Automot Ind Mag 9(3):83
2. Kučera L, Gajdáč I, Mruzek M (2016) Výskum a vývoj energetického asistenta vodiča elektromobilu. Automot Ind Mag 9(5):62–63
3. Málik L, Hrček S (2016) Všeobecný model výpočtu prevodových mechanizmov so stupňovou a plynulou zmenou prevodového pomeru. ŽU v Žiline, Žilina

Contact Analysis of Selected Toothed Contact of the Two-Stage Front Gearbox

Stanislav Gramblička, Róbert Kohár, Maroš Majchrák and Michal Vrabec

Abstract The article deals with the contact analysis of selected toothed gears of the two-stage front gearbox. They are extensively used throughout the machine industry, but also in other mechanical devices where input parameters change to the required working values. The aim of this work is to point out the effect of deformations of the individual gear components on the correctness of the toothed engagement. The work deals with the contact analysis of selected gears of the two-stage front gearbox. The subject of the study is to determine the resulting coefficient of unevenness of the tooth load over the width of the contact for deformation of individual parts of the gearbox using the KISSsoft computerized software. The pc software KISSsoft has several ways to define a coefficient $K_{H\beta}$. The method using the ISO 6336 standard was used in this case. Due to load, the teeth of the wheels resiliently deform. This deformation has an effect in an enlargement of the coefficient of the toothed contact. The result is the processing of data and their evaluation for individual types of deformations of the parts of the gearbox unit and the subsequent evaluation of the overall impact on the life of the gearbox.

Keywords Analysis · Deformation · Coefficient

1 Introduction

The toothed gears have the ability to transmit maximum performance then, when the shafts of the toothed gears are perfectly arranged, and the load transmitted is evenly distributed across the entire active tooth width. The so how as the load decomposes on the surface the side of the tooth; it greatly affects the life of the gearing. Incorrect load distribution results in inception of a place on the surface of the tooth with high contact stress, resulting in fatigue wear or the fracture fatigue. Manufacturing and assembly inaccuracies, elastic deformations of individual gearbox components, thermal deformations and external influences can cause tooth misalignment, which affects the load distribution on the sides of the teeth. During the assembling parts of

S. Gramblička · R. Kohár (✉) · M. Majchrák · M. Vrabec
University of Žilina, Univerzitná, 8215/1, 01026 Žilina, Slovak Republic
e-mail: robert.kohar@fstroj.uniza.sk

Š. Medvecký et al. (eds.), *Current Methods of Construction Design*, Lecture Notes in Mechanical Engineering, https://doi.org/10.1007/978-3-030-33146-7_30

the gearbox, a check should be made to avoid unexpected damage during operation. An important consideration of the contact analysis is the so-called coefficient of unevenness of the tooth load after the width in contact $K_{H\beta}$ [1–5].

2 Methods and Materials

2.1 *Definition of the Coefficient* $K_{H\beta}$ *Without Shafts Deformation*

The software KISSsoft includes several ways of defining a coefficient $K_{H\beta}$. If the value is known, it can be entered manually, which is the first method. Another method of defining the coefficient is, depending of the used the calculating norm, in this case it is the specific norm DIN 3990 method B (method C), which is relate from the norm ISO 6336 method B, consequently it contains the approximate procedure. For the contact analysis, a third gearing unit was selected, which consists of a z4 a pinion gear in the middle shaft and a gear a z6 output shaft wheel. When determining the $K_{H\beta}$ coefficient according to the DIN norm, the dimensions of the pinion shaft are included, but the driven shaft geometry in this case is not included. The specified pinion shaft values are the distance between the bearings l, the distance of the pinion from the center of the distance between the bearings *s*, and the outer diameter of the shaft between the bearings *d*. Another parameter that the norm takes into account is the pinion shaft type. That is defined based on the norm ISO 6336 (type *a* as much as *e*). The values given for the calculation of the $K_{H\beta}$ coefficient for the third gear are $l = 243$ mm, $s = 70.5$ mm, and $d = 62.256$ mm. With this the values so defined the coefficient is defined $K_{H\beta} = 1.1525$. In this calculation, only the geometry of the pinion shaft is taken into account and the geometry of the output shaft is not included. Due to load, the teeth of the wheels resiliently deform [2, 6, 7].

Figure 1 shows the distribution of the applied load on the width of the tooth, with the red color showing the location with the highest contact stress. Places with the maximum stress occur at the beginning of the gear engagement and at the center of

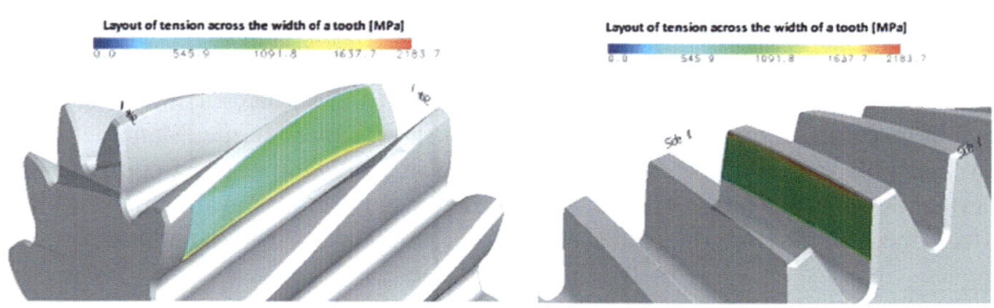

Fig. 1 Layout of tension across the width of tooth at left of pinion and at right of a wheel

the gear engagement, while the maximum tension on the pinion occurs at the root of the tooth. The wheel has maximum tension on the head of the tooth, at the beginning, and at the center of the tooth width. Such a stress distribution is not ideal because the blue color area, the stress 545 MPa, should be located outside the spacing circle. In this case, the medium contact stress, a green color with a voltage of about 1091 MPa, is located in the head and the root of the tooth from the center to the end of the toothed engagement.

2.2 *Definition of the Coefficient* $\mathrm{K}_{H\beta}$ *with Shafts Deformation*

The toothed gears transfer performance-causing stress and torque load on the shafts what caused their elastic deformations. These deviations may affect the toothed wheel engagement and hence the effect on load distribution on the side of the tooth. A more precise definition of the $K_{\mathrm{H}\beta}$ coefficient is the application of the norm ISO 6336-1 Annex E. Annex E of this norm takes into account the load distribution on the tooth width between parallel axes of the cogwheels. This method focuses on the most important deviations, namely bends and twisting of the shafts and the elastic deformation of the wheel teeth. The determination of the $K_{\mathrm{H}\beta}$ coefficient was carried out based on the norm ISO 6336-1 Annex E with by considering deflections of shafts of a pinion, wheels bending, and tooth bending. The shaft deflection of the pinion under the tooth gear has value 0.0852 mm, and the output shaft deflection of the pinion under the tooth gear has value 0.0634 mm. The size of twisting angle in the pinion shaft in the place toothing is 0.0016°, and the size of twisting angle in the output shaft in the place toothing is 0.0126°. This calculation does not take into account manufacturing tolerances. The value of the coefficient of the $K_{\mathrm{H}\beta}$ in this case is 1.2155 [2, 6, 7].

Figure 2 shows the distribution of the applied load along the width of the tooth, with a considerable difference in results compared to the result without considering of the shafts deflection. The contact area of the pinion its contents a larger area of pale

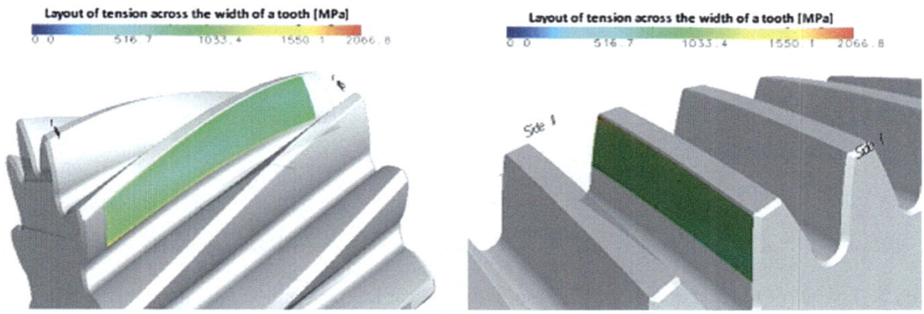

Fig. 2 Layout of tension across the width of the tooth at left of the pinion and at right of the wheel

blue color, meaning a contact voltage of about 517 MPa and a distinctive green color area, representing a voltage of approximately 1033 MPa, mostly at the beginning of the engagement. The range of the maximum stress is again at the bottom at the beginning of the toothed engagement.

2.3 Definition of the Coefficient $K_{H\beta}$ with a Shafts Deformation and the Gearbox Cabinet Deformation

The software KISSsys is into computing of the $K_{H\beta}$ coefficient is able to incorporate the stiffness of a particular the gearbox cabinet. The stiffness of the cabinet is described by a reduced stiffness matrix. The stiffness matrix of the cabinet is reduced in this case to six points representing the center of the bearings. Subsequently, these points are loaded by the calculated reaction forces in the bearings, and thus, the displaçements of the bearings in the direction of the axes x, y, and z are obtained. The offset values obtained are applied into the calculation of the $K_{H\beta}$ coefficient, while will be considered with in the calculation of the shaft deformation and of the wheels teeth bending [1, 4, 6].

Reduced stiffness matrix of the gearbox cabinet The software ANSYS Workbench was used to obtain the reduced stiffness matrix of the gearbox cabinet. The model of the gearbox is built up than simplified, i.e., without screw holes and without internal preload [3] (Fig. 3).

The coefficient of the tooth load after the width, as it after the analysis without deformations of the gearbox cabinet, has changed from value 1.2155 to the value 1.2261. This change in the coefficient does not have a significant effect on the distribution of the tension over the width of the tooth, which implies that the gearbox cabinet is sufficiently stiff and there is no significant deformation due to the operating loads.

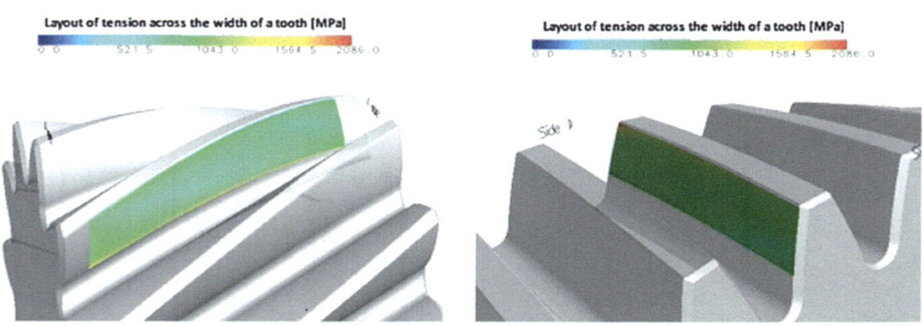

Fig. 3 Layout of tension across the width of a tooth to at left of a pinion and at right of a wheel

2.4 *The Optimization of the Third Gear, the Toothed Modification*

To optimize the toothed engagement of double-type tooth modification is used in practice. The first type is a modification the evolvent of teeth, for example a linearly chamfer and a rounding in the area of the head and heel of the tooth, a changing the pressure angle α and their combinations. The advantages of this type are, for example, quiet running, smoother a toothed engagement, and higher load stress. The second type is used to improve the load distribution over the width of the tooth, i.e., the $K_{H\beta}$ coefficient. There are various types of these modifications, for example, the linear grinding of tooth start and end, rounded grinding of the teeth start and the teeth end, the modification of angle of the inclination tooth, the crowning of the teeth, the eccentric crowning of the teeth, contortion of the teeth, and others. To improve the $K_{H\beta}$ coefficient, the modification across the width of the teeth was used, namely in the concrete in the crowning teeth to the norm ISO 6336-1 Annex B. Once the calculation has been completed, we have gained a value of the coefficient was $K_{H\beta} = 1.0914$. To remind, the difference of the $K_{H\beta}$ coefficient without modifying the tooth and considering the same deformations was 1.2261 [2, 6, 7].

Figure 4 shows the layout of tension across the width of the tooth of the pinion and the wheel, while modified was only the teeth of the wheel. In this case, the stress is distributed over the whole surface evenly, the size of which is about 880 MPa. The maximum stress is predominantly located in the heel of the tooth of the pinion and in the head of the tooth wheel in the beginning and center of the toothed engagement.

3 Results and Discussion

For comparison of the calculations, a brief overview of the resulting $K_{H\beta}$ coefficients for selected types of calculations is given. Table 1 describes how the $K_{H\beta}$ coefficient

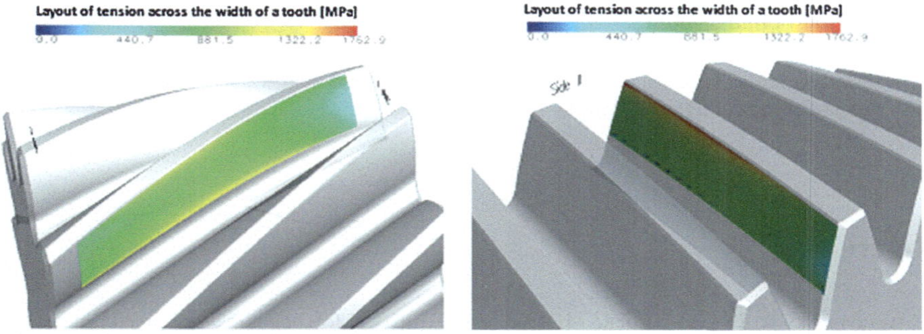

Fig. 4 Layout of tension across the width of the tooth at left of the pinion and at right of a wheel

Table 1 Overview of the results of the individual calculations of the $K_{H\beta}$ coefficient

Considered deformations	Norm	Hertz of contact stress (MPa)	Coefficient $K_{H\beta}$
The bends of teeth	DIN 3990	2183.664	1.1525
The bends of teeth – Deformations of shafts	ISO 6336-1 annex E	2066.821	1.2155
The bends of teeth – Deformations of shafts – Deformations of gearbox cabinet	ISO 6336-1 annex E	2086.018	1.2261
The bends of teeth – Deformations of shafts – Deformations of gearbox cabinet – Modification of wheel tooth	ISO 6336-1 annex E modification ISO 6336-1 annex B	1762.94	1.0914

under the influence of deformations of the various parts of the gearbox, as well as the size of contact stress.

4 Conclusion

At the conclusion, it is pointed out that from the mentioned results of the contact analyses, they have a great influence on the correctness of the toothed engagement of the shafts deformation, while increase of a coefficient with consideration deformations of the gearbox cabinet has changed slightly; i.e., the gearbox has sufficient stiffness at the specified operational loads and material used. It can be said with certainty that the biggest impact on the overall life of the gearbox being examined has a toothed engagement.

Acknowledgements This study was supported by Cultural and Educational Grant Agency MŠVVaŠ under the contract no. 040ŽU-4/2016.

References

1. Bašťovansky R, Tropp M, Lukáč M, Brumerčík F (2017) Molybdenum sheet metal test device. Commun Sci Lett Univ Žilina 19(2):124–127
2. Jedlinski L, Caban J, Krzywonos L, Wierzbicki S, Brumerčík F (2015) Application of vibration signal in the diagnosis of IC engine valve clearance. J Vibro Eng 17(1):175–187

3. Weis P, Kučera Ľ, Pecháč P, Močilán M (2017) Modal analysis of gearbox housing with applied load Elsevier. TRANSCOM 2017. International scientific conference on sustainable, modern and safe transport, high Tatras, Slovakia. University to Žilina, Žilina, EDIS, pp 953–958
4. Klebanov B, Barlam D, Nystrom F (2008) Machine elements life and design. CRC Press, USA
5. Malik L, Medvecký Š et al (2003) Časti a mechanizmy strojov. University of Žilina, Žilina, EDIS
6. Lukáč M, Brumerčík F, Krzywonos L, Krzysiak Z (2017) Transmission system power flow model. Commun Sci Lett Univ Žilina 19(2):27–31
7. ISO 6336:1996 (1996) The strength calculation of the front and cone wheels

Evaluation of the Energy Balance on Test Bench with Open Power Flow and Closed Power Flow for Testing of Transmission

Michal Herda, Ĺuboš Kučera, Tomáš Gajdošík and Milan Benko

Abstract The article deals with the diagram of the disposition of components and the creation of kinematic bonds of the test benches for transmission systems. The source of the drive on test benches is an electric motor. The article describes the configuration of components and the creation of kinematic bonds for the test bench with an open flow of mechanical power. The test bench also includes a dynamometer in addition to the electric motor. The article is also described an assembly procedure and creation kinematic bonds for test bench with a closed flow of mechanical power. In this, test bench used only an electric motor without a dynamometer. The electric motor is the same for both test benches. The kinematics of test benches is modeled and calculated in the Swiss program KissSys/KissSoft. As a result, the article deals with evaluation the energy balance of the individual test benches, especially by evaluation the energy balance of the electric motor.

Keywords Transmission · Energy balance · Open test bench · Closed test bench

1 Introduction

Nowadays, there are very developed simulation programs for various types of analyses and simulations. This also applies to the analysis and simulation of gearboxes and transmissions [1]. However, simulations and analysis in programs are mostly performed under ideal conditions with the addition of boundary conditions replacing the real state. These results may not correspond to reality. Therefore, it is still important to test the designed gearboxes and transmission systems on real test benches. Mechanical testing of transmission systems is major power consumption, especially in test benches with open mechanical power flow. Therefore, the main objective of the article is the calculation and evaluation of the energy balance of test benches for transmission systems. This is a comparison of the test bench with an open flow of the mechanical power and the test bench with a closed flow of mechanical power. Today, it is important to pay attention to energy consumption. With today's focus on the

M. Herda (✉) · Ĺ. Kučera · T. Gajdošík · M. Benko
University of Žilina, Univerzitná, 8215/1, 01026 Žilina, Slovak Republic
e-mail: Michal.Herda@fstroj.uniza.sk

Š. Medvecký et al. (eds.), *Current Methods of Construction Design*, Lecture Notes in Mechanical Engineering, https://doi.org/10.1007/978-3-030-33146-7_31

economy of development and production, energy consumption is one of the decisive economic indicators. For the analytical calculation, the same two-stage planetary gearboxes A2000 are used. The two-pole asynchronous engine with output of 50 kW is used as a drive. For both test cases, the same components (planetary gearbox A2000 and electric motor) were used.

2 Design of the Kinematic Scheme for the Test Bench with an Open Power Flow

During assembling a kinematic scheme, the knowledge of testing of the transmission systems has been applied. It was also based on tests performed at the University of Žilina's testing laboratory [2–5]. The test bench with an open flow of mechanical power has a simple schematic on the components. (see Fig. 1).

In the test bench were used components: a two-pole asynchronous electric motor with a 50 kW output controlled by a phase transducer. In addition, there are two A2000 two-stage planetary gearboxes in the circuit. At the end of the circuit, the dynamometer is connected. As shown (Fig. 1), at the beginning of the test bench, the electric motor is the source of the drive. Behind it, two planetary gearboxes are connected in series. These gearboxes are connected in a mirror position, so that approximately the same values are measured at the output shaft from the second gearbox as the input values are. The difference between input and output values is due to losses in gearboxes. After we retrieve output power and losses in planetary gearboxes, we get the input power value. The test bench closes the dynamometer. It produces electricity from the output power by recuperation, which then returns to the electricity grid. To connect all components in the test bench, cardan shafts were used. To calculate the total losses in the test circuit, the losses incurred on the joint of the cardan shafts do not count. In our case, they are negligible for calculation.

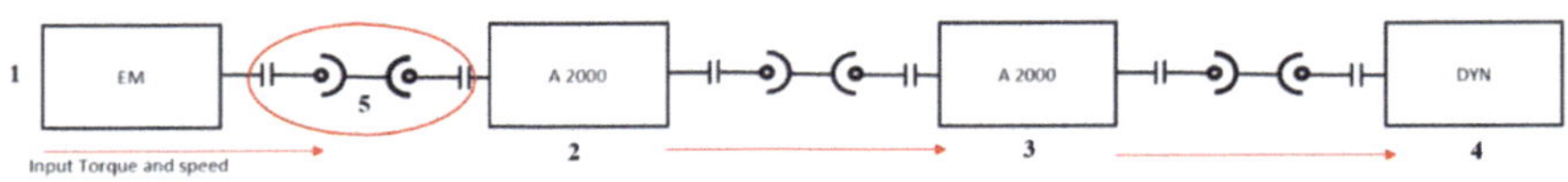

Fig. 1 Kinematic scheme of the test bench with an open flow of mechanical power 1 EM—Electric motor; 2, 3 A2000—Planetary gearbox A2000; 4 DYN—Dynamometer; 5 Kinematic connection; arrowhead show power flow

3 Modeling and Calculation of an Open Test Bench in KissSys

For modeling and calculating total losses in a bench with open flow of mechanical power, the Swiss calculation program KissSoft/KissSys [6], for the design and simulation transmissions, was used. The advantage of the program is its high precision in modeling and calculation of transmission systems. As a precondition for overall efficiency calculation, the A2000 gearbox in KissSys was first modeled and saved as a template to be used repeatedly and did not have to be reworked. After the planet gear has been correctly modeled, a new project has been launched in the program. After the launch of the new project, two planetary gearboxes were inserted and placed in series. But they have been mirrored against each other. The reason was, to connect the output shaft of the carrier shafts. After the gearboxes were placed, a direct kinematic connection was established between the gearboxes. The connection the output shaft of the carriers shafts, thereby forming the transmission system with a transmission ratio of $i = 1$. The schema and 3D model in KissSys (see Figs. 2 and 3).

After creating a kinematic link between planetary gearboxes, the boundary conditions were entered. The input and output were defined, and also the braking torques for the ring shafts, to ensure the passage of power from the input shaft to the central

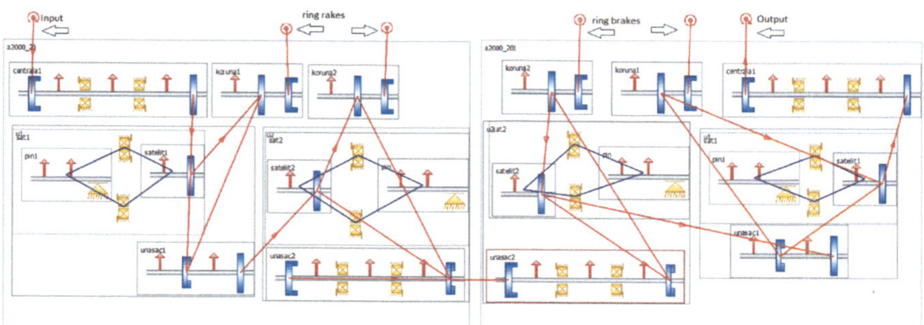

Fig. 2 Technical scheme in KissSys

Fig. 3 Three-dimensional model in KissSys

Table 1 Results from calculation of the bench with an open flow of mechanical power

Title	Value	Units
Input power	45,914.4	W
Output power	43,911.2	W
Total loss	2003.2	W
Total efficiency	95.64	%

wheel of the first planetary gear, and the output was the center wheel of the second planetary gearbox. The entered inputs to the system are 1500 rpm and a torque of 292.3 Nm. The system has already calculated the required input power that was 45,914.4 W. After entering the boundary conditions, the calculation of the kinematics for the transmission system was started. Calculation of kinematics in KissSys is necessary for further calculations (see Table 1). The next step was to calculate the efficiency of the whole circuit. After entering all the conditions for calculation of efficiency, the calculation method was chosen according to ISO TR 14179-2. Efficiency calculations did not take into account the contact pattern analysis. Calculations are performed in an iterative manner.

The total calculated efficiency is divided into multiple partial losses. All partial losses are delineated in the calculation report. For simplicity, all partial losses were divided into two points, namely tooth loss, which was 1696.3 W and other losses that were 306.9 W.

4 Design of the Kinematic Scheme for the Test Bench with a Closed Power Flow

Design of a test bench with a closed power flow proceeds similarly to the test bench with an open power flow, but here it had to be ensured that the power circulates. In the kinematic scheme (Fig. 4), the distribution of components in the test bench with a closed power flow of mechanical power is shown.

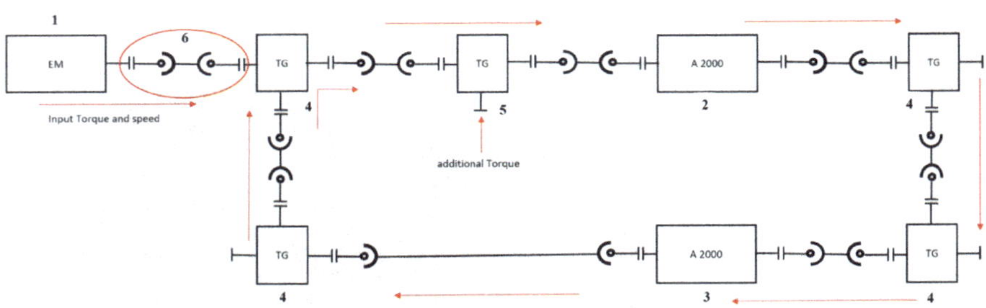

Fig. 4 Kinematic scheme of the test bench with a closed flow of mechanical power 1 EM—Electric motor; 2, 3 A2000—Planetary gearbox A2000; 4 TG—transmission gear units; 5 TG—Technological gearbox; 6 Kinematic connection; arrowhead show power flow

The same electric motor is used as the drive as is for a test bench with an open flow of mechanical power. Similarly, there are two A2000 planetary gearboxes in the circuit. Unlike the first test bench, there are auxiliary gear hubs in the circuit, also a technological gearbox and a dynamometer absent. The gear hubs (see Fig. 3), with the number 4, serve to close the test circuit. Several alternatives have been considered when designing gear hubs. The most preferred variant was a variant with a bevel gearbox. When deciding, the condition with the greatest weight was to maintain the gear ratio $i = 1$ in order to avoid additional calculations or compensation for the power flow. The next step was to design a technological gearbox. For the design of the technological gearbox, the differential, cylindrical gearboxes, and bevel gearboxes were considered. As a technological gearbox, bevel gearboxes were selected to provide a simple power connection and also maintain the $i = 1$ gear ratio. All parts of the circuit are coupled with cardan shafts.

5 Modeling and Calculation of a Closed Test Bench in KissSys

As with open test bench, KissSys was used to assemble a test bench with a closed flow of mechanical power. Before the start, bevel gearboxes have been modeled and saved as a template. Since the planet gearbox A2000 has already been modeled from the previous case, it did not have to model again. After the launch of the new project, two planetary gearboxes and five bevel gearboxes were added. The schema and 3D model in KissSys (see Figs. 5 and 6).

After the deployment, the closed test bench components, links were created between the individual parts. For the correct definition of links, it was necessary

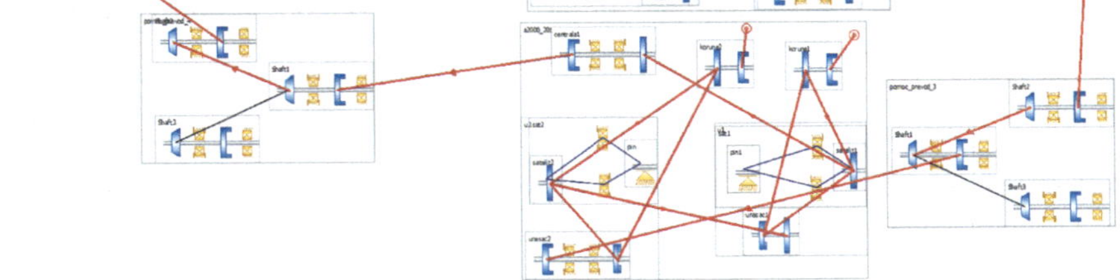

Fig. 5 Technical scheme in KissSys

Fig. 6 Three-dimensional model in KissSys

to determine the flow of power, where the power will be divided and where it will be joined. If the power division function was incorrectly defined, the program would not be able to calculate the kinematics and other calculations would not be possible. Subsequently, the boundary conditions have been entered. These were identical to the open test bench, 1500 rpm and a torque of 292.3 Nm. After defining the power distribution and entering the boundary conditions for the input, the kinematics could be calculated (see Table 2). After kinematics, the calculation of the efficiency of the whole bench was continued. As in the first case, the system has calculated the input power of the engine. Even in this case, it was 45914.4 W. Efficiency was also calculated according to ISO TR 14179-2.

After completing the calculation, we have redefined the input conditions to confirm by the following calculation the correctness of the claim that if the input speed [1500 rpm] and input power equal to [7502.5 W] losses in the circuit, and prestressed technology gearbox with the required torque [292.3 Nm], maintains the same power flow output.

Table 2 Results from calculation of the bench with a closed flow of mechanical power

Title	Value	Units
Input power	45,914.4	W
Output power	38,411.9	W
Total loss	7502.5	W
Total efficiency	83.66	%

Table 3 Results from calculation of the bench with an open flow of mechanical power

Title	Test bench with an open flow of mechanical power	Test bench with an closed flow of mechanical power
Input rpm (1/min)	1500	1500
Input torque (Nm)	292.3	292.3
Input power (kW)	45.91	45.91
Output rpm (1/min)	1500	1500
Output torque (Nm)	279.6	244.58
Output power (kW)	43.91	38.41
Total losses (kW)	2	7.5
Efficiency (%)	95.64	83.66

There was no full torque defined at the input, but a moment was added, which in the previous calculation was equal to the loss in the test bench. So, for the input, the torque was defined only 47.77 Nm, while maintaining the initial speed [1500 rpm]. In order to achieve a similar power flow in the test bench, the technological gearbox was prestressed with a torque of 292.3 Nm. After the calculation was completed, it was found that the assumption was correct and the desired effect of the closed flow of mechanical power was achieved.

6 Evaluation

After the calculations, the values obtained were evaluated (see Table 3).

From calculated data follows, when the power flow is open, the loss in the measuring state is 4.36%, which corresponds to a power of 2 kW. The engine consumption per test hour is therefore 45.9 kW/h. At the end of the test bench, the dynamometer works with a power of 43.91 kW, recuperating approximately 40% of this power, which in our case is 17.56 kW. The total power loss in this case is 28.35 kW. It is clear from the calculation that the open test circuit in the test conditions mentioned above is an hourly energy consumption of 28.35 kW/h. In the closed test bench where the dynamometer is absent and the desired torque is achieved by mechanical prestressing, the power consumption of the electric motor is 7.5 kW/h, which represents the coverage of torque losses throughout the test bench.

7 Conclusion

It is therefore clear that the power consumption of a closed test bench for the given test parameters is almost 4 times lower than that of the open test bench. However, the initial financial difficulty in constructing this circuit (the use of five bevel gearboxes and a higher number of cardan shafts) is eliminated with the number of hours tested, and these technology gearboxes can also be used to test other types of transmission. It follows from the above that a closed test bench is more suitable for long-term transmission tests, where it appears to be financially less demanding than the open bench.

Acknowledgements This study was supported by Cultural and Educational Grant Agency MŠVVaŠ under the contract no. 040ŽU-4/2016.

References

1. Li JY, Gou ZJ, Li Y (2005) A dynamic simulation of meshing force in gear meshing process based on ADAMS. Machinery 3
2. Gramblička S, Kohár R, Stopka M (2017) Dynamic analysis of mechanical conveyor drive system. Procedia Eng 192:259–264
3. Tomášiková M, Gajdošík T, Lukáč M, Brumerčík F (2017) Simulation of planetary gearbox. Commun, Sci Lett Univ Žilina 19(2A):48–53
4. Lukáč M, Brumerčík F, Krzywonos L, Krzysiak Z (2017) Transmission system power flow model. Commun, Sci Lett Univ Žilina 19(2):27–31
5. Liang X, Zuo MJ, Feng Z (2018) Dynamic modeling of gearbox faults, a review. Mech Syst Signal Process 98:852–876
6. KissSoft. Calculation software for machine design—design, analyse, optimize; online help

Dynamic Analysis of Vibrating Sorter and Description of Optical Methods for Experimental Verification

Jan Hoidekr and Vojtěch Dynybyl

Abstract The aim of this work is to determine the system of equations of motion for laboratory vibrating sorter mechanism. Description of motion and experimental verification is useful for designing of industrial vibrating sorters. These sorters are mostly used for sorting oil seeds in oil pressing industry. The mechanism which is described in this work is part of a laboratory testing circuit, where the methodology of designing of industrial machines is being developed. In this case, the sorter is placed on linear compression springs for easier dynamic modeling. The model of vibrating sorter is considered as a nonlinear two bodies system with 6 degrees of freedom and its solution leads to application of numerical solver using software MATLAB. In the second part of this work, the experimental measurement using optical methods is described. There are placed markers to be captured with the camera. With the help of software MATLAB and its image processing toolbox, the displacement of markers is determined and the correctness of dynamic analysis can be verified.

Keywords Dynamic · Equations of motion · Optical methods · MATLAB · Vibration sorting

1 Introduction

Vibrating sorters are abundantly used in oil processing industry in pretreatment section for cleaning oil seeds or sorting specimen according its size. The sorter described below is a part of a laboratory testing circuit where hulled seeds are sorted into five categories: large seeds, medium-large seeds, medium-small seeds, small seeds, and dust. The first four categories are further processed in cyclone machine where hulls are separated from kernels. The largest kernels are further processed in optical sorter to determine which are fully hulled and which are not.

J. Hoidekr (✉) · V. Dynybyl
Czech Technical University in Prague, Technická 4, 16607 Prague, Czech Republic
e-mail: jan.hoidekr@fs.cvut.cz

V. Dynybyl
e-mail: vojtech.dynybyl@fs.cvut.cz

Š. Medvecký et al. (eds.), *Current Methods of Construction Design*, Lecture Notes in Mechanical Engineering, https://doi.org/10.1007/978-3-030-33146-7_32

The vibration sorting machine is simplified laboratory copy of industrial machines. It consists of two bodies, the mainframe and the sieves placed in wooden frames. Normally, the mainframe is placed on four silent blocks. For the first modeling step, the silent blocks were replaced with linear compression springs. Motion analysis of this laboratory vibrating sorter helps in the development of industrial machines.

2 Methods

2.1 Dynamic Analysis

The dynamic model is considered according to Fig. 1. For the first step of the analysis, the dynamic system is reduced to symmetric planar problem. The horizontal springs labeled *kx* represent lateral stiffness of springs labeled *ky*. For simplification, lateral stiffness in this model is considered as a constant, nonetheless in further analysis, usage of the lateral stiffness dependent on the axial is expected, e.g., according to [1].

The sieves are kinematically excited through eccentric pin that generates dynamic effects. Equations of motion are determined in static equilibrium position; therefore, forces representing mass of bodies are not considered in Fig. 2. Either, the friction and horizontal forces in place of contact between the frame and the sieves are not considered due to lubrication of a polished sliding rod (Fig. 3).

In real construction, the eccentric pin is not oriented exactly according Fig. 1, but the axis of rotation is oriented vertically. Kinematic excitation is described in

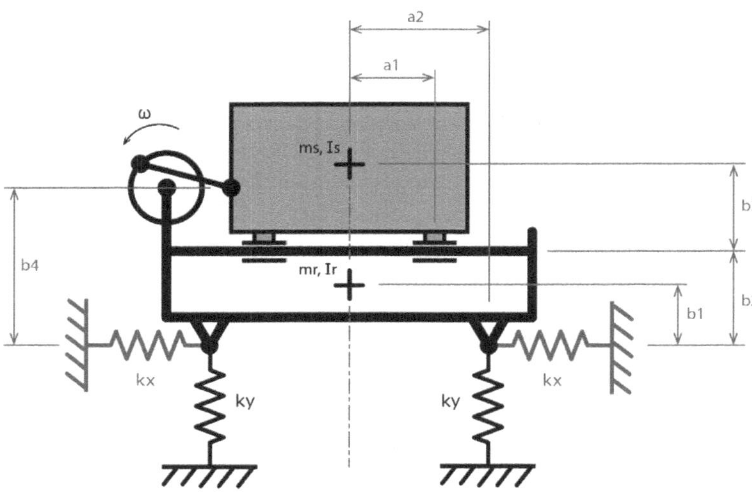

Fig. 1 Dynamic model of vibrating sorter

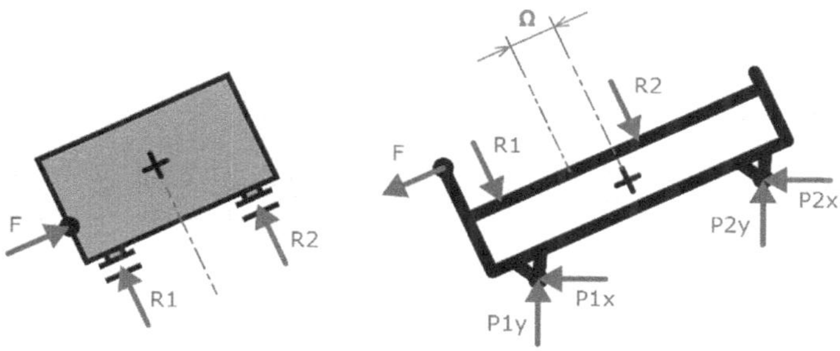

Fig. 2 Free body diagram of vibrating sorter

Eqs. (1) and (2) according Fig. 4.

$$\Omega = \sqrt{c_1^2 - r_e^2} - \sqrt{c_1^2 - (r_e \cdot \cos(\varphi_e))^2} + r_e \cdot \sin(\varphi_e) \tag{1}$$

$$\varphi_e = \omega \cdot t \tag{2}$$

Coordinates of displacement are defined in Fig. 2. The Eqs. (3)–(8) are equations of motion and they are determined for the center of gravity for both bodies. After using simple transformation, it is not difficult to detect the displacement of any point

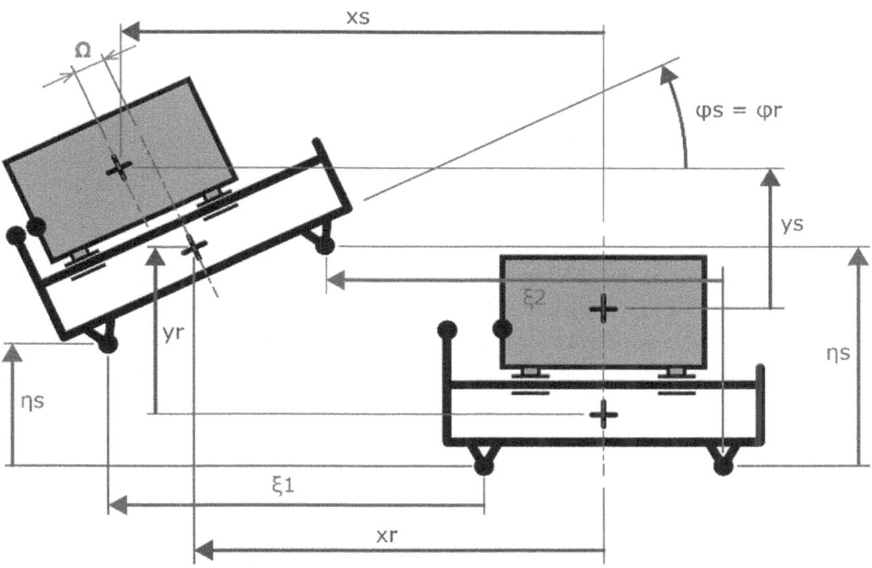

Fig. 3 Coordinate system considered in dynamic model

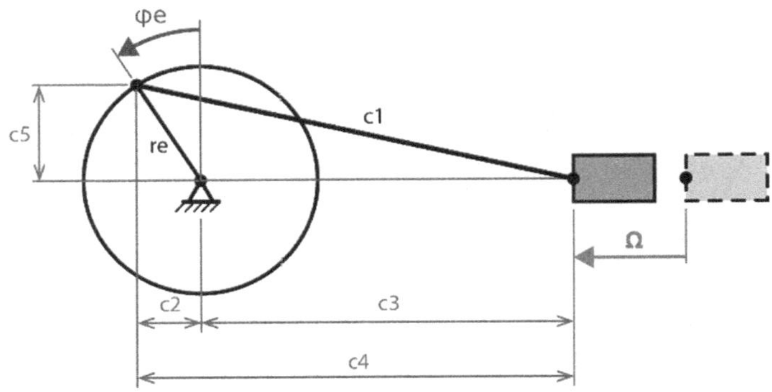

Fig. 4 Kinematic excitation of mechanism through eccentric pin

of vibrating sorter.

$$m_R \cdot \ddot{x}_R = F \cdot \cos(\varphi_R) - R_1 \cdot \sin(\varphi_R) - R_2 \cdot \sin(\varphi_R) + P_{1x} + P_{2x} \tag{3}$$

$$m_R \cdot \ddot{y}_R = -F \cdot \sin(\varphi_R) - R_1 \cdot \cos(\varphi_R) - R_2 \cdot \cos(\varphi_R) + P_{1y} + P_{2y} \tag{4}$$

$$\begin{aligned} I_R \cdot \ddot{\varphi}_R = {} & F \cdot (b_4 - b_1) + R_1 \cdot (a_1 + \Omega) - R_2 \cdot (a_1 - \Omega) \\ & + \cos(\varphi_R) \cdot \left[-P_{1x} \cdot b_1 - P_{1y} \cdot a_2 - P_{2x} \cdot b_1 + P_{2y} \cdot a_2\right] \\ & + \sin(\varphi_R) \cdot \left[-P_{1x} \cdot a_2 + P_{1y} \cdot b_1 + P_{2x} \cdot a_2 + P_{2y} \cdot b_1\right] \end{aligned} \tag{5}$$

$$m_S \cdot \ddot{x}_S = R_1 \cdot \sin(\varphi_S) + R_2 \cdot \sin(\varphi_S) - F \cdot \cos(\varphi_S) \tag{6}$$

$$m_S \cdot \ddot{y}_S = F \cdot \sin(\varphi_S) + R_1 \cdot \cos(\varphi_S) + R_2 \cdot \cos(\varphi_S) \tag{7}$$

$$I_S \cdot \ddot{\varphi}_S = F \cdot (b_3 + b_2 - b_4) + R_1 \cdot a_1 + R_2 \cdot a_1 \tag{8}$$

There are nine variables in Eqs. (3)–(8); so, the relationships between coordinates of displacement of frame and sieves have to be defined according to [2]. By second derivative of these relationships determined in Eqs. (9)–(11), the acceleration of sieves is expressed and can be substituted in Eqs. (6)–(8). That reduces the system to six nonlinear second-order algebraic differential equations which can be solved using numeric methods. The forces P_{1x}, P_{2x}, P_{1y}, and P_{2y} are reaction forces caused by deformation of linear compression springs out of static equilibrium position.

$$\varphi_S = \varphi_R \tag{9}$$

$$x_S = x_R + (b_3 + b_2 - b_1) \cdot \sin(\varphi_R) + \Omega \cdot \cos(\varphi_R) \tag{10}$$

$$y_S = y_R + (b_3 + b_2 - b_1) \cdot \cos(\varphi_R) - \Omega \cdot \sin(\varphi_R) - (b_3 + b_2 - b_1) \quad (11)$$

This system of equations is solved with the help of MATLAB software using numeric solver ode15i for solving fully implicit differential equations.

2.2 Experimental Measurement Using Optical Methods

There are several ways how to measure displacement during vibrations. According to [3], one of the options is an optical method. Canon 6D DSLR camera and 50 mm $f/1.4$ lens were used to capture motion on video with resolution 1280 × 720 px and 1/50 s framerate.

To determine exact description of motion, the camera should be as far as possible from sorter to avoid perspective deformation. It is convenient to use circle markers since MATLAB software image processing toolbox can easily find the centers of circles. The markers were printed in dark color on white background paper and stuck on body of sorter frame according to Fig. 5.

Actually, there are only two markers needed. One circle represents the described point and the second circle is used to determine angle of rotation φ.

Each frame of video can be processed in MATLAB like independent static image. All pixels are loaded into matrix dimensionally equaling to resolution of image. Listing and processing all frames of video gives a trajectory of displacement of certain point. By simple transformation, the displacement of any point can be calculated.

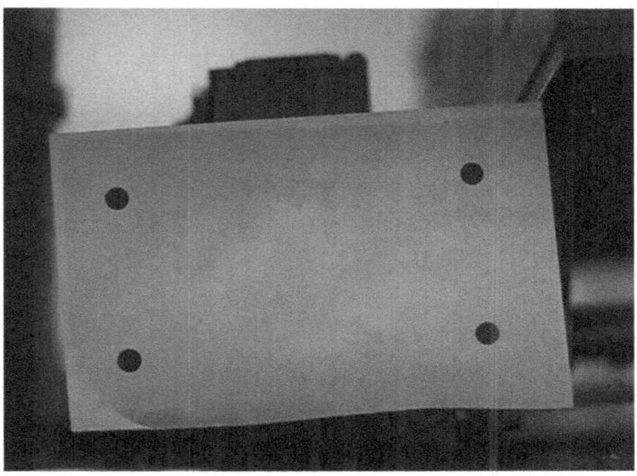

Fig. 5 Markers stuck on frame body

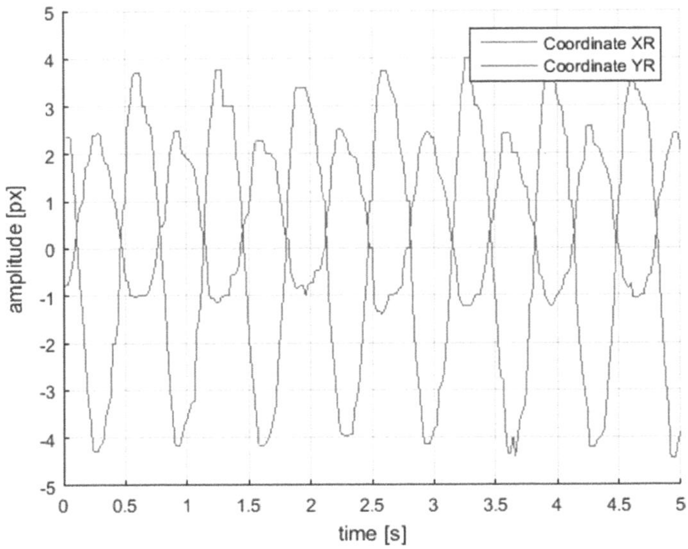

Fig. 6 Trajectory of certain point of mainframe measured experimentally by optical method

3 Results

Results of experimental measurement are shown in Fig. 6. There were several measurements with different frequencies of vibrating, starting from $f = 1$ Hz up to $f = 3$ Hz with 0.5 Hz step. An example in Fig. 6 shows measurement with frequency $f = 1.5$ Hz.

The vertical coordinate in graph in Fig. 6 is plotted in pixels. One pixel equals approximately 0.36 mm.

4 Discussion

Results from experimental measurement match with expectations and corresponds with real sorter vibration. The laboratory sorter used for experimental measurements does not fully correspond to dynamic model listed above; therefore, there is no bigger sense to directly compare this result. Several simplifications were mainly used to show the base of methods of dynamic analysis for similarly working devices.

The further step of modeling is customizing dynamic system according to real vibrating sorter so experimental measurement can be compared.

5 Conclusion

In this work were described basic methods of dynamic analysis of vibration sorter working according to Fig. 1. Based on these methods, the more sophisticated and more detailed models can be defined; so, these models could be directly compared to experimental measurement data. Simplifying of dynamic model in this work consisted mostly of ignoring antisymmetric dimensions and considering this model fully symmetric. Further detail, which is possible to include into model, is friction between mainframe and sieves.

In the methodology of using optical methods for experimental measurement would be recommended the usage of a camera with higher framerate than 50 fps to capture higher frequencies correctly. In this project, only camera with 50 fps framerate was available.

Acknowledgements This work was supported by SGS ČVUT 2018 No: SGS18/126/OHK2/2T/12 "Continualization of process of sorting oil seeds."

References

1. Stejskal V (2002) Kmitání s Matlabem, 1st edn. Vydavatelství ČVUT, Prague
2. Turek O, Hoidekr J (2018) Popis kmitání vibrační třídičky s vice stupni volnosti pomocí numerických metod. In: Moravec J (ed) Konference studentské tvůrčí činnosti, STČ 2018. ČVUT Prague, Prague
3. Machač M, Hoidekr J (2018) Vyhodnocení experimentálního měření kmitání vibrační třídičky pomocí optické metody. In: Moravec J (ed) Konference studentské tvůrčí činnosti, STČ 2018. ČVUT Prague, Prague

Modeling of Cornea Applanation When Measuring Eye Pressure

Branislav Hučko, Ĺuboš Kučera, Stanislav Ďuriš, Peter Pavlásek, Jan Rybář and Juraj Hodál

Abstract This paper deals with the numerical and experimental modeling of a cornea applanation during pressure loading–air puff. The determination of cornea applanation plays a key role in measuring inner eye pressure. A raised level of inner eye pressure is one of the most obvious symptoms of glaucoma. The numerical model for the cornea was established in consideration of its geometrical and physical properties using the finite element method. The primary applanation pressure was determined from obtained numerical results. The experimental model of the cornea was built from silicone. The experimental cornea was subjected to inner pressure that was measured by traditional medical contact and non-contact tonometers applied outside the cornea. Subsequently, the results for this model were used as the referential ones.

Keywords Applanation · Eye pressure · Model · Glaucoma · Measurement

1 Introduction

The intraocular pressure (IOP) is the fluid pressure inside the eye. The IOP is determined by the coupling of the production of aqueous humor and the drainage of aqueous humor mainly through the trabecular meshwork located in the anterior chamber angle [1]. The reason for this is because the vitreous humor in the posterior segment has a relatively fixed volume and thus does not affect intraocular pressure regulation. The IOP preserves the integrity of the eyeball. Its maintenance is the ability of an aqueous humor that is secreted from the ciliary epithelium—the structure supporting the lens. It depends on two factors: the rate of fluid formation and the ability to drain it. If the first one is greater than the second one, the intraocular pressure may be increased.

B. Hučko (✉) · S. Ďuriš · P. Pavlásek · J. Rybář · J. Hodál
Slovak University of Technology in Bratislava, Nam. Slobody 17, 81231 Bratislava, Slovak Republic
e-mail: branislav.hucko@stuba.sk

Ĺ. Kučera
University of Žilina Univerzitná, 8215/1, 01026 Žilina, Slovak Republic

Š. Medvecký et al. (eds.), *Current Methods of Construction Design*, Lecture Notes in Mechanical Engineering, https://doi.org/10.1007/978-3-030-33146-7_33

Tonometry is the method of measuring IOP by tonometer [2]. The IOP measurement is an important aspect in the evaluation of patients at risk of glaucoma [3]. Most tonometers are calibrated to measure the pressure in millimeters of mercury (mmHg).

The intraocular pressure is a dynamical variable like the blood pressure. The IOP varies not only throughout the night and day but also during longer period. The diurnal variation for normal eyes is between 3 and 6 mmHg and the variation may increase in glaucomatous eyes. During the night, intraocular pressure may not decrease despite the slower production of aqueous humor. It was observed that 95% of population had an average IOP of 15.7 $\pm$ 2 SD (standard deviation) mm Hg. This results into the interval from 10 to 21 mm Hg. The normal distribution of IOP is not the Gaussian one and is distorted to the right. This distortion suggests that more than 2.5% of population should have the IOP greater than 21 mmHg [4].

Current tonometers are based on the contact and non-contact technologies of measuring the cornea deformation [2]. The application of contact tonometers requires the direct unpleasant contact of measuring device, probe, with the human cornea. Because the probe makes contact with the cornea, a topical anesthetic is introduced on to the surface of the eye in the form of an eye drop. The non-contact tonometry utilizes the air puff for deforming the cornea without any anesthetic.

Both technologies are calibrating to the measurement of cornea applanation—the loaded curved cornea surface becomes flatten-applanated. The calibration of measuring devices requires to build the referential cornea: numerical and experimental. These corneas should be applicable in calibration process of any tonometers.

The aim of this paper is to present numerical and experimental models of cornea in the application of non-contact measuring of IOP.

2 Principles of non-contact tonometry

The non-contact tonometry is system consists of a central air chamber lined from one side by a light emitter (infrared light emitter) and a second detector (see Fig. 1).

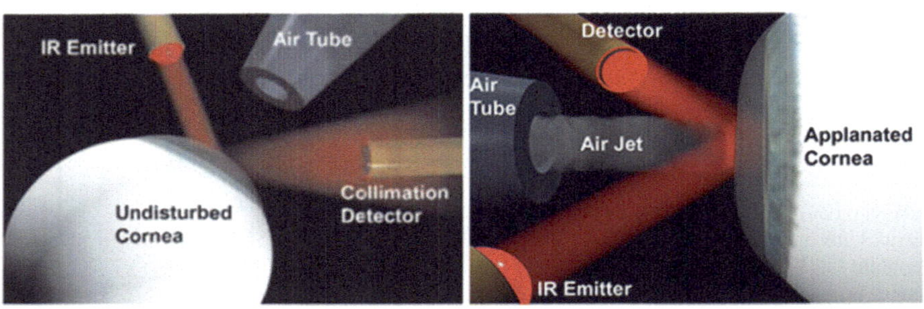

Fig. 1 The principle non-contact cornea applanation [5, 6]

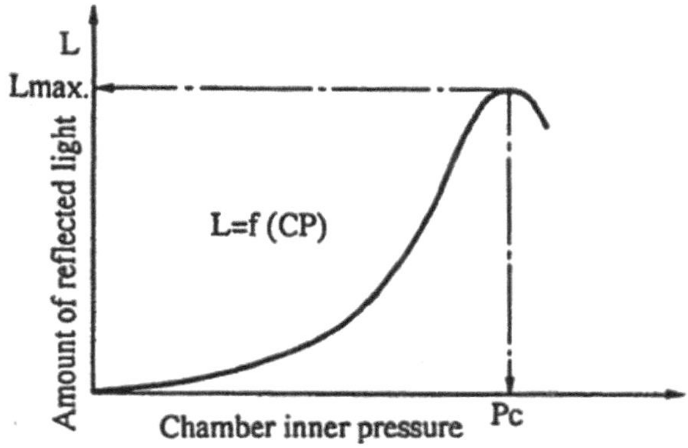

Fig. 2 Intensity of reflected light versus chamber inner pressure [7]

In the unloaded state of cornea, the light emitted is scattered by its convex surface. As the pressure of the air impulse directed to the cornea increases to deform the cornea, the surface of the cornea behaves like a plane mirror reflecting the IR light into the detector. The IR light beam is reflected by physical law: the angle of incidence is equal to the reflection angle. At the point of maximum light detection, when the cornea is completely flattened, the device switches off the air pressure pulse. In other words, the device will terminate the measurement if the field of application reaches at least the size required to redirect the entire light beam. The emitted beam is the same at each measurement, it has a constant cross-section. Only the applied pressure and the duration are changed.

The pressure of the air flow increases with the pressure of the anterior chamber, aqueous humor (CP—chamber inner pressure), which is continuously measured by the pressure sensor. The intensity of the light reflected (L) on the cornea is detected as a function of CP (see Fig. 2). The applanation pressure (Pc) is determined from the maximum reflected light intensity (Lmax) and subsequently, the corresponding IOP can be determined.

The determination of IOP is based on the assumption of equality of pressures on both sides of the flattened corneal apex, i.e., the cornea reaches the equilibrium position, similarly [8]. In this case, the external applanation pressure Pc caused by measuring air impulse equals to the IOP:

$$\text{IOP} = Pc \tag{1}$$

3 Numerical model of cornea applanation

The numerical model of cornea presented in this paper has been built from basic geometrical parameters [9] (see Fig. 3), mechanical properties [10] and corresponding boundary conditions.

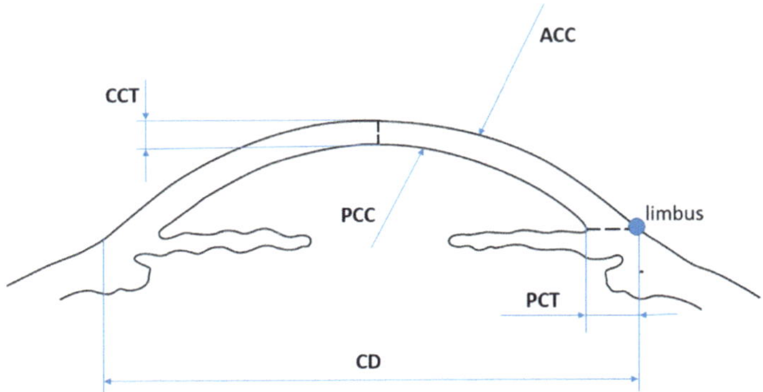

Fig. 3 Geometrical parameters of cornea, modified from [10]

The corneal parameters presented in [9] such as the corneal diameter (CD), the anterior corneal curvature (ACC), the posterior corneal curvature (PCC), the peripheral corneal thickness (PCT) and the central corneal thickness (CCT) provide information about the healthy cornea and possible changes associated with ocular diseases or other factors such as aging gender, ethnicity, etc. Normal values for the CD ranges from 10.50 to 12.75 mm, the ACC ranges from 7.06 to 8.66 mm, the PCC ranges from 6.15 to 7.38 mm, the PCT reaches value of 1.2 mm at limbus and the CCT ranges from 512 to 569.5 μm. These corneal parameters vary with age, gender, ethnicity, refractive state, stature and anthropometric factors. In our model, we applied the average values of all geometrical parameters.

There exist many different mechanical models of cornea properties [10]. The simplest mechanical model is the elastic one. The basic mechanical parameters are Young's modulus and Poisson's ratio. Young's modulus ranges from 0.054 to 0.359 MPa, Poisson's ratio has the constant value of 0.48 [10, 11]. The average value of Young's modulus (0.207 MPa) should be applied in mechanical model.

The boundary conditions also contain the loading conditions. We have applied the impulsive load—the air puff of maximum value 5 kPa that corresponds to 40 mmHg. The distribution function due to the free airflow is presented in Fig. 4.

The time duration of the air puff was 15 ms. The relation between the applied load and time is linear–ramped function from the zero value to the maximum value. Then, the final numerical model using the finite element method (FEM) using ANSYS [12] can be created (see Fig. 5). Due the axial symmetry of cornea model, we simplified it to one half of plane model.

4 Experimental model

The experimental model of cornea has been built like the verification and referential device (see Fig. 6).

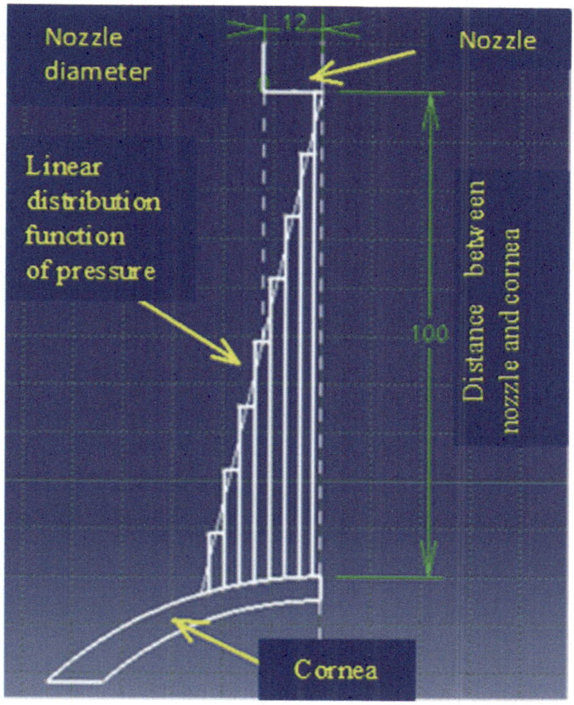

Fig. 4 Distribution function of applied pressure

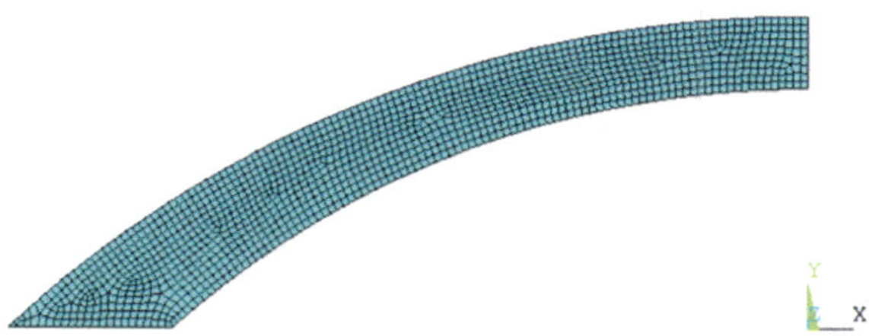

Fig. 5 FEM model of cornea

The experimental model of cornea consists of the cylindrical chamber ended with the silicone model of cornea. The mechanical properties of silicone are different in comparison with a human cornea. Young's modulus of silicone ranges from 0.1 to 1.4 MPa. This silicone cornea is subjected to the constant inner pressure by the air. The inner pressure is stabilized and calibrated to the value of 3 kPa = 22.5 mmHg. This value is less than applied pressure in the numerical model. Subsequently the IOP has been measured with the contact and non-contact tonometers.

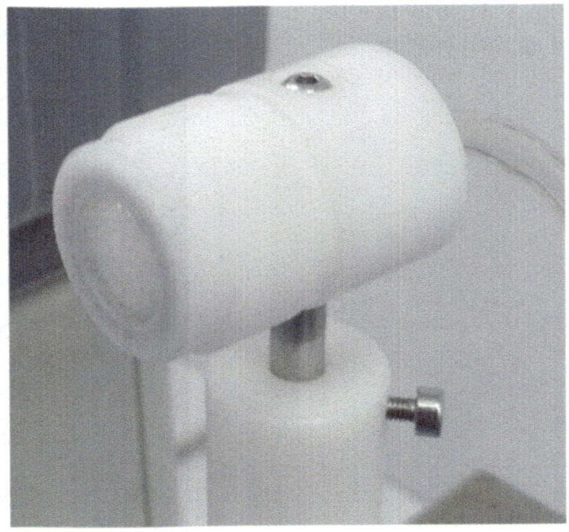

Fig. 6 Experimental model of cornea

5 Results and Discussion

Due to the fact that our experimental model contains the silicone cornea we used Young's modulus of 0.8 MPa—more less the average value, instead of values presented in Chap. 3. Another important fact is that the producer of silicone cornea could not guarantee the value of Young's modulus.

The deformation of numerical model is presented in Fig. 7. It is clear that the deformation of cornea is behind the applanation state.

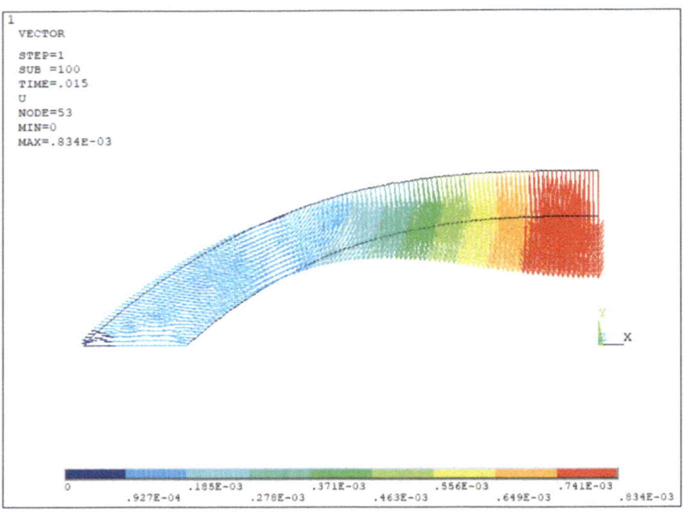

Fig. 7 Deformation of cornea—vector plot

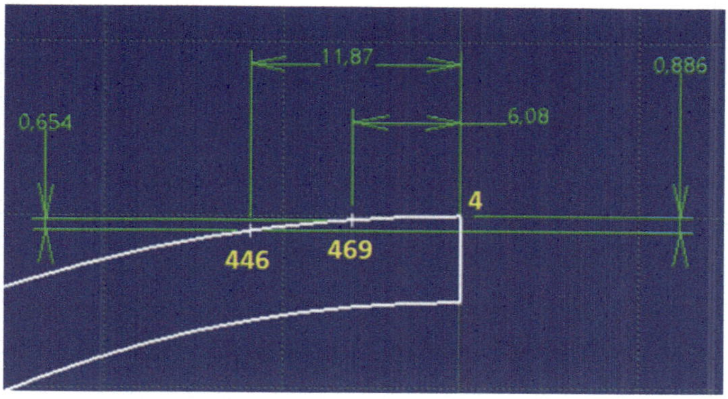

Fig. 8 Selected nodes for applanation pressure estimation

Therefore, we proposed the simple algorithm of estimating Pc from the displacement field. We select three nodes on the place of load application (see Fig. 8).

We found out the vertical relative displacements of those nodes. Because the response is elastic and linear one can estimate

$$UY_4 - UY_{446} = 0.8270 - 0.6267 = 0.2003 \text{ mm} \quad (2)$$

$$UY_{04} - UY_{0446} = 0.0866 - 0 = 0.0866 \text{ mm} \quad (3)$$

$$P_C = 5000 * \frac{0.0886}{0.2003} = 2211.68\ Pa \cong 2200\ Pa \quad (4)$$

$$t_C = 15 * \frac{2200}{5000} = 6.6 \text{ ms} = 0.0066 \text{ s} \quad (5)$$

where UY_i, resp.UY_{0i} is the vertical displacement and the initial vertical coordinate of *i*th node. The applanation pressure is $P_C = 2.2$ kPa = 16.7 mmHg and the applanation time is 6.6 ms. Using Eq. (1) we get IOP = 16.7 mmHg. The results of experimental measurements are presented in the following table, see Table 1.

The exact value of IOP is 22.5 mmHg. The experimental results reached the good correlation with the exact solution. The numerical results can be acceptable the difference is less than 30% that can be caused by the unreliable input of mechanical properties as we mentioned before. But both approaches have the potential to be simple referential and calibrating methods for measuring IOP devices.

Table 1 Experimental measurements of IOP

Measurement number	Non-contact tonometer NIDEK NT-2000 (mmHg)	Contact tonometer Rebound ICARE (mmHg)
1	21	23.5
2	21	23.6
3	23	23.8
4	–	24.1
5	–	23.9
Average value	21.7	23.78

Acknowledgement The authors hereby express their gratitude for the financial supports of KEGA 060STU-4/2016 and EMPIR 16PT03-InTENSE projects.

References

1. Aptel F, Weinreb RN, Chiquet Ch, Mansouri K (2016) 24-h monitoring devices and nyctohemeral rhythms of intraocular pressure. Prog Retin Eye Research 55:108–148
2. Garway-Heath DF (2007) Measurement of intraocular pressure, Kugler Publications
3. Farandos NM, Yetisen AK, Monteiro MJ, Lowe CR, Yun SH (2014) Contact lens sensors in ocular diagnostics. Adv Healthc Mater 4(6):792–810
4. Nema HV, Nema N (2014) Diagnostic Procedures in Ophthalmology. JP Medical Ltd, India
5. Realini T (2008) The ocular response analyzer, Glaucoma today, 27–30, July/Aug 2008
6. http://www.devinemedihealth.com/Ocular_Response_Analyser.asp?Compid=2&pid=17
7. Essilor F (2011) Document CT-80A_RM.pdf. Last accessed 11 Apr 2011
8. Srodka W (2009) Biomechanical model of human eyeball and its applications. Optica Applicata, (XXXIX) 2:401–413
9. Mashige KP (2013) A review of corneal diameter, curvature and thickness values and influencing factors. S Afr Optometrist 72(4):185–194
10. Guarnieri FA (2015) Corneal biomechanics and refractive surgery, New York
11. Shin PJ, et al (2017) Estimation of the corneal Young's Modulus in vivo based on a fluid-filled Spherical-shell model with Scheimpflug imagin. J Ophthalmol. Article ID 5410143, 11 pages
12. ANSYS, Version 18

Screw Connections with Application of Modern Insurance Sealants and Adhesives Analysis

Jan Kanaval, Eliška Cézová and František Starý

Abstract This contribution describes analytical evaluation and experimental verification of the friction and force conditions in the pre-stressed bolted connections. In the case of using standard fasteners (bolts and nuts), there were evaluated mainly influence of advanced coatings using and lubricants and modern insurance sealants and adhesives into the threaded surfaces of bolted joints. To achieve the set goals, modular design-testing stand was designed. Achieved analytical findings and experimental verification will allow to refine the technological processes of assembly of bolted joints, which will have a major impact on the strength, durability, and reliability of the screw connections (fasteners). Discovered knowledge is not currently available for design engineers during the designing of bolted joints. Design engineers typically use only recommendation of retailers and manufacturers when designing of bolted joints. In industrial practice, it is now very often required to calculate the screws according to the standard VDI 2230. Standard VDI 2230 specifies a procedure for the strength calculation of pre-stressed screw connections. The calculation is usually done on a personal computer using the KissSoft program.

Keywords Pre-stressed bolted connection · Advanced coatings and lubricants · Sealants · Adhesives · Standard VDI 2230

1 Introduction

The process of tightening threaded fastener assemblies, especially for critical bolted joints, involves controlling both input torque and angle of turn to achieve the desired result of proper preload of the bolted assembly.

Understanding the role of friction in both the underhead and threaded contact zones is the key to defining the relationship between torque, angle, and tension. There can be as many as 200 or more factors that affect the tension created in a bolt when tightening torque is applied. Fortunately, torque–angle signature curves can be obtained for most bolted joints.

J. Kanaval (✉) · E. Cézová · F. Starý
Czech Technical University in Prague, Technická 4, 16607 Prague, Dejvice, Czech Republic
e-mail: jan.kanaval@fs.cvut.cz

Š. Medvecký et al. (eds.), *Current Methods of Construction Design*, Lecture Notes in Mechanical Engineering, https://doi.org/10.1007/978-3-030-33146-7_34

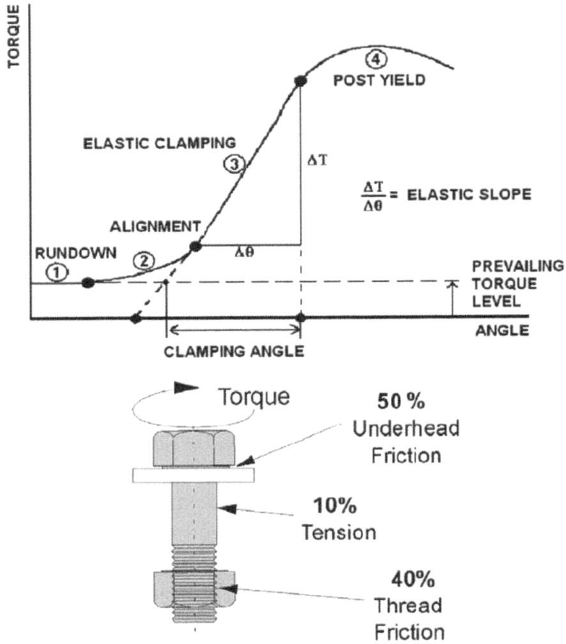

Fig. 1 Measured torque–angle curve. Four zones of the tightening process [1]

By combining the torque–angle curves with a few simple calculations and a basic understanding of the engineering mechanics of threaded fasteners, you can obtain the practical information needed to evaluate the characteristics of individual fastener tightening processes. The torque–angle curves can also provide the necessary information to properly qualify the capability of tightening tools to properly tighten a given fastener (see Fig. 1) [1].

There is shown a pre-stressed bolted connection loaded with a centric axial force and diagram of pre-stressed bolted connection force—deformation (see Fig. 2).

In industrial practice, it is now very often required to calculate the screws according to the standard VDI 2230. Standard VDI 2230 specifies a procedure for the strength calculation of pre-stressed screw connections [2]. The calculation is usually done on a personal computer using the KissSoft program.

2 Methods and Materials

The simple testing equipment for evaluation of threaded contact friction between the bolt and the nut was designed and manufactured at Department Designing and Machine Components of the Czech Technical University in Prague and simple experiments there were realized too (see Figs. 3 and 4).

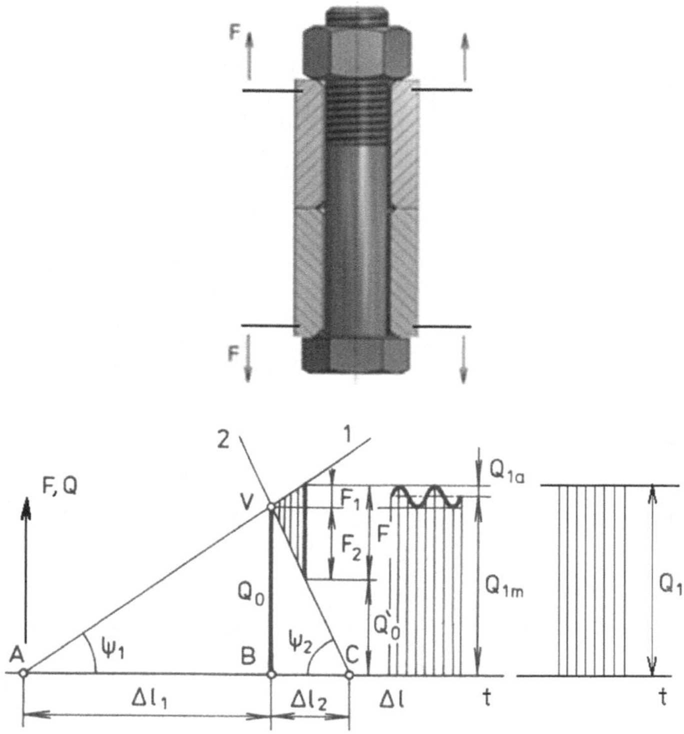

Fig. 2 Pre-stressed bolted connection loaded with a centric axial force; diagram of pre-stressed bolted connection force—deformation [3]

In the case of using standard fasteners (bolts and nuts), there were evaluated mainly influence of advanced coatings using and lubricants and modern insurance sealants and adhesives into the threaded surfaces of bolted joints.

There were also tested M10 screws with different surface treatment (galvanized zinc, black oxide finish and stainless steel A2-70) without grease, and with different types of lubricants (Glattol, Molyka) and different insulating adhesives (Loctite 243 [4], Loctite 270 [5]) in the laboratory of the Department Designing and Machine Components of the Czech Technical University in Prague. The results of these experiments are shown in Tables 1 and 2, and see Fig. 5.

On this newly designed testing equipment (see Fig. 4) can be detected:

- the value of the friction coefficient in the thread;
- the value of the friction coefficient under screw head (under nut);
- the tension axial pre-stress reducing after the screw connection is assembled;
- to evaluate the characteristics of individual fastener tightening processes (Torque–angle curves);
- verifying the strength of the screw connection.

Fig. 3 Simple testing equipment for evaluation of threaded contact friction (for torque measurement) HBM torque sensor T20WN, 20 Nm, was used

Technical parameters of the testing equipment:

- maximum testing torque moment 185 Nm;
- maximum testing tension axial force 100 kN;
- maximum testing rotation speed 97 min^{-1};
- maximum testing stroke 300 mm.

3 Results

The results of realized experiments are shown in Tables 1 and 2, and in Fig. 5.

There are presented (as an example) calculated, and measured values for bolt size M10 × 60, strength class 8.8, galvanized (Zn) and nut M10, strength 8, galvanized (Zn) when using insulating adhesive Loctite 270 and when measured upwards in Table 1.

There are presented experimentally determined thread friction coefficients for all realized measurements in Table 2.

There is made a graphical comparison of all experimentally determined thread friction coefficients for all realized measurements in Fig. 5.

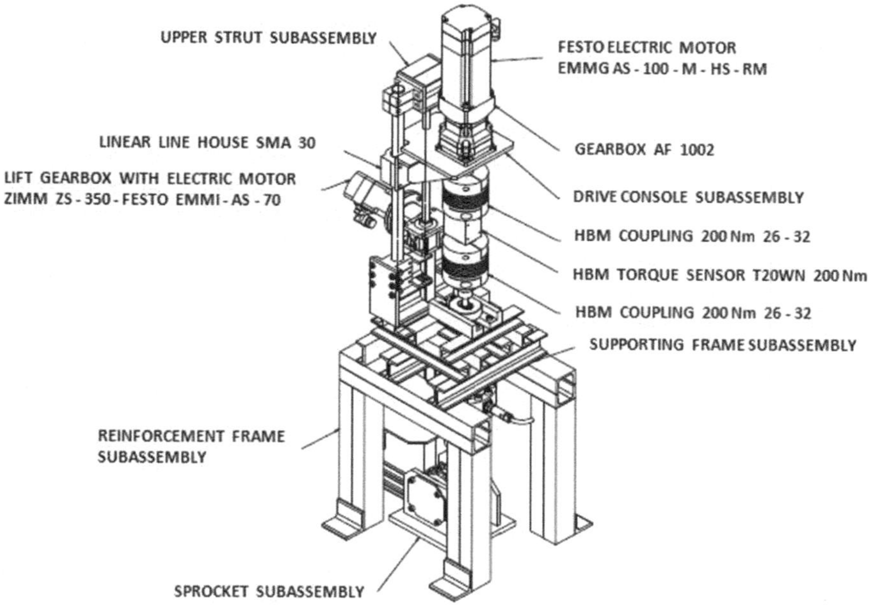

Fig. 4 Testing equipment for complex analysis of screw fasteners

Table 1 Table presenting measured and calculated values for bolt size M10 × 60. Strength class 8.8, galvanized (Zn), nut: M10, strength class 8, galvanized (Zn), Loctite 270, direction up. Hinge mass 2.63 kg, nut friction moment Mp: 0 Nmm

Weight mass (including m [kg])	Tensile force Qo [N]	Wrench torque moment Mkk [N.mm]	Wrench torque moment without nut Mkk [N.mm]	Calculated thread friction angle φ'[°]	Calculated thread friction coefficient f' [1]	Calculated friction coefficient f [1]
82.23	806.68	1900	1900	24.532	0.456	0.395
161.28	1582.16	3400	3400	22.434	0.413	0.358
200.48	1966.71	4100	4100	21.766	0.399	0.346
249.62	2448.77	4900	4900	20.884	0.382	0.330
Statistical evaluation of experiment		Sample average μ		22.404	0.413	0.357
		Sample standard deviation σ		1.554	0.032	0.028

4 Discussion

Aim of the last experiments was to determine the magnitude of the friction coefficient in the thread depending on the different surface treatment (galvanized zinc, black

Table 2 Table presenting experimentally detected thread friction coefficient [f'] for Tested M10 × 70 (M10 × 80) screws with different surface treatment (galvanized zinc, black oxide finish, and stainless steel A2-70) without grease, and with different types of lubricants (Glattol, Molyka) and different insulating adhesives (Loctite 243, Loctite 270)

Screw connection specification		Bolt M10 × 60, strength class 8.8, galvanized (Zn), nut M10, strength class 8, galvanized (Zn)	Bolt M10 × 80, strength class 5.8, black oxide finish, nut M10, strength class 05, black oxide finish	Bolt M10 × 60, stainless steel, strength class A4-70, nut M10, stainless steel, strength class A4-80
Experimentally detected thread friction coefficient [f']				
Without lubricant	Direction down	0.268	0.177	0.346
	Direction up	0.484	0.178	0.560
	Average value	0.376	0.178	0.453
With grease Glattol	Direction down	0.186	0.187	0.175
	Direction up	0.174	0.174	0.171
	Average value	0.180	0.181	0.173
With grease Molyka	Direction down	0.128	0.178	0.167
	Direction up	0.140	0.161	0.166
	Average value	0.134	0.170	0.167
With insulating adhesives Loctite 243	Direction down	0.241	0.142	0.224
	Direction up	0.295	0.138	0.274
	Average value	0.268	0.140	0.249
With insulating adhesives Loctite 270	Direction down	0.440	0.349	0.343
	Direction up	0.413	0.389	0.334
	Average value	0.427	0.369	0.339

oxide finish, and stainless steel A2-70) without grease, and with different types of lubricants (Glattol, Molyka) and different insulating adhesives (Loctite 243, Loctite 270) of the screw and the nut. The experimentally determined values of the thread friction coefficient correspond to the values reported in the technical literature.

It was demonstrated that the using of lubricants when assembling the screw connections decreases the coefficient of friction in the thread. And also was demonstrated that the using modern insulating adhesives (Loctite 243, Loctite 270) when assembling the screw connections increases the coefficient of friction in the thread.

5 Conclusions

The simple experiments performed in the past verified the chosen measurement methodology for measuring the coefficient of friction in the threads. The experience

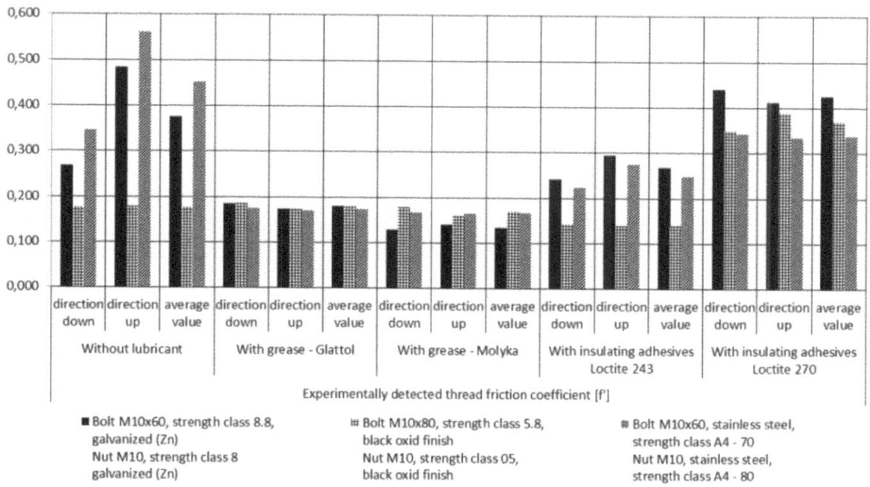

Fig. 5 Graph presenting experimentally detected thread friction coefficient for tested M10 × 70 (M10 × 80) screws with different surface treatment (galvanized zinc, black oxide finish, and stainless steel A2-70) without grease, and with different types of lubricants (Glattol, Molyka) and different insulating adhesives (Loctite 243, Loctite 270)

gained in this way was used to design new testing equipment for complex screw connection analysis (when using modern screw fasteners with coatings; when using different types of lubricants; when using different insulating adhesives).

It has been shown that the using of modern insulating adhesives (Loctite 243, Loctite 270) when assembling the screw connections increases the coefficient of friction in the thread.

Acknowledgements This study was supported by Student grants competition of the Czech Technical University in Prague, Faculty of Engineering, no. SGS16/145/OHK2/2T/12.

References

1. Shoberg R (2010) Engineering fundamentals of threaded fastener design and analysis, MI 48335, USA: RS technologies. A division of PCB load and torque, Inc
2. Guideline VDI 2230 (2003) Systematic calculation of high duty bolted joints: Part 1, joints with one cylindrical bolt
3. Švec V (2008) Parts and mechanisms of machines, joints and connecting parts. CTU, Prague
4. Prumex SRO (2018) Technical sheet, information on http://proprumysl.cz/soubory/20/243.pdf. Hrušovany nad Jevišovkou. Last accessed 11 June 2018
5. Prumex SRO (2018) Technical sheet, information on https://proprumysl.cz/soubory/33/270.pdf. Hrušovany nad Jevišovkou. Last accessed 11 June 2018

Measuring Mass Moment of Inertia of a Rotor—Two Simple Methods Using no Special Equipment

Peter Kaššay and **Robert Grega**

Abstract The paper presents simple methods for measuring mass moment of inertia of rotors without need of dismantling the rotor or any special laboratory equipment. Only simple mechanical components (hub, threaded rod, weights, screws, nuts, etc.) and a common compact digital camera capable of taking 30 fps VGA video recording were used. First method is based on unwinding a specific mass (weight) hanged on string which is wound on cylindrical hub, where the times of consecutive rotations of the shaft are recorded. The second one is based on oscillation of pendulum with specified properties added to the rotor. In this case, the times of extreme positions of the pendulum are recorded. The evaluation of recorded time values is based on simple physical principles. The influence of various parameters as weight of added mass, length of pendulum and friction is examined. These methods are suitable for determining mass moment for machines without relevant technical data, when dismounting of rotor would not be an appropriate solution or special equipment is not available.

Keywords Mass moment of inertia · Rotor · Experimental determination · Measurement · Unwinding mass · Added pendulum

1 Introduction

Currently, ever increasing demands on performance, safety, operational reliability, and durability of machinery are imposed. Therefore, the amount of technical problems which need to be effectively solved in the design process of new design and the complexity of performed calculations keep growing [1]. For the realization of these calculations and design, the high level of knowledge about mechanical properties of the components of mechanical systems and physical processes which arise during the operation of mechanical systems is necessary [2, 3].

P. Kaššay (✉) · R. Grega
Technical University of Košice, Letná 9, 40200 Košice, Slovak Republic
e-mail: Peter.Kassay@tuke.sk

R. Grega
e-mail: Robert.Grega@tuke.sk

Š. Medvecký et al. (eds.), *Current Methods of Construction Design*, Lecture Notes in Mechanical Engineering, https://doi.org/10.1007/978-3-030-33146-7_35

In the field of mechanical drives, problems of controlling driving elements (motors) according to desired operating parameters [4], problems of torsional vibration [5–7], selecting proper drive elements (e.g., mechanical transmissions, shaft couplings, etc.) [8, 9], noisiness [10], and fault diagnostics [11] are encountered very often. In all of the mentioned problems, the knowledge of specific mechanical parameters is essential to solving them properly [12–14].

One of the basic parameters of rotary machines is the mass moment of inertia of the rotor. This is usually provided by the manufacturer of the given device. But sometimes, in case of certain manufacturers, older machines, prototypes, or with additional equipment mounted, etc. we have to determine the moment of inertia by ourselves.

In literature dealing with dynamics and vibration, several methods for measuring the mass moment of inertia can be found [15, 16]. But usually, the dismantling of the rotor is required. Some frequency converters offer an option to measure mass moment of inertia of whole drive during acceleration or deceleration by measuring the motor currents. These currents can be also measured directly [17]. But in this case, a frequency converter and electric motor with precisely known parameters, eventually equipment for measuring motor currents is needed.

The paper presents simple methods for measuring mass moment of inertia of rotors without need of dismantling the rotor or any special laboratory equipment. First method is based on unwinding a mass hanged on string wound on cylindrical hub, and the second one is based on oscillation of pendulum with specified properties added to the rotor. Presented methods are suitable for determining mass moment for machines without relevant technical data, when dismounting of rotor would not be an appropriate solution or special equipment is not available. The methods are presented on an example of specific DC electric motor.

2 Measuring Mass Moment of Inertia of Electric Motor

Proposed methods for measuring mass moment of inertia are presented on a SM 160 L type DC electric motor shown in Fig. 1. This motor is a part of our laboratory equipment. Previously, it was many times used for torsional vibration measurements and analyses [1]. Presently, we use other electric motors for these purposes. So this older motor now serves as equipment for testing methods for measuring moments of inertia, bearing resistance, for student laboratory practices, etc.

The motor has a hub mounted on shaft end. The brushes of electric motor were removed in order to minimize the friction during measurements. The axis of motor is lifted approximately 1.5 m above the ground and the horizontal position of axis was checked. For both presented methods, a common compact digital camera capable of taking 30 fps VGA video recording was used.

Fig. 1 SM 160 L type DC electric motor prepared for measurements

2.1 *Measuring Mass Moment of Inertia by Unwinding a Weight*

This method uses a mass hanged on string which is wound on cylindrical hub mounted on the rotor shaft as it is shown in Fig. 2. The motor axis has to be in horizontal position. The string has to be flexible but not too stretchy, with uniform cross-section, and the windings may not cross. By releasing the rotor, the weight acts on the rotor with

Fig. 2 Measuring mass moment of inertia by unwinding a weight

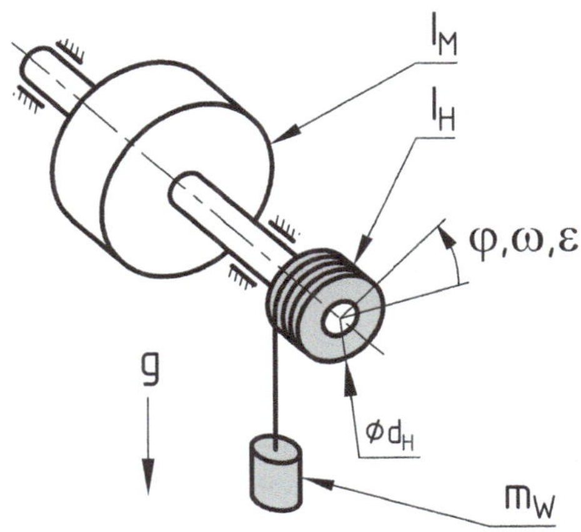

a constant torque which gives the rotor a constant angular acceleration (considering only constant friction torque).

The mass moment of inertia of the whole setup I consists of mass moment of inertia of the motor I_{M}, hub I_{H} and equivalent mass moment of inertia of the weight m_{W}:

$$I = I_{\mathrm{M}} + I_{\mathrm{H}} + m_{\mathrm{W}} \cdot \frac{d_{\mathrm{H}}^2}{4} \tag{1}$$

where $d_{\mathrm{H}} = 0.05825$ m is the diameter of string's medium fiber.

The torque from external forces acting on the hub can be computed as:

$$M_{\mathrm{K}} = m_{\mathrm{W}} \cdot g \cdot \frac{d_{\mathrm{H}}}{2} - M_{\mathrm{f}} \tag{2}$$

where:

- $g = 9.81$ m s^{-2} is gravitational acceleration,
- M_{f}—friction torque.

Then, the acceleration of the rotor can be computed as:

$$\varepsilon = \frac{M_{\mathrm{K}}}{I} \tag{3}$$

Then, the mass moment of inertia of motor I_{M} can be expressed from previous formulas as:

$$I_{\mathrm{M}} = \left(\frac{m_{\mathrm{W}} \cdot g \cdot \frac{d_{\mathrm{H}}}{2} - M_{\mathrm{f}}}{\varepsilon} \right) - I_{\mathrm{H}} - m_{\mathrm{W}} \cdot \frac{d_{\mathrm{H}}^2}{4} \tag{4}$$

The twist angle φ depending on time can be described by formula:

$$\varphi = \varphi_0 + \frac{\varepsilon}{2} \cdot (t - t_{\mathrm{START}})^2 \tag{5}$$

where:

- φ_0 is initial twist angle (by zero speed),
- t_{START} is start time (by zero speed).

This formula can be expressed in general form of second-order polynomial as:

$$\varphi = a_0 + a_1 \cdot t + \frac{\varepsilon}{2} \cdot t^2 \tag{6}$$

where a_0, a_1 and $^{\varepsilon}/_2$ are coefficients of polynomial. The process of unwinding is recorded and the time values t_i, when the mark on the hub is passing the reference on the motor housing, are read from the recording. The angles of twist corresponding to times t_i can be computed as:

$$\varphi_i = 2\pi \cdot i \tag{7}$$

Through the measured values, a regression curves corresponding to Formula (6) were fitted, where the quadratic coefficients are equal to $^{\varepsilon}/_{2}$.

The measured values and results of computations are listed in Table 1 in numerical form. Values of mass moment of inertia in column $^{*}I_{\mathrm{M}}$ are computed considering zero friction torque M_{f}, and for computed values in column $^{**}I_{\mathrm{M}}$ an optimized value of constant friction torque M_{f} was used.

For illustration, measured values with evaluated regression curves for $m_{\mathrm{W}} = 2$ kg are shown in Fig. 3.

In Fig. 4 (left) are shown the values of computed mass moment inertia I_{M} depending on used weight m_{W}. The values of I_{M} lie in a wide range. They are slightly rising with decreasing weight from maximum weight of $m_{\mathrm{W}} = 3$ kg until $m_{\mathrm{W}} \approx 1$ kg, then rapidly increasing from $m_{\mathrm{W}} \approx 0.5$ kg. The computed values of I_{M} for weight $m_{\mathrm{W}} =$ 0.25 and 0.15 kg have also a considerable dispersion. Mentioned phenomena can be attributed to increasing influence of friction torque M_{f} for low weight values.

In Fig. 4 (right) are shown result where constant friction torque was used for computing mass moment of inertia values. As friction has increasing impact with decreasing weight, only measured data for $m_{\mathrm{W}} = 1$ kg and higher were used for moment of inertia computations. The value of constant friction torque M_{f} was optimized so that the sum of squared residuals between average moment of inertia $\overline{I_{\mathrm{M}}}$ and computed values of I_{M} is minimum. The average computed moment of inertia is $\overline{I_{\mathrm{M}}} = 0.09426\ \mathrm{kg\,m^2}$. The standard sample deviation is $s_{I_{\mathrm{M}}} = 0.001488\ \mathrm{kg\,m^2}$ which is equal to 1.58% of the average value. The optimized value of friction torque is $M_{\mathrm{f}} = 0.06681$ N m. For practical purposes, the moment of inertia can be considered as $I_{\mathrm{M}} = 0.094 \pm 0.0015\ \mathrm{kg\,m^2}$.

2.2 *Measuring Mass Moment of Inertia by an Added Pendulum*

This method uses an added pendulum mounted on the motor shaft as it is schematically shown in Fig. 5. The motor axis, as for previous method, has to be in horizontal position. The mass parameters of pendulum (weight, position of center of gravity—*CG*, mass moment of inertia around motor axis) have to be determined with a sufficient precision [18].

The mass moment of inertia of the whole setup I consists of mass moment of inertia of the motor I_{M} and pendulum I_{P}:

$$I = I_{\mathrm{M}} + I_{\mathrm{P}} \tag{8}$$

Table 1 Measured and computed values for measurements with unwinding weight

No.	m_W [kg]	t_0 [s]	t_1 [s]	t_2 [s]	t_3 [s]	t_4 [s]	t_5 [s]	t_6 [s]	$^{\varepsilon}/_2$ [rad s^{-2}]	$^*I_M$ [kg m^2]	$^{**}I_M$ [kg m^2]
1	3	2.701	3.542	4.023	4.408	4.744	5.041	–	3.880	0.1029	0.09427
2		2.804	3.578	4.062	4.445	4.771	5.056	–	3.977	0.1002	0.09179
3		3.212	4.186	4.707	5.089	5.439	5.733	–	3.886	0.1027	0.09411
4	2.5	2.896	4.090	4.660	5.101	5.471	5.802	6.105	3.205	0.1043	0.09386
5		3.100	4.242	4.801	5.235	5.631	5.954	6.246	3.181	0.1051	0.09462
6		2.105	2.669	3.121	3.504	3.838	4.134	–	3.081	0.1088	0.09792
7	2	3.163	4.474	5.108	5.615	6.048	6.407	6.738	2.530	0.1062	0.09300
8		0.393	1.399	1.995	2.469	2.877	3.236	3.573	2.547	0.1055	0.09233
9		1.702	2.798	3.435	3.888	4.335	4.700	5.031	2.483	0.1083	0.09489
10	1.5	2.347	3.873	4.602	5.187	5.691	6.129	6.534	1.793	0.1132	0.09458
11		2.425	3.809	4.539	5.106	5.600	6.049	6.452	1.779	0.1142	0.09537
12		2.444	3.672	4.381	4.952	5.436	5.872	6.276	1.776	0.1144	0.09554
13	1	2.221	3.904	4.837	5.570	6.185	6.752	7.263	1.101	0.1239	0.09353
14		2.844	4.673	5.608	6.343	6.981	7.538	8.051	1.106	0.1233	0.09308
15		2.773	4.185	5.081	5.787	6.408	6.959	7.460	1.085	0.1258	0.09500
16	0.75	3.944	5.544	6.571	7.404	8.136	8.803	9.405	0.7433	0.1385	–
17		3.736	5.323	6.357	7.191	7.922	8.574	9.187	0.7468	0.1378	–
18		2.869	4.636	5.709	6.554	7.300	7.959	8.568	0.7622	0.1349	–
19	0.5	2.488	4.806	6.187	7.309	8.267	9.163	9.971	0.4327	0.1596	–
20		2.982	5.144	6.486	7.605	8.572	9.443	10.238	0.4314	0.1601	–
21		2.912	5.005	6.344	7.433	8.401	9.270	10.082	0.4245	0.1628	–

(continued)

Table 1 (continued)

No.	m_W [kg]	t_0 [s]	t_1 [s]	t_2 [s]	t_3 [s]	t_4 [s]	t_5 [s]	t_6 [s]	$^{\varepsilon}/_{2}$ [rad s^{-2}]	$^{*}I_M$ [kg m^2]	$^{**}I_M$ [kg m^2]
22	0.25	4.274	8.010	10.315	12.246	13.937	15.502	16.953	0.1342	0.2609	–
23		7.542	9.983	11.972	13.720	15.310	16.806	–	0.1142	0.3075	–
24		5.201	8.298	10.479	12.326	13.991	15.533	16.955	0.1257	0.2789	–
25	0.15	4.729	10.824	14.691	18.076	21.138	24.021	26.783	0.03648	0.5823	–
26		4.603	10.896	14.833	18.232	21.306	24.214	26.970	0.03704	0.5734	–
27		1.002	8.137	12.078	15.463	18.532	21.404	24.171	0.04053	0.5236	–

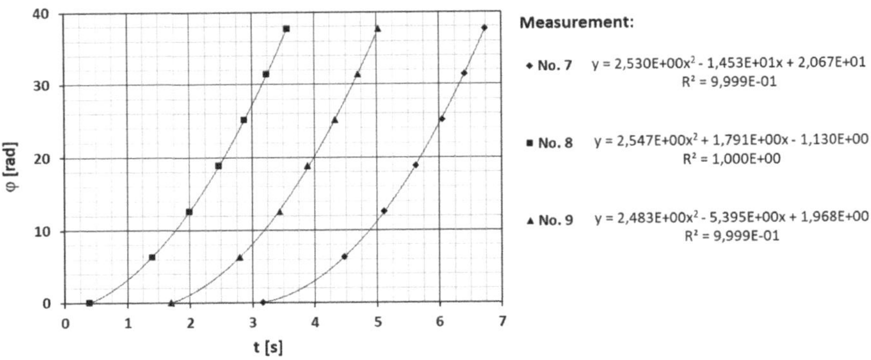

Fig. 3 Selected measured values with evaluated regression curves for weight $m_W = 2$ kg

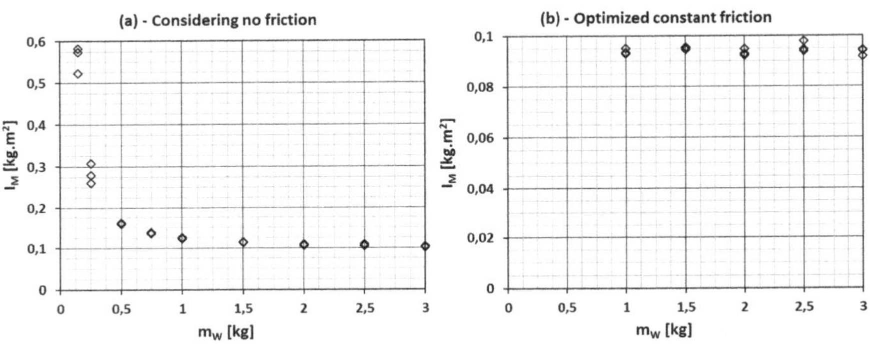

Fig. 4 Moment of inertia—computed values for unwinding weight method

Fig. 5 Measuring mass moment of inertia by added pendulum

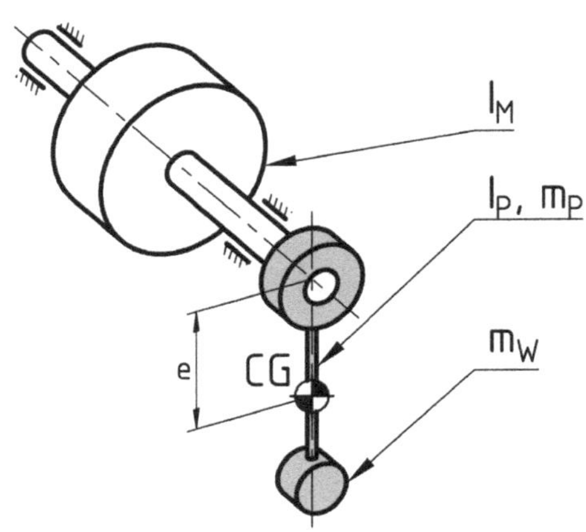

The approximate equivalent torsional stiffness k resulting from pendulum mass m_P with center of gravity in distance e from motor axis is:

$$k \approx m_P \cdot g \cdot e \tag{9}$$

Then, the natural frequency of free oscillation can be computed as:

$$f_0 = \frac{1}{2\pi} \cdot \sqrt{\frac{k}{I}} \tag{10}$$

And finally, according to previous formulas, the mass moment of inertia of motor can be expressed as:

$$I_M = \frac{m_P \cdot g \cdot e}{\pi^2 \cdot (2 f_0)^2} - I_P \tag{11}$$

The design of pendulum is shown as 3D model in Fig. 6. The weight (or a set of weights) is mounted on a threaded rod by nuts and washers so that the weights and their position can be easily changed and measured. The parameters of pendulum (weight, position of the center of gravity and the mass moment of inertia) were determined from its 3D model.

Different weights and position of weights were used. The pendulum was tilt on side, then released and the oscillation was recorded. For the first measurements, the times in the moment of reaching extreme positions for both sides were read from the video recording. Thus, the difference between two consecutive times are half-periods of oscillation $T_0/2$ giving the double of the oscillation frequency $2{\cdot}f_0$.

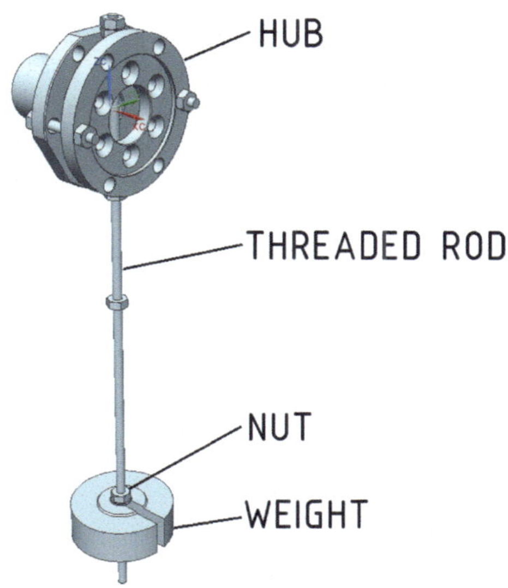

Fig. 6 Design of the pendulum used for experiments

Theoretically, for friction damping the period of oscillation should be a constant value [15, 16]. For our case, the maximum recorded differences between half-periods were no greater than ±1/30 of second corresponding to the frame rate of used digital camera. Therefore, we considered the frequency of oscillation as constant and only the average value of $2{\cdot}f_0$ for selected n half-periods was computed (as difference between the times of initial and final amplitude divided by number of amplitudes). The measured values and results of computations are listed in Table 2 in numerical form.

We also tried to measure the oscillation frequency by a laser optical speed sensor connected to a portable vibration analyzer. As reference for the sensor, a reflexive mark was attached on an additional nut on the threaded rod. Unfortunately, the evaluated value of frequency was constantly changing. It was due to fact that the reflexive mark has no zero width and the difference between the moments of entering and leaving the laser beam is rising with decreasing amplitude. Thus, this method of frequency measurement cannot be directly used for evaluating the frequency of free oscillating pendulum with friction damping. The optical speed sensor could be used

Table 2 Measured and computed values for measurements with added pendulum

No.	m_W [kg]	m_P [kg]	e [m]	I_P [kg m^2]	n [–]	$T_0/2$ [s]	$2{\cdot}f_0$ [Hz]	I_M [kg m^2]
1	1	2.685	0.2132	0.2794	23	0.8101	1.234	0.09403
2	0.5	2.185	0.1498	0.1588	22	0.8772	1.140	0.09153
3	2	3.685	0.2853	0.5095	59	0.7616	1.313	0.09667
4	3	4.685	0.3227	0.7222	35	0.7391	1.353	0.09859
5	0.5	2.185	0.1337	0.1259	15	0.8667	1.154	0.09220
6	1	2.685	0.1884	0.2177	39	0.7897	1.266	0.09587
7	2	3.685	0.2487	0.3872	32	0.7281	1.373	0.09570
8	3	4.685	0.2800	0.5439	46	0.7001	1.428	0.09524
9	0.5	2.185	0.1220	0.1049	18	0.8592	1.164	0.09071
10	1	2.685	0.1695	0.1764	30	0.7735	1.293	0.09425
11	2	3.685	0.2226	0.3103	50	0.7073	1.414	0.09763
12	3	4.685	0.2489	0.4306	50	0.6741	1.483	0.09609
13	0.5	2.185	0.1097	0.08556	16	0.8687	1.151	0.09423
14	1	2.685	0.1502	0.1394	25	0.7615	1.313	0.09303
15	2	3.685	0.1948	0.239	50	0.6860	1.458	0.09677
16	3	4.685	0.2164	0.3273	50	0.6493	1.540	0.09749
17	0.5	2.185	0.09765	0.06913	22	0.8787	1.138	0.09463
18	1	2.685	0.1310	0.1078	25	0.7614	1.313	0.09490
19	2	3.685	0.1673	0.1785	40	0.6699	1.493	0.09651
20	3	4.685	0.1840	0.2396	50	0.6286	1.591	0.09899

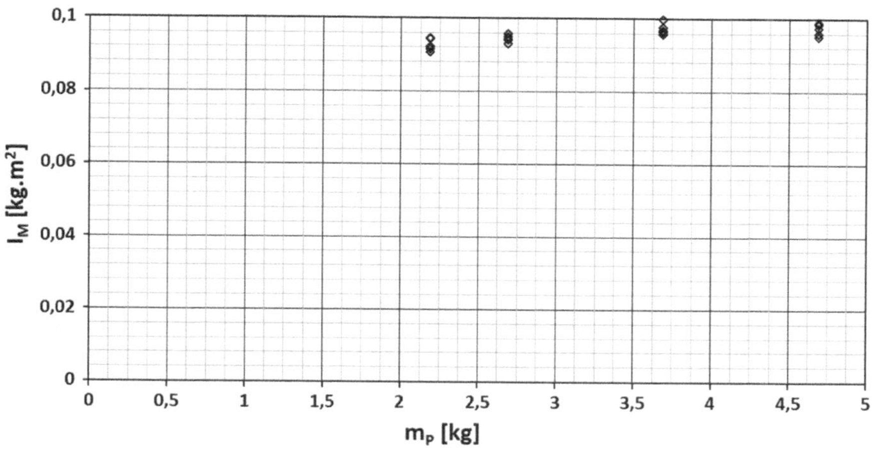

Fig. 7 Moment of inertia—computed values for added pendulum method

in case when the acquired signal is processed with an algorithm adapted to given conditions.

The dependence of computed mass moment inertia on pendulum weight is shown in Fig. 7. It can be seen that the computed values of moment of inertia are slightly increasing with increasing weight. Which could be caused by the presence of other than dry friction, dependence of dry friction on radial forces acting on the rotor bearings, using imprecise mass parameters of pendulum [18], etc.

The average computed moment of inertia for this method is $\overline{I_M} = 0.09547$ kg m^2. The standard sample deviation is $s_{I_M} = 0.002403$ kg m^2 which is equal to 2.52% of the average value. For practical purposes, the moment of inertia can be considered as $I_M = 0.095 \pm 0.0024$ kg m^2.

3 Conclusion

Based on presented data, both of used methods are giving very similar results for the rotor mass moment of inertia for the used SM 160 L type DC electric motor:

- $I_M = 0.094 \pm 0.0015$ kg m^2 ($\pm 1.58\%$) determined by unwinding weight method versus,
- $I_M = 0.095 \pm 0.0024$ kg m^2 ($\pm 2.52\%$) determined by added pendulum method.

The main advantage of presented methods is the usage of very simple equipment as weights, standard mechanic components, and only a common digital camera as recording device.

However, the precision of obtained mass moment of inertia values should be confirmed by a more precise method unaffected by friction, e.g., bifilar torsional pendulum with unassembled rotor.

Also the assumption of constant friction torque of rotor bearings should be closely examined by additional measurements. If the friction torque depends, e.g., on radial force or speed, the presented formulas for evaluating mass moment of inertia should be adapted to that fact.

But the above-mentioned problems will be part of further research.

Acknowledgements This chapter was created with support from the Ph.D. students and young researchers project Solution of a Control System Element for Mechanical Systems' Continuous Tuning.

References

1. Kaššay P (2014) Modeling, analysis and optimization of torsional oscillating mechanical systems: habilitation thesis. Technical University of Košice, Košice
2. Homišin J (2016) Characteristics of pneumatic tuners of torsional oscillation as a result of patent activity. Acta Mech Automat 10(4):316–323
3. Urbanský M (2014) Theoretic and experimental determination of the flow resistance coefficient at gaseous medium flow into and out of the pneumatic coupling. Sci J Silesian Univ Technol Ser Transp 85:119–125
4. Kyslan K, Slapak V, Lacko M, Durovsky F (2015) Cost functions in finite control set model predictive control of permanent magnet DC machine. In: Fedak V, Dudrik J, Jakopovic Z, Kolonic F (eds) 2015 international conference on electrical drives and power electronics (EDPE). IEEE, New York, pp 124–129
5. Sapietová A, Dekýš V (2016) Dynamic analysis of rotating machines in MSC.ADAMS. In: Vavro J, Ondrušová D (eds) MMS 2015. Elsevier Science BV, Amsterdam, pp 143–149
6. Urbanský M, Homišin J, Čopan P (2013) Examination of mechanical system response to gaseous media pressure changes in the pneumatic coupling. Sci J Silesian Univ Technol Ser Transp 81:143–149
7. Urbanský M (2017) Harmonic analysis of torsional vibration force excitation. Sci J Silesian Univ Technol Ser Transp 97:181–187
8. Łazarz B, Wojnar G, Madej H, Czech P (2009) Evaluation of gear power losses from experimental test data and analytical methods. Mechanika 80(6):56–63
9. Vojtková J (2017) Determining and verifying the geometric characteristics of helical grooves in the worm in planetary toroidal drives in a more effective way. Sci J Silesian Univ Technol Ser Transp 96:197–204
10. Liptai P, Moravec M, Lumnitzer E, Gergeľová M (2017) Proposal of the sound insulating measures for a vibrational sorter and verification of the measured effectiveness. Adv Sci Technol Res J 11(3):196–203
11. Czech P, Wojnar G, Burdzik R, Konieczny Ł, Warczek J (2014) Application of the discrete wavelet transform and probabilistic neural networks in IC engine fault diagnostics. J VibroEng 16:1619–1639
12. Homišin J, Čopan P, Urbanský M (2013) Experimental determination of characteristic properties of selected types of flexible shaft couplings. Sci J Silesian Univ Technol Ser Transp 81:51–57

13. Homišin J, Urbanský M (2011) Results of measurement of transitional effects at air pressure changes in pneumatic coupling during mechanical system operation. Sci J Silesian Univ Technol Ser Transp 72:41–50
14. Krajňák J (2014) Influence of gaseous media on damping of pneumatic couplings. Sci J Silesian Univ Technol Ser Transp 85:83–87
15. Beards CF (1996) Structural vibration analysis and damping. Arnold, London
16. Brepta R, Půst L, Turek F (1994) Mechanical oscillation. Sobotáles, Praha
17. Poliak F, Fedák V, Zboray L (1987) Electric drives. SNTL, Praha
18. Mantič M, Kuľka J, Krajňák J, Kopas M, Schneider M (2016) Influence of selected digitization methods on final accuracy of 3D model. In: Majerník M, Daneshjo N (eds) ICESPM 2015, PMES. CRC Press, Boca Raton, pp 475–480

The Influence Long-Term Operating Load to the Riveted Join

Václav Kubec, Josef Hodek, Antonín Prantl and Petr Votápek

Abstract The paper deals with the life of the joints in the long-term operational load. For designing and dimensioning is important the feedback that the design was correctly designed and the component was not over-sized or undersized. This feedback becomes increasingly important, because it is now a trend to design lightweight structures with maximum use of the material. The consequence of this suggestion method is that the established safety margin does not allow for any errors or inaccuracies. The danger and weak side of this approach is to estimate the long-term life of the designed structure. In the research, we had the opportunity to analyze riveted bridge structure, which has been in operation since 1905. An analysis of the residual stress is published in the paper and is compared with the assumed bias in the newly made construction. The state of the rivet joint after a century operation is judged on the basis of this comparison.

Keywords Joining · Riveted joint · Long-term load · Residual stress

1 Introduction

The problem of how to join materials together emerged with the first human activities that could be described as manufacturing. Although this problem may seem to have been solved already, research into and development of joining processes are still relevant, particularly due to the ever-increasing number of new materials.

In terms of their lifetime, prestressed structures tend to exhibit better properties than those without prestressing.

This year, a reconstruction of bridges (see Fig. 1) in a railway station in the Czech city of Pilsen began. Under this project, their riveted steel frames will be replaced with reinforced concrete structures. The bridges have been in use since 1905, over

V. Kubec (✉) · J. Hodek · A. Prantl
COMTES FHT a.s., Průmyslová 995, 33401 Dobřany, Czech Republic
e-mail: vkubec@comtesfht.cz

P. Votápek
University of West Bohemia, Univerzitní 22, 30614 Plzeň, Czech Republic
e-mail: pvotapek@kks.zcu.cz

Š. Medvecký et al. (eds.), *Current Methods of Construction Design*, Lecture Notes in Mechanical Engineering, https://doi.org/10.1007/978-3-030-33146-7_36

Fig. 1 Bridge before reconstruction [1]. Test sample—on the right

a century. Their original riveted frames were to be scrapped. We were thus able to arrange with their owner to take samples for investigation. These samples will be studied using non-destructive as well as destructive methods. The purpose is to evaluate effects of service loads and time on this structure and its material. One of the samples, which were used for analysis, is shown in Fig. 1.

This contribution concerns a riveted joint in the bridge and focuses on comparing residual stress values determined by the contour method and by FE analysis.

2 Theoretical Background

Two methods were used to determine the condition of the joint. One was the contour method for determining residual stresses in components and the other was an ordinary finite element (FE) calculation. The purpose was to assess the condition of the joints after 100 years in service by comparing the results of the contour method and the FE analysis.

2.1 Contour Method

The contour method [2–4] is based on the elastic superposition principle: if parts of a divided body are subjected to an external force or to a defined displacement in the cut area such that the shape change resulting from the division is eliminated, then these forces must be equivalent to the stress present in the intact body before cutting. This principle is shown schematically in Fig. 2.

The component in which residual stress is present, Fig. 2a, is cut into two parts. Normal and tangential components of the released stress cause the surface deformation shown in Fig. 2b. By imposing external forces, the original stress conditions can be restored in Fig. 2c.

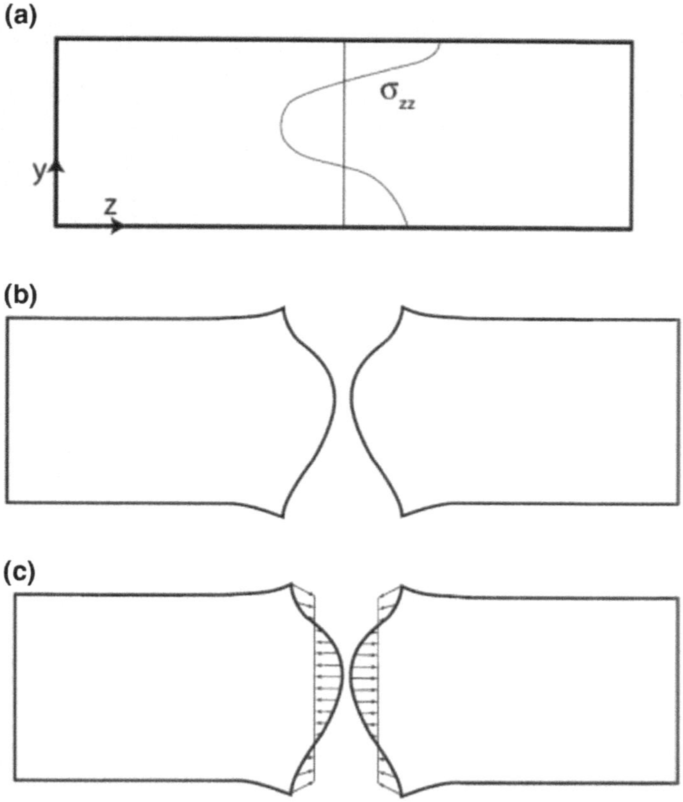

Fig. 2 Schematic representation of the contour method

The following principles must apply:

- all stresses after cutting are purely elastic,
- the cutting process must not have a significant effect on the resulting parts and
- the cut surface must be a plane.

After cutting, the new surface becomes distorted due to the normal and tangential stresses released. In practice, only normal stresses can be measured. If, however, the measurement is carried out on both cut surfaces and the values are averaged, the tangential stresses cancel out.

3 Application of Contour Method to Bridge Rivet

A bridge rivet was cut out from a fragment of the bridge structure using a table saw, as illustrated in Fig. 3.

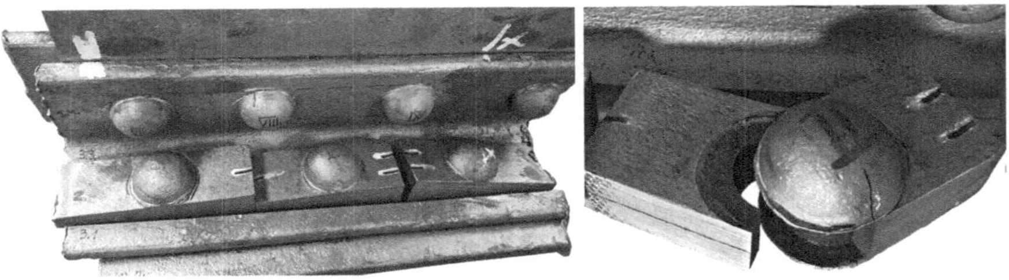

Fig. 3 Rivets cut out from a fragment of the bridge structure on the left. Disassembly of the joint—on the right

The materials fastened together by the rivet were carefully removed after a small cut had been made in the rivet head with a wire EDM machine, as shown in Fig. 3.

The rivet was cut using a wire EDM machine (see Fig. 4, on the left). The contour method was applied to the resultant cut surfaces (see Fig. 4, on the right) to determine residual stresses perpendicular to these surfaces which existed after the rivet had been removed from the structure.

The cutting operation released the residual stresses in the rivet, resulting in the cut surface distortion. The distortion was measured with a Mitutoyo coordinate measuring machine (CMM) fitted with the Renishaw PH10T measuring head. The readings were adjusted, summed, smoothed and inverted by a user script in the octave environment (see Fig. 5, on the left). The resultant data were input as a boundary condition into the software MSC MARC (see Fig. 5, on the right).

The simulation was formulated as a linear elastic problem, using the modulus of elasticity $E = 211$ GPa and Poisson's ratio $\nu = 0.29$. The mesh of the model consisted of 4130 linear quadrilateral elements.

By applying the boundary condition to the cut surface of the rivet, one can determine the stress state in the rivet before cutting. Figure 6 shows the distribution of calculated stress normal to the plane of cutting. It is non-uniform. It appears that cutting with a wire EDM machine probably affected the stress distribution because it is symmetrical with respect to the cutting axis.

Stress profiles along line segments oriented in the cutting direction and perpendicularly to the cutting direction (see Fig. 7).

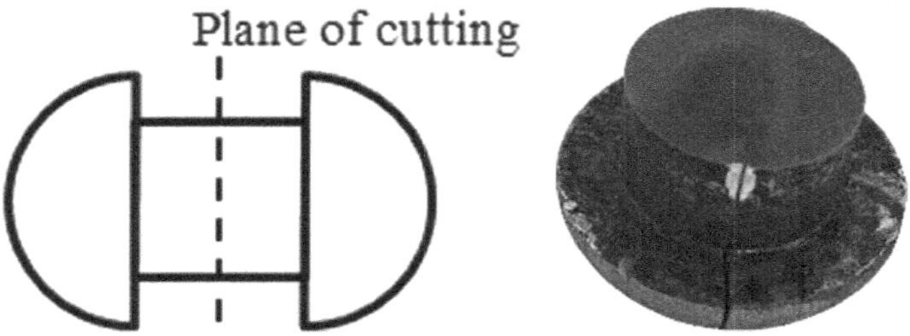

Fig. 4 Method of cutting the rivet—on the left. Rivet after cutting—on the right

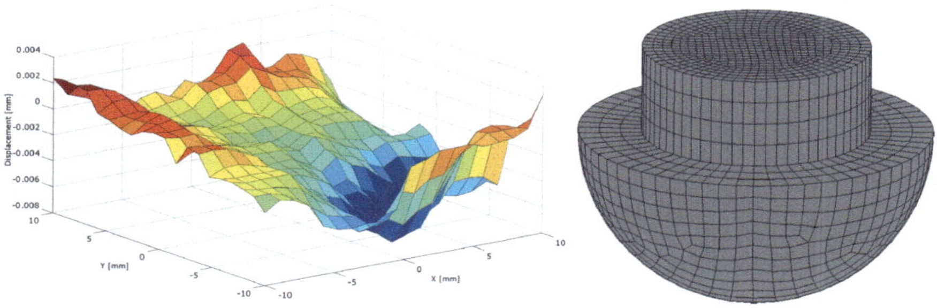

Fig. 5 Output generated by octave script—in left, FE model in the MSC MARC software in right

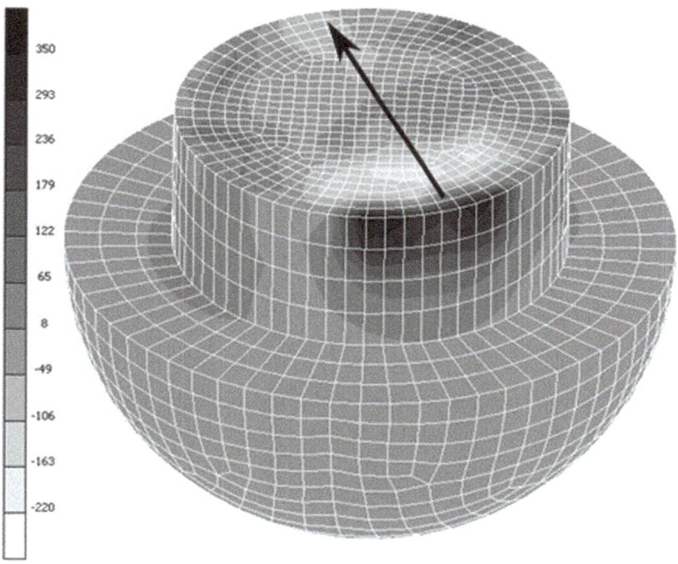

Fig. 6 Stress in the rivet normal to the plane of cutting [MPa] and indicated the direction of cutting in a wire EDM machine

4 Calculation of Post-assembly Condition

In the first step, input data for materials of the joint were determined. After considering the year of manufacture of the joint, we decided to use the JMatPro software for finding the materials characteristics. Using this software, thermomechanical properties of materials can be calculated from their chemical composition alone. Table 1 describes the chemical composition of the rivet material.

This way, all data for elastic behaviour are obtained (among them the modulus of elasticity and Poisson's ratio), as well as thermal properties (including thermal conductivity and heat capacity) and their temperature dependences. The data for

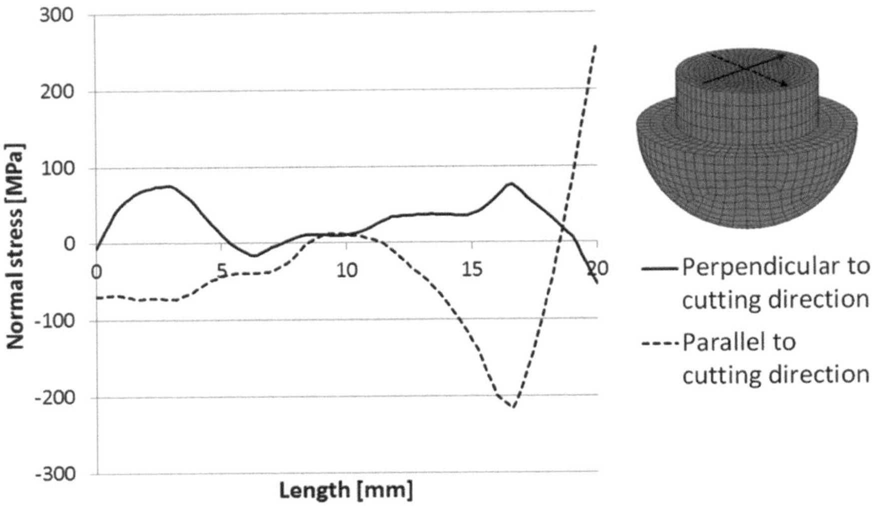

Fig. 7 Profiles of stress on the cut surface normal to the plane of cutting [MPa]

Table 1 Chemical composition of rivet, in [%]

C	Si	Mn	S	Cr	Ni	Cu	N	Fe
0.136	0.025	0.306	0.040	0.056	0.039	0.084	0.0167	99.205

plastic behaviour is generated in a tabular form for the full range of temperatures the full spectrum of strain rates. A portion of this data is shown in Fig. 8.

4.1 *Parameters of the Riveted Joint*

At the time of construction of this bridge, riveted joints for bridges were specified in applicable standards. These included load calculations, which cannot be replicated today because the input information for the bridge design is no longer available.

In this particular case, it was important to find standardized dimensions of rivets and assembly specifications. The rivets were brought to bright red heat, i.e. to approximately 800 °C [5].

The exact rivet size was determined from the thickness of the parts being joined. With rivets chosen for this experimental investigation, the thickness of plates was $s = 10$ mm.

$$d = 2 \cdot s = 2 \cdot 10 = 20\,\text{mm} \tag{1}$$

$$D = 1.5 \cdot d = 1.5 \cdot 20 = 30\,\text{mm} \tag{2}$$

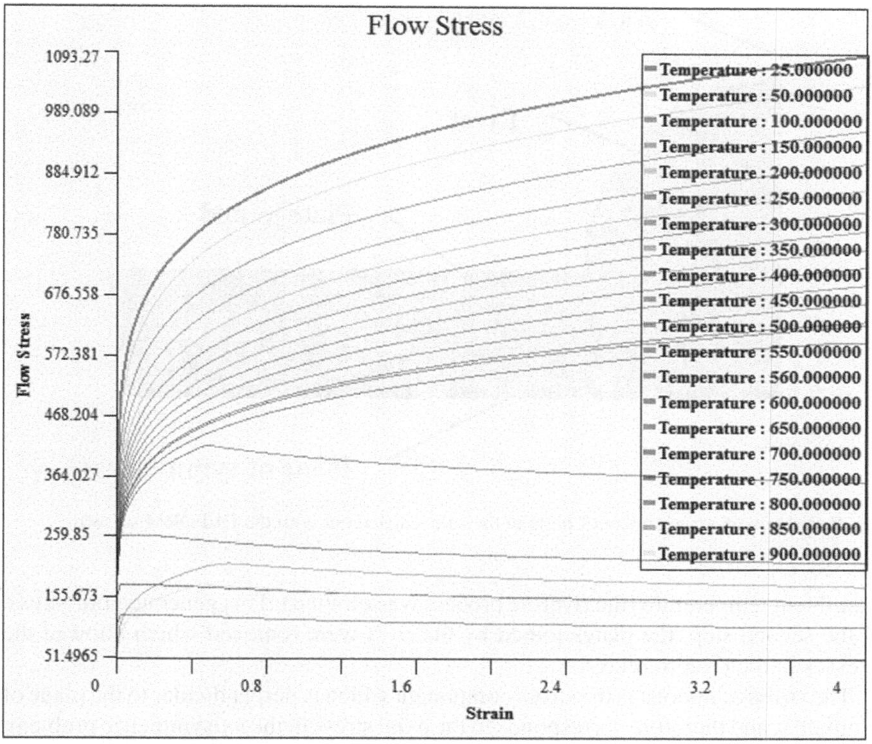

Fig. 8 Portion of plastic behaviour data for the material of the rivet

$$v = 0.6 \cdot d = 0.6 \cdot 20 = 12\,\mathrm{m} \tag{3}$$

$$r = {}^{d}\!/_{1.3} = {}^{20}\!/_{1.3} = 15\,\mathrm{m} \tag{4}$$

The above dimensions were input into finite element calculations. The purpose was to find the stress distribution in the rivet after assembling and the rivet distortion after the joint was disassembled.

4.2 *Numerical Calculation of Assembly*

The simulation was carried out as an axisymmetric problem. Only one half of the joint was included in the model. The calculation had two steps, which simulated the process of assembly and the disassembly of the joint (see Fig. 9).

For the first step, a model was constructed for calculating stresses in the riveted joint in the bridge structure. The rivet cooled from a riveting temperature of 800 °C

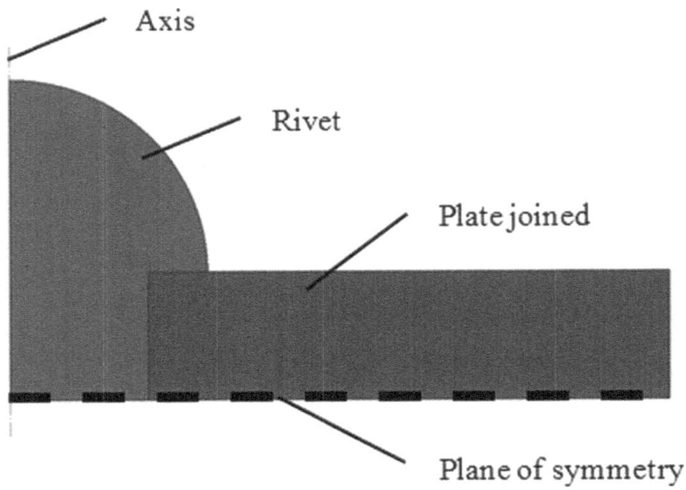

Fig. 9 Schematic arrangement of parts in the calc. carried out with the DEFORM software

to ambient temperature (the riveting process was omitted). This generated pre-stress. In the second step, the plates joined by the rivet were removed which allowed the stresses in the rivet to relax.

The stress of interest is the stress component which is perpendicular to the plane of symmetry, and therefore corresponds to the axial stress in the axisymmetric problem. The stress distribution is shown in Fig. 10.

For the sake of comparison, axial stress values in the plane of symmetry were plotted in a graph Fig. 11.

Fig. 10 Axial stress calculated with the DEFORM software

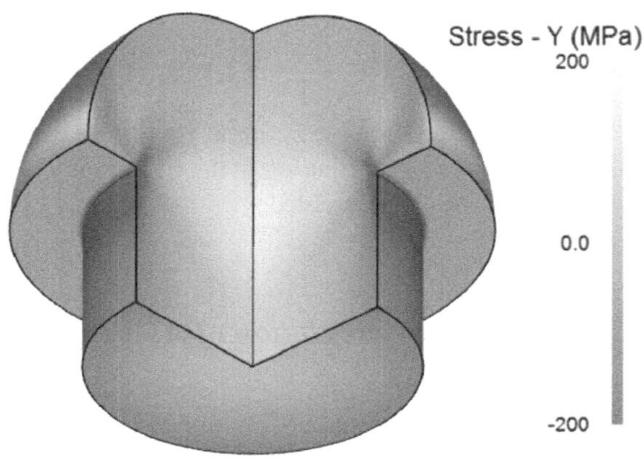

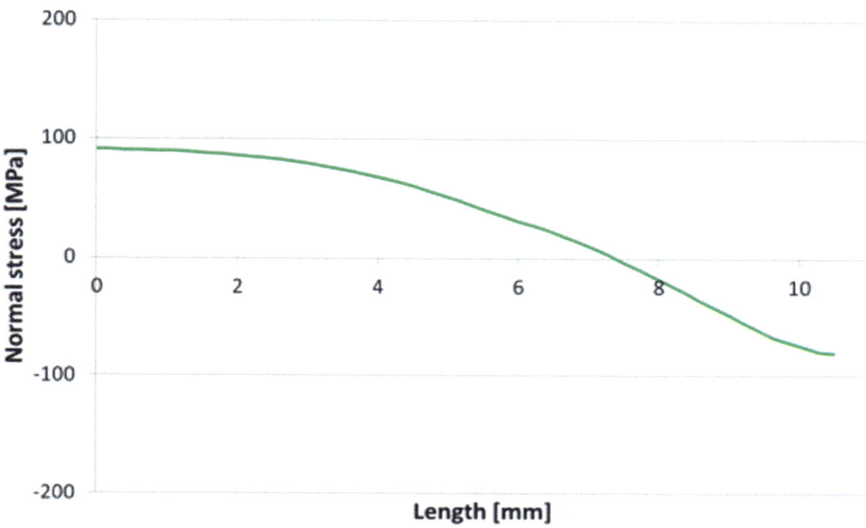

Fig. 11 Axial stress in the plane of symmetry as calculated with the DEFORM software

5 Conclusion

One of the crucial aspects of designing structures involves predicting their life. Calculations are often based on experience-based assumptions. When a particular structure reaches its predicted life, the operator faces the question whether to continue using it or replace it.

This paper deals with an FE calculation of residual stress in a rivet in a bridge structure and compares the result with values found by the contour method. It is the first step in a series of investigations into effects of operation overtime on the structure of the bridge.

It is clear from the graph in Fig. 12 that in the sub-surface regions of the rivet shank, the magnitude of residual stress found by the contour method is in relatively good agreement with calculated values. Differences in stress levels can be seen in the centre of the rivet, where the contour method yields lower stress values. One of the possible reasons is the effect of a long-term load on the rivet in the bridge, which the FE simulation might be unable to reflect accurately. Another limitation of this experimental investigation is that only one rivet was examined.

Nevertheless, these initial results suggest that the condition of the bridge joints is good. The authors will continue to examine more rivets, carry out metallographic characterization and perform additional tests to obtain as accurate description of their material as possible. With such information, it will be possible to refine the FE model of the rivet and perhaps identify the cause of the discrepancy reported in this paper.

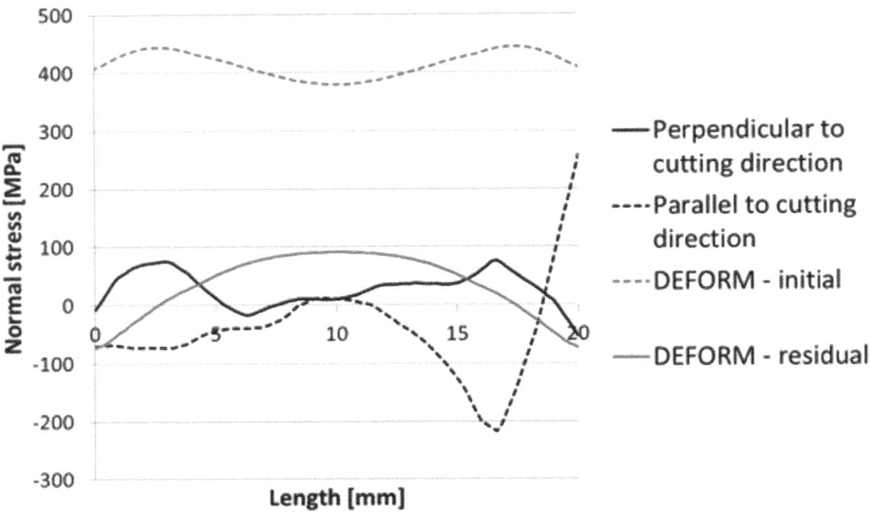

Fig. 12 Stress values compared

Acknowledgements This study was created by project Development of West-Bohemian Centre of Materials and Metallurgy No.: LO1412, financed by the MEYS of the Czech Republic.

References

1. Zahn M (2017) Kolem nádraží. http://www.oplzni.eu/?p=6644. Accessed 2018/05/31
2. Prime MB, DeWald AT (2013) The contour method. In: Schajer GS (ed) Practical residual stress measurement methods. Wiley, Chichester, pp 109–138
3. Hosseinzadeh F, Kowal J, Bouchard PJ (2014) Towards good practice guidelines for the contour method of residual stress measurement. J Eng
4. Prime MB (2001) Cross-sectional mapping of residual stresses by measuring the surface contour after a cut. J Eng Mater Technol 123(2):162–168
5. Grim L, Zvoníček J (1932) Technický průvodnce pro inženýry a stavitele: 6. číslo, ČÁSTI STROJŮ (Technical guide for engineering specialists and master builders: sixth volume MACHINE PARTS). Česká matice technická, Praha

Monitoring of the Measurement Process Capability by Using Capability Indices

Ľuboš Kučera, Jakub Palenčár, Rudolf Palenčár, Stanislav Ďuriš, Ján Vachálek and Jan Rybář

Abstract Requirements related to accuracy of measurement and other metrological properties are required for measurement. Expression of the quality of the measurement is generally solved by the uncertainty of measurement. For measurements in industry, in addition to uncertainty increasingly also used capability index. We take advantage of experiences when determining the capability of production processes. There are some specifics related to the probability of their validity and in particular due to the uncertainty of the check measurement standard used. With help of this standard, we determine capability of measurement process. This contribution is addressed mainly to the impact of the uncertainty of a check standard for the capability index calculation. In practice, it is generally assumed that the uncertainty of the check standard in the assessment of the capability indices of measurement process is negligible. However, there may be cases that it is not negligible. In this paper, we examine how uncertainty of the check standard affects capability indices. This may be particularly important for lower index values, when data from the uncertainty of the check standard indicate that the process is satisfactory and in fact it may not be. The contribution also includes correction factors that allow correct the empirical value of the capability index and thus avoiding a possible mistaken assessment of the measurement process capability.

Keywords Capability index · Check measurement standard · Check measurement standard uncertainty

Ľ. Kučera (✉)
University of Žilina, Univerzitná 8215/1, Žilina, Slovak Republic
e-mail: lubos.kucera@fstroj.uniza.sk

J. Palenčár · R. Palenčár · S. Ďuriš · J. Vachálek · J. Rybář
Slovak University of Technology in Bratislava, Nam. Slobody 17, 81231 Bratislava 1, Slovak Republic
e-mail: jakub.palencar@stuba.sk

Š. Medvecký et al. (eds.), *Current Methods of Construction Design*, Lecture Notes in Mechanical Engineering, https://doi.org/10.1007/978-3-030-33146-7_37

1 Introduction

Capability indices compare the required (prescribed) precision of the measurement process with real process variability and deflection (bias). In praxis, the so-called first generation of the indices: C_g and C_{gk} are used. The calculated values of these indices should be higher than 1.33 to claim that the measurement process is suitable for the process it has been created for.

The use of capability indices makes it possible to avoid controlling the suitability of the measurement process by means of uncertainties, which is challenging since the sources of uncertainty and their valuation need to be determined. The use of capability indices is conditioned using so-called check standard (CS). The exact "true" nominal value of the CS is not known. We can only estimate its value with a certain uncertainty. This CS uncertainty under 10% of the overall process uncertainty is usually neglected, similarly like the measurement uncertainty in manufacturing processes [1–8]. This assumption assumes that the error of CS during a repeated measurement is within the limits of the expanded uncertainty and is therefore random. Unfortunately, during the repeated measurements, the check standard can develop a bias under some specific conditions. The influence of the CS uncertainty was analyzed in [9, 10].

There was thesis, which proposes also usage of the capability indices of the next generations in measurement process analysis [10, 11]. In the upcoming sections of this publication, the effect of such bias will be shown together with the evidence that a CS uncertainty with a value less than 10% can have a considerable effect on the final value of the capability indices and consequently on the final assessment of the capability of the measurement process.

2 Theory

What we know about the CS, it is true value around the declared value within $\pm U_{cs}$, where U_{cs} is extended CS uncertainty. Thus, the CS error can be in range of $\pm U_{cs}$. When we assess the measurement process capability, we need the most unfavorable situation to consider. This is when the CS error is equal the value of the U_{cs} and at the same time will be aiming opposite direction like the true deflection of the measured process.

Let us assume that the measurement process has a distribution:

$$X : N\left(\mu, \sigma^2\right) \tag{1}$$

the check standard has a distribution:

$$X_{CS} : N(\mu_{CS}, 0) \tag{2}$$

and the measurement result y originating from the CS will have the following distribution:

$$Y : N(\mu_Y + \mu_{CS}, \sigma_Y^2 = \sigma^2) \tag{3}$$

assumption of normality is not strict, practically speaking.

3 Capability Index C_{gk} and the CS Uncertainty

We have defined the capability index according to the following equation:

$$C_{gk} = \frac{U - |\delta|}{2\sigma} = \frac{U}{2\sigma} - \frac{|\delta|}{2\sigma} = C_g - \frac{|\delta|}{U} C_g(1 - \nu) \tag{4}$$

Capability index C_g is defined:

$$C_g = \frac{U}{2\sigma} \tag{5}$$

The parameter ν is the proportional deflection of the measurement process:

$$\nu = \frac{|\delta|}{U} \tag{6}$$

where $\delta = \mu - m$ and $m = \mu_{cs}$.

In this case, μ_{cs} represents the "true" value of the CS, which is the desired final value.

In this scenario, we will obtain the empirical capability index C_{gk}^{Y} by swapping the μ for μ_Y and σ for σ_Y. If we apply all the previously mentioned point the relationship between the "real" capability index C_{gk} and empirical capability index C_{gk}^{Y} will have the following form:

$$\begin{aligned} C_{gk}^{\zeta} &= \frac{U - |\delta| - \delta_{CS}}{2\sigma} = \frac{U - |\delta|}{2\sigma} - \frac{\delta_{CS}}{2\sigma} = C_{gk} - \frac{\delta_{CS}}{2\sigma} = C_{gk} - \frac{U}{2\sigma}\frac{\delta_{CS}}{U} \\ &= C_{gk} - C_g\gamma = C_{gk} - C_{gk}\frac{\gamma}{1 - \upsilon} = C_{gk}\left(1 - \frac{\gamma}{1 - \upsilon}\right) \end{aligned} \tag{7}$$

where

$$\gamma = \frac{\delta_{CS}}{U}, \tag{8}$$

$$\delta_{CS} = \mu_{CS} - x_{CS} \tag{9}$$

While x_{cs} is the nominal value of CS (indicated values by CS). This means that the empirical capability index can have a smaller or higher value than the "true" capability index. With the increase of λA (which is a consequence of increasing CS data fluctuations based on the overall uncertainty of CS), C_{gk} increases as well. On the other hand, with the increase of CS systematic error, the C_{gk} has an increasing tendency as well. This is valid for the case where the systematic error of CS (δ_{cs}) deflects in the same direction as the measurement process δ. Then:

$$C_{gk} = C_{gk}^{\zeta} \frac{1}{1 - \frac{\gamma}{1-\upsilon}} \tag{10}$$

In the case that the systematic error CS (δ_{cs}) has an opposite deflection direction as the measuring process δ, the γ will be negative. In this case, we must consider the worst-case scenario and multiply the empirical capability index with a factor:

$$1/\left(1 + \frac{|\gamma|}{1 - \upsilon}\right). \tag{11}$$

In order to apply, this correction knowledge of the CS uncertainty sources is required. This means to know the contribution to the uncertainty originating from the fluctuations of the values generated by the CS and the systematic error of the CS. To gather and define this information can be in some cases complicated and time consuming. In most cases, we assumed that considerable part of the CS uncertainty would manifest itself by the fluctuation of its generated values. If the considerable part of the CS uncertainty is manifested by in the systematic error, and that the γ (relative deflection of CS) is represented by the whole relative uncertainty of CS.

The possible differences between the calculated and "real" probability index that are dependent of the CS uncertainty values can be seen in Fig. 1.

For the measured values Y_i the probability index $\widehat{C_{Cg}^{Y}}$ will be calculated form empirical data according to the following equation:

$$\hat{C}_{gk}^{\zeta} = \frac{U - \left|\overline{Y} - x_{CS}\right|}{2s} \tag{12}$$

where

$$s^2 = \frac{1}{n}\sum_{i=1}^{n}\left(Y_i - \overline{Y}\right)^2 \tag{13}$$

$$\overline{Y} = \frac{1}{n}\sum_{i=1}^{n} Y_i \tag{14}$$

are the most credible estimates of σ_Y^2 and μ_y.

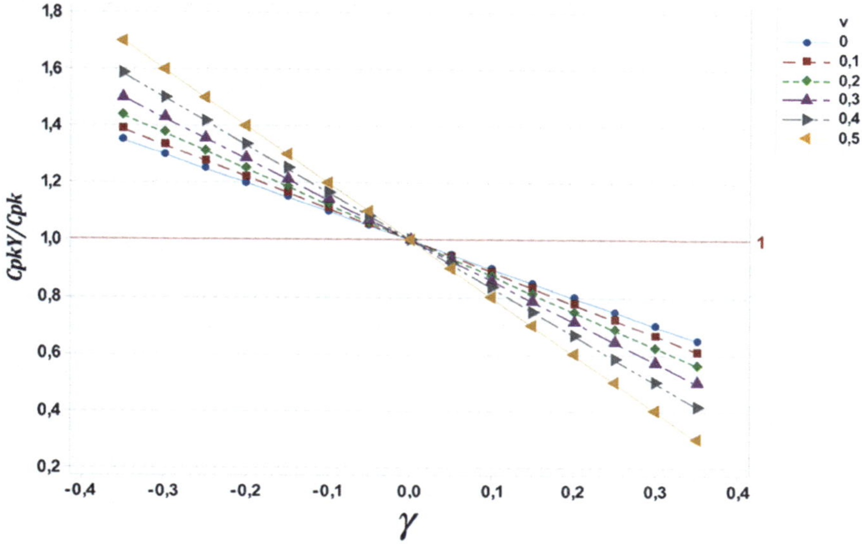

Fig. 1 Share of experimental indices to "real"

4 Conclusion

In practice, it is generally assumed that the uncertainty of the check standard is negligible when it comes to the assessment of probability indices. Although there can be cases, in which this assumption is not correct. Within this publication, the effects of check standards uncertainty on the probability indices were investigated. Knowledge of check standard uncertainty effect can be important at lower index values when the data can indicate that the process is unsuitable, when in reality, it is suitable. As we can see from Fig. 1 that not considering the CS uncertainty even at values lower than 10% (especially, by small components of type A uncertainties) can result in false conclusions. This is especially true for the probability index values C_{kg} that are close to a number 1.33. Specifically, in these cases, neglecting of CS uncertainty can result in false evaluation of suitability of the measurement process. The introduction of a correction factor that was presented in the previous sections enables the correction of the empirical value, which results in the avoidance of a possible incorrect assessment of the suitability of the measurement process.

Acknowledgements The authors would like to thank the Slovak University of Technology in Bratislava and the Slovak Institute of metrology and the grand agency APVV project no.: 15-0295, 15-0164, agency VEGA grant no.: 1/0098/18, 1/0610/17 a 1/0556/19 and KEGA projects no.: 006STU-4/2018, 039STU-4/2017.

References

1. Pearn WL, Kotz S (2006) Encyclopedia and handbook of process capability indices. World Scientific Co. Pte. Ltd., Singapore
2. Pearn WL, Liao MY (2006) One-sided process capability assessment in the presence of measurement errors. Qual Reliab Eng Int 22(7):771–785
3. Baral AK, Anis MZ (2015) Assessment of Cpm in the presence of measurement errors. J Stat Theory Appl 14(1):13–27
4. Grau D (2013) Testing capability indices for one-sided processes with measurement errors. Int J Metrol Qual Eng 4:71–80
5. Haq A, Brown J, Moltchanova E et al (2015) Effect of measurement error on exponentially weighted moving average control charts under ranked set sampling schemes. J Stat Comput Simul 85:1224–1246
6. Maleki MR, Amiri A, Castagliola P (2017) Measurement errors in statistical process monitoring: a literature review. Comput Ind Eng 103:316–329
7. NIST/SEMATECH e-handbook of statistical methods (2018). http://www.itl.nist.gov. Accessed 2018/11/3
8. Bordignon S, Scagliarini M (2002) Statistical analysis of process capability indices with measurement errors. Qual Reliab Eng Int 18(4):321–332
9. Palencar J, Palencar R, Šooš L, Schreier Z (2016) Impact of the check measurement standard uncertainty on the calculation of the capability indices of the measurement process. Metrol Test 22(2):10–13
10. Palencar J (2017) Measurement process monitoring using the capability indices. Dissertation thesis, Slovak University of Technology in Bratislava, Bratislava
11. Kurekova E (2017) Measurement process capability—trends and approaches. Meas Sci Rev 1(1):43–46

The Digital Twin of a Measuring Process Within the Industry 4.0 Concept

Ĺuboš Kučera, Ján Vachálek, Markus Melicher, Pavol Vašek and Juraj Slovák

Abstract This paper approaches the creation of a digital twin based on a real measuring process in order to convert it into digital form as a virtual model within a selected software-based virtualization environment. Subsequently, a real-world data comparative flow is obtained from sensors; afterward, analysis within their calculated uncertainties and the digital twin data values are made. Such treated real measuring process data are inserted into a database, from where they are compared with the virtual digital twin data, where necessary trending-based analysis is made, for the purpose of proactive maintenance and the measuring process optimization. Such collected data are also stored for later-planned analysis within the Industry 4.0 concept in the form of big data databases.

Keywords Industry 4.0 · Digital twin · Measurements uncertainties · Proactive maintenance · Ultrasonic sensor

1 Introduction

The basic idea, which this paper approaches, is the creation of a digital twin to a real measuring process in order to compare the real measured values, gathered from the physically available system, with the ideal values gathered from a virtually created model called a digital twin. Such a comparison of data and the discovery of discrepancies between the measured and ideal values have great justification for the use of proactive maintenance. Afterward, it is possible to set the limits where the measured values are already outside the set optimal interval, but still within the interval where the system has no fault, which would affect the function of the process [1].

Ĺ. Kučera (✉)
University of Žilina, Univerzitná 8215/1, 01026 Žilina, Slovak Republic
e-mail: lubos.kucera@fstroj.uniza.sk

J. Vachálek · M. Melicher · P. Vašek · J. Slovák
Slovak University of Technology in Bratislava,
Nam. Slobody 17, 81231 Bratislava 1, Slovak Republic
e-mail: jan.vachalek@stuba.sk

Š. Medvecký et al. (eds.), *Current Methods of Construction Design*, Lecture Notes in Mechanical Engineering, https://doi.org/10.1007/978-3-030-33146-7_38

This interval between the still-usable limit and the imminent fault is just ideal from the point of carrying out proactive maintenance on the system. Saving funds from the long-term perspective will be achieved by such precise performed actions to repair or calibrate the device within the described interval, because the time window for performing such proactive maintenance will be visible and known in advance from the graphical representations and the maintenance department can be prepared for the 2 outages sufficiently in advance. Otherwise, it is a common practice that such systems are repaired or calibrated at predetermined intervals, which are oversized due to the consideration of security coefficients, or once the fault has occurred, and the system is forced to underlay maintenance. In both the described cases, higher device operating costs are caused by early intervention or unexpected downtime.

2 Proactive Maintenance

The proactive type of maintenance should keep the serviceability of the object under control and continuously increasing the level of it. Data from diagnostics and the measurements are used as the source of information for the purpose of estimating the levels of proactive maintenance [2].

One of the barriers for the expansion of proactive maintenance was the difficulty of gathering consistent and relevant operating conditions and the technical state of the equipment. Such information was difficult to obtain in the past, as reports on working conditions were depended in general on human input, which can be considered at best as incomplete or inaccurate [2].

With the increasing amount of automation, current information from the devices can rely heavily on information provided by software tools to analyze their technical state with modern monitoring systems. These systems usually generate alarm messages in cases where signs of an approaching failure appear suddenly in the monitored elements so that a corrective action can be activated immediately, before any signs of faults appear [3].

The main goals of proactive maintenance are [2]:

- Further reduction of maintenance and operation costs of the object.
- Preventing faults by extending the technical lifespan of the objects.
- Statistical control of random and systematic influences affecting the serviceability of the object.

A proactive approach is the one that not only monitors and evaluates the technical state, but also primarily performs interventions that prevent or at least delay the occurrence of faults and damage.

3 Uncertainties in the Measurement Process

Industry 4.0 concept uses a large amount of information called "big data." In the measurement process, big data are formed by measured values in real time. Measured values have minimal informative value for the customer if uncertainties of the measurement are unknown. The reason why uncertainties are necessary for proactive maintenance stems from using proactive maintenance to optimize technical processes. Finding and removing the causes of an undesired state of the system requires reading values from sensors and defining the range of values for which the system is still operating in optimal state. The main problem is the fact that the measured values do not represent real values which are not taken into consideration. In practice, the source of such errors may be a human factor, incorrectly calibrated sensor, inappropriate measurement method, etc., [4].

Consider that the measurement interval in which a system still works optimally is (a, b). No measurement is accurate, and therefore, when measuring this interval, there is either a random or a systematic error. There may be situations where the condition $X \in (a, b)$ does not apply because $X = \pm x$, where x represents uncertainty of measurement and X measured value. To avoid these situations, it is necessary to add the measurement uncertainty to the measured value and thus avoid exceeding the allowed interval in which the system is working properly. The secure interval will be in the form of: $a < X \pm x < b, X, x \in R$ [5].

Uncertainties can be evaluated by two basic methods depending on whether the causes of these uncertainties are known. In case of unknown causes, standard uncertainties type A are evaluated using statistical analysis. According to the new calculation method, standard uncertainty type A is calculated as [4–6]:

$$u_{\mathrm{A}} = \sqrt{\frac{n-1}{n-3}\sqrt{\frac{1}{n(n-1)}\sum_{i=1}^{n}(x_i - \bar{x})(x_i - \bar{x})^2}} \tag{1}$$

where n is number of measurements, x_i is the measured value, and $\bar{x}$ is the arithmetic mean of measured values. Standard uncertainties type B are linked to known, identified sources. This type of uncertainties is based on partial uncertainties of individual sources (2).

$$u_{\mathrm{b}} = \sqrt{u_{\mathrm{B1}}^2 + u_{\mathrm{B2}}^2 + \ldots + u_{\mathrm{B}k}^2} \tag{2}$$

where $u_{\mathrm{B}i}^2$ are individual components of the uncertainties measured by method B and calculated as:

$$u_{\mathrm{B}} = k_{x\Delta} s_{\Delta} \tag{3}$$

$$s_\Delta = \frac{\Delta}{\sqrt{3}} \tag{4}$$

Parameter k represents the expansion coefficient with a value based on the probability distribution. The Δ is the magnitude estimate of changes from the true value of the variable. Using a combined standard measurement uncertainty (5), an interval is formed which contains the actual value of measured variable with the probability of approximately 68%.

$$u_c = \sqrt{u_A^2 + U_B^2} \tag{5}$$

The combined standard uncertainty of measurement and cover coefficient arise expanded uncertainty of measurement (6). Cover coefficient k depends on the probability distribution with a value greater than 1.

$$U = k \cdot u_c \tag{6}$$

4 Case Study

The main goal was to create an interface, which is user-friendly and allows the user to easily change the parameters of the measuring instrument. The main benefit of the interface is that the user is able to test more sensors, only by changing the parameters without the need to reprogram. For the creation of the interface, the Siemens PLM solution plant simulation platform was used. The part of the interface used for setting parameters is shown in Fig. 1.

For the testing purpose of our interface, the Sonar-Bero 3RG6113-3BF00 sensor by Siemens was used. The measuring range of this sensor is from 20 to 130 cm. For

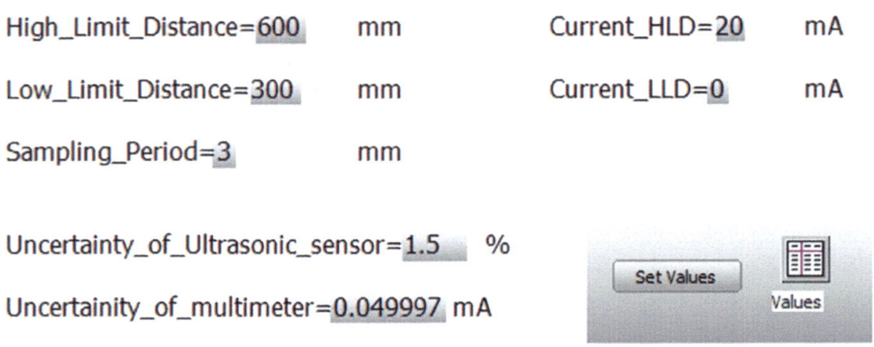

Fig. 1 Sensor settings within the user interface

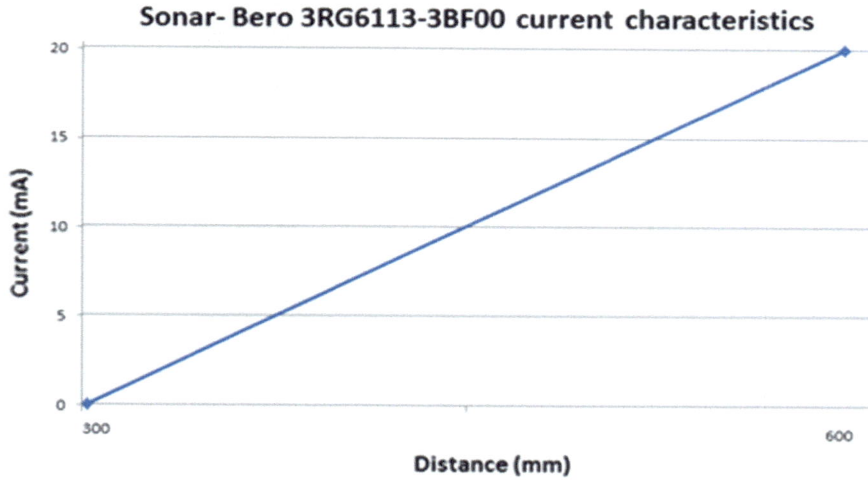

Fig. 2 The ideal current characteristic for the Sonar-Bero ultrasonic sensor

our purpose, the low limit distance was programmed to 30 cm, which corresponds with the current 0 mA and the high limit distance to 60 cm which correspond with the current 20 mA.

Since the output current of the sensor is proportional to measured distance, the current characteristics of the sensor are a straight line as shown in Fig. 2.

The sampling period for the measurement was set to 3 mm, and the real value of the current in measured points was counted using the equation of a straight line. Then, the calculation was programmed into the interface, which allowed the interface to compute the real values for any sampling period and limits [7].

The dataset used for the testing consists of 50 whole range measurements with the same sampling period of 3 mm. The dataset was acquired using Fluke 287 RMS multimeters connected to a database. Our calculations also take into consideration two individual components of uncertainties, the uncertainty of the multimeters and the uncertainty of the ultrasonic sensor. These parameters are also adjustable in the interface to suit the currently used components.

The interface operates in two different modes (see Fig. 3). The first one is the manual mode. The manual mode is controlled by buttons and allows the user to import data from excel or other database to compute uncertainties and export the data. The automatic mode contains a generator, which updates the connected database after a preset period. After the update, the program computes uncertainties and examines if the values measured by the sensor are within the limits. If the values are located outside of the permitted range bordered by the high limit current and low limit current for the measured distance, the program prompts a message for the user to let him know there is a need for proactive maintenance.

The figures show measured values obtained using Sonar-Bero 3RG6113-3BF00 sensors. The figures are divided into three zones. The first zone is located between

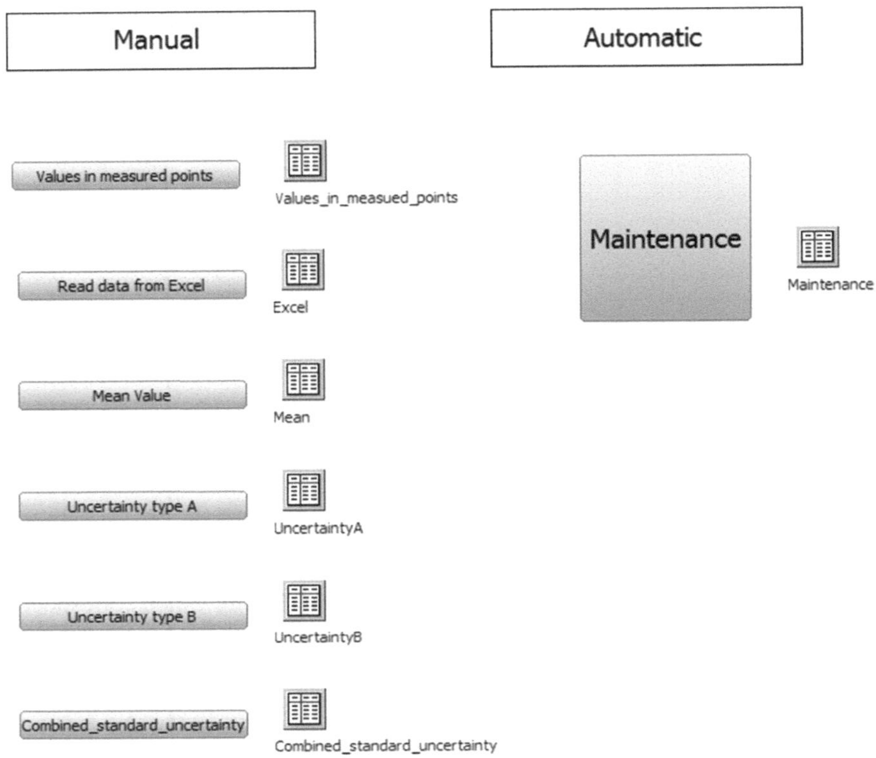

Fig. 3 Operation mode selection within the user interface

the high limit and the low limit line. The high limit line represents the sum of mean value computed from the measurements in measured points and combined standard uncertainty in measured points. The low limit line represents mean value minus the combined standard uncertainty. The real value of current in measured points is computed using the equation of a straight line. The size of the proactive maintenance zone depends on the needs of the process. For our purpose, the zone is bounded by two lines, which represent the ideal value in given measurement point plus or minus 10% of its value [8].

In Fig. 4, all the values received from the sensor are located in between the high and the low limit. In this case, the sensor is measuring correctly and there is no need for any kind of maintenance.

In Fig. 5, one of the values is located above the high limit but inside of the proactive maintenance zone. If this case occurs, the program saves the date and time of the occurrence into a table together with the ID of the sensor and informs the user there might be a need for proactive maintenance. The proactive maintenance zone signalizes there might be a problem with the sensor or the process, but for now, it is not a danger for the process.

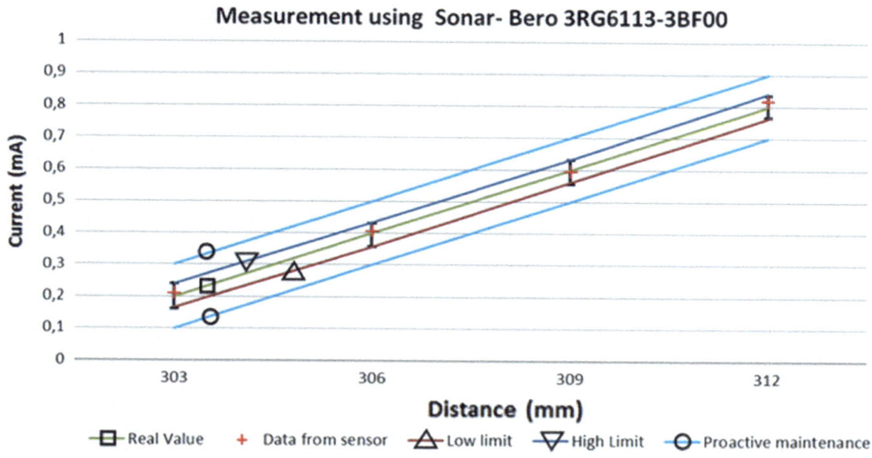

Fig. 4 Measurement sample of correct data

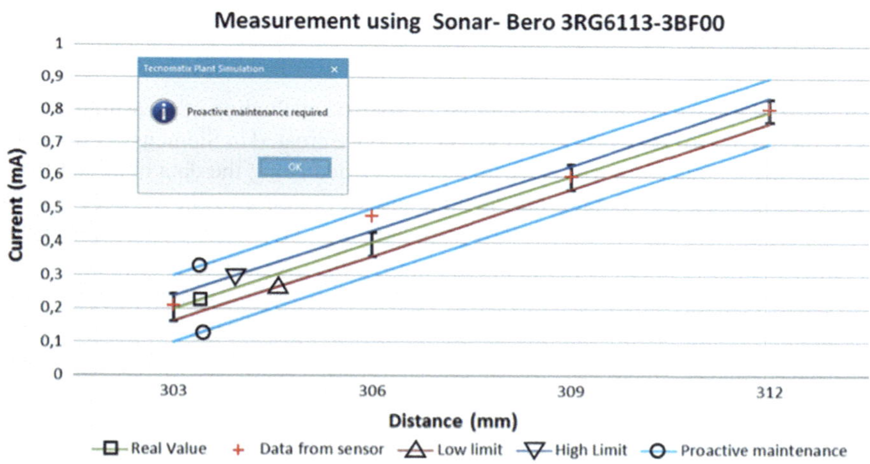

Fig. 5 Measurement sample of data within the proactive maintenance limits

In Fig. 6, some of the values are located outside of the proactive maintenance zone. In this case, the operator receives a warning that the value is out of limits. It indicates that the process is no longer working properly and has to be terminated before it causes damage to instruments or products.

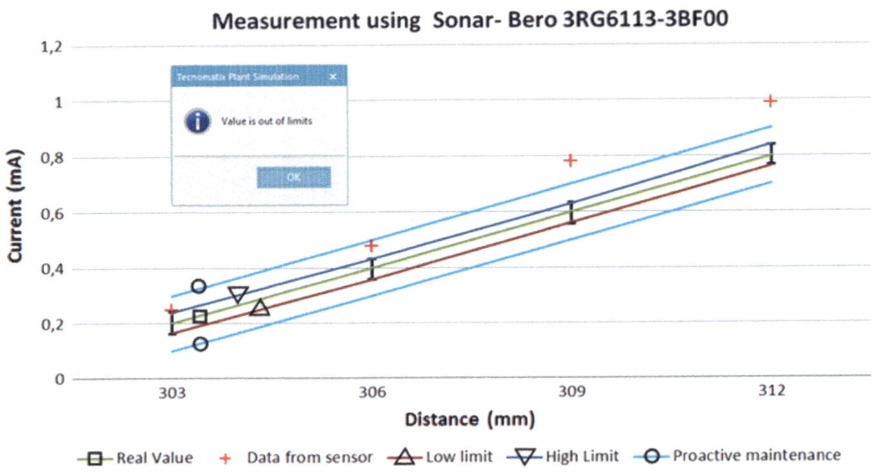

Fig. 6 Measurement sample of data outside the proactive maintenance limits

5 Conclusion

The paper introduces progressive methods of statistic evaluation of measured values for the needs of proactive maintenance. The interface created in Siemens Tecnomatix plant simulation platform was experimentally verified using the data obtained from ultrasonic sensors. Based on trends tracking and deviation growth, we were able to set the parameters required for proactive maintenance. After setting up the interface, the program is able to justify the credibility of collected data and to detect potential error of the measuring device. The program is also able to detect the need for proactive maintenance of the sensor, and based on this knowledge, plan the proactive maintenance of the chosen technology. Our future goal is to extend the created concept and test the interface on a real industrial process.

Acknowledgements The authors would like to thank the Slovak University of Technology in Bratislava, the Slovak Institute of Metrology, and the grand agency APVV project no.: 15-0164, 15-0295 agency VEGA grant no.: 1/0317/17, and KEGA project no.: 027STU-4/2017. 9.

References

1. Vachálek J, Lokšík M, Morháč M, Bartalský L, Rovný O, Šišmišová D (2017) The digital twin of an industrial production line within the Industry 4.0 concept. In: Fikar M, Kvasnica M (eds) 21st international conference on process control (PC). IEEE, Red Hook USA, pp 258–262
2. Podolsky J (2010) Prediktívna údržba ako súčasť celkovej stratégie optimalizácie technických prostriedkov. AT&P J 10:34
3. Duchoň F (2015) Výrobné systémy a CIM, 1st edn. Nakladateľstvo STU, Bratislava

4. Palenčár R, Ďuriš S, Dovica M, Palenčár J (2015) Application of Monte Carlo method for evaluation of uncertainties of temperature measurement by SPRT. In: 21st world congress of the international measurement confederation, Prague. IMEKO, Budapest, pp 1194–1200
5. Palenčár R, Sopkuliak P, Palenčár J, Ďuriš S, Suroviak E, Halaj M (2017) Application of Monte Carlo Method for evaluation of uncertainties of ITS-90 by Standard Platinum Resistance Thermometer. Meas Sci Rev 17(3):108–116
6. JCGM 100—evaluation of measurement data—guide to the expression of uncertainty in measurement (GUM). GUM 1995 with minor corrections. JCGM, BIPM, Paris (2008). http://www.bipm.org/utils/common/documents/jcgm/JCGM_100_2008_E.pdf. Accessed 2018/04/12
7. Vítečková M, Víteček A, Babiuch M (2011) Unified approach to analog and digital two-degree-of-freedom PI controller tuning for integrating plants with time delay. Acta Montan Slovaca 1:89–94
8. Vasek V, Dostalek P, Dolinay J, Janacova D, Kolomazník K (2008) Microcontrollers and modern control methods. In: Proceedings of the 2nd WSEAS international conference on computer engineering and applications. WSEAS Press [IT], Mexico, pp 195–198

FE Analysis of Load Distribution in Clinched Joints Array

Pavel Maly and Frantisek Lopot

Abstract Clinching represents one of the modern technologies applied in industrial production for connecting sheet-metal components into complex assemblies considering desired functionality and parameters. This article describes the use of FE method for simulation of connection of two sheet-metal plates using three clinching joints. The simulation model uses 3D solid bodies of connected sheet-metal plates and 3D model of the clinched joint. The clinched joint is modeled as a shaped interlock of sheet-metal plates with help of the tie connection and hard contact with penalty friction. The uniaxial stress in the vicinity of each clinched joints is used for analysis of the simulation results. This specific place corresponds with the placement of strain gauges during the experimental measurements on real specimens. The set of five simulation model and tested specimens with different joints spacing was used. Therefore, the results of simulation and experimental testing can be compared and used for general simplified analysis of load distribution between three joints depending on their spacing.

Keywords Clinched joint · FE simulation · FEM · Array of joints · Experimental testing · Strain gauge

1 Introduction

There is an effort to produce lightweight constructions in these days, because engineering structural, together with economic and environmental reasons, is becoming more important. These goals can be achieved using thin-walled parts joined together into a final assembly. Therefore, suitable joining method of sheet-metal plates is also of great importance especially in many industrial branches such as automotive, aerospace, or production of electric appliances and furniture. When the parts are connected together, more simple joints are used to achieve desired properties of the final assemblies.

P. Maly (✉) · F. Lopot
Czech Technical University in Prague, Technická 4, 16607 Prague 6, Czech Republic
e-mail: pavel.maly@fs.cvut.cz

Š. Medvecký et al. (eds.), *Current Methods of Construction Design*, Lecture Notes in Mechanical Engineering, https://doi.org/10.1007/978-3-030-33146-7_39

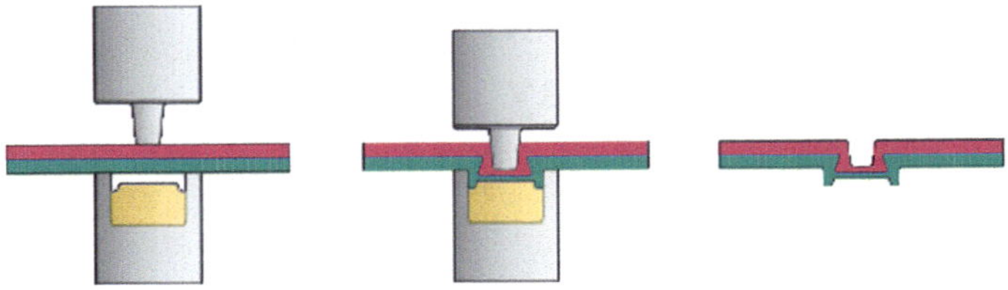

Fig. 1 Clinched round joint [1]

1.1 Clinching Technology

The clinching technology is a method of joining sheet-metal plates by localized cold forming of materials. There are many worldwide suppliers of this technology (e.g., TOX PRESSOTECHNIK [1], BTM Corporation, Böllhoff, Eckold clinching, etc.). The example of the clinching technology, specifically for a round joint, is shown in Fig. 1. The main principle of joint creation is to form an interlock between connected plates with help of special tooling, i.e., shaped punch and die.

1.2 Methods

All components and products have different parameters that describe the main properties. Moreover, they often specify the functionality and feasibility. There are various methods on how to obtain the desired parameters, and simulation and experimental methods are one of the most used. The general application of these methods can be found in [2, 3].

Some sources can be found describing different approaches regarding clinched joints and joints in general. The previous study was done also with help of analytic methods. The articles [4, 5] describe the concept for load distribution in connection with joints arrays focusing on the analytic approach. The general analysis of load distribution between more rivets in the row using analytical method is described in [6]. Experimental approach for load transfer and fatigue life for multiple-row joints is given in [7]. Another research on FE modeling of clinched joints in assemblies can be found in [8].

2 Finite Element Simulation

Previous research of clinched joint FE simulations was done using the model with one round joint and is summarized in [9, 10]. The simulations with one clinched joint model considered different influences and parameters, and the conclusions were therefore used in the simulations of specimens with three joints in the row. The engineering software Abaqus CAE was used for the FE analysis.

2.1 FE Simulation Model

The simulation 3D model consists of two plates connected by the three joints. Scheme of the simulation model is shown in Fig. 2 where the boundary conditions are also shown. The assumed cross-section of the plates is 40 × 3, and the length of plates corresponds to the real specimens used for experimental testing. Half of the model was used in simulations because the symmetry of the specimen in plane *XY* is obvious.

The used material is steel with isotropic elastic behavior (Young's modulus 210 GPa and Poisson's ratio 0.3), and the mesh uses 8-node linear bricks with reduced integration (C3D8R).

The loading force was applied at the end of the bottom plate in the *X* direction and corresponds to the value of 3000 N. The force is lower than shear strength of the joint, so the plastic deformations are not assumed.

Three paths and places of interest near each joint were defined to analyze the results. Each place of interest is located in the vicinity of the joint, i.e., 5 mm from the edge of the punched hole, and corresponds to the position of the uniaxial strain gauge on the upper plate of the experimental specimens. The place of interest and the paths are shown in Fig. 2.

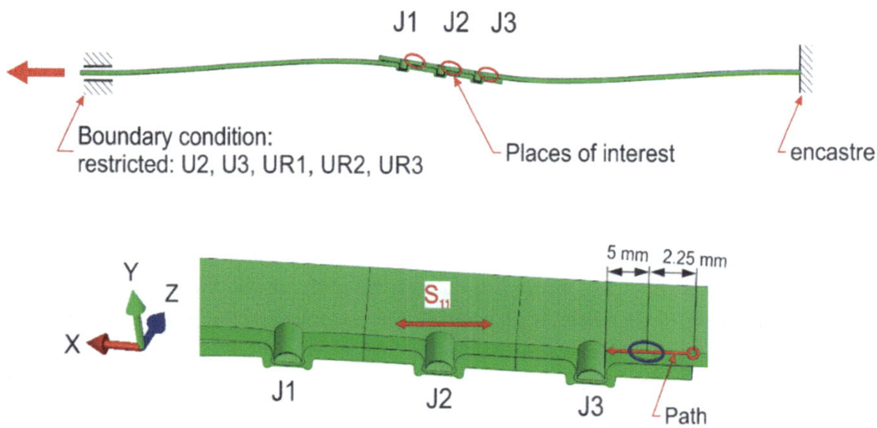

Fig. 2 Simulation FE model

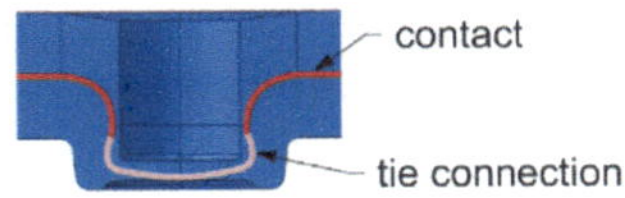

Fig. 3 Clinched joint model

Table 1 Set of specimens

Specimen	Cross-section [mm]	Joints spacing [mm]
03-40-20	40 × 3	20
03-40-30	40 × 3	30
03-40-40	40 × 3	40
03-40-50	40 × 3	50
03-40-60	40 × 3	60

2.2 *Clinched Joint Model*

Each joint is an interlock of upper and bottom plates and is created using tie connection in the bottom part of the joint and hard surface-to-surface contact with friction for other surfaces (see Fig. 3). The characteristic dimension of the used round clinched joint, i.e., the outer diameter, is 10 mm.

2.3 *Simulations*

The set of six simulation models was created and subjected to the calculation. The spacing between the three joints was the varying parameter. The set of models with the notation and joints spacing parameter is given in Table 1.

2.4 *Results*

The resultant stress in *X* direction (S11) along the paths in the vicinity of the joints is used for analysis of the simulation results. The example of the results is shown in Fig. 4. The place of interest is also marked in this graph.

The simulation results, i.e., the calculated stress S11 in the place of interest, are given in Table 2. The proportionally distributed total load between the joints is also given, and the sum of the stress of all joints is assumed for the determination. The proportional load distribution for the specimen with joint spacing of 20 mm is not included because the stress near the first joint was positive.

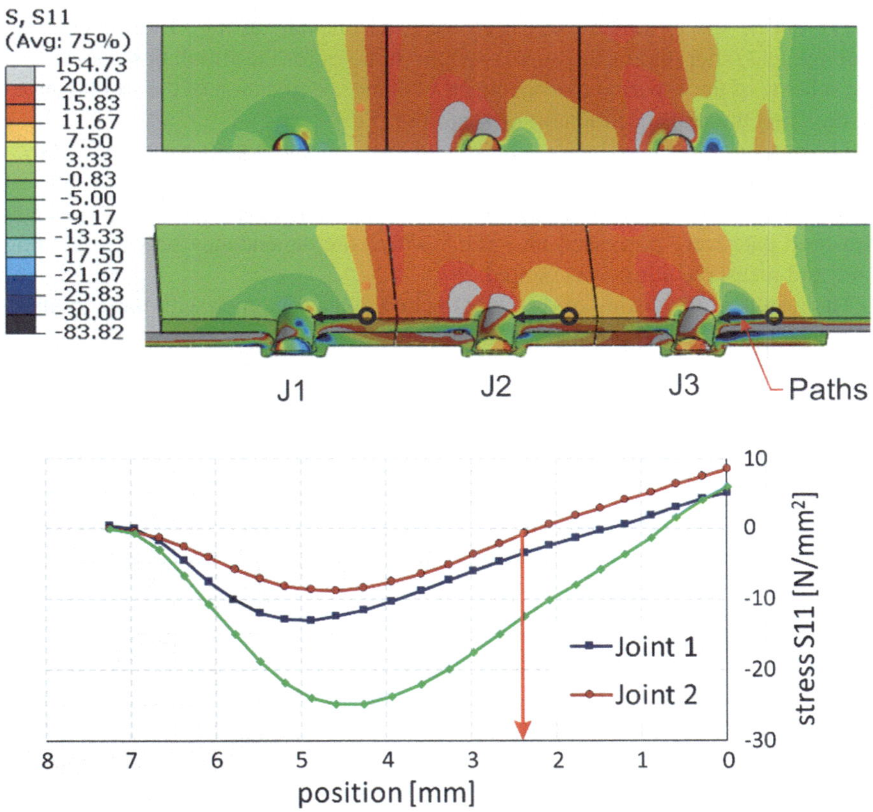

Fig. 4 Resultant stress in *X* direction (S11)

Table 2 Summary of results

Specimen	Joint 1 [N/mm^2]	Joint 2 [N/mm^2]	Joint 3 [N/mm^2]	Joint 1 [%]	Joint 2 [%]	Joint 3 [%]
03-40-20	5.09	−3.14	−17.01	NA	NA	NA
03-40-30	−0.91	−1.46	−14.66	5	9	86
03-40-40	−3.58	−0.86	−12.64	21	5	74
03-40-50	−4.81	−0.81	−10.89	29	5	66
03-40-60	−5.93	−1.06	−9.41	36	6	57

3 Experimental Testing

Detailed description of experimental testing of the real clinched joint specimens with strain gauges is described in [9–11].

The set of six specimens was used for experimental testing. The dimensions, material, and other parameters were used the same as for the simulation specimens. The scheme of the experimental specimen with the specification of the strain gauges position and the real specimen is shown in Fig. 5.

The measurement was done on the special testing stand, loading was applied by a hydraulic cylinder, actual loading was measured by a force sensor, and the uniaxial stress near the joints was obtained with help of strain gauges. The time-dependent measured data were processed into graph where the dependency between loading force and measured stress was plotted.

The results of the testing of the real clinched joints specimen with three joints in the row are given in Table 3 and given in [11].

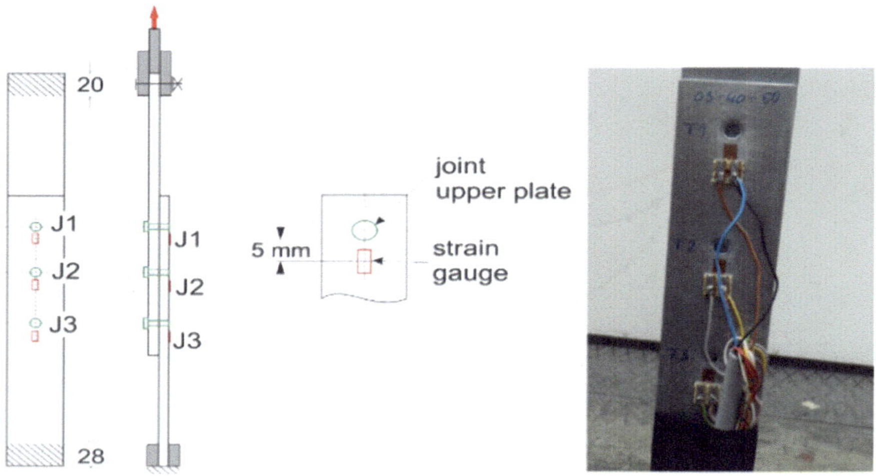

Fig. 5 Specimen for experimental measurement

Table 3 Results of experimental measurement

Specimen	Joint 1 [%]	Joint 2 [%]	Joint 3 [%]	Total loading
03-40-20	40.13	19.75	40.13	100
03-40-30	39.75	20.49	39.75	100
03-40-40	36.72	26.56	36.72	100
03-40-50	36.58	26.85	36.58	100
03-40-60	33.97	32.05	33.97	100

4 Conclusion

The set of five FE 3D solid models of joined two thin-walled plates with three clinched joints in the row was simulated and analyzed. The joint modeled as an interlock of the plates and the symmetry of the specimen were used. The courses of the uniaxial stress near the joints and their values in the places of interest were used for the analysis of the results. They can be used for determination of load distribution between the joints.

In general, the loading of the middle joint is the lowest and the side joints transfer higher loads. It corresponds to the analytical approach of the load distribution calculation of three joints in the row [4, 5]. However, according to the analytical approach, the side joints transfer the same load for the specific case.

If the results of experimental testing and simulations are compared, it is evident that they differ quite significantly. This can be explained by the influence of the simulation model, e.g., boundary condition, and by the used experimental method, e.g., position of the strain gauge or additional bending of the specimen.

Therefore, there is a vast scope for further work and improvements. The research on the clinched joints continues focusing on the detailed analysis and improvement of the experimental and simulation methods.

Acknowledgements This study was supported by the cooperation with the TOX®PRESSOTECHNIK, Weingarten, Germany.

References

1. Tox-Pressotechnik. https://cz.tox-pressotechnik.com. Accessed 28/04/02
2. Dub M, Kolar J, Lopot F, Dynybyl V, Berka O (2014) Experimental analysis of stress state and motion of tram gearbox hinge. In: EAN 2014—52nd international conference on experimental stress analysis. Czech Society for Mechanics, Praha, p 22
3. Berka O, Dub M, Lopot F, Dynybyl V (2015) Analysis of satellites loading in planetary transmission. In: Padevet P, Bittnar P (eds) EAN 2015, 53rd conference on experimental stress analysis. Czech Society for Mechanics, Praha, pp 16–19
4. Maly P, Lopot F, Sojka J (2016) Properties of the clinched joints array in the row. In: Kubec V, Kubec Krónerová E, Trapp Cajthamlová Š (eds) Book of proceeding 57th international conference of machine design departments (ICMD 2016). University of West Bohemia, Pilsen, pp 247–250
5. Maly P, Dynybyl V, Sojka J (2015) Design and verification of connection realized by the array of clinched joints. In: Bošanský M, Bucha J, Kadnár M, Nemčeková M, Rusnák J, Vereš M (eds) ICMD 2015, book of proceedings of the 56th international conference of machine design departments. Slovak University of Agriculture in Nitra, Nitra, pp 195–198
6. Cvekl Z, Dražan F et al (1976) Theoretical basis of transportation equipment (in Czech). SNTL Praha, Praha
7. Huth H (1986) Influence of fastener flexibility on the prediction of load transfer and fatigue life for multiple-row joints. Fatigue in mechanically fastened composite and metallic joints, vol 927. ASTM Special Technical Publication, Philadelphia, pp 221–250

8. Breda A, Coppieters S, Debruyne D (2016) Modeling strategy for clinched joints in assemblies. J Phys Conf Ser 734:032001
9. Maly P, Lopot F, Sojka J (2017) FEM model and experimental measurement of clinched joint. IOP Conf Ser Mater Sci Eng 179:012051
10. Maly P, Lopot F, Stary F, Sojka J (2017) Strain gauge measurement on clinched joints specimens. In: Trebuňa F, Bocko J, Frankovský P, Huňady R, Kostka J, Kula T (eds) EAN 2017, 55th conference on experimental stress analysis. Czech Society for Mechanics, Praha, pp 139–142
11. Maly P, Lopot F, Sojka J (2017) Load distribution analysis in the connection of three clinched joints array in the row. In: Herak D (ed) ICMD 2017, 58th international conference of machine design departments. Czech University of Life Sciences Prague, Prague, pp 218–221

Transmission Mechanical Vibrations in the Car Seat in the Laboratory Conditions

Rudolf Martonka and Vítězslav Pfliegel

Abstract The comfort of seating on car seats depends on many parameters. People differ in sex, body, weight, quantity and quality of muscle tissue, and mental and nerve constitution. Such differences may, according to the actual situation, be a real mood, a sense of hunger or superstitions, levels of fatigue. In the list of differences we could continue, e.g. such as a people's age, weight, corpulence, and so health, everything has a great effect on the comfort of sitting. Each individual person is different, behaves differently in the car, and feels different on the car seats while driving. Such an original person will sit on a unique car seat and demand it to suit everyone without distinction. We all understand that this is not possible. Without the personalization of the car seat we can not get anywhere. For the design of the right car seat, we will first have to carefully examine the person. Then we can ask the car seat to adapt to the person who sits on it, preferably automatically. Next we should explore, what kind of sitting more harms health of human, static sitting, or influence of mechanical vibrations (mechanical shacking) on human, or so damage to the human nervous system under the influence of driving (stress from complicated traffic, fast driving, etc.) A sitting position for a person is not entirely natural. Lying is definitely a better position, which is not good for the long term, but probably not in the car is good enough. So we have to stay at the seating. We will examine all the possible parameters and we may come up with some objective conclusions about the innovation of car seats.

Keywords Car seat · Human · 3-D dummy · Sitting comfort

1 Introduction

In order to perform human testing in laboratory conditions, there are a number of limitations to be observed, such as the perfect health status of the test person, a test signal with an acceleration lower than 0.5 g, the possibility of terminating the test with the test subject immediately, without limitation, the presence of a physician, and

R. Martonka (✉) · V. Pfliegel
Technical University of Liberec, Studentska 2, 46117 Liberec, Czech Republic
e-mail: rudolf.martonka@tul.cz

Š. Medvecký et al. (eds.), *Current Methods of Construction Design*, Lecture Notes in Mechanical Engineering, https://doi.org/10.1007/978-3-030-33146-7_40

the presence of a person specifically supervising the test. Interestingly, no restrictions apply to perform the same human testing under real conditions. However, we have decided to perform testing under laboratory conditions because of the good repeatability of the tests, the possibility of setting the same test conditions independently of the season [1, 2]. The test persons were selected from a number of volunteers with a valid medical check-up and written approval of the tests.

An interview with the test person was performed before each test; the course and method of the test were explained. The test person was instructed about the safety mode and the use of the stop button to immediately terminate the test. The first tests focused on examining the two main comfort parameters of the car seats, the first pressure distribution in the seat area (static comfort) and the seat cushion (dynamic comfort). During the test, the person under test did not perform any psychosomatic tasks.

2 Experiment

Testing was performed at the hydrodynamic laboratory on hexapod. An X sensor for measuring pressure maps and an acceleration sensor for acceleration measurement were used. The testing was carried out according to the following plan [3]: first, after stacking of the pressure maps, followed by static seating without motion for 5 min, again measuring the pressure maps, loading with a dynamic signal of 0.1 g sin for 3 min, followed by load Ehra for 7 min, followed by dynamic signal 0.1 g for 3 min, and again measuring the pressure maps. The experiment with dummy is shown in Fig. 1. Characteristic parameters were defined by international standards, for example [4–6].

In same time was measured input acceleration on seat rails and output acceleration on dummy butt. After the mathematical processing of the FFT by transforming the measured acceleration signals, the signals were mutually divided. From these ratios, we obtained the transmission characteristics. A set of automobile seats was chosen to measure and calculate the transmission characteristics of each automobile seat. The obtained graphs are shown in Fig. 4.

Before and after the dynamic test, measurements of contact pressure distribution between dummy and car seat were performed. We have compared the change in contact area size and contact pressure between individual measured images.

3 Results

The measured result is shown in Figs. 2 and 3. By comparing time variations of measured acceleration and contact pressure field sizes, it is possible to determine the correlation between dynamic and static parameters that characterize comfort seats.

Fig. 1 Experiment with dummy

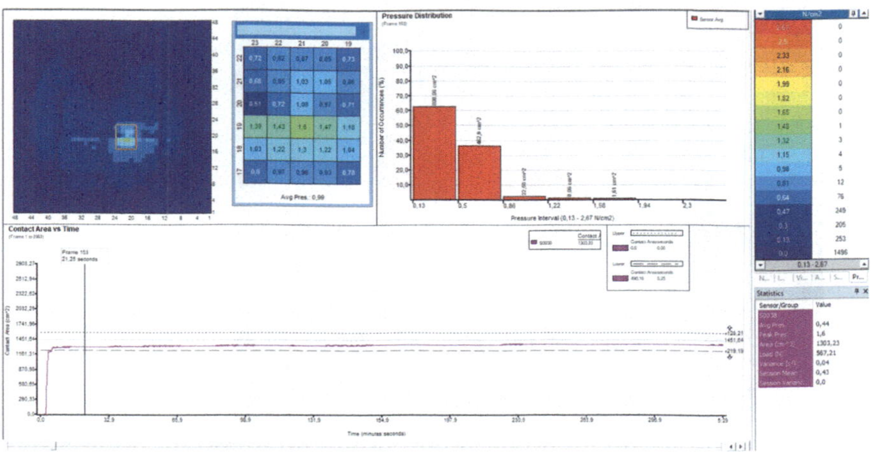

Fig. 2 Contact pressure field of the car seat

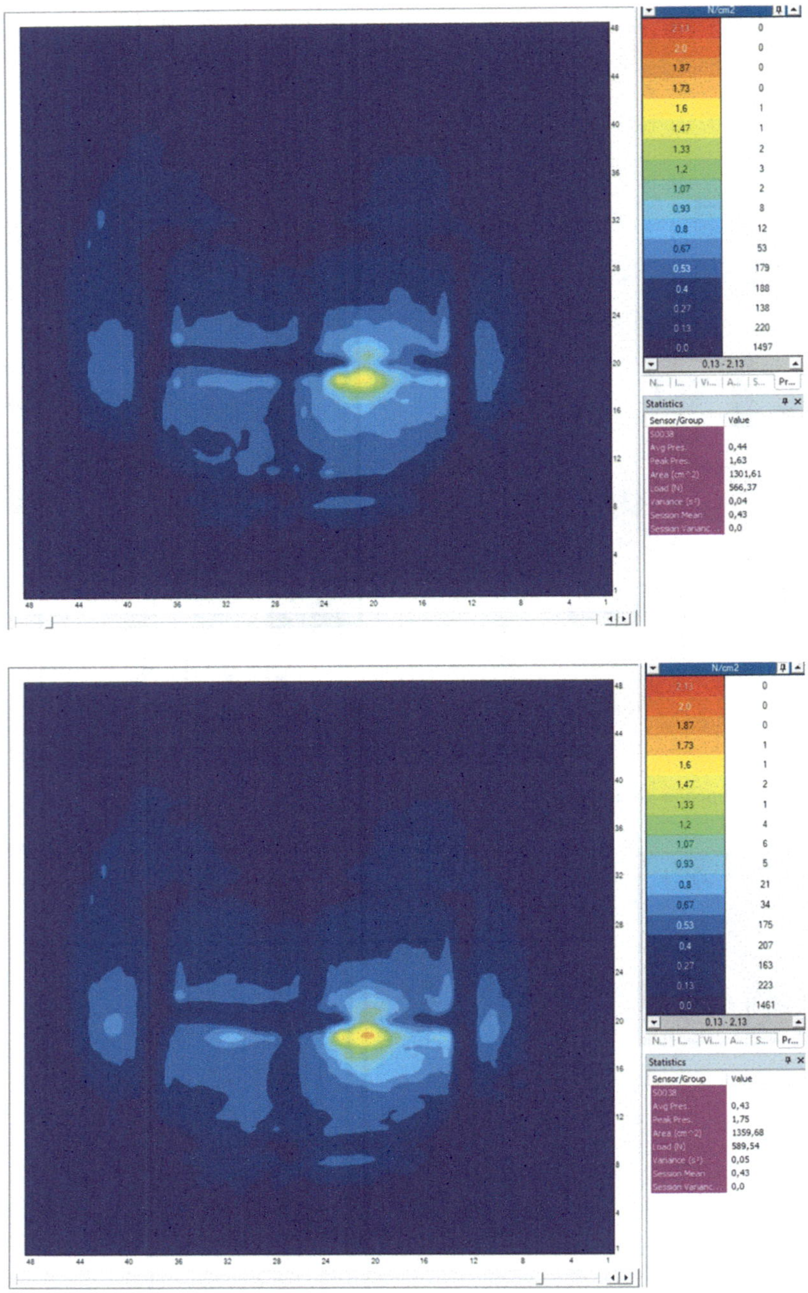

Fig. 3 Contact pressure field before and after dynamic test

On the two images—Fig. 3 shows the change in contact area size and contact pressure distribution before and after the dynamic test. At first sight, it can be seen that the two controlled parameters increased after the test—area and contact pressure. This is due to the upsetting of the PUR foam in the seat filler and thereby its solidification. This can also be seen on dynamic transmission characteristics. On the transmission characteristics, this effect is manifested by reducing the peak of the transmission characteristic and moving it to a higher frequency. This can be seen between a pair of 1 and 2 curves, the same for 3 and 4 curves (see Fig. 4).

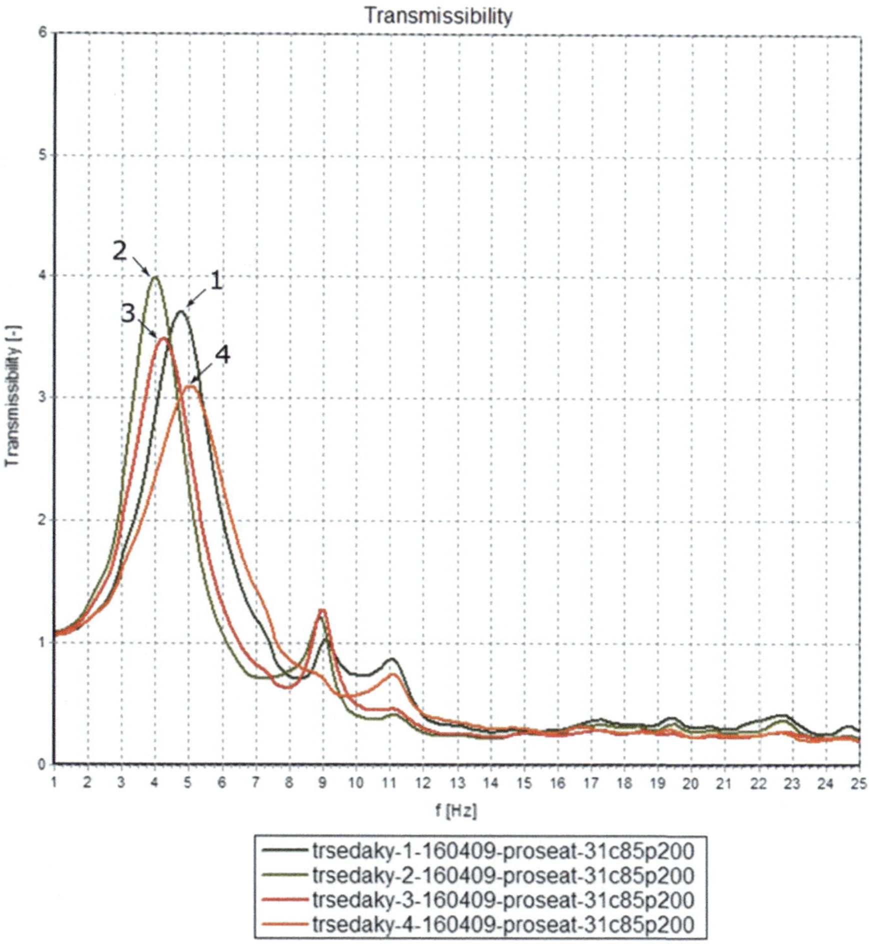

Fig. 4 Transmissibility of the car seat

4 Conclusion

The following conclusions can be drawn from the measured data. With dynamic loading of the car seat, the PUR filler is compacted and thus solidified. In this case, the solidification of the foam results in an increase in contact area and contact pressure. Thus, it is possible to perform faster and less time-consuming measurement of contact pressure distribution, from which it can be concluded that the selected seat has a large contact area with high contact pressure and thus will have bitter dynamic properties or dynamic comfort over a small seat. With the larger contact area the contact pressure is lower. These measurements can also be measured over a period of time on a dynamically loaded seat and compare changes in dynamic comfort during seat use. This gives us information on the stability of the mechanical properties of the PUR filler and the entire seat over time.

Acknowledgements This publication was written at the Technical University of Liberec as part of the project "Innovation of products and equipment in engineering practice" with the support of the Specific University Research Grant, as provided by the Ministry of Education, Youth and Sports of the Czech Republic in the year 2018. This work was supported by the Ministry of Education, Youth and Sports of the Czech Republic and the European Union—European Structural and Investment Funds in the frames of Operational Programme Research, Development and Education—project Hybrid Materials for Hierarchical Structures (HyHi, Reg. No. CZ.02.1.01/0.0/0.0/16_019/0000843).

References

1. Vink P, Lips D (2017) Sensitivity of the human back and buttocks: the missing link in comfort seat design. Appl Ergon 58:287–292
2. Park S, Lee N, Lee S (1993) An investigation of the relationship between subjective comfort evaluation and body pressure distribution of vehicle seats. J Ergon Soc Korea 12(1):3–15
3. Wijaya AR, Jönsson P, Johansson O (2003) The effect of seat design on vibration comfort. Int J Occup Saf Ergon 9(2):193–210
4. Standard: ASTM D3574-11 (2003) Standard test methods for flexible cellular materials-slab, bonded, and molded urethane foams
5. Standard JASO B407-87 (1987) Test code of seating comfort for automobile seats
6. Standard DIN EN ISO 3385 (2017) Flexible cellular polymeric materials-determination of fatigue by constant-load pounding

Safety Models of Systems in the Engine Room with Renewable Elements

Zbigniew Matuszak

Abstract The idea of technical safety system had been characterised. Presented were basic models of technical safety systems regarded to the highest measures of safety. Chosen measures of safety were included. Safety measures were illustrated with Kolmogorov equations. A safety analysis based on the engine room technical system with renewable elements was presented. It was illustrated, among others, with Kolmogorov equations. Measures of readiness and unreadiness as well as other measures of unreliability were indexed. In the complicated system, such as the engine room, its service depends on its accident-free functioning, as well and often on the safety of the environment. In estimating of the reliability of the technical system, it is necessary to take into consideration not only the system's technical damages, but also the possibility of errors made by the technical personnel of the engine room. These errors cause various consequences of safety unreliability in the engine room system, and identifying these errors should be taken into consideration in the safety models and indicators of safety reliability and usable. Presented models do not include all of the possible areas. The presented way includes a narrow possible portion of safety analysis of exploitation in engine room.

Keywords Safety models · Measures of safety · Measurements of readiness and unreadiness · Engine room

1 Introduction

The notion of the object's safety came into use in numerous technical branches. Significantly, earlier, when technology was related to structures raised by man, safety was related only to danger, which is posed to the life of its users. In the thirties, a study technique was developed which was called a theory of safety, which regarded mainly to building structures. In research works of that era, it is customary to operate the idea of safety of the object and to search for various measures of safety; however, clear and concise definition of safety is not provided, identifying it generally with

Z. Matuszak (✉)
Maritime Academy in Szczecin, ul. Wały Chrobrego 1-2, 70500 Szczecin, Poland
e-mail: z.matuszak@am.szczecin.pl

Š. Medvecký et al. (eds.), *Current Methods of Construction Design*, Lecture Notes in Mechanical Engineering, https://doi.org/10.1007/978-3-030-33146-7_41

idea of non-destruction of the object (or control point) and sometimes an idea of non-emergency [1, 2].

Presently, the idea of safety is utilised in relation to objects to which damage creates a realistic danger to the life of health of human being, as well as in relation to technical objects, where damage causes smaller or greater economic losses caused by pauses in utilisation and the necessity of repair. The category of these technical objects includes all floating objects and particularly vessels [3, 4].

Under the idea of engine room safety, it is necessary to comprehend its quality characterised by resistance to the formation of dangerous situations. Two states of safety were highlighted, in which:

- danger to safety appears,
- there does not appear danger to safety.

The measures of safety may be [5]:

- reliability of safety R_B, being resistance of engine room systems to errors of its function causing danger to the system, systems interacting with it, the environment and human life;
- unreliability of safety Q_B, i.e. susceptibility of engine room systems, which cause danger to safety.

Between above-mentioned measures of safety there appears to be an obvious dependence:

$$R_B + Q_B = 1 \tag{1}$$

Additionally to the presented basic indicators of safety, the probability of the appearance in engine room damages may be included, which causes a pause or incomplete functioning of the engine room. The indicator is identified as unreliability to safety Q_Z or unreliability of functioning [2].

If the events causing unreliability to safety Q_B or the unreliability of usable Q_Z exclude one another, then the overall unreliability of engine room Q is the total of indicators Q_B and Q_Z:

$$Q = Q_B + Q_Z \tag{2}$$

the meaning:

$$R_B = 1 - Q + Q_Z \tag{3}$$

and measures R_B, Q_B and Q_Z have the character of conditional probability.

There existing general dangers in the functioning of engine room are:

- danger to safety, such as due to the state of engine room, which results in a secondary damage or other undesirable processes accompanying the damages;

- danger of safety to objects interacting with engine room such as being the state of engine room in which the damage could indirectly or directly cause danger to safety of the equipment and elements interacting therewith;
- danger to the safety of environment, such as the state of the engine room, in which damage could cause indirect or direct damage is the natural human environment;
- danger to human life, such as the state of engine room, in which damage of its elements could cause serious bodily injuries or the loss of human life.

2 Methods and Materials

2.1 Selected Safety Models of Renewable Safety Elements and Systems in the Engine Room

The vast majority of equipment included in the component of particular installations in the engine room are renewable elements, or as results of repair, or the replacement for new. Selected renewable models are presented in the example of greasing of the turbochargers of the series B-591 vessel, of which the simplified structure of reliability is presented in Fig. 1.

In the engine room installations, two types of elements can be pointed out:

- following the damage to any installation element, the remaining undamaged elements stop functioning, and during its renewal, they do not become damaged—this applies to the damaged elements which create a series reliability structure;
- following the damage to any installation element, the remaining undamaged elements still function and can become damaged during the time of its renewal—this applies to the damaged elements creating parallel or type "k of n" reliability structure.

Figure 2 presents the system's graphs of the reliability series structure called full renewal, based on a fact that during the renewal process of the damaged element, the technical condition of the entire system is inspected.

System elements in quantity n are characterised by intensity damages λ_i intensity and renewal μ_i. For such a system, it is accepted that the probability of simultaneous damage of two or more elements is unlikely. The described model can be described by a set of Kolmogorov differential equations:

$$G'(t) = -G(t) \cdot \sum_{i=1}^{n} \lambda_i + \sum_{i=1}^{n} g_i(t) \cdot \mu_i \tag{4}$$

$$g_i'(t) = \lambda_i \cdot G_i(t) - \mu_i \cdot g_i(t) \tag{5}$$

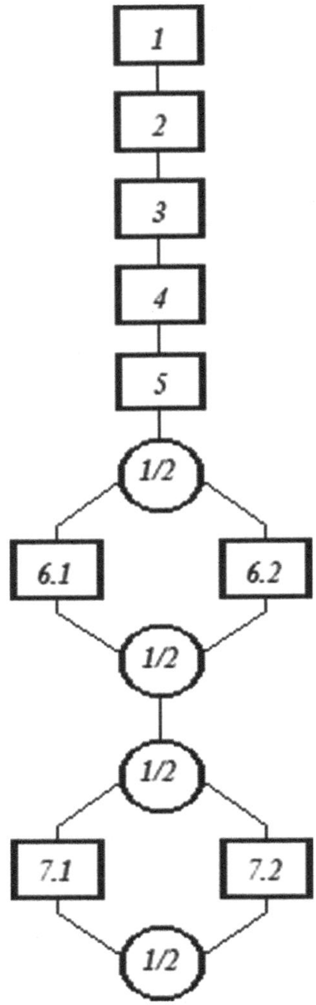

Fig. 1 Infallibility structure of the lubrication installation of turbocompressors: 1—lubrication oil cooler; 2—oil filter; 3—lower circulation oil tank; 4—upper circulation oil tank; 5—reserve circulation oil tank; 6.1, 6.2—oil filters before turbocompressors; 7.1—main circulation oil pump; 7.2—reserve circulation oil pump

$$G(t) + \sum_{i=1}^{n} g_i(t) = 1 \tag{6}$$

$$G(0) = 1;\ \ g_i(0) = 0;\ \ i = 1, 2, \ldots n \tag{7}$$

where: $G(t)$—measure of system's readiness; $g_i(t)$—measure of system's unreadiness.

The Laplace transformations of the above measures have the following form:

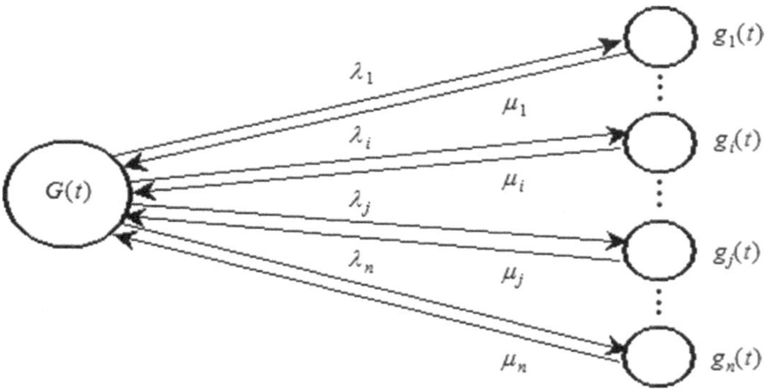

Fig. 2 Graphs of a system with parallel reliability structure with full renewal

$$\tilde{G}(s) = \frac{\prod_{l=1}^{n}(s+\mu_l)}{s\cdot\left[\prod_{l=1}^{n}(s+\mu_l)+\sum_{j=1}^{n}\lambda_j\cdot\prod_{l=1,l\neq 1}^{n}(s+\mu_l)\right]} \tag{8}$$

$$\tilde{g}_i(s) = \frac{\lambda_i\cdot\prod_{\substack{j=1\\ j\neq i}}^{n}(s+\mu_j)}{s\cdot\left[\prod_{l=1}^{n}(s+\mu_l)+\sum_{j=1}^{n}\lambda_j\cdot\prod_{\substack{l=1\\ l\neq j}}^{n}(s+\mu_l)\right]} \tag{9}$$

Based on the above dependencies, the stationary measures of G readiness and g_i unreadiness are indicated on the dependence of:

$$G = \lim_{t\to\infty} G(t) = \lim_{t\to\infty} s\cdot\tilde{G}(s) = \frac{\prod_{l=1}^{n}\mu_l}{\prod_{l=1}^{n}\mu_l+\sum_{j=1}^{n}\lambda_j\cdot\prod_{\substack{l=1\\ l\neq j}}^{n}\mu_l} \tag{10}$$

$$g_i = \lim_{t\to\infty} g_i(t) = \lim_{t\to\infty} s\cdot\tilde{g}_i(s) = \frac{\lambda_i\cdot\prod_{\substack{j=1\\ j\neq i}}^{n}\mu_j}{\prod_{l=1}^{n}\mu_l+\sum_{j=1}^{n}\lambda_j\cdot\prod_{\substack{l=1\\ l\neq j}}^{n}\mu_l} \tag{11}$$

Technical equipment of the engine room installation being elements of the analysed system, in particular case, could possess identical intensities of renewal ($\mu_1 = \mu_2 = \ldots = \mu_n = \mu$), and in such a case, dependencies (10) and (11) have a form of:

$$G = \frac{\mu}{\mu+\lambda_O} \tag{12}$$

$$g_i = \frac{\lambda_i}{\mu+\lambda_O} \tag{13}$$

$$g = \sum_{i=1}^{n} g_i = \frac{\lambda_O}{\mu+\lambda_O} \tag{14}$$

where

$$\lambda_O = \sum_{i=1}^{n} \lambda_i. \tag{15}$$

For the above-described principles, the Laplace transformations of a simplified format as well as of their opposite transformation are available.

Analysing the described system from the safety point, the result of its damage can be an appearance of usable unreliability, in which event, the system is renewed with the μ_i intensity or safety unreliability, which can appear with λ_{Bi} ($i = 1, 2, \ldots, n$) intensity. An appearance of the last condition causes a withdrawal of the system from exploitation (a pause in exploitation for the renewal activity or for the system exchange—installation).

Based on graphs of the renewable system with taking into consideration the possibility of the appearance of safety unreliability is presented in Fig. 3. A set of Kolmogorov equations was developed, which later on was useful in indicating, among others, such measures of reliability of the system as the functioning of system's unreliability usable $Q_i(t)$, functioning of safety reliability $R_B(t)$ and functioning of safety unreliability $Q_B(t)$:

$$R'_O(t) = -\lambda_O \cdot R_O(t) + \sum_{i=1}^{n} \mu_i \cdot Q_{Zi}(t) \tag{16}$$

$$Q'_{Zi}(t) = \lambda_i \cdot R_O(t) - (\mu_i + \mu_{Bi}) \cdot Q_{Zi} \tag{17}$$

$$q'_{Bi}(t) = \lambda_{Bi} \cdot Q_{Zi}(t) \tag{18}$$

with: $i = 1, 2, 3 \ldots, n$; $R_O(0) = 1$; $Q_{Zi}(0) = 0$; $Q_{Bi}(0) = 0$; where:

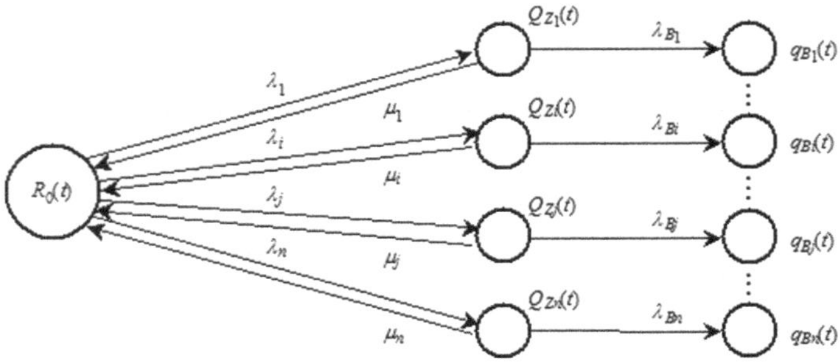

Fig. 3 Graphs of a renewable system regarding the possibility of unreliable safety state

$$\lambda_O = \sum_{i=1}^{n} \lambda_i. \tag{19}$$

Utilising the Laplace transformations as well as opposite transformations, indicated can be the previously presented measures of reliability:

$$Q_Z(t) = \sum_{i=1}^{n} Q_{Z_i}(t) \tag{20}$$

$$R_B(t) = R_O(t) + Q_Z(t) \tag{21}$$

$$Q_B(t) = \sum_{i=1}^{n} q_{Bi}(t) = 1 - R_B(t) \tag{22}$$

However, the stationary conditional safety unreliability is described by the dependency of:

$$q_{Bi} = \lim_{t\to\infty} q_{Bi}(t) = \frac{\lambda_{Bi} \cdot \lambda_i \cdot \prod_{\substack{j=1 \\ i\neq j}}^{n} (\mu_j + \lambda_{Bj})}{\lambda_O \cdot \prod_{l=1}^{n} (\mu_l + \mu_{Bl}) - \sum_{l=1}^{n} \lambda_l \cdot \mu_l \cdot \prod_{\substack{l=1 \\ j\neq l}}^{n} (\mu_j + \lambda_{Bj})} \tag{23}$$

Installations of the engine room or the same installations can be in a state of functioning (working), state of repair, state of awaiting (standby), instituting a hot or cold reserve. The quantity of these conditions equals to an n, and the time duration in a given condition is equivalent to lot time of A_i $(i = 1, 2, \ldots, n)$; otherwise, the system functions until an appearance of damage through time $T_{i,\,n+1}$. Following an appearance of damage, the system is renewed through time equivalent to T_{n+1}.

The process of passing to respective conditions is applicator as one creating a Markov chain with passing probabilities of a_{ij} $(i, j = 1, 2, \ldots, n)$.

Marking distribution and function failure probability variable by random time $A_i(t)$ respective to $A_i(t)$ and $a_i(t)$—Laplace–Stieltjes transformation function $A_i(t)$ as $A_i(s)$, and following assuming random variable T_i $(i = 1, 2, \ldots, n)$ exponential distribution of intensity of damages λ_i and for random variable T_{n+1} the same intensity time-table of renewal μ, selected measures of reliability and safety of installation in engine room (systems) have a form:

time of duration of system i form:

$$T_i = \min\,(A_i i,\ T_{i,\,n+1}) \tag{24}$$

distribution random variable T_i:

$$F_i(t) = 1 - \exp(-\lambda_i \cdot t)[1 - A_i(t)] \tag{25}$$

transformation Laplace–Stieltjes distribution $F_i(t)$:

$$\tilde{F}_l(s) = \int_0^t e^{-st} dF_i(t) = \frac{s \cdot \tilde{A}_l \cdot (s + \lambda_i) + \lambda_i}{\lambda_i} \tag{26}$$

value of awaiting:

$$\overline{T}_i = E[T_i] = \left[\frac{1}{\lambda_i} \cdot \tilde{A}_i(\lambda_i)\right] \tag{27}$$

probability of passing of the system from state i to j in time shorter than t:

$$P_{ij}(t) = a_{ij} \cdot \int_0^t \exp(-\lambda_i \cdot u) dF_i(u) \tag{28}$$

probability of passing of the system from the state i to the state of damage $(n + 1)$ in time shorter than t:

$$P_{i,n+1}(t) = \int_0^t \left[1 - \exp(-\lambda_i \cdot u)\right] dF_i(u) \tag{29}$$

If possible, n conditions of installation of engine room (systems) will be acknowledged as conditions of usable, and then the time of duration in such conditions is equivalent to time of duration in a condition of usable.

3 Conclusions

The engine room, such as most technical systems, is a structure of a human-machine type. In the complicated system, such as the engine room, its service depends on its accident-free functioning, as well and often on the safety of the environment. In estimating of the reliability of the technical system, it is necessary to take into consideration not only the system's technical damages, but also the possibility of errors made by the technical personnel of the engine room. These errors cause various consequences of safety unreliability in the engine room system, and identifying these errors should be taken into consideration in the safety models and indicators of safety reliability and usable [6–8]. Presented models do not include all of the possible areas. The presented way includes a narrow possible portion of safety analysis of exploitation in engine room. A wider and more detailed is included in the presentation [9].

References

1. Kołodziejski M, Matuszak Z (1996) Importance assessment of ship power plant system components. Pol Marit Res 2(8):3, 27–30
2. Kołodziejski M, Matuszak Z (2017) Reliability centred maintenance (RCM)—basics of implementation and general characteristics. J Mach Constr Maint Probl Eksploatacji 4:99–105
3. Matuszak Z (2002) Modeli otkazow i prinadleznost dannych ob otkazach k generalnoj sowokupnosti na primierie sudowych energeticzeskich ustanowok. Kaliningradskij gosudarstwiennyj techniczeskij uniwersytet, Kaliningrad
4. Matuszak Z (2003) Kompozicii raspredelenij charakteristik nadioznosti i modeli otkazow sistem sudowych energeticzeskich ustanowok. Kaliningradskij gosudarstwiennyj techniczeskij uniwersytet, Kaliningrad
5. Jaźwiński J, Borgoń J (1989) Niezawodność eksploatacyjna i bezpieczeństwo lotów. WKiŁ, Warszawa
6. Bergman B (1985) On reliability theory and its applications. Scand J Stat 12:1–42
7. Natvig B (1979) A suggestion of a new measure of importance of importance of system components. Stoch Process Appl 9:319–330
8. Natvig B (1985) New light on measures of importance of system components. Scand J Stat 12:43–54
9. Matuszak Z (1996) Wybrane zagadnienia bezpieczeństwa siłowni okrętowej. Szczecin, Wyższa Szkoła Morska w Szczecinie, Studia nr. 25, Szczecin

Low-Speed Bearings Diagnostic Equipment

Michael Oeljeklaus and Lubomír Pešík

Abstract The subject of paper is the design of control system of slow running bearings including strength analysis and load capacity measurement with the intention to apply the results to the manufacturing operation by the chain conveyor in the paint shop. The diagnostics of low-speed bearings is very desirable in a wide range of industrial sectors. Mainly, in the transport technologies is necessary to identify the damaged bearing before a production failure and large economic losses become. Chain conveyors are used in the bodywork painting process and chain wheels are mounted on shaft, which is currently mounted on two roller bearings. Shaft speed is relatively slow, ones of rpm. There is no doubt that the reliability of this construction has a main influence to the volume of production, especially since the paint shop is the bottleneck of whole production process of cars. The diagnostics of high-speed bearings is based on vibration measurement that is reliably used for many years (Sturm and Förster in Einführung in die Theorie der Technischen Diagnostik. Vieweg+Teubner Verlag, Wiesbaden, pp. 18–92, 1990 [1]; Hauptmann in Sensoren—Prinzipien und Anwendungen. Verlag C. Hanser, München, Wien, 1991 [2]). For the diagnostics of low-speed bearings, there is no appropriate physical method, which has been able to identify the bearing damage.

Keywords Vibration diagnostics · Low-speed bearings · Bearing diagnostics

1 Introduction

The topic of the article is focused on solving problem of diagnostics of low-speed bearings with intention to apply obtained results to the manufacturing process in the paint shop.

M. Oeljeklaus
Škoda Auto a.s. Mladá Boleslav, tř. Václava Klementa 869, 29301 Mladá Boleslav, Czech Republic

L. Pešík (✉)
Technical University of Liberec, Studentská 1402/2, 46117 Liberec 1, Czech Republic
e-mail: lubomir.pesik@tul.cz

Š. Medvecký et al. (eds.), *Current Methods of Construction Design*, Lecture Notes in Mechanical Engineering, https://doi.org/10.1007/978-3-030-33146-7_42

2 Used Methods

The design of control system of slow running roller bearings is based on analyzation of the current state. The basis is to determine a load in all manufacturing conditions and to determine the force of the chain in the most exposed position. With knowledge of manufacturing operation force of the chain, it is possible to calculate the load of shaft and roller bearings. The results may be used for FEM analysis, calculation stresses and deformations of each part of the current state as a basis of design solution.

The design solution of a system of diagnostics of slow running roller bearings can be based on the principle of rolling resistance or vibration measurement. However, effectivity of the vibration detection depends on a sufficient level of measured acceleration, which cannot be achieved in slow rotational movements. It means for the current state to dismount chain and then measure the rolling resistance or increase the rpm by external power and measure the vibration.

Pursuant to demand for the reliability of chain conveyors and the shafts and chain wheels mounting, was designed a system consists of two pairs of roller bearings—shaft and frame. These are connected to each other by the so-called reference part that is the freely rotable part. This solution allows identification of bearings damage with two methods. The first method is depending on change of rolling resistance during forced rotation of reference part. The other method uses vibration measurement by increasing the rpm with properly connected external power. This innovative solution is protected by patent [3, 4].

Kinematic link between shaft and reference part allows distribution of dynamic load between shaft and frame bearings, which is convenient for manufacturing operational matters [5]. Based on this kinematic link, shaft and reference part are in forced rotation. Due to kinematic link of shaft, reference part and frame, the design solution leads to integration of planetary mechanism with spur or bevel cogwheels. This solution is protected by patent as well.

After strength analysis of the designed solution, the prototype was made with the intention to experimentally measure the deformation by nominal and maximum value of load. The fatigue test is already prepared.

3 Description of Current State

Current state of shaft mounting of chain wheel of conveyor (see Fig. 1) is made of two spherical roller bearings mounted to the frame by bearing units. Chain wheel is connected to free end of shaft and is separated by front plate of frame, which is connected to the frame of paint shop production line.

The advantage of current shaft mounting of chain wheel of conveyor is its relative simplicity. Significant disadvantage is complicated dismounting and changing of bearings. Because of that, the vibrodiagnostics of bearings was implemented by using latest technologies. Sensors of low-frequency vibration are used for identification of damaged bearings. The difficulties consist of insufficient intensity of vibration of

Fig. 1 Current state of shaft mounting of chain wheel

slow running bearing. Its value merges with vibrational background of paint shop production line. The reliability of identification of damaged bearing is insufficient.

4 Design Solution of Diagnostic System

Reliable design of control system of slow running bearings can be based on identification of change of the rolling resistance of bearing or its vibrations during the sufficient rpm [6]. Both these methods are difficult to be realized during the manufacturing operation. The change of rolling resistance can be identified only with increased power load, which is usually too late. The sufficient rpm value of bearing can happen only if the standard, low-speed operation is off.

The design solution of design of control system of slow running bearings (see Fig. 2), which is the topic of this paper, comes from the idea, that if one ring of the roller bearing is forced low speed, the second ring of the roller bearing is freely rotable [3, 4]. This diagnostic solution allows identification of change of rolling resistance of shaft bearing or sufficient rpm increase on its freely rotable ring that allows identification of damage of bearings using known methods of vibrodiagnostics. The freely rotable ring of the shaft bearing is rotably mounted to frame using frame bearing. The reference part is mounted between shaft and frame bearing (see Fig. 2).

The change of bearings is quite a difficult operation, and because of that, the shaft is divided between chain wheel and bearing. This design allows to repair or change of bearing outside of the paint shop production.

This solution provides a number of design variants, one of which was chosen as optimal and then analyzed.

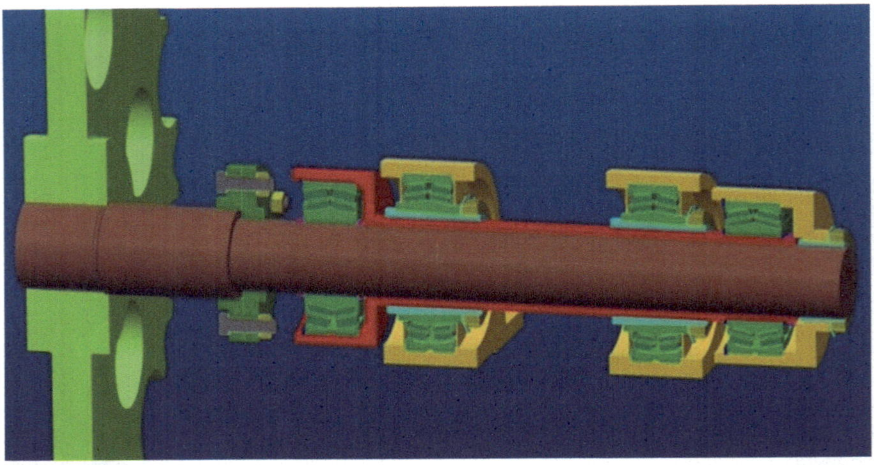

Fig. 2 Drafted solution of shaft mounting of chain wheel

Fig. 3 Equivalent stress (von Misses) of designed solution

5 Strength Analysis

Strength analysis (see Figs. 3 and 4) is an important field of every design solution. With this shaft mounting solution, the strength analysis is mainly focused on stress and deformation calculations. In all cases, the Ansys software for FEM analysis was used.

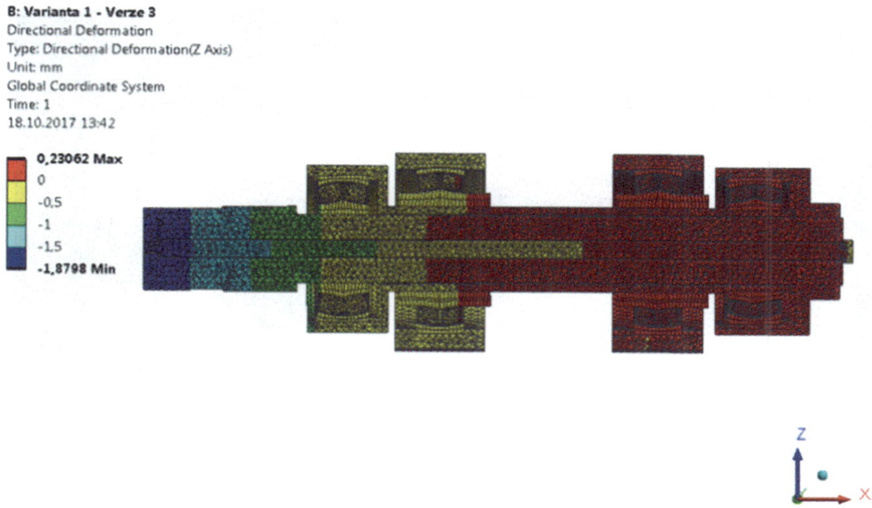

Fig. 4 Deformation in *z*-axis direction of designed solution

6 Damage Identification System

The identification of damaged shaft or frame bearing can be made based on the change of rolling resistance. For this purpose, the kinematic and moment link of shaft, reference part and frame that consists of planetary gear with simple planets and bevel cogwheels was designed. Shaft cogwheel with main axis of rotation is tightly connected to shaft. Frame cogwheel with main axis of rotation is connected to frame by adjustable frictional moment. The link between these cogwheels is made by satellites that are mounted to the pins of reference part.

The motion of reference part is determined by kinematic and moment link of planetary gear. If the rolling resistance of shaft bearing increases out of load capacity limit, the frictional moment between frame cogwheel and frame is exceeded. The frame cogwheel then starts to rotate in the same shaft direction.

If the rolling resistance of frame bearing increases out of load capacity limit, the frictional moment between frame cogwheel and frame is exceeded. The frame cogwheel then starts to rotate in the counter shaft direction.

Both rotations of frame cogwheel against the frame can be easily identified by connected sensor (see Fig. 5).

7 Conclusion

The paper is focused on actual issue of diagnostic of low-speed roller bearings. The designed solution uses reference part between shaft and frame bearing. Change of rolling resistance, which leads to bearing damage, can be identified using reference

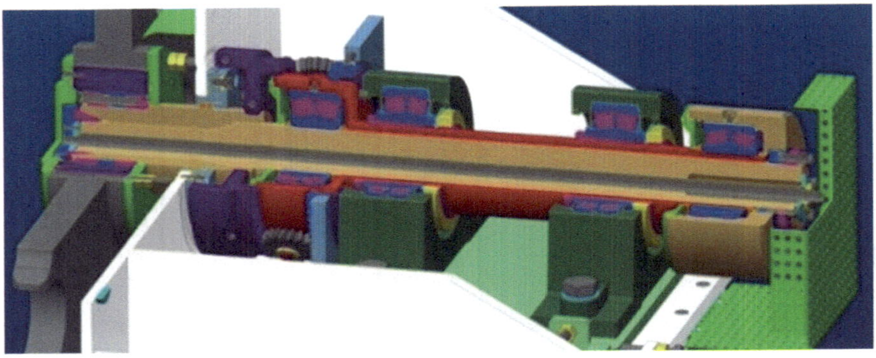

Fig. 5 Designed solution of shaft mounting of chain wheel

part. This can be realized during the manufacturing operation. The designed diagnostic system consists of planetary gear with bevel gears. Considering this is an original solution, two applications for patent were filed. One of them has been granted.

Acknowledgements This publication was written at the Technical University of Liberec as part of the project "Innovation of products and equipment in engineering practice" with the support of the Specific University Research Grant, as provided by the Ministry of Education, Youth and Sports of the Czech Republic in the year 2018.

References

1. Sturm A, Förster R (1990) Einführung in die Theorie der Technischen Diagnostik. Vieweg+Teubner Verlag, Wiesbaden, pp 18–92
2. Hauptmann P (1991) Sensoren—Prinzipien und Anwendungen, 1st edn. Verlag C. Hanser, München, Wien
3. Oeljeklaus M, Pešík L, Jancák M (2017) ŠKODA AUTO a. s. Zařízení pro diagnostiku poruch pomaloběžných ložisek. Česká Republika. Patent č. 306656, 13 Mar 2017
4. Oeljeklaus M, Pešík L, Jancák M (2018) ŠKODA AUTO a. s. Vorrichtung zum Überwachen wenigstens einer Lagereinrichtung sowie Maschine mit wenigstens einer solchen Vorrichtung. Česká Republika. Přihláška vynálezu PV 2018-86, 21 Feb 2018
5. Oeljeklaus M, Pešík L (2017) Kinematic similarity of ball and roller bearings with planetary gears. In: Herák D (ed) Proceedings of 58th international conference of machine design departments—ICMD 2017. Czech University of Life Sciences Prague, Prague, pp 276–279
6. Tomeh E (2007) Diagnostics methodology of rolling element and journal bearings, 1st edn. Technická univerzita v Liberci, Liberec

Calculating Strength of Truck-Trailer Frames by Using Software MSC Adams and the Finite Element Method

Adam Przemyk, Andrzej Harlecki and Szymon Tengler

Abstract A method of calculating strength of frames of truck trailers with a central axle (axles) by using interface of MSC Adams software, for a dynamic analysis of mechanical systems, and NX Nastran/Femap software, for the finite element method (FEM) analysis, is presented in the paper. The transfer of information between these both programs is realized within this interface which is presented in detail in the paper. In the approach proposed, a method of reduction of degrees of freedom of the FEM model has been developed based on Craig-Bampton Method belonging to the group of methods known as component mode synthesis. At the beginning of the proceeding geometrical models of components of the truck and the trailer by using program SolidWorks have been worked out. In the analysis, it has been assumed that the modeled truck-trailer combination can drive over the road with unevenness of a specific shape. Some computing results were verified experimentally. The model of truck-trailer combination can be treated as its virtual prototype useful in a design process.

Keywords Truck · Trailer · Frame · Dynamics · MSC Adams · FEM · Component mode synthesis · Craig-Bampton method

1 Introduction

The subject matter of the considerations presented in this paper is a trailer with so-called a central axle (or two central axles) towed by a truck (Fig. 1). The trailer axle is near its gravity center what provides that at an even load distribution on its surface only a small part of the static load of the trailer is carried through its drawbar to the towing truck. For these reasons, vehicles equipped with that kind of trailers are characterized by high maneuverability. Since there is a lot of loading space available the trailers with the central axle are widely used for transporting goods of a large

A. Przemyk (✉) · A. Harlecki · S. Tengler
University of Bielsko-Biala, Willowa 2 Street, 43309 Bielsko-Biala, Poland
e-mail: a.przemyk@gmail.com

A. Harlecki
e-mail: aharlecki@ath.bielsko.pl

Š. Medvecký et al. (eds.), *Current Methods of Construction Design*, Lecture Notes in Mechanical Engineering, https://doi.org/10.1007/978-3-030-33146-7_43

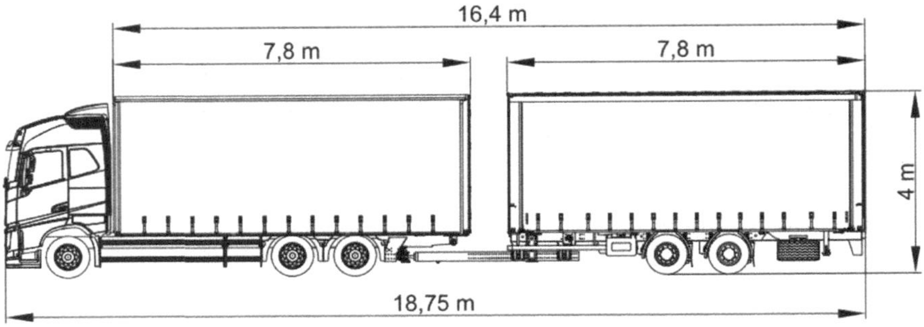

Fig. 1 View of the truck-trailer combination

volume. Practice shows that the trailers with the central axle are exposed to loss of stability. It means that so-called snaking motion (also termed as yaw motion or lateral motion) can occur while driving. The authors of this paper developed a mathematical model of the truck-trailer combination, which allows analyzing this issue. It was presented by them in paper [1]. In this paper, they do not refer to this issue, concentrating on an analysis of strength of the trailer frame. For this purpose, driving of the combination in question through a speed bump was simulated and courses of accelerations of the selected trailer points and Huber-von Mises equivalent stresses occurring in its frame were determined.

2 Modeling of the Trailer

2.1 A Geometrical Model of the Trailer

For the needs of this analysis, a geometrical model of the truck-trailer combination was developed using program SolidWorks. The model is a composition of the geometrical models of the elementary sub-assemblies of the vehicle and the trailer (Fig. 2), including the trailer frame, being its basic part. The frame model is built essentially of two longitudinal frame rails joined with cross members. This model also includes a drawbar connecting the trailer with the vehicle. A board in a form of plywood, making loading space, was placed on the frame.

2.2 A FEM Model of the Trailer Frame

On basis of the geometrical model of the trailer frame its FEM model was built in environment Femap [2]. This model contains about 145,000 finite elements, mainly shell and solid ones, including beam ones, and it has about 413,000 nodes, and a

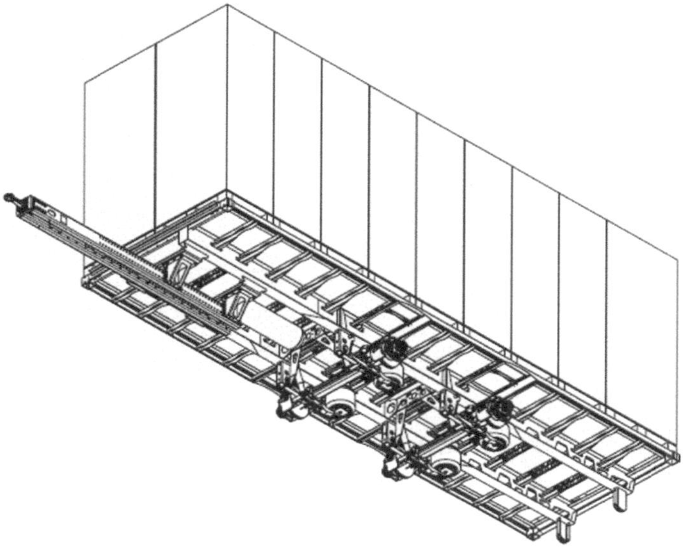

Fig. 2 Geometrical model of the trailer

number of its degrees of freedom is about 2,478,000. Finite elements modeling load in a form of rectangular bodies placed on the standard loading pallets (Fig. 3) were also considered. In the places, where in the environment MSC Adams the frame in question is connected with the components adjacent to it additional nodes, which were deprived of all degrees of freedom, were generated. Those nodes were joined

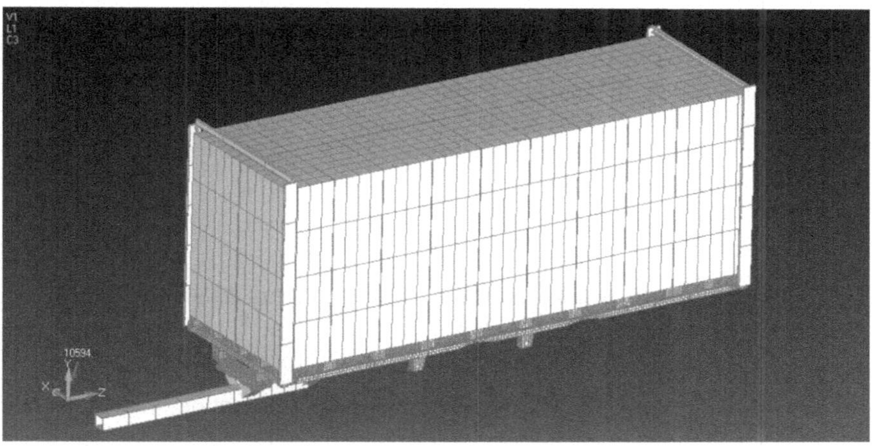

Fig. 3 FEM model of the trailer

Fig. 4 Model of the truck-trailer combination in the environment MSC Adams

with over a dozen nodes of the FEM grid adjacent to them by stiff RBE2 beam elements.

2.3 A Model of the Truck-Trailer Combination in the Environment MSC Adams

A model of the analyzed combination developed in the environment MSC Adams [3] is presented in Fig. 4. The vehicle frame, unlike the trailer frame, was modeled as a stiff body. The trailer was connected with the vehicle by a drawbar with a spherical joint at the end. Tire parameters of wheels of the vehicle and the trailer were selected on basis of a tire model offered in the environment MSC Adams, based on the "Magic Formula" developed by Pacejka [4].

3 An Application of an Interface of Programs MSC Adams and FEM

Beginning with general considerations it can be stated that motion of a deformable body which is subjected to an action of constraints from the adjacent bodies can be considered as motion of this body treated as non-deformable, on which its deformations, being a result of its free vibrations, are superimposed [5]. This way of an approach, which is termed as "floating frame of reference method" in the literature, is also used in modeling issues with use of FEM of deformable moving bodies. It should be borne in mind that a number of degrees of freedom of the FEM models may reach a few millions. In order to make a computing process feasible, a special procedure allowing a significant reduction in a number of degrees of freedom of the FEM model in question should be used and it is named the modal reduction technique in the literature. In the case of the adopted interface, Craig-Bampton method [6] is frequently used. It belongs to a group of methods referred to as component mode synthesis. More information about the interface discussed can be found in the paper [7].

A transfer of information between particular computing environments, performed within the interface of programs MSC Adams and NX Nastran/Femap, is presented in Fig. 5. The geometrical models of the sub-assemblies of the analyzed combination, developed in the environment SolidWorks (it may be also another program of type CAD 3D), are exported to a universal file of type STEP, and then they can be imported to environment MSC Adams or Femap. In this environment (that is a pre-processor), a FEM model of the trailer frame is made, and a modal analysis is performed by a solver NX Nastran [8]. During this analysis file *.mnf is made. It contains information on natural frequencies and mode shapes of vibrations of the FEM model. This information is transferred to environment MSC Adams, and in particular to its module MSC Adams/View, in which a model of the analyzed combinations is built. A simulation of the combination motion by use of module MSC Adams/Solver is a next stage. The obtained results in a form of deformations of the modeled frame changing with time are exported to environment NX Nastran/Femap, in which calculations concerning, e.g., determining contours of the Huber-von Mises equivalent stresses or deformations are performed by solver NX Nastran. The results of this analysis are presented by use of post-processor Femap, whereas the results of the dynamic analysis made in environment MSC Adams are presented by module MSC Adams/Post-processor.

4 Some Results of the Simulations

The truck-trailer combination model prepared in such a way was set into motion and covered an obstacle in a form of a 50 mm high-speed bump with all wheels. The driving took place at 4.4 m/s speed (that is about 16 km/h).

Courses "A" of vertical accelerations of the mass centers of the geometrical models of the frame and the front axle of the trailer determined appropriately by computing as functions of time are presented in Figs. 6 and 7. Those courses were verified experimentally by making measurements of those accelerations by piezoelectric sensors which resulted in determining courses "B". As it can be seen clearly vibrations appeared at moment $t = 4.5$ s, when the rear wheels of the towing truck began to drive over the speed bump. Intensification of those vibrations took place in the time interval from the moment at $t = 6.4$ s to the moment at $t = 6.8$ s, when the trailer wheels were driving over the bump.

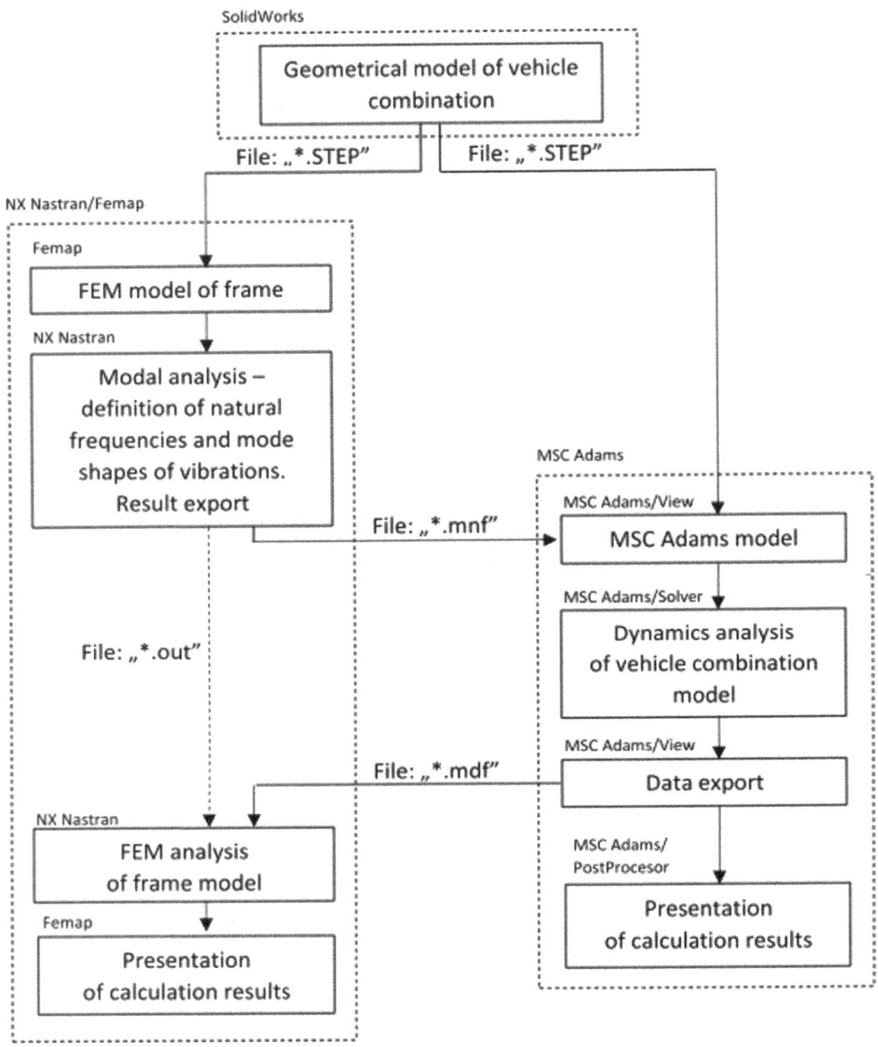

Fig. 5 Interface of programs MSC Adams and NX Nastran/Femap—information transfer

Contours of the Huber-von Mises equivalent stresses in the trailer frame at the selected moment when the rear wheels of the truck were driving over the speed bump are presented in Fig. 8. Stresses of higher values occurring in the area of mounting the drawbar to the trailer frame should be noted.

Contours of the Huber-von Mises equivalent stresses in the trailer frame when the vertical acceleration of its mass center reaches the highest value are presented in Fig. 9. An increased level of stresses in the rear part of the trailer frame is clearly visible—in that place holes in the frame rails were made to reduce the mass.

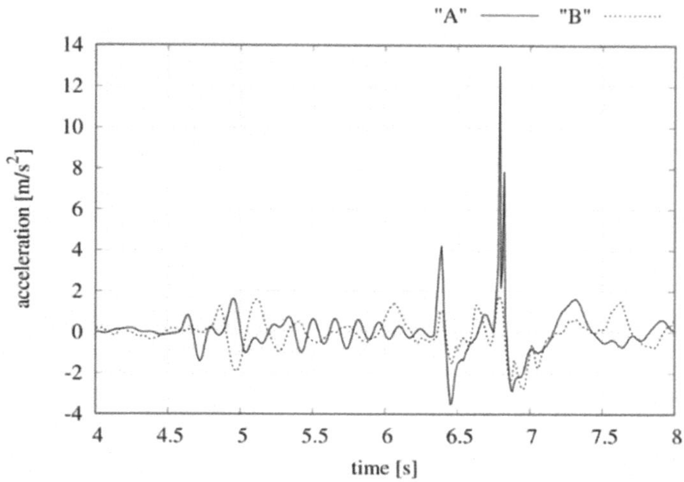

Fig. 6 Course of the vertical acceleration of the mass center of the trailer frame

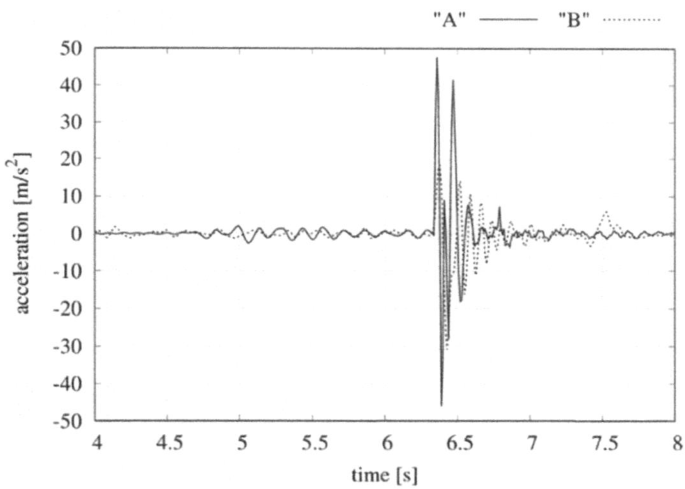

Fig. 7 Course of the vertical acceleration of the mass center of the front axle of the trailer

Transported load is protected against falling onto the road by the front and back walls visible in the figure. The walls are screwed to the frame and they may not be joined together, e.g., by roof beams. This case was analyzed while making calculations. It turns out that as a result of driving of the combination in question over the obstacle vibrations of those walls can be excited what leads to strong load of the frame areas in the places of connections with them. Therefore, it can result in occurring fatigue cracks in these areas.

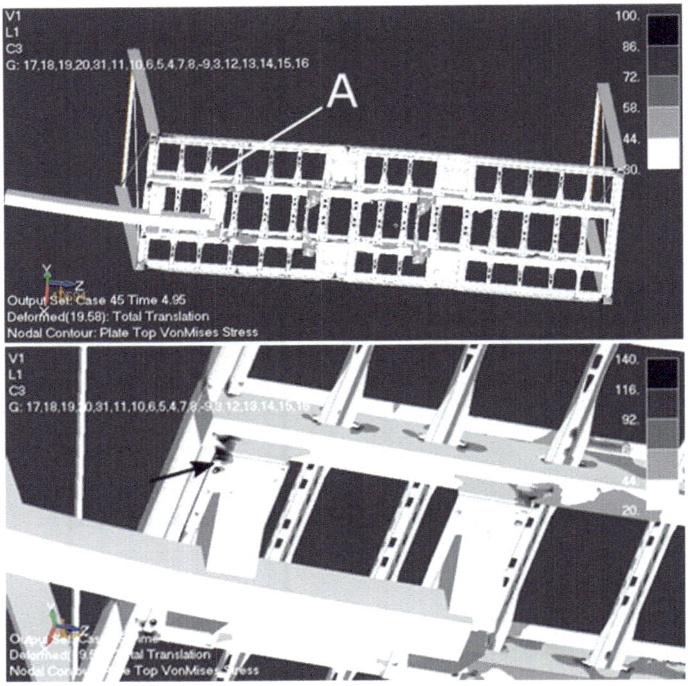

Fig. 8 Contours of the Huber-von Mises equivalent stresses in the trailer frame at the selected moment of motion, detail A is shown down (arrow = 140 MPa)

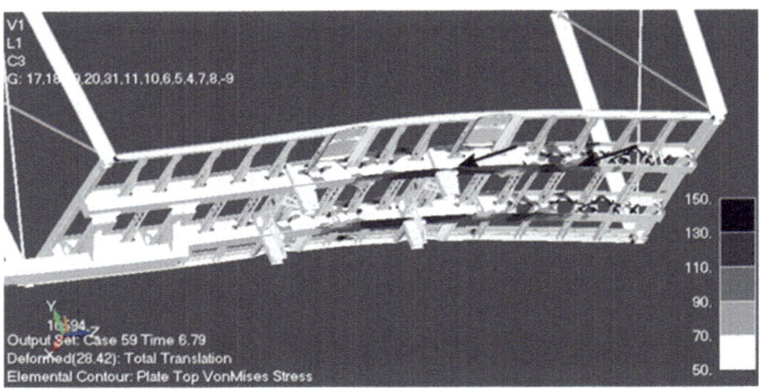

Fig. 9 Contours of the Huber-von Mises equivalent stresses in the trailer frame at the selected moment of motion (arrows = 150 MPa)

5 Conclusions

According to the authors, the presented method can be of interest for engineers dealing with designing frames of truck trailers. The calculation model of the frame developed by use of interface of programs MSC Adams and NX Nastran/Femap, understood as a prototype, enables to perform any number of calculations in the design process. Their results should provide an image of loads of its sub-assemblies, including primarily the frame, near to real one. An advantage of the developed trailer model is that its flexible frame loaded is exposed to an action of flexible and damping element co-operating with it reflecting closely real conditions of its load. An analysis of dynamics of a properly developed virtual prototype of a trailer may become basis for preparing its real prototype and only conclusions drawn from its testing will complete the design process.

References

1. Przemyk A, Tengler S, Harlecki A (2017) An analysis of dynamics of a truck with a trailer. In: Proceedings of international conference, dynamical systems—theory and applications DSTA 2017, Lodz, Poland
2. Siemens PLM Software (2012) Femap v11 Help File
3. MSC Software Corp. (2013) MSC Adams Documentation
4. Pacejka HB, Bakker E (1992) The magic formula tyre model. Vehicle system dynamics. Int J Veh Mech Mobility 21:1–18
5. Shabana AA (2005) Dynamics of multibody systems. Cambridge University Press
6. Craig RR, Bampton MC (1968) Coupling of substructures for dynamic analyses. Am Inst Aeronaut Astronaut AIAA J 6(7):1313–1319
7. Przemyk A, Harlecki A, Tengler S (2017) Method of strength calculations of truck frames (in Polish). Autobusy Tech Eksploat Syst Transp 12:1–6
8. UGS Corp. (2007) NX Nastran 5 Release Guide

Solution of Damped Oscillations by Coulomb Friction at the Karakuri Mechanism Using MAPLE Software

Tomáš Riegr and Ivan Mašín

Abstract Calculation of the time course and speed of the oscillation motion dampened by centripetal force typically represented by sliding friction is rather complicated mathematical task, which is unnecessary to solve repeatedly for individual stages of the given oscillations until they are completely subdued. Using classical mathematics is a rather lengthy calculation when the results obtained for one of the examined oscillations are simultaneously the initial conditions for the following oscillations. The fact that said damping friction force changes its direction of action depending on the direction of movement of the mechanism considerably complicates given solution. It is rather complicated to solve the given calculation by classical programming in the available mathematical software. To solve the motion equation of such a mechanism, the mathematical software MAPLE using the signum function is being used. The values obtained by calculation are further verified on a model of low-energy device for the transfer of objects in a material flow designed on the principle of Karakuri mechanism.

Keywords Karakuri · MAPLE · Damped oscillations · Coulomb friction

1 Introduction

The matter of calculating oscillations with constant (Coulomb) friction is not too widespread in practice, in contrast to oscillations attenuated by linear (viscous) damping proportional to the first power velocity. With the need for a solution of this type of oscillatory motion, we encounter during the design of Karakuri spring mechanisms, which are a perspective solution of differently usable structures capable of its function with the need of minimal supply of external energy [1], they are working in a so-called energy mode. According to [2], the Karakuri mechanisms use their basic physical phenomena such a gravity, magnetism or energy accumulation

T. Riegr · I. Mašín (✉)
Technical University of Liberec, Studentská 2, 46117 Liberec, Czech Republic
e-mail: ivan.masin@tul.cz

T. Riegr
e-mail: tomas.riegr@tiscali.cz

Š. Medvecký et al. (eds.), *Current Methods of Construction Design*, Lecture Notes in Mechanical Engineering, https://doi.org/10.1007/978-3-030-33146-7_44

into mechanical springs. A spring system that accumulates a portion of the potential energy of the transported body to be reused for reversing motion, and it is also used in the below-described design of a trolley for low-energy transfer of objects in the material flow. To construct the resulting motion equation of the trolley mechanism, first, the method of reducing the mass and force quantities of the whole structure on the frame of the trolley is used and the equation thus created is further solved by the MAPLE software.

2 The Standard Mathematical Solution of the Motion Equation of Oscillatory Motion Damped by Sliding (Coulomb) Friction

The movement of the point must be described by the two differential equations according to the velocity sign. The resistive force is in this case constant size and it is directed against the sense of speed as mentioned in [3],

$$m\ddot{x} = -kx - Nf \qquad \text{pro}\, \dot{x} > 0 \tag{1}$$

$$m\ddot{x} = -kx + Nf \qquad \text{pro}\, \dot{x} < 0 \tag{2}$$

which can also be written as stated in [4],

$$\ddot{x} + \Omega^2 x = \pm \frac{Nf}{m} \tag{3}$$

where the top sign is for $\dot{x} > 0$ and bottom for $\dot{x} < 0$. This reality considerably complicates the solution because the task has be to solved in stages, during which time individual stages are being limited by the extreme positions of the course curve when $\dot{x} = 0$ (points x_{1K}, x_{2K}, x_{3K}). From the character of Eq. (3) results that in the individual stages, the course $x_i(t)$ of the same type as the undamped oscillatory motion with the circular frequency Ω with the effect of the frictional force being expressed by displacement of the centre either left or right by the value p:

$$p = \frac{Nf}{k} \tag{4}$$

in general, for the individual stages applies motion equation:

$$x_i = C_i \sin(\Omega t_i + \varphi_{\mathrm{oi}}) \pm p \tag{5}$$

where index i denotes the interval of the movement of the given stage of the course $x(t)$, with the values x_i, C_i, φ_{oi} and time t_i is measured from the beginning of the ith stage of the movement and displacement of the centre. Figure 1 shows the time

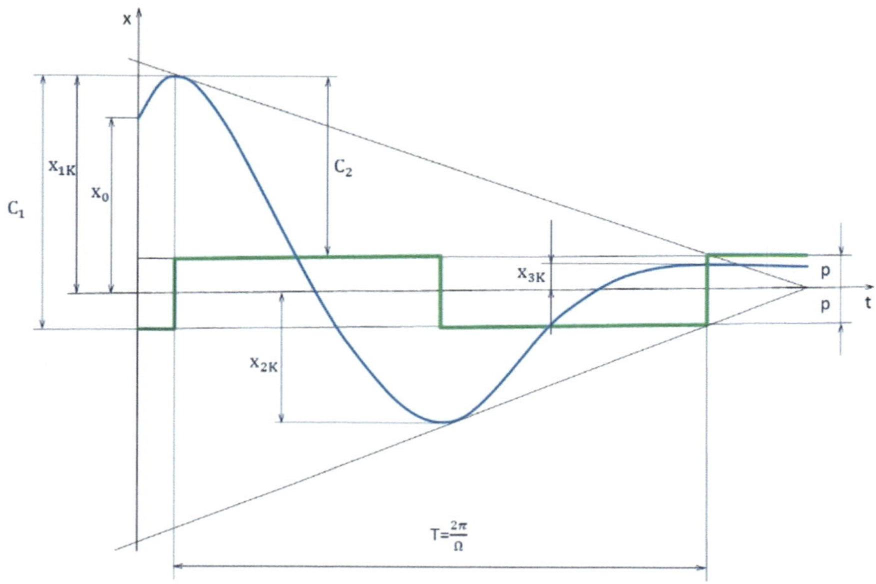

Fig. 1 Time course of oscillating motion variations with friction damping

course of the damped oscillation where the variance of the individual oscillations is reduced by $2p$ every half-period $T/2$.

3 Solution of the Motion Movement Equation Using the MAPLE Software and the Signum Function

It is possible to solve differential Eqs. (1) and (2) by means of the commonly available and practically very much used mathematical software MAPLE described in [5], but the problem is how to correctly enter the sign change of the friction force value to the equations.

The sign change of the friction force value depending on the direction of movement of the system can be solved using the signum mathematical function.

The signum function (see Fig. 2) is a mathematical function that assigns to any number x a value one ($x \neq 0$) or zero (for $x = 0$). Any real number can be expressed as the product of the sign and absolute values:

$$x = \operatorname{sign}(x)\ |x|, \quad x \neq 0 \tag{6}$$

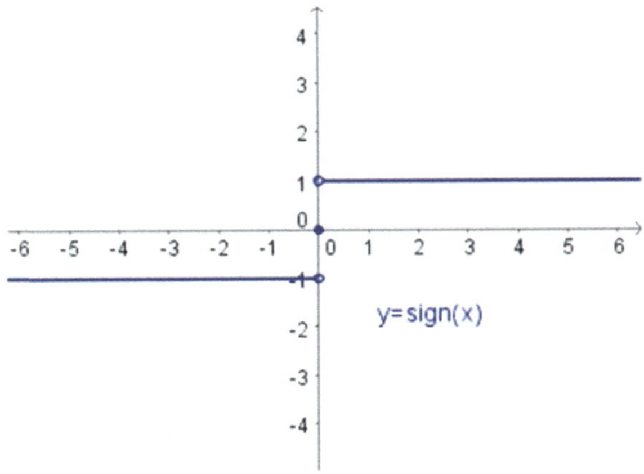

Fig. 2 Graph of signum function

4 Experiment and Calculations

The use of the signum function in the differential motion equation has been verified on calculations of dynamic behaviour of the trolley for transport of object between two points. The test trolley was simply designed on the principles of the Karakuri mechanisms as a low-energy device used principles mentioned in [6] for the transfer of objects in the material flow. For its propulsion, it uses gravitational forces of the transported object with an accumulation of part of the potential energy of the transported object into the compression springs during forward movement. This accumulated energy is then used to return the trolley back to the starting point.

From Fig. 3, it is obvious the oscillation motion initiator after insertion of the transported object with mass m_z into the transport box of mass m_b. Then the existing balance of the gravitational force from the box and the force of the spring:

$$m_b g = k l_b \tag{7}$$

is disturbed and thus to disruption of the equilibrium state. The mechanism moves and vibrates until the system dampens, where applies

$$(m_b + m_z)g = k(l_b - x_{st}) \tag{8}$$

Using the signum function Eq. (7) has this form:

$$m_r^* \ddot{x} + 4k\left(\frac{r_2 r_4}{r_3 r_k}\right)^2 x + \mathrm{sign}(\dot{x})\left(4F_t \frac{r_2 r_4}{r_3 r_k}\right) = (m_b + m_z)g\frac{r_2 r_4}{r_3 r_k} - m_{total} g \frac{\xi}{r_k} \tag{9}$$

where

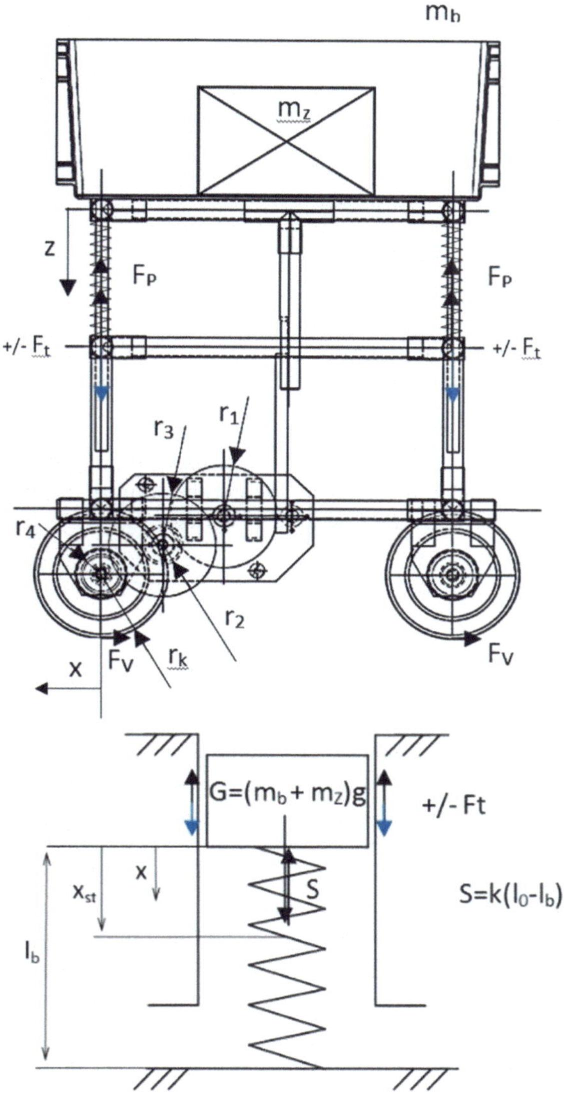

Fig. 3 Experimental Karakuri trolley and replacement scheme of geometric force and mass quantities

m_r^* reduced weight of trolley on its frame [kg]
m_b upper part of trolley weight, box weight and rack weight [kg]
m_z transferred object weight [kg]
m_{total} total weight of trolley [kg]
x axes of trolley moving [m]
r_1 radius of gearwheel Nr. 1 in gearbox [m]
r_2 radius of gearwheel Nr. 2 in gearbox [m]
r_3 radius of gearwheel Nr. 3 in gearbox [m]
r_4 radius of gearwheel Nr. 4 in gearbox [m]
r_k radius of ground wheel [m]
F_t friction force in linear guide of upper frame [N]
ξ arm of rolling resistance [m]
g gravitational acceleration [m/s^2]
k stiffness of the spring [N/m].

By entering Eq. (9) into the MAPLE software using the signum function [5] and defined parameters of trolley mechanism, we get the solution of the travelled paths (distance) and the speed depending on time as per (see Fig. 4).

5 Conclusion

Experiment and calculations showed the following facts:

- The course of the speed and distance calculated by the time corresponds with its shape to the theoretical curves obtained by the classical calculation of the individual stages of the oscillation motion curve according to the Equation relation (5).
- With currently used kinematic bonds and geometry of individual parts, the trolley is able to transfer the object in graph shown distance 1.45 m where it stops after vibration damping.
- Maximal reached distance from the start is 2.35 m—the maximal distance point before the oscillation movement is dampened. When the trolley is locked in this position by additional mechanism, it is possible to increase the driving distance by up to +62% compared to the friction stop distance!
- The rate of the course corresponds to the time course of the path and at its vertices is equal $\dot{x} = 0$.

Acknowledgements The research work and outcome reported here have been achieved as part of studying the Karakuri mechanisms and their use for the transfer of objects in the material flow of production, which is also the subject of the dissertation work of the author.

```
> restart;
  ode := m[R]*diff(x(t), t, t) + K·x(t) + F[0]·signum(diff(x(t), t)) = F;
```

$$ode := m_R \left(\frac{d^2}{dt^2} x(t) \right) + K\, x(t) + F_0 \operatorname{signum}\left(\frac{d}{dt} x(t) \right) = F \quad (1)$$

```
> params := {m[R] = 17.993, K = 19.16, F[0] = 4.5, F = 27.14};
```

$$params := \{F = 27.14, K = 19.16, F_0 = 4.5, m_R = 17.993\} \quad (2)$$

```
> res := dsolve( {eval(ode, params), x(0) = 0, D(x)(0) = 0}, numeric, maxfun = 0);
```

res := **proc**(x_rkf45) ... **end proc** (3)

```
> plots[odeplot](res, [[t, x(t)], [t, D(x)(t)]], t = 0..15, numpoints = 100, color = [red, blue],
  legend = ["distance", "velocity"]);
```

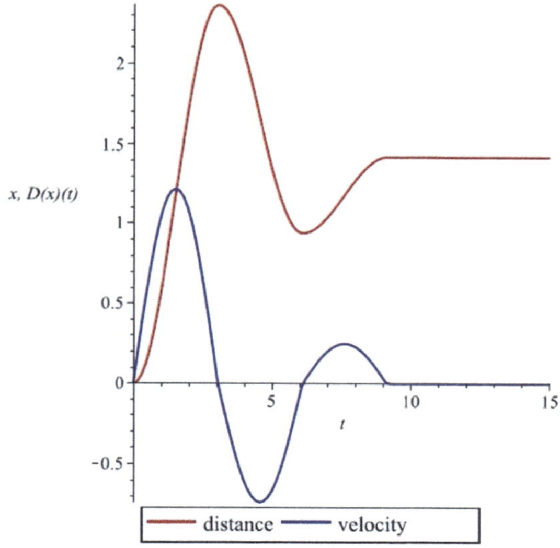

Fig. 4 Graph of distance and velocity movement of trolley over time

References

1. Hatcher D (2004) The free-energy device handbook, 1st edn. Adventures Unlimited Press, Kempton, Illinois, USA
2. Katayama H, Sawa K, Hwang R, Ishiwatari N, Hayashi N (2014) Analysis and classification of Karakuri technologies for reinforcement of their visibility, improvement and transferability: an attempt for enhancing lean management. In: PICMET 2014—Portland international center for management of engineering and technology, proceedings: infrastructure and service integration, Institute of Electrical and Electronics Engineers Inc., pp 1895–1906. Institute of Electrical and Electronics Engineers Inc., Kanazawa, Japan
3. Beitz W, Küttner KH (1983) Taschenbuch für den Maschinenbau, 15th edn. Springer-Verlag, Berlin, Heidelberg
4. Brát V, Brousil J (1988) Dynamika, 1st edn. ČVUT Praha, Czech Republic

5. Maple user's manual. https://www.maplesoft.com/documentation_center/maple18/usermanual.pdf. Accessed 2015/5/2
6. Brown HT (2009) 507 mechanical movements mechanism and devices. BN Publishing, Milton Keynes, UK

The Procedure for Reduction of the Acoustic Performance Level of the Spinning Cycle in the Washing Machine

Ján Šteininger, Ján Ďungel, Slavomír Hrček and Igor Gajdáč

Abstract The development of automatic washing machines and their gradual upgrade of already sold models is a very challenging task, mainly due to the changing operating conditions of the washing and subsequent spinning cycle. Increasing demands of consumers to buy the appliance constantly force manufacturers to continually innovate and improve their products to be able to succeed in the enormous competition. The trend for the centrifuging cycle lies in continual spin increase, resulting in greater forces and thus load on the overall machine construction. In general, it is known that with increasing spins more water is removed from the laundry. This is a significant benefit for the consumer as the drying time of the laundry is considerably shortened. However, the increase in spins can also bring the negative phenomena. One of them is a louder noise of the washing machine. Therefore, it is necessary to continually improve the procedures that can reduce the noise. For their correct design, experiments have been carried out. They have consisted of the measurement of the declaration, technical and subsequent measurement using an acoustic camera. Its using will determine the proportion of acoustic energy spread from individual construction components. These measurements and simulations will be used to set engineer theories so that it can be possible to transform the appliance, resulting in an effective reduction of the acoustic performance level. The theories will be verified by subsequent stated measurements and compared with the original values.

Keywords Noise reduction · Washing machine

J. Šteininger (✉) · S. Hrček · I. Gajdáč
University of Žilina, Univerzitná 8215/1, 01026 Žilina, Slovak Republic
e-mail: jan.steininger@fstroj.uniza.sk

J. Ďungel
CEIT Engineering Services, s.r.o, Univerzitná 8413/6, 01008 Žilina, Slovak Republic

Š. Medvecký et al. (eds.), *Current Methods of Construction Design*, Lecture Notes in Mechanical Engineering, https://doi.org/10.1007/978-3-030-33146-7_45

1 The Project of Experiments for Identifying Noise Sources

Measurements were performed in a certified acoustic laboratory using the spin programme at 1 400 RPM. In order to prevent a change in the distribution of the laundry in the drum and its imbalance, an evenly distributed 3 × 3 kg rubber was used in the drum of the washing machine to ensure the repeatability of the individual compared measurements. The rubber imbalance was oriented at the centre of the 0.5 kg drum [1].

1.1 Technical Parameters of Measurement

Measurement was done in anechoic chamber, (see Fig. 1) and chosen measuring system was PULSE system B&K 3053-B-120. The type of microphones used for measurement was B&K 4190 and calibrator B&K 4231. Microphones worked with acoustic camera. Selected washing programme during measurement was spin. 3D scan was made on measuring device, type FARO® Design ScanArm 2.0.

Before the measurements for the washing machine were performed, the errors that were made during the production and the lack of control were eliminated. They consisted of adding screws on individual coverings of the washing machine and inserting a gasket between the cabinet and the side panels [2].

The spinning cycle of the washing machine consists of four speeds, and the highest acoustic power was recorded at the maximum speed revolutions of 1 400 RPM. From this section, a 50 s interval is taken and its total acoustic performance is used to calculate the value written on the energy label. For other speeds, the 50 s section

Fig. 1 Position of the appliance in an anechoic chamber

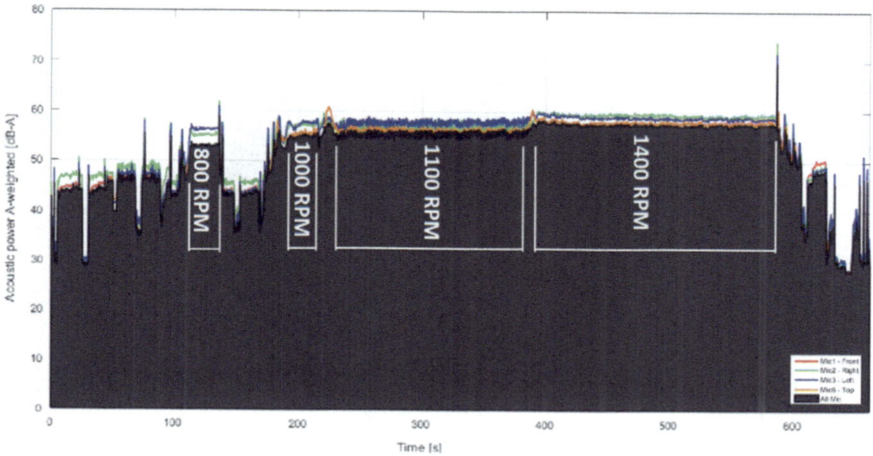

Fig. 2 Profile of the acoustic power of the washing machine during spinning at different speeds

of the 1 100 RPM and 20 s sections for 800 and 1000 RPMs were taken from the acoustic performance profile [3]; (see Fig. 2).

3D measurement method that using the FARO system was used to construct the model, and consequently, a model for individual components of the washing machine was created from the point cloud (see Figs. 3 and 4). These models were used in the COMSOL Multiphysics simulation software to identify the actual shapes of individual components using the finite element method; (see Fig. 5). Here we were interested in the frequencies recorded in the declarative measurements so that the individual components of the washing machine were not exposed to resonant frequencies during each spinning cycle, and it was possible to effectively reduce the noise level of the appliance [1, 3].

The shape complexity of the whole appliance is very demanding for the calculation of the FEM method. Because of this fact, the hammer resonance method (on the cabin, the drum, the water tank of the washing machine) was used together with all the components to identify the resonant frequencies of the whole appliance. From

Fig. 3 Scanning of the washing machine components point cloud scanning

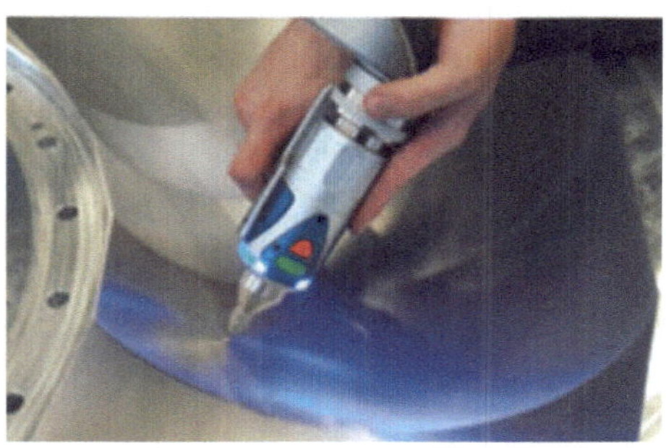

Fig. 4 Point cloud scanning

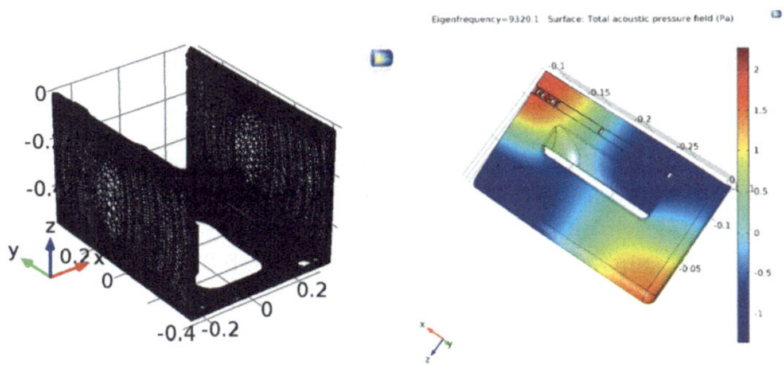

Fig. 5 Calculation of actual shapes of the washing machine component in COMSOL multiphysics simulation software

a physical point of view, the impact of the hammer on the component will create different kinds of waves that go through the tested component at different speeds [1, 3] (see Fig. 6).

After subsequent modifications of the washing machine components, repeated declarative measurements were performed in three repetitions. Here, a decrease in the acoustic level performance of the appliance was detected, and it ranged from 0.8 to 1.7 dB (A); see Table 1.

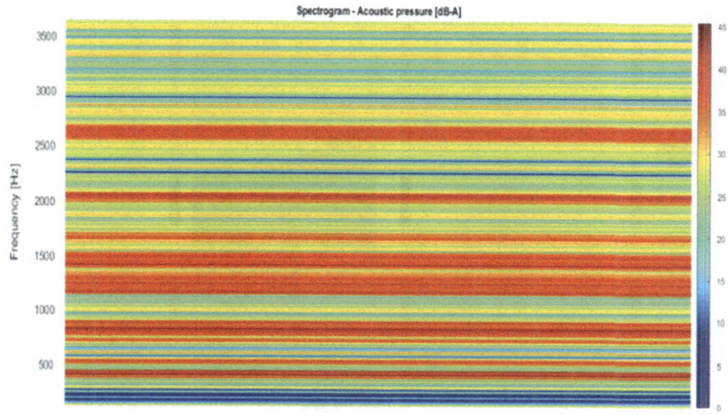

Fig. 6 Spectrogram of the hammer method (the cabinet of the washing machine)

Table 1 Average improvements in acoustic performance after shape modifications of the washing machine components

Improvement of acoustic power (dB A)	Drum speed (rpm)	800	0.8
		1000	1.5
		1100	1.3
		1400	1.7

1.2 Design of Noise Insulation for the Modified Appliance

For the correct design of position and type of insulating material, measurements were made using an acoustic camera. The assumption is that the acoustic energy spread from individual construction components should be correctly determined so that the highest possible acoustic pressure level reduction is reached by using the smallest amount of acoustic material [4].

As shown above (see Fig. 7), a significant proportion of the emitted acoustic energy at 315 Hz and spinning is spread from the bottom of the washing machine to its surroundings. Therefore, the insulation was also located at the bottom part of the appliance. From the frequency analysis, a suitable type of moderating material was designed to be inserted from the bottom part of the washing machine along the perimeter before the placement of the removable sheet metal cover with flood control function [4].

The measurement was carried out as in the case of declarative measurements for three washing machines with three repetitions with (see Fig. 8) and without insulating material. After the washing machine placing and its feet setting, the appliance was no more moved and the insulation was removed so that the washing machine stayed at its place to maintain the best measurement repeatability [1, 4].

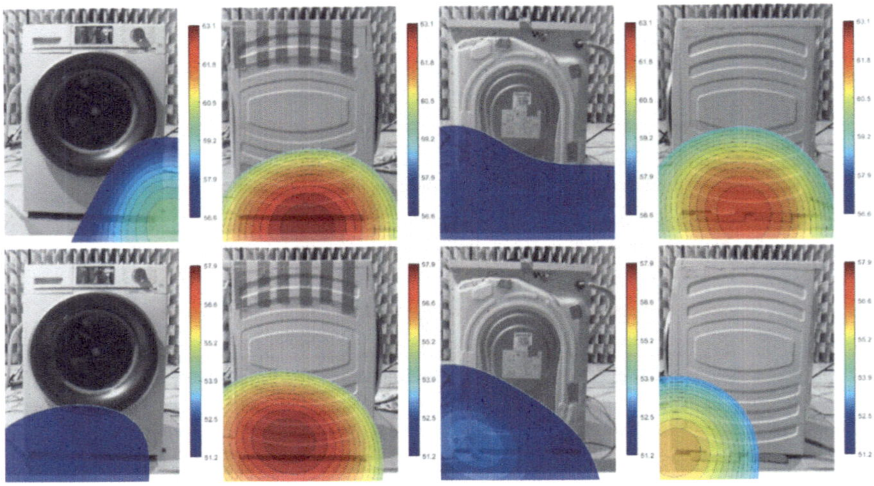

Fig. 7 Recording from the 315 Hz acoustic camera in the washing machine at 1400 RPM spinning

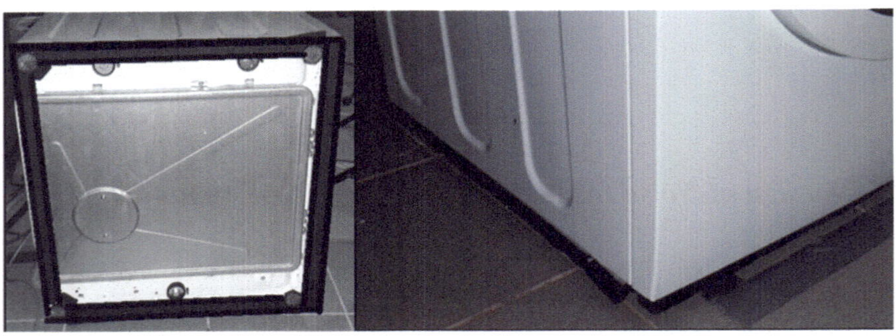

Fig. 8 Placement of the acoustic insulation at the bottom part of the washing machine

For repeated measurements, when the unbalance distribution is the same and the appliance was not moved, there are different acoustic powers and they can move up to 2.7 dB (A). For the washing machine 3 and 800 RPM, the acoustic power value is significantly higher than for the washing machines 1 and 2 [4]; (see Table 2).

2 Conclusion

Using the most modern methods of noise, elimination has an enormous impact on effective noisiness reduction, where with appropriate modifications of individual components, it has a positive impact on reducing the overall acoustic power of the

Table 2 Acoustic power at different revolutions during spinning cycle

	The acoustic power (dB A)		Rubble load 3 × 3 kg + UB 500 g middle								
			Washing machine 1 measurement			Washing machine 2 measurement			Washing machine 3 measurement		
			M 1	M 2	M 3	M 1	M 2	M 3	M 1	M 2	M 3
With insulation	Drum speed (rpm)	800	54.7	56.4	56.3	55.8	55.1	55.0	57.4	61.3	63.1
		1000	55.9	57.4	57.8	57.4	57.3	57.0	58.5	58.2	57.2
		1100	57.3	58.5	60.0	59.3	60.5	61.0	57.4	57.9	58.1
		1400	62.7	63.2	63.5	61.8	63.5	62.7	62.7	60.8	62.2
No insulation	Drum speed (rpm)	800	57.1	5619	56.9	56.8	56.3	58.2	63.2	63.0	62.0
		1000	58.7	57.6	59.0	59.6	59.3	58.7	58.7	59.7	59.1
		1100	62.5	62.9	63.7	62.6	63.7	63.7	59.9	61.3	60.5
		1400	64.1	63.9	65.1	66.1	65.0	65.5	64.0	64.3	64.2

Table 3 Average improvements of the acoustic power of the washing machines for different revolutions of the drum

Improvement of acoustic power (dB A)	Drum speed (rpm)	800	1.7
		1000	1.5
		1100	3.4
		1400	2.1

appliance. Also the using of a suitable acoustic material by an acoustic camera precisely localized place, it can save the financial resources during the production of the household appliances. Another criterion is the improvement rate, which is reflected in the energy label and it can help to distinguish the appliance from the competition and improves its marketability. Average acoustic improvements (see Table 3) ranged from 1.5 to 3.4 dB (A), and the most noticeable values were recorded at 1100 RPM, and it was the value 3.4 dB (A). For the highest acoustic power values at maximum revolutions of 1400 RPM, the average improvement was 2.1 dB (A). As the revolutions of 1100 and 1400 have a total of 70% of the time duration of the 12-min spinning cycle, there is an assumption that the overall equivalent noise level emitted from the appliance throughout the whole process of spinning will be significantly lower.

Acknowledgements This publication is the result of the project implementation: Competency Centre for Knowledge technologies applied in Innovation of Production Systems in Industry and Services, ITMS: 26220220155, supported by the Research & Development Operational Programme funded by the ERDF.

References

1. STN EN 60704-1, Akustika (2010) Household and similar electrical appliances. Test code for the determination of airborne noise. Part 1: general requirements
2. Zvolenský P, Kašiar Ľ, Volna P, Barta D (2017) Simulated computation of the acoustic energy transfer through the structure of porous media in application of passenger carriage body. Procedia Eng 187:100–109
3. Zvolenský P, Grenčík J, Pultznerová A, Kašiar Ľ (2017) Research of noise emission sources in railway transport and effective ways of their reduction. In: MATEC web of conferences, dynamics of civil engineering and transport structures and wind engineering—DYN-WIND 2017, Trstená, Slovak Republic, 21–25 May 2017, vol 107, article no 00073
4. Zvolenský P, Pultznerová A, Grenčík J (2016) The simulation calculation of acoustics energy transfer through the material structure. In: MATEC web of conferences, 5th international scientific conference, integration, partnership and innovation in construction science and education, vol 86, article no 04001

Load Analysis of Ship Generating Sets During the Maneuvers of the Vessel

Dariusz Tarnapowicz

Abstract The specificity of ship power plants is connected with the vessel's working mode and, above all, with operational safety. The synchronous generator powered by a combustion engine (diesel generator) is widely used in shipbuilding. During maneuvers and on "difficult" waters regions (sailing down the river, canal, straits, lakes, etc.), in order to ensure continuity of power supply, the parallel operation of generating sets is required. Safety reasons during maneuvers have impact on the parallel operation, rather than power demand. Therefore, auxiliary engines are often underloaded. These conditions cause that the fuel is not fully burned. It has an effect on operational problems of auxiliary engines and an increase in emissions of air pollutants. The article presents empirical researches of loads in ship generating sets in operational conditions requiring the parallel operation of sets for different types of ships. Moreover, it shows solutions that enable the limitation of unfavorable phenomenon of underloading of auxiliary engines so-called wet stacking.

Keywords Ship generating sets · Wet stacking of diesel generator · Electrical load balance

1 Introduction

On each ship, regardless of size, the ship's power plant is the basic system deciding on the safety of shipping—operational safety and security of the crew.

Determination of the power for a ship's power plant and the number of generating sets is crucial during the design of vessels. There are several ways to select generating sets.

The first method is related to the comparison with power plants on similar or the same types of ships. This simple method is burdened with errors reproduced from other vessels. The experience of the designer and operational reports from twin vessels plays a huge role.

D. Tarnapowicz (✉)
Maritime University of Szczecin, ul. Wały Chrobrego 1-2, 70500 Szczecin, Poland
e-mail: d.tarnapowicz@am.szczecin.pl

Š. Medvecký et al. (eds.), *Current Methods of Construction Design*, Lecture Notes in Mechanical Engineering, https://doi.org/10.1007/978-3-030-33146-7_46

The second method is based on statistical analyses predicting the maximum load of power plants in various operational states [1]. This method uses the probability of switching on a given receiver (or a group of receivers).

The third method is to determine the power of the power plant on the basis of the power of the main propulsion engine, and this is typically about 10%. This method must take into account the type of vessel. It is burdened with a big error and is rarely used.

A commonly used method for the selection of the quantity and power of generating sets is an analysis based on the electrical balance of the vessel [2].

The ship's power balance is carried out for various ship operating conditions: sea travel, stoppage in the port, and maneuvers. The power balance begins with the determination of all electricity receivers in a given operational state. Then, the loading factor, peak power, and simultaneity factor of receptions are determined.

On the basis of the balance, data for the selection of power in generating sets is obtained. The number of sets is selected in a way that during the operational state of the ship lasting for the longest period (sea travel), the generating sets can operate as economically as possible.

It is recommended that one generating should operate during the sea travel. In accordance with the regulations [3], the vessel must be equipped with at least two generating sets. Usually, there are three of four sets of the same power of the ship.

During maneuvers and on "difficult" waters (sailing down the river, canal, straits, lakes, etc.), in order to ensure the continuity of power supply, parallel operation of generating sets is required. During maneuvers and on "difficult" wasters, safety reasons, not power demand, are decisive for parallel operation. Therefore, auxiliary engines are often underloaded.

2 Methods and Materials

Measurements of loads for diesel generator sets (D-G) on dozens of different types, different sizes, and different production date of ships (21 vessels) were carried out. The examinations were performed in all working conditions of the ship, i.e., sea voyage, stoppage in the port and during maneuvers. The measurements were made at 10-min time intervals for hours, reading the indications of active power given by the generators of generating sets.

The load of ship generating sets read on power meters is not equivalent to the load of auxiliary engines. By determining the load of the auxiliary engine, the coefficient of excess power of the auxiliary engine to the generator power α_{NM} and generator efficiency η_G was taken into account [4].

$$\mathrm{LF_{AE}} = \frac{\mathrm{LF_{GS}}}{\eta_G \alpha_{NM}} \tag{1}$$

where

LF_{AE} load factor of auxiliary engine;
LF_{GS} load factor of generator;
η_G efficiency of generator;
α_{NM} auxiliary engine excess power factor to generator.

The average value obtained from active power measurements for one generating set was taken as the generator load factor.

The tests performed during the sea voyage presented the load of generating sets at the level of 60–75%, and during the stay in the port—at the level of 40–60% (depending on whether the ship is transshipped).

3 Results

The lowest load was during maneuvers and sailing in "difficult" wasters. The results of measurements made during maneuvers are presented in Figs. 1, 2, 3, and 4.

The presented researches show that the load of auxiliary engines of generating sets during maneuvers is less than 40% and for the majority of ships—less than 30%.

Such a low load of internal combustion engines is the reason for the occurrence of the phenomenon, which is called wet stacking.

The methods of reducing the wet stacking phenomenon in D-G power generating sets are becoming very important.

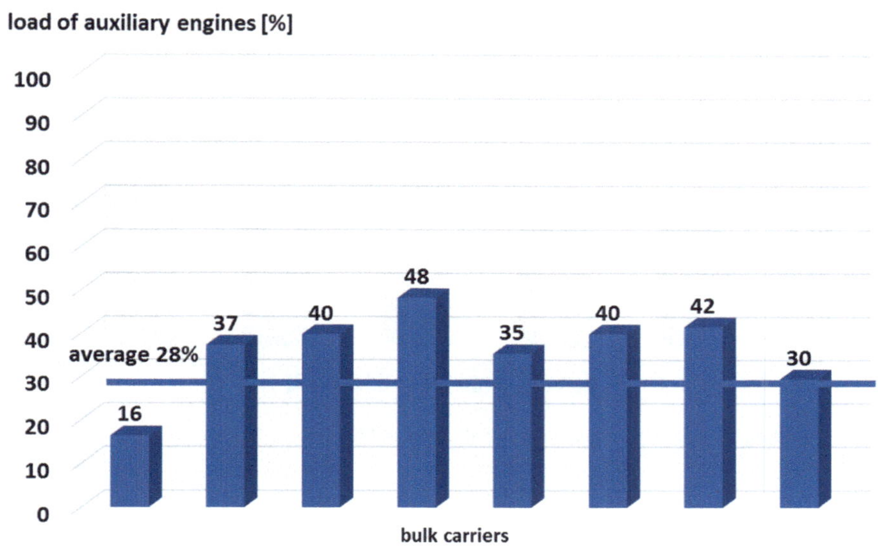

Fig. 1 Loading of auxiliary engines of bulk carriers

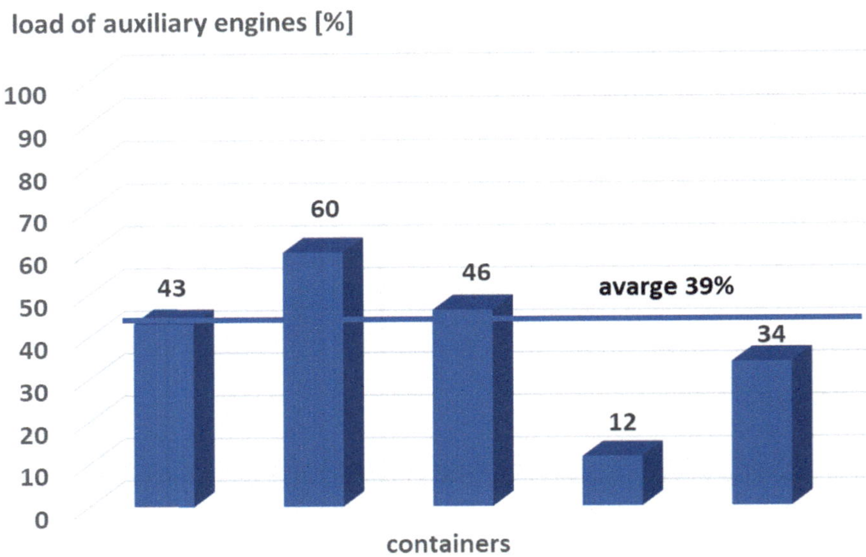

Fig. 2 Loading of auxiliary engines of containers

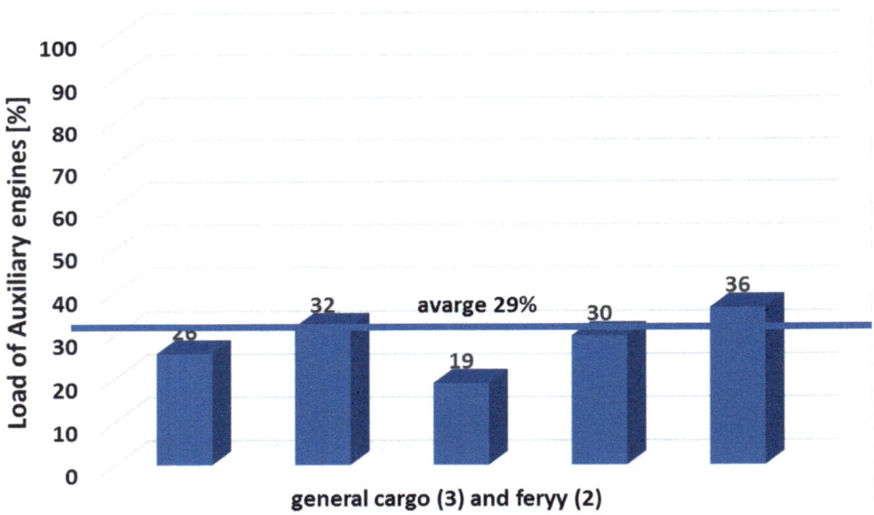

Fig. 3 Loading of auxiliary engines of general cargo and ferry

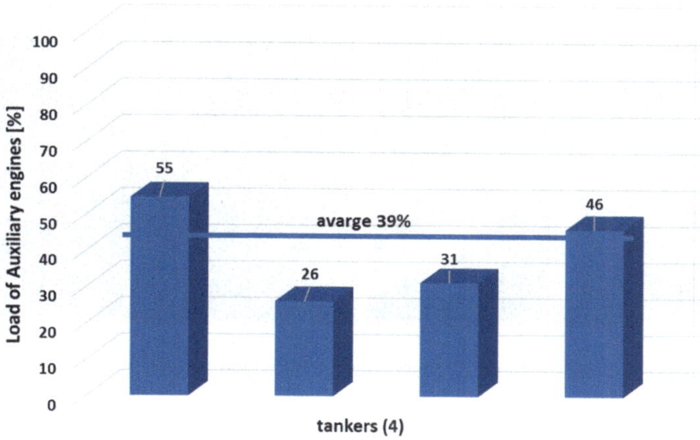

Fig. 4 Loading of auxiliary engines of tankers

4 Discussion

4.1 The Effects of the Occurrence of Wet Stacking Phenomenon

When the D-G sets are underloaded, there is a lower temperature of fuel combustion and thus combustion is not total. Unburnt fuel will be ejected from the combustion chamber, and it begins to accumulate in the exhaust part of the engine, causing the accumulation of carbon deposits on the exhaust valves, turbocharger and exhaust. Figure 5 shows the effects of the wet stacking phenomenon on the D-G exhaust pipe.

When the D-G sets work in the underloaded mode, piston rings do not expand sufficiently to properly seal the space between the piston and the cylinder wall. This causes the burning of unburned fuel and gases into the oil sump and diluting the lubricating properties of the oil, leading to premature wear of the engine. Wet stacking also causes damage to the injectors, which start to inject fuel poorly (Fig. 6)

Fig. 5 Wet stacking—"extreme clogging" of exhaust pipe [5]

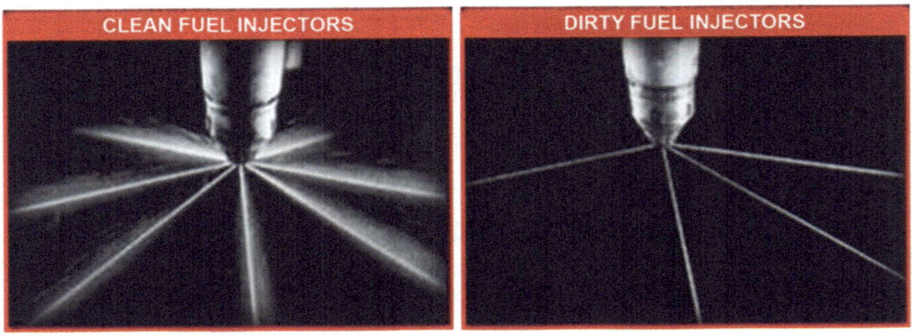

Fig. 6 Loading of auxiliary engines of general cargo, ferry, and tankers [6]

[6].

The effect of wet stacking:

- reduces the engine life for many years;
- increases exhaust emissions;
- leads to the lower maximum power;
- leads to costly repair and maintenance.

4.2 Asymmetrical Operation of Generating Sets

One of the methods of limiting the effects of the occurrence of the wet stacking phenomenon is the asymmetrical operation of generating sets. On computerized vessels, where the supervision over the operation of a ship's power plant is realized by power management system (PMS), the asymmetrical load of generating sets with active power [7] is possible. After a few hours of asymmetrical operation of generating sets, PMS decides to change the load distribution, so for about 1 h one of the sets is loaded with almost all active power, and the other practically works without load (Fig. 7).

After burning the accumulated residues, there is a change in the load of sets in order to "clean" the other set. Then, there is a return to symmetrical work of generating sets.

4.3 Optimal Power Selection of Generating Sets

As mentioned in the introduction, the selection of generating sets is usually determined on the basis of the power demand during the operation of the vessel on the

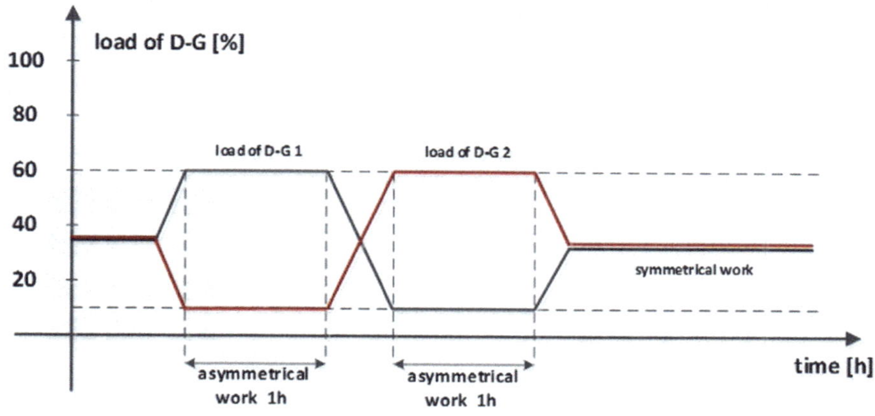

Fig. 7 Asymmetrical operation of generating sets

voyage. The D-G set is selected in a manner to ensure the operation of one set during the "sea" voyage. Other sets usually have the same power.

In order to reduce the wet stacking phenomenon, generating sets for the designed vessel can be selected to optimize their operation both "in the sea" and during "maneuvers." It is possible to choose one D-G set with higher power for operation "in the sea" and two sets with lower power for "maneuvers."

For example, the D-G set was selected from the 1 MVA electric power balance (optimal for "seal travel"), and two D-G sets with a power of 600 kVA were selected for the maneuvers.

New power electronics technologies allow for any power distribution between two generating sets with different or the same powers.

4.4 The Use of Load Banks

The obvious solution to the problem associated with the underloading of D-G ship generating sets is the temporary loading in order to burn away accumulated soot and "purging" of the fuel system. For this purpose, it is possible to use an additional receiver, which is switched on for a short time, e.g., 40 min.

The offer of generator set manufacturers includes the so-called load banks, which are selected for the power of D-G sets [8]. Load banks range from a few 10 kW to 3 MW.

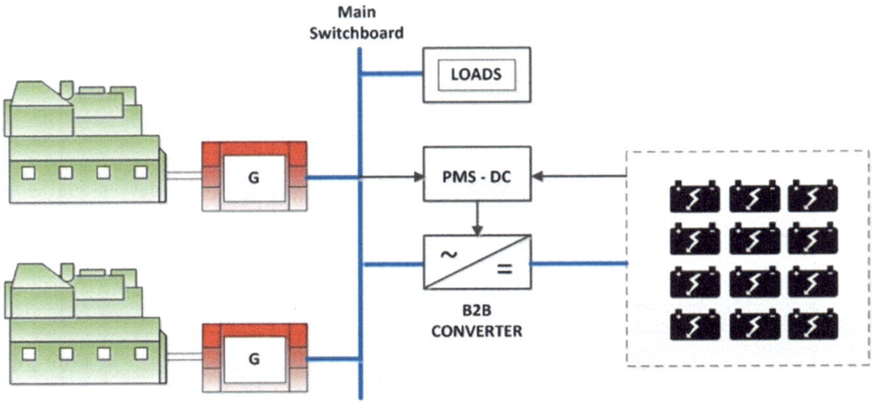

Fig. 8 Battery-powered ecological system

4.5 Battery-Powered Ecological System

A very interesting solution is the use of electricity storages that can be used in the port. Acidic batteries can be used as the energy storage. The power of these batteries is chosen to meet the vessel's demand in the port. Some types of ships, e.g., bulk carriers, require a small amount of power in the port, even during transshipment. Figure 8 shows the battery-powered ecological system.

The system works in such a way that during maneuvers, D-G sets are additionally loaded by charging the battery matrix. After docking, D-G sets are stopped and the batteries supply the receivers. The energy flow in both directions is provided by the B2B (back to back) converter. The convertor's operation is controlled by PMS–DC (Power Management System–DC).

Literature [9] confirms that vessels in the port are the largest source of pollution. The suggested solution will help to reduce air pollution caused by vessels mooring in the port while limiting the effects of underloading of generating sets.

5 Conclusions

The empirical researches of load for ship generating sets presented in this article show that D-G sets during ship's maneuvers are underloaded, and there is the so-called wet stacking phenomenon. Wet stacking causes the consumption of combustion engines, more frequent repairs, as well as an increase in the emission of air pollutants. The article proposes preventive solutions that limit the negative phenomenon.

Acknowledgements This research outcome has been achieved under the research project: Nowoczesne technologie w systemach "Shore to Ship" No 2/S/IEiAO/16 financed from a subsidy of the Ministry of Science and Higher Education for statutory activities.

References

1. Kostyszyn R, Nowak T (2016) Elektroenergetyka okrętowa. Akademia Morska w Gdyni
2. Wyszkowski S (1991) Elektrotechnika okretowa. Tom1. Wydawnictwo Morskie Gdańsk
3. Polski Rejestr Statków (2018) Przepisy klasyfikacji i budowy statków morskich część VIII. Copyright by Polski Rejestr Statków S.A, Gdańsk
4. Nicewicz G, Sosinski M, Tarnapowicz D (2014) Identification of power factor in marine electrical grid. In: Geoconference on energy and clean technologies, vol II, 14th international multidisciplinary scientific geoconference (SGEM), pp 391–398. Albena, BULGARIA
5. MQ POWER Homepage. http://www.mqpower.com/wet-stacking.html. Last accessed 2018/04/21
6. Multiquip Homepage. http://www.wincogen.com/wp-content/uploads/PD/Engines/wet_stacking_info_prevention.pdf. Last accessed 2018/04/21
7. Śmierzchalski R (2004) Automatyzacja systemu elektroenergetycznego statku. Gdańsk
8. Assocpower Homepage. http://www.assocpower.com/publicdocs/getpdf-04.php?file=Avtron-60Hz-Buyers-Guide. Last accessed 2018/04/21
9. Tarnapowicz D, German-Galkin S (2018) International standardization in the design of "shore to ship"—power supply systems of ships in port. Manag Syst Prod Eng 26(1):9–13

Measurement of Torque and Axial Force on Cardan Shaft

Miroslav Trochta and Jiří Začal

Abstract Cardan shafts are commonly used when the drive and driven shafts are not coaxial but are still parallel. If the distance between the shaft changes, then the length compensation is moved under the influence of torque, and the resultant friction causes an axial force to be generated between the tooth flanks in the spline. This movement can be caused by the deformation of the frame or misalignment of shafts. The magnitude of this axial force is important for the design of other machine parts in terms of service life. This article deals with the measurement of torque and axial force on cardan shaft located on machine drive. This was done to verify the magnitude of the axial forces and torque at the run of the machine due to the fatigue fracture of the bolts at the cardan flange. Article describes the methodology of measurement, used equipment, and evaluation method. Measured values of the axial force are compared with theoretical calculations of axial force according to the manufacture of cardan shaft.

Keywords Cardan · Shaft · Measurement · Force · Torque

1 Introduction

Measurement of load conditions during operation is a very important matter for the determination of actual loads on measured parts. Cardan shafts are one of the main parts in machines, and they are used to transmit torque between not coaxial shafts. That means that they are mainly used in machines drives, typically they can be found on heavy trucks between gearbox and axle differential cardan usually consists of two flanges: two universal joints and shaft with length compensation (see Fig. 1).

The aim of measurement was to determinate real values of axial force and torque. That was because from the theoretical calculation provided from the manufacture of cardan shaft, maximal axial force should be calculated with Eq. (1) [1] with use of friction coefficient 0.03 for Rilsan-coated involute spline. Drive and its parts were designed with the use of this maximal force, and result was fatigue fracture of the bolts

M. Trochta (✉) · J. Začal
Technical University of Ostrava, 17. listopadu 15/2172, 70833 Ostrava-Poruba, Czech Republic
e-mail: miroslav.trochta@vsb.cz

Š. Medvecký et al. (eds.), *Current Methods of Construction Design*, Lecture Notes in Mechanical Engineering, https://doi.org/10.1007/978-3-030-33146-7_47

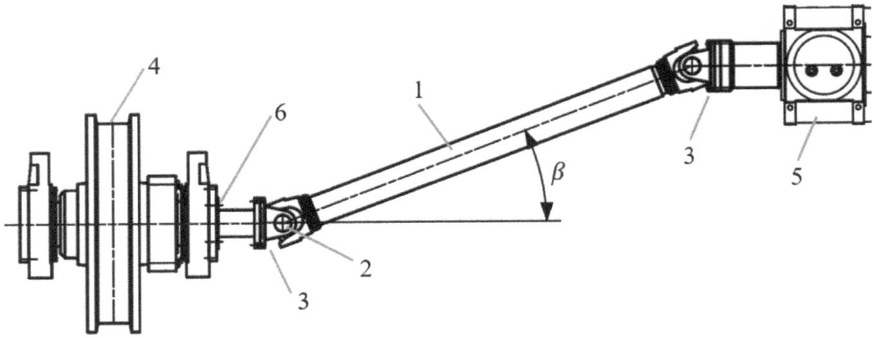

Fig. 1 Cardan shaft. 1—Shaft with length compensation, 2—universal joint, 3—flange, 4—wheel, 5—drive, 6—shaft extension

on shaft extension (position 6, see Fig. 1). After customer eliminated the possibility of design error at his side, he asked us to measure real values of torque and axial force. To compare them it with theoretical forces,

$$F_{\mathrm{ax}} = 2 \cdot T \cdot \mu \cdot \left(\frac{1}{d_{\mathrm{t}}} + \frac{\sin \beta}{L} \right), \tag{1}$$

where

F_{ax} Axial force,
T Nominal torque,
μ Friction coefficient,
d_{t} Pitch cycle diameter of the spline,
β Working angle,
L Profile coverage.

2 Measurement Arrangement

To measure axial force and torque, we had first chosen method of measurement. This choice was simple because our workplace is well equipped for strain gauge measurement; we have chosen this method due to the request for the measurement of axial force and torque. Two independent measurement spots next to each other had to be prepared. For both we used full-bridge configuration of connection in Wheatstone bridge. The strain gauges were glued to the tubular part of cardan shaft shown in Fig. 2. Strain gauges were connected to Wheatstone bridges, then to wireless strain gauge apparatus, and signals from the apparatus were digitized using A/D converter.

Because we were not allowed to dismount cardan shaft and move it to direct calibration. Torque and axial force had to be calculated from strain that was calculated as followed:

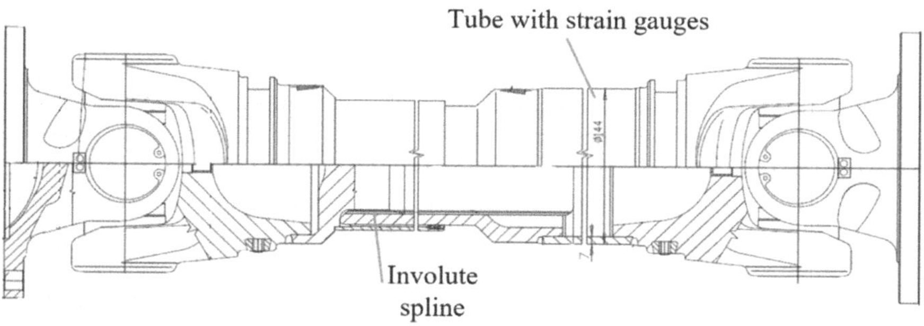

Fig. 2 Cardan shaft drawing

$$\varepsilon_{\sigma} = \frac{4 \cdot C}{k \cdot n} \cdot U_{\mathrm{m}} = \frac{4 \times 0.2}{2 \times 2.6} \cdot U_{\mathrm{m}} = 0.153846 \cdot U_{\mathrm{m}}, \quad (2)$$

$$\varepsilon_{\tau} = \frac{4 \cdot C}{k \cdot n} \cdot U_{\mathrm{m}} = \frac{4 \times 0.2}{2 \times 4} \cdot U_{\mathrm{m}} = 0.1 \cdot U_{\mathrm{m}}, \quad (3)$$

where

C Sensitivity of the bridge,
U_{m} Measured voltage,
k Gauge factor,
n Bridge factor.

Then torque was calculated from Eq. (4) and axial force from Eq. (5).

$$T = \varepsilon_{\tau} \cdot \frac{J_{\mathrm{z}}}{r} \cdot G = \varepsilon_{\tau} \cdot \frac{\pi \cdot \left(D^4 - d^4\right)}{16 \cdot D} \cdot G = 318.9 \cdot U_{\mathrm{m}}\,[\mathrm{Nm}], \quad (4)$$

$$F_{\mathrm{a}} = \varepsilon_{\sigma} \cdot S \cdot E = \varepsilon_{\tau} \cdot \frac{\pi \cdot \left(D^2 - d^2\right)}{4} \cdot E = 97,336.2 \cdot U_{\mathrm{m}}\,[\mathrm{N}], \quad (5)$$

where

J_{k} Second moment of area,
r Distance between the rotational axis and the farthest point in the section,
G Shear modulus of the material,
D Outside diameter of tube,
d Inner diameter of tube,
S Working angle,
E Elastic modulus of the material.

3 Measurement Results

Due to the real scope of measurement, axial force and torque could not be measured simultaneously but had to be measured separately. Because of this fact, measurement was performed one after another and on the same driving path with the same driving speed and load conditions. Graph of measured torque and axial force can be seen in Figs. 3 and 4. Graphs are only for one load condition, specifically for slow driving speed with no load. Measurements were made for three load states, namely

- slow speed without load,
- maximum speed without load,
- maximum speed with load (normal workload).

Measured torque values were within the expectation of our customer, but values of axial force were unexpectedly high. Because of this, a deeper analysis of the problem was needed. Direct cooperation of torque and axial force is not possible due to the separated measurement of torque and axial force.

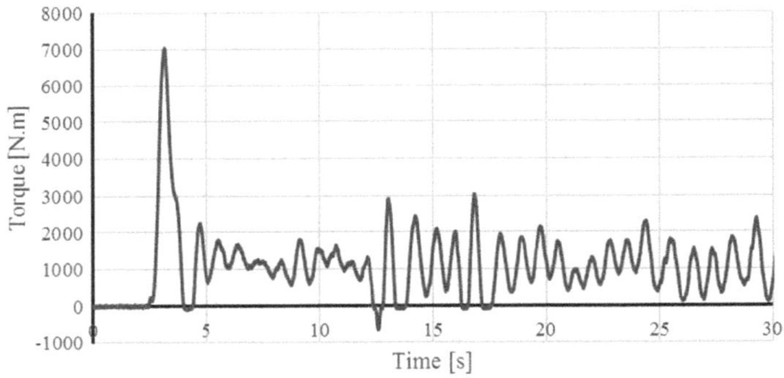

Fig. 3 Measured torque

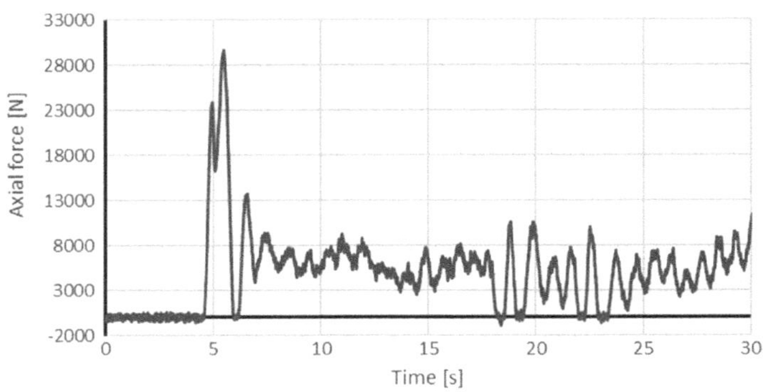

Fig. 4 Measured axial force

Table 1 Comparison of measured and calculated values

		Nominal values			Maximal values		
		1	2	3	1	2	3
Measured torque [N m]		1263	966	2627	7053	16,906	19,978
Measured axial force [N]		5869	3236	10,866	29,626	69,877	81,524
Calculated axial force [N]	$\mu = 0.03$	2305	1763	4796	12,874	30,855	36,461
	$\mu = 0.07$	5378	4114	11,190	30,039	71,997	85,077
Axial force deviation [%]	$\mu = 0.03$	−60.7	−45.5	−55.9	−56.5	−55.8	−55.3
	$\mu = 0.07$	−8.4	27.1	3.0	1.4	3.0	4.4

4 Comparison of Results

For comparison of measured axial force and calculated axial force, we used maximal values and then some nominal values from parts of signal where torque is constant. These values can be seen in Table 1. Values in column 1 are for slow speed without load, column 2 for maximum speed without load, and column 3 for maximum speed with load. It can be seen from the graph of (see Fig. 3) that torque never has a constant value; that is, normal behavior of the cardan shaft [2, 3] for the evaluation, we used the mean value from the steady part of the signal.

Due to the large difference between measured and computed axial force (up to 60.7%), we began to investigate where the problem is. The result was that the manufacturer declared coefficient of friction 0.03 does not reflect the reality. The coefficient of friction from another manufacturer, which gives the coefficient of friction for the same lubricant and surface finish, is more than doubled, namely 0.07 [1]. When this coefficient of friction was used, the deviation of the measured axial force from the calculated axial force varied to ten percent (with one exception). Exact values of deviation can be seen in Table 1. This remote value is due to the measurement method, specifically, splitting the measurement into torque measurement and axial force apart. Because of that conditions for torque and axial force measurement could be little bit different, even if the machine moved along the same track section, the control unit behaves every time a little differently.

5 Conclusion

When fatigue fracture of some part occurs it's in most cases because of either calculation error or different load condition that expected. At first in most cases, calculations are checked. If calculation is correct, measurement of the real load conditions is basically only way how to check real values of loads in system. In our case, the problem was in the wrong coefficient of friction, which did not correspond to reality. Due to this designer of machine that calculated necessary dimensions of parts have wrong input data and that caused fatigue fracture of designed bolts. This case shows how hard job designers can sometimes have as they must rely on the partner to deliver them reliable date inputs for the work. In our case, problem was that real force was more than two times bigger than force calculated during the design phase. When this big difference occurs in most cases on parts with variable loads, fatigue fractures are most likely to occur. By comparing the calculated and measuring axial force in the cardan shaft, it can be seen that if the correct input data are available, theoretical calculations correspond to the real load condition.

References

1. Wichmann, May 2015. http://www.wichmann-os.de/fileadmin/redaktion/englisch/Downloads/Industrie/WiCHMANN_Overview_Industry.pdf. Accessed 14 June 2018
2. Yanying G, Zhonghui S, Zhonghong S (2008) The kinematics analysis on single cross universal joint. In: 2008 IEEE, vehicle power and propulsion conference. IEEE, pp 1–3
3. Porat I (1980) Moment transmission by a universal joint. Mech Mach Theory 15(4):245–254

AHP-Based Evaluation of Vertical Gardens Design

Tomáš Vach and Ivan Mašín

Abstract The article deals with the analysis of design solutions used in the implementation of vertical gardens. Besides the typological elaboration of the realized design solutions, the authors also deal with the assessment of typical concepts using the multi-criteria decision-making method analytic hierarchy process (AHP) that helps decision-makers and engineers solve a complex problem with multiple conflicting and subjective criteria. Initially, three typical vertical garden designs are isolated from existing solutions available on the market, which are then subjected to AHP. Within the AHP, the following criteria were chosen: modularity, capacity, comfort, and design. In the analysis, the complex unstructured situation of the individual components, the arrangement of the components in the hierarchical order, the allocation of the numerical value to the subjective evaluation of the relative importance for each chosen variable, and the evaluation of the highest priority were evaluated using the synthesis itself.

Keywords Vertical garden · Design · Analytic hierarchy process

1 Introduction

With the growing population of the Earth logically growing the need for food, which today is a human being's lack of food on our plate. One possible solution for solving the problem of food shortages or minimizing these problems can be the development and cultivation of crops in vertical gardens (decenter food source) combined with normal horticultural crops such as fields [1]. A similar trend in decentralization is currently seen in the field of renewable energy. For the mass use of vertical gardens and their use in already arrested areas, which are not used in any construction, it is necessary to consider the construction of so-called vertical garden system itself. Vertical garden technology itself is not a breakthrough, historical roots can be found for thousands of years in the history of mankind [2]. The construction of these vertical gardens can be equipped with a variety of subsystems—light technics, nutrient supply

T. Vach (✉) · I. Mašín
Technical University of Liberec, Studentská 2, 46117 Liberec, Czech Republic
e-mail: vach@modus.cz

Š. Medvecký et al. (eds.), *Current Methods of Construction Design*, Lecture Notes in Mechanical Engineering, https://doi.org/10.1007/978-3-030-33146-7_48

systems, growth tracking, and so on [3]. At present, there are possibilities to use different types of composite materials, and other possibilities can be brought by nanotechnologists and others. The paper focuses only on the initial evaluation of the basic typical design of vertical gardens that are currently available in the current market.

2 Typical Design of Vertical Gardens

Before AHP-based evaluation, three design alternatives were formulated. The absolute majority of vertical garden systems use modular architecture. Typical design solutions can be used both for domestic use for small-scale plant cultivation and for large-scale industrial cultivation. Typical solutions can also be suited to manage environments. At present, there is a wide variety of materials used, from commercially available polymers to stainless steel or aluminum, or the specific coating of the support materials. Greening elements are used, or the trend is to use vertical products that are already used at the end of the life cycle—typically PET bottles and the like.

The first typical design solution (within AHP marked as the A concept) is a modular vertical system located on the building wall (see Fig. 1). This system can be used for both indoor and outdoor applications. Thanks to flexible use of different modules, it can often be seen in various apartment

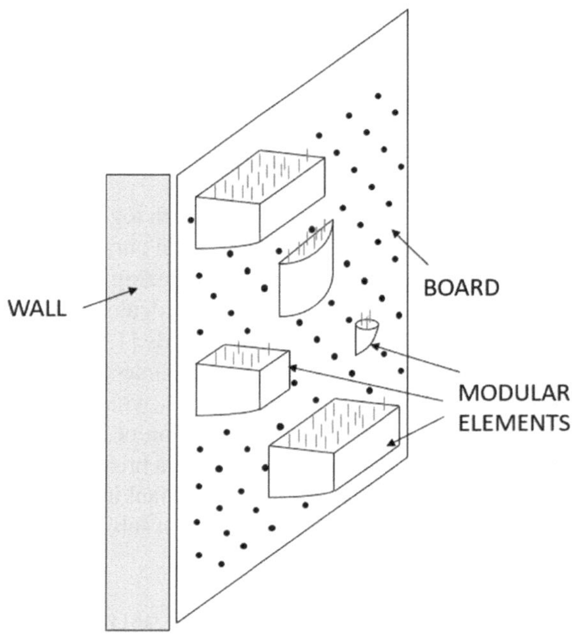

Fig. 1 Schematic representation of modular vertical system for wall mounting (Concept A)

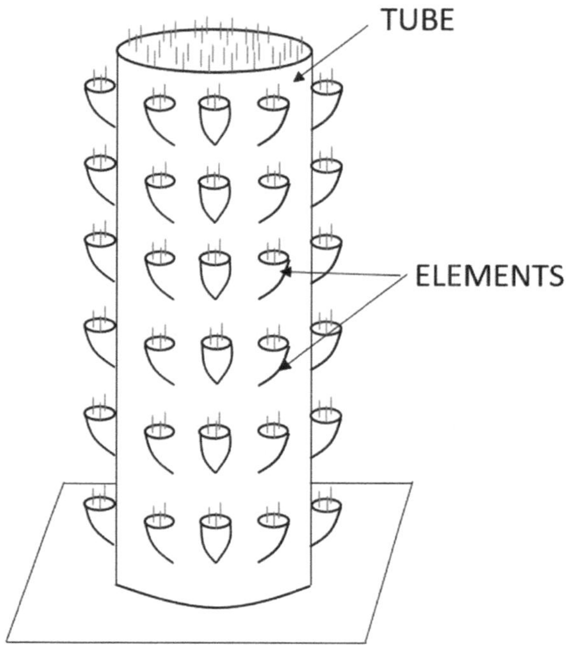

Fig. 2 Schematic representation of vertical garden based on a pillar structure (Concept B)

buildings, balconies, lodges, and other similar surfaces. This typical design is also used for the growth of ornamental plants.

A second typical design solution (within AHP marked as the B concept) is a modular vertical garden system based on a column structure whereby the plants themselves are attached laterally to the tube or small tubular modules and the cylinder is used (see Fig. 2). As a rule, the subsystem provides nutrient and plant subsystems for the plants. The construction is at the limit of application for use for home and industrial use. A typical construction is also used for indoor applications with controlled environments such as cellars, halls, or other non-residential spaces even without daylight access under the condition of artificial illumination. This typical design thanks to its shape of the main carrier, combined with other subsystems, can also be used to rotate and achieve better plant illumination.

A third typical design solution (within AHP marked as the C concept) is a modular vertical system based on a shelf structure (see Fig. 3). The plants themselves are mounted in horizontal boxes or boxes that are arranged vertically above each other. This typical vertical garden design is primarily for mass cultivation. The design is very suitable for deploying various subsystems to support plant growth. Due to the large capacities of the planted plant, various profiles, metal, or composite materials are used within the structure. Within the design, lifetime requirements (durability with repeatable use) are laid down. The structure itself uses elements of the modular system, but in combination with size and wiring, it may not be too user-friendly to use—handling the operator.

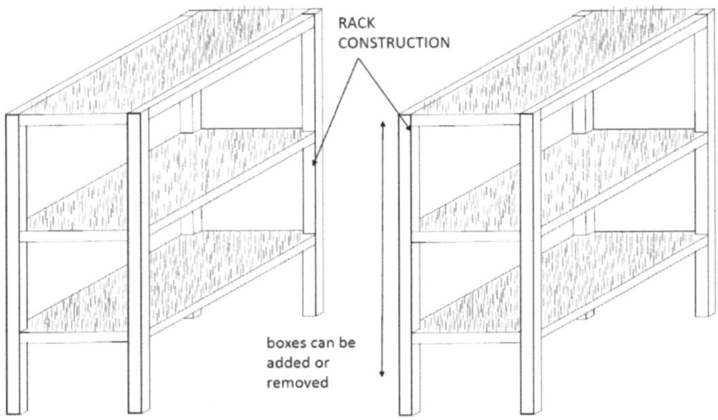

Fig. 3 Schematic representation of vertical garden based on a rack structure (Concept C)

3 AHP-Based Evaluation of Design Alternatives

The AHP procedure is according to [4] summarized as:

- Model the problem as a hierarchy containing the decision goal, the alternatives for reaching it, and the criteria for evaluating the alternatives.
- Establish priorities among the elements of the hierarchy by making a series of judgments based on pairwise comparisons of the elements. For example, when comparing potential purchases of commercial real estate, the investors might say they prefer location over price and price over timing.
- Synthesize these judgments to yield a set of overall priorities for the hierarchy. This would combine the investors' judgments about location, price, and timing for properties A, B, C, and D into overall priorities for each property.
- Check the consistency of the judgments.
- Come to a final decision based on the results of this process [4].

The following design parameters were selected as criteria:

- Modularity—a parameter that reflects flexibility of use and customization.
- Capacity—planting of as many plants of the same species (while maintaining the required space for each plant).
- Comfort—a parameter that reflects the easy planting, access to plants, or easy harvesting of crops.
- Design—a parameter that reflects simplicity of design and production.

The resulting hierarchy is shown in Fig. 4.

In order to select the best concept of vertical garden design, a proven traditional procedure of the analytical hierarchy process in the design concept selection was used [5–8].

Performed standard AHP process included among others the pairwise comparison of criteria (see Table 1), calculation of relative priority vectors (see Table 2), and par-

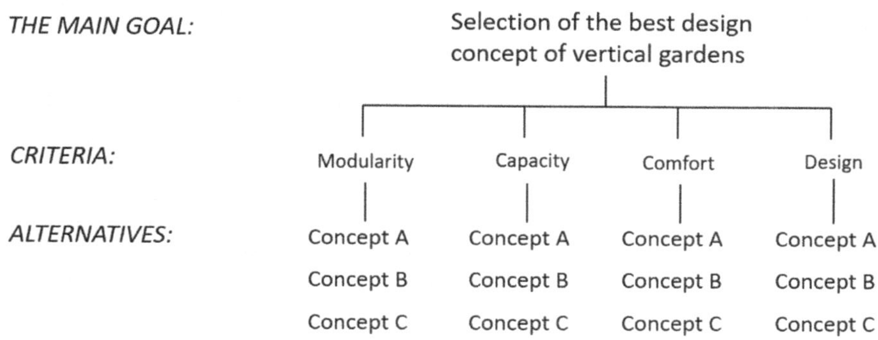

Fig. 4 AHP hierarchy for selecting the best vertical garden design concept

Table 1 Pairwise comparison of criteria

Criterion	Modularity	Capacity	Comfort	Design
Modularity	1	4	2	4
Capacity	¼	1	¼	¼
Comfort	½	4	1	½
Design	¼	4	2	1

Table 2 Relative priority vectors matrix

Concept	Relative priorities according to criteria				Criterion	Relative priority
	Modularity	Capacity	Comfort	Design	Modularity	0.471
A	0.164	0.589	0.200	0.524	Capacity	0.073
B	0.297	0.128	0.400	0.172	Comfort	0.209
C	0.539	0.283	0.400	0.304	Design	0.247

tial priorities calculation (see Table 3). Ratings were provided by a multi-professional team.

Table 3 shows that, taking into account mentioned criteria, the highest relative partial priority received the Concept C (0.433). On the other hand, this concept

Table 3 Partial relative priorities

Concept (Design)	Partial priority
A	0.291
B	0.276
C	0.433
Sum	1.000

Fig. 5 Benefit–cost diagram

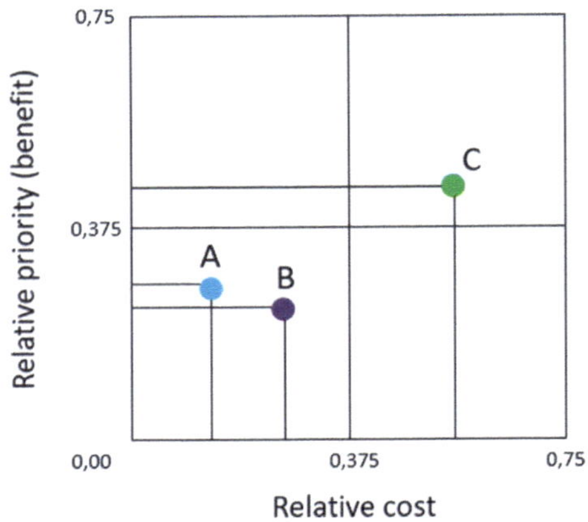

generally achieves a higher (double) benefit at the cost of double unit relative costs (Fig. 5).

Based on AHP's results, the research planned in connection with the prepared dissertation will use a C-based design solution.

4 Conclusion

Analytic hierarchy process helps designers to choose the best solution from several options and selection criteria. AHP has broad applications in the initial steps of innovation process when the concept of innovated product is generated. In the case of the above-mentioned research, the AHP method was used to determine the direction of the innovative design of vertical gardens that will use the shelf or tank character of the structure in conjunction with modern light sources and other subsystems.

References

1. Trendov MN (2018) Comparative study on the motivations that drive urban community gardens in Central Eastern Europe. Ann Agrar Sci 16:85–89
2. Zaid SM, Perisamy E, Hussein H, Myeda NF, Zainon N (2018) Vertical Greenery System in urban tropical climate and its carbon sequestration potential: a review. Ecol Ind 91:57–70
3. Davis MJM, Ramirez F, Pérez ME (2016) More than just a Green Façade: vertical gardens as active air-conditioning units. Procedia Eng 145:1250–1257
4. Saaty TL (2008) Decision making for leaders: the analytic hierarchy process for decisions in a complex world. RWS Publications, Pittsburgh
5. Ulloa C, Nunez NM, Lin C, Rey G (2018) AHP-based design method of a lightweight, portable and flexible air-based PV-T module for UAV shelter hangars. Renew Energy 123:767–780

6. Hambali A, Sapuan SM, Ismail N, Nukman Y (2009) Application of analytical hierarchy process in the design concept selection of automotive composite bumper beam during the conceptual design stage. Sci Res Essay 4:198–211
7. Zhu G, Hu J, Qi J, Gu C, Peng Y (2015) An integrated AHP and VIKOR for design concept evaluation based on rough number. Adv Eng Inform 29:408–418
8. Balo F, Şağbanşua L (2016) The selection of the best solar panel for the photovoltaic system design by using AHP. Energy Procedia 100:50–53

Examining the Modal Characteristics of the Prestressed Gearbox

Peter Weis, Milan Sapieta and Peter Bezák

Abstract The subject of the paper is to examine the modal characteristics of the conveyor gearbox. On gearbox were detected excessive vibrations during operation. Due to the fact that the gearbox operates in continuous operation, it was an unacceptable condition. For this reason, experimental measurements were made using the bump test to identify the cause of excessive vibrations. In addition, a finite element analysis was solved in the ANSYS Workbench software to compare results with experimental analysis results. Load was considered from torque on the input shaft, the mass, and inertia effects of the electric motor and prestressed bolt connections. This means, we solved modal analysis of prestressed system. This approach has enabled us to better understand the dynamic behavior of the gearbox under real operating loads. Output from the finite element analysis also confirmed the oscillation tendency occurs mainly on the input gear shaft at certain natural frequencies. From analyzes results, we can conclude that resonance is caused by the matching of gear mesh frequency of the input gear shaft with one of the natural frequencies of gearbox housing.

Keywords Modal analysis · Gearbox · Natural frequencies

1 Modal Analyses of Prestressed Gearbox

For the purpose of examining, the modal characteristics were selected gearbox designed by the company Transmisie Engineering a.s. Martin, SK. On gearbox were detected increased vibrations during operation, that could cause resonance. For this reason, a bump test was performed to identify the cause of these vibrations. Table 1

P. Weis (✉) · M. Sapieta
University of Žilina, Univerzitná 8215/1, 01026 Žilina, Slovak Republic
e-mail: peter.weis@fstroj.uniza.sk

P. Bezák
University of Žilina, Institute of Competitiveness and Innovations, Univerzitná 8215/1, 01026 Žilina, Slovak Republic

Š. Medvecký et al. (eds.), *Current Methods of Construction Design*, Lecture Notes in Mechanical Engineering, https://doi.org/10.1007/978-3-030-33146-7_49

Table 1 Data of the gearbox

Type of the gearbox	TE-KP1-12,5-U2-P4P5P7/1E-X1
Type of electromotor	1PH8226-1HF12-2BC1
Power output of electromotor (kW)	30
Gear ratio (–)	50.46
Rotation of input shaft (min^{-1})	1470
Rotation of output shaft (min^{-1})	29.23
Weight of gearbox and motor without oil filling and toque reaction arm (kg)	865
Weight of electromotor (kg)	280
Weight of brake (kg)	68

Table 2 Gear mesh frequencies

Gear mesh frequencies → rotation of input shaft $n = 1470\ \text{min}^{-1}$						
Gear	Number of teeth	Gear mesh frequency and their harmonic frequencies (Hz)				
		$1x$	$2x$	$3x$	$4x$	$5x$
Bevel 1	11	$f_{Z1} = 269.5$	$f_{Z1} = 539$	$f_{Z1} = 808.5$	$f_{Z1} = 1078$	$f_{Z1} = 1347$
Bevel 1	34					
Sun 1.row	32	$f_{Z2} = 3.179$	$f_{Z2} = 6.358$	$f_{Z2} = 9.537$	$f_{Z2} = 12.716$	$f_{Z2} = 15.895$
Planet 1.row	32					
Sun 2.row	32	$f_{Z3} = 0.788$	$f_{Z3} = 1.576$	$f_{Z3} = 2.364$	$f_{Z3} = 3.12$	$f_{Z3} = 3.94$
Planet 2.row	32					
Ring	97					

presents the main data of the gearbox. Table 2 shows gear mesh frequencies and their harmonic frequencies of individual gears. It is necessary to know these frequencies in comparison with the natural frequencies of the gearbox due to the resonance problem. More information can be found in the literature [1–3].

1.1 Experimental Modal Analysis

To measure vibration response, the SPM Leonova Diamond, S/N 1241080, was used. Three tests were performed. They were different in place and direction of impact. In Fig. 1, the place of measurement is marked by red dot. Places and directions of impact are marked by red arrows.

The measurements showed that frequency response is near the gear mesh frequency of input gear shaft. This match causes resonance which further goes to the gearbox housing. Figure 2 shows frequency response from test 1 and impact 2.

Fig. 1 Test 1, Test 2, Test 3 (from left to right)

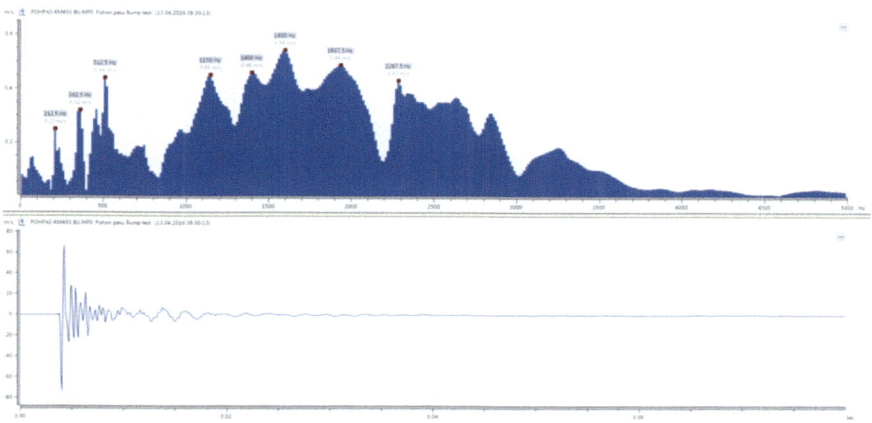

Fig. 2 Frequency response, test 1, and impact 2

1.2 *Modal Analysis Using Finite Element Method*

Analysis was solved in software ANSYS Workbench, which is primarily used to solve complex and robust 3D models. Real operational load was considered. The load on the input shaft was defined by torque with nominal value of 274.72 Nm. On the surface of the inlet, flange was defined load from mass of electromotor and brake. First was solved static analysis, which gave the new stiffness matrix due to new contact status. Then modal analysis was solved. Before importing a model into ANSYS Workbench, complex geometry modification has been made. For this geometry modification, Creo Parametric software was used. We have removed all components that are unnecessary from the point of view of the finite element analysis such as sealing and bearing elements. We have also removed all small holes, chamfers, fillets. Geometry modification continues in design modeler module in ANSYS Workbench. The complete model consists of a number of components. Those that behave as a single component in terms of analysis are added together using Boolean

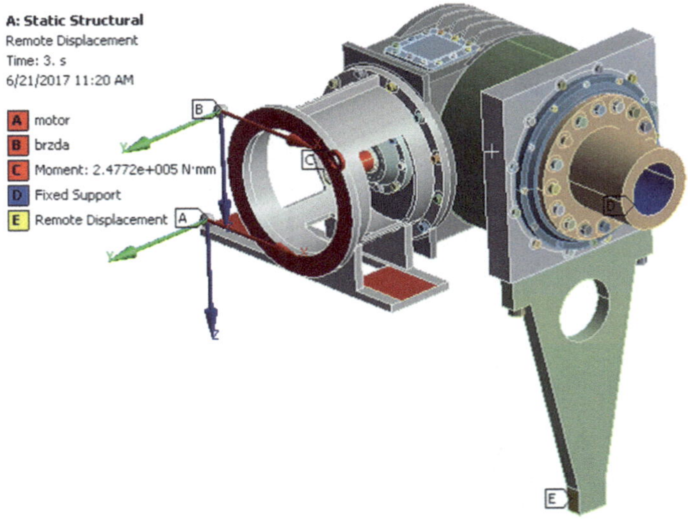

Fig. 3 Boundary conditions

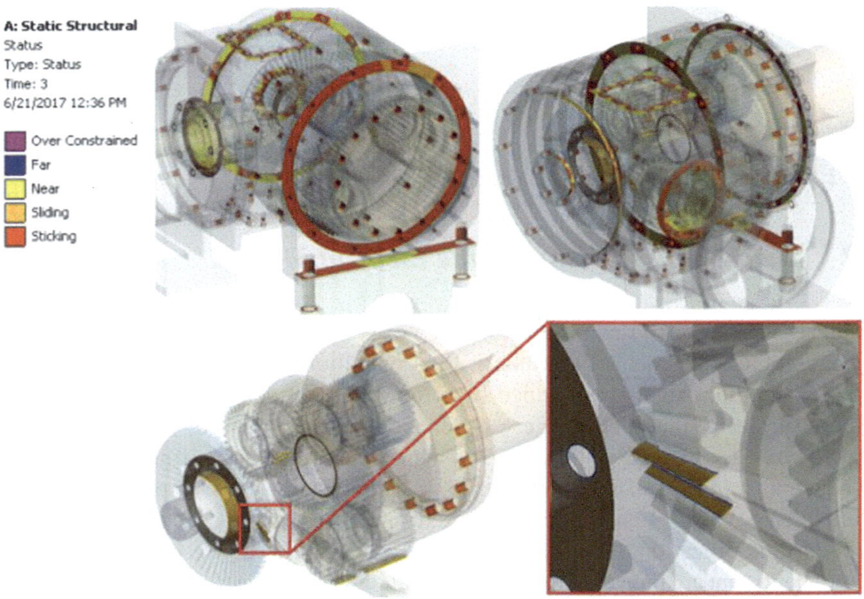

Fig. 4 Contact status

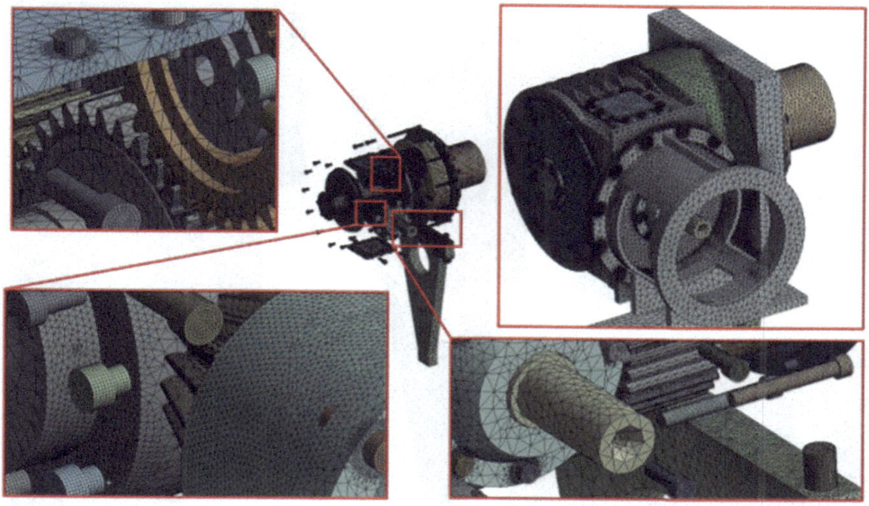

Fig. 5 Finite element mesh

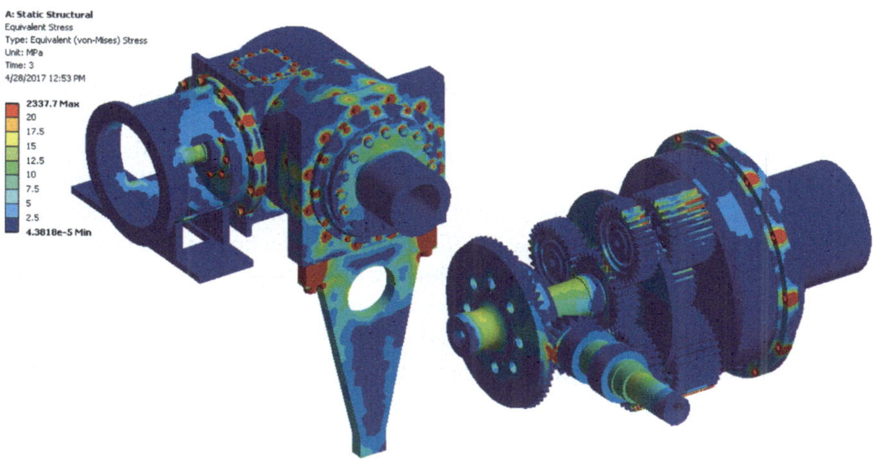

Fig. 6 Equivalent stress distribution

operations. We have reduced the number of components and eliminated the need to couple individual parts using contacts.

Boundary Conditions. The first load step was defined as prestress in bolted connections (see Fig. 3). The weight of the electric motor and a brake is represented by point mass attached to the gearbox housing. The torque was defined in the second load step to value $Mk = 1$ Nm and in the third load step to value $Mk = 274.72$ Nm. On the

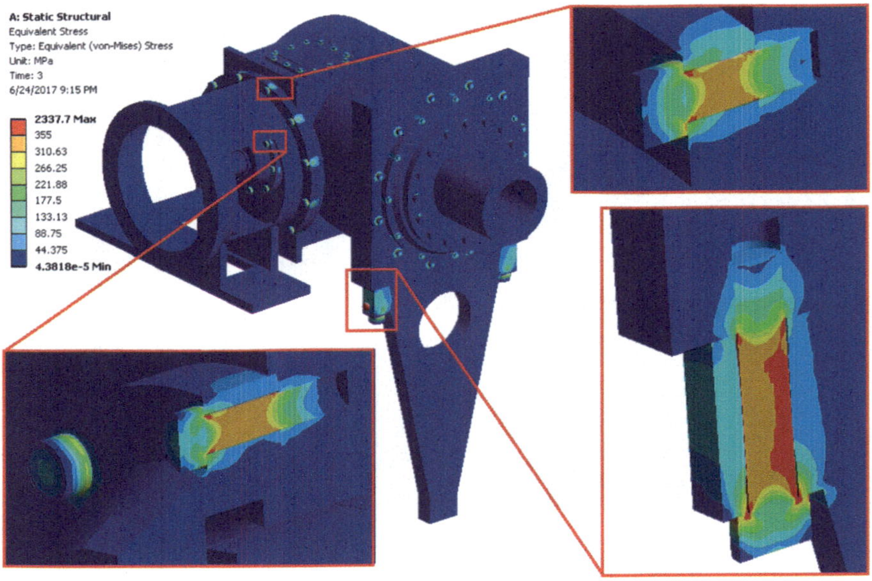

Fig. 7 Equivalent stress distribution in bolted connections

cylindrical surface of the output shaft, all degrees of freedom have been removed. The reaction force was captured at the lower part of the reaction arm.

Contacts and Bearing. From a contact point of view, this was a challenging task. There were contacts in the gears, between the housing parts and all bolted connections. Contacts in gears were defined as no separation type. Frictional contacts with a friction coefficient of 0.1 were defined between the housing parts. In bolted connections was contact type set as Frictional with friction coefficient of 0.1 with the Augmented Lagrange formula and Gauss points detection. Bonded contact type simulated threaded connection was defined on the cylindrical surfaces of the bolt and on the threaded holes in gearbox housing parts. The contact status is shown in Fig. 4.

Substitution of bearings was performed by function joint in ANSYS Workbench. There are two bearing types in the gearbox. The spherical roller bearings have been replaced by spherical joint. This joint has all the rotational degrees of freedom (around x, y, z). Tapered roller bearings do not allow the tilting of the outer ring against the inner bearing ring, and thus, they have been replaced by revolute joint.

Finite Element Mesh. The complete finite element mesh (see Fig. 5) consisted of 7,939,280 linear tetrahedrons, 1,957,958 nodal points, representing 5,873,874 solved equations. Linear elements were chosen mainly due to a large number of elements and using quadratic elements would mean a very long computation time.

The most important was mainly the size of elements on contact areas where the element size was set to 1.5 mm. For contact meshing, ANSYS Workbench automatically uses elements type CONTA174 and TARGE170. The CONTA174 element is

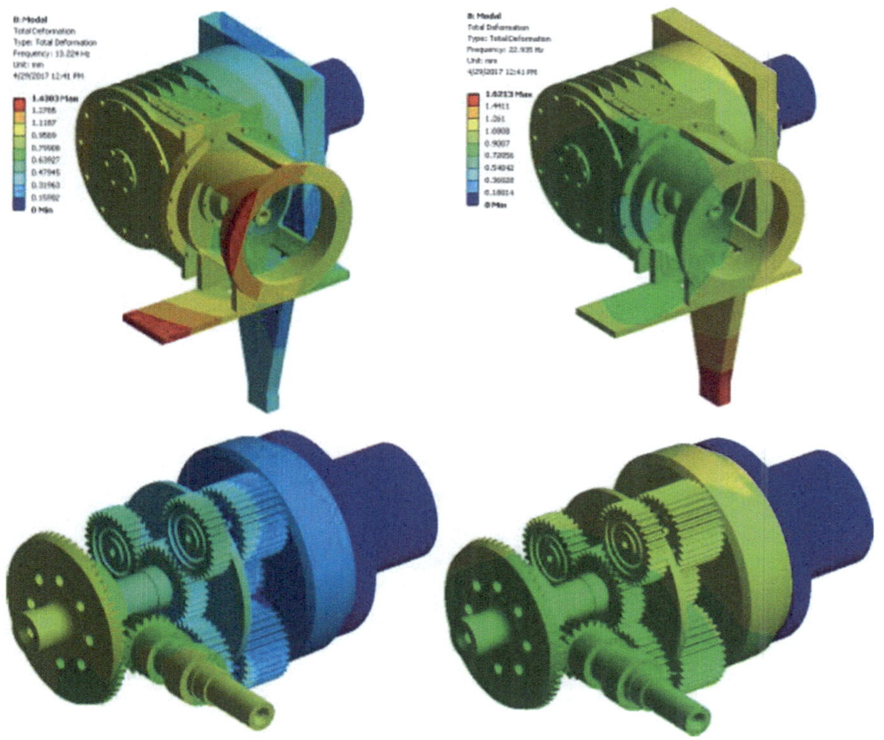

Fig. 8 Mode shape 2-left, Mode shape 3-right

the eight-node quadratic element used on the contact surface, and the TARGE170 element is a linear triangle element used on the target surface of the contact pair. When generating a finite element mesh, attempting to use the smallest number of elements, but too few could affect the accuracy of the solution.

2 Results

Figure 6 shows an equivalent stress distribution. Maximum values are found on the gear teeth. The stress concentrators are due to the geometry of the bolted connections where sharp edges are presented to simplify the geometry. Figure 7 presents equivalent stress distribution in bolted connections.

Table 3 presents the first forty lowest natural frequencies, which are significantly closer together with respect to a number of components, since all the bolted connections have been included in the model. There are also some frequencies near to

Table 3 Natural frequencies

Mode shape	Natural frequency (Hz)	Mode shape	Natural frequency (Hz)
1	9.15	21	455.53
2	13.22	22	491.45
3	22.93	23	500.66
4	31.37	24	524.52
5	44.69	25	531.93
6	61.74	26	553.12
7	84.63	27	575.15
8	92.80	28	587.30
9	109.56	29	594.55
10	121.03	30	602.15
11	190.12	31	667.60
12	194.01	32	699.80
13	253.74	33	830.62
14	269.96	34	883.27
15	298.22	35	896.74
16	303.29	36	911.29
17	362.51	37	934.87
18	367.14	38	983.52
19	401.78	39	1015.40
20	427.91	40	1031.80

gear mesh frequency of input gear shaft, which is 269.5 Hz and their harmonic frequencies. This frequency match causes the resonance on the gearbox. The oscillation tendency occurs mainly in the area of the input gear shaft, which further goes to gearbox housing. Figure 8 shows mode shape 2 with maximum value of displacement 1.4303 mm and mode shape 3 with maximum value of displacement 1.6203 mm. Mode shape 39 has maximum value 6.426 mm, and mode shape 40 has maximum value of displacement 9.891 mm (see Fig. 9).

3 Conclusions

This paper enables practical application of the knowledge gained during the solution of this problem in the development of similar complex gearbox. By solving a modal analysis with real-operating conditions, many problems can be avoided before they are included in production. For this reason, it is also important to understand the modal characteristics in order to eliminate possible resonance. Resonance does not

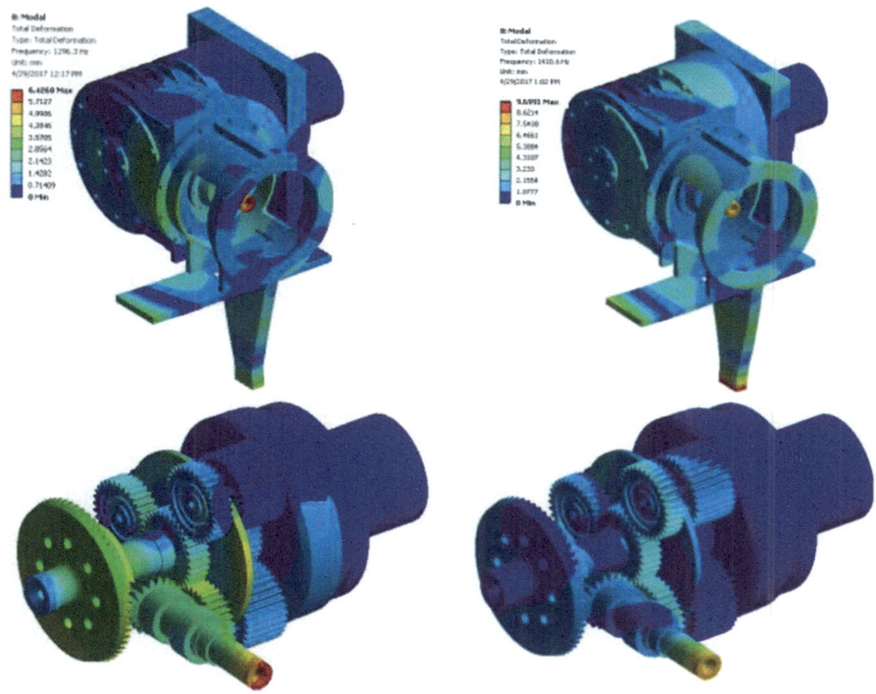

Fig. 9 Mode shape 39-left, Mode shape 40-right

immediately lead to damage the device but also to a significant reduction in service life, increased vibration, or high noise.

Acknowledgements The research is supported by the Cultural and Educational Grant Agency of the Ministry of Education, Science, Research, and Sport of the Slovak Republic under the project no. KEGA 040ŽU-4/2016.

References

1. Kohár R, Hrček S (2014) Dynamic analysis of a rolling bearing cage with respect to the elastic properties of the cage for the axial and radial load cases. Communications 3A(16):74–81
2. Hrček S, Kohár R, Medvecký Š (2012) Determination on the maximum roller bearing load with regards to durability there of using FEM analysis. Communications 3(14):55–61
3. Kohár R, Hrček S, Medvecký Š (2012) Usage of dynamic analysis to determine force interactions between components of rolling bearings. Communications 3(14):62–67

Effect of Temperature on Bolt Working Load of Pressure Vessels

Jiří Začal and Lukáš Jančar

Abstract The area of sealed flange connections of pressure vessels often encounters critical operating states that are defined by particular load conditions. These loads are exhibited from assembly stage, through pressure testing, to operation of the pressure vessel. The issues arising in assembly conditions alone are complex and significant. However, this article discusses the problematics of flange joints in subsequent conditions. Specifically, it is examining the effect of temperature on the bolt pretension. Stiffness calculations for machine parts which are constantly exposed to high temperatures at continuous exertion of force pose a significant level of complexity even nowadays. For calculations of pre-stressed bolt connections, the computations are even more complicated. This is given by the fact these are highly complex components that are almost always extremely stressed. However, the requirements for critical flange joints are even higher. There are not only bolts strength requirements, but strict leak rate requirements of gasket joint as well. This article provides a basic description of the effect of temperature on bolt pretension. Its theoretical basis is complemented by concrete examples from practice. Subsequently, one of the examples is subjected to FEM analysis. Finally, the possible ways of partial elimination of negative thermal influences on flange connections of pressure vessels are demonstrated.

Keywords Flange joints · Gasket · Bolt pretension · Pressure vessel · Circular flange · Temperature

1 Introduction

Critical flange joints of pressure vessels (see Fig. 1) are often working under high operation temperatures. The stiffness calculations of the pressure vessel parts are thus considerably complicated. It is not correct to rely on the knowledge of the mechanical

J. Začal (✉) · L. Jančar
Technical University of Ostrava, 17. listopadu 15/2172, 70833 Ostrava, Poruba, Czech Republic
e-mail: jiri.zacal@vsb.cz

L. Jančar
e-mail: lukas.jancar@vsb.cz

Š. Medvecký et al. (eds.), *Current Methods of Construction Design*, Lecture Notes in Mechanical Engineering, https://doi.org/10.1007/978-3-030-33146-7_50

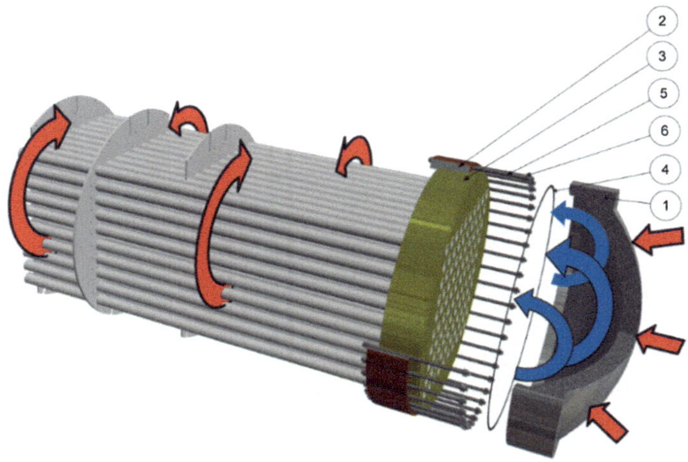

Fig. 1 Tube bundle with cover of floating head [2]

properties of materials that are inherent in normal temperature conditions [1]. Such results may be too conservative. Sufficiently it would not correspond with reality. A high reliability of results can only be guaranteed if it comes from an experiment whose conditions best describe the reality of the load on a particular component during the fatigue life. However, creating such tests is very difficult and expensive. In addition, the basic idea of the experiment is denied. This is why in practice it is often based only on experience.

2 Basic Description of the Effect of Temperature on Flange Joints with Gasket

For the correct design of the thermally stressed flange connection, the key knowledge of operating temperatures and thermal expansion coefficients, for components of the joint, is crucial. Due to the increased operating temperature, Young's modulus of the materials elasticity is changing, and therefore, the stiffness of the joint is changing too [1]. Knowledge of this parameter is also important.

It may not seem difficult to include these influences. The calculation procedure is given, for example, in CSN EN 1591-1 [3, p. 31]. The simplified Eq. (1) is given in the following [1 p. 224].

$$F_{0t} = \frac{\alpha_p(t_p - t_o) - \alpha_s(t_s - t_o) + F_0 \cdot \left(\frac{1}{S_p \cdot E_{pt}} + \frac{1}{S_s \cdot E_{st}}\right)}{\frac{1}{S_p \cdot E_{pt}} - \frac{1}{S_s \cdot E_{st}}} \tag{1}$$

However, only steady temperatures are often inserted inside this equation. Only values of steady-state temperature are substituted into the mathematical relation. Nevertheless, the most significant temperature differences are observed during the period of approach to the working condition, and not during the steady state itself.

2.1 Effect of High Temperature on Bolt Pretension and Leakage

What effect the operating temperature will have on the bolt pretension will depend primarily on the thermal expansion of the flange joint parts. Rate of this deformation depends on coefficient of thermal linear expansion and temperature. In practice, the temperature of the bolt is usually lower than the temperature of the flanges. Due to this behaviour, the bolt pretension increases.

The gasket of pressure vessel is another important component, which is affected by temperature. The leakage is strongly dependent on the gasket stress. The gasket properties are determined by the tests according to CSN EN 13,555 [4]. This standard specifies leakage graphs (see Fig. 2), which depict the leak rate requirement dependence on used gasket stress. Furthermore, it shows the graphs of dependence of gasket pressure on gasket closure. Also important are the linear or nonlinear unloading slope graphs that define the degree of plastic and elastic deformation of the gasket after unloading. Based on all these dependencies, it is possible in practice to determine the optimal gasket pressure for any load conditions. However, if the bolt working load increases due to thermal expansion of flanges, the gasket stress increases too. After the temperature has reached equilibrium, both the bolt load and gasket stress decrease, but the large gasket plasticity does not allow full recovery of sealing joint. Therefore, leakage may occur under the operation conditions. This situation can occur already at the beginning of the working cycle (when the operating temperature is approaching) or after several work cycles. These negative effects are

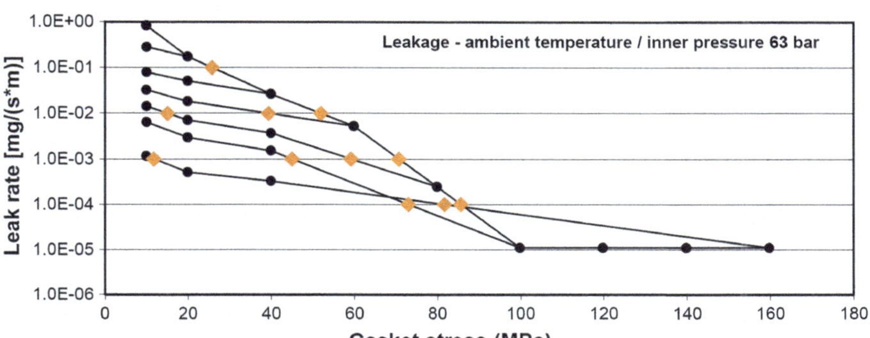

Fig. 2 Graph of leak rate requirement

most apparent in floating-type joints (FLT) [5]. This type of joint does not allow the flanges to metal-to-metal contact (MMC) [5] around sealing surface; thus, the whole gasket pressure is exerted on effective sealing surface. The greater effect of inappropriate heat transfer is on loose flange with collar or stub. Temperature differences can be significant in this case, even over 100 °C.

2.2 Long-Term Thermal Effect on the Flange Connection

Many critical flange joints are in practice loaded with temperature over a long period of time. Partial or total loss of bolt pretension is due to the relaxation and creep deformation of material. These two terms cannot be confused. While during relaxing, the stress in the component decreases due to the increase of the plastic deformation to the detriment of the elastic deformation, and the length of the component is unchanged. During the creep deformation, the component changes its length. Creep deformation can be defined as a continuous, very slow, and permanent deformation of material under the influence of mechanical stresses during the ambient or higher temperature. It is plastic deformation only.

It is very difficult to determine the loss of preload at the creep-affected joint. The nature of the flange and bolt tension is too complicated. Creep parameters from experimental measurements (e.g., ultimate strength limit), which correspond to the smooth test specimens, cannot be used for the calculation of the bolt connection. The results would not correspond to reality [1]. The only reliable method is an experiment performed directly on a bolt flange connection. But there are not many such examinations that would match the length in the thousands of hours. Similar tests have proven that greatest loss of bolt pretension will occur over the first six to ten hours. This is because both the creep rate and the relaxation rate depend on the magnitude of the stress that is highest in the component at the start of the test.

In practice, the issue of creep and relaxation is mainly resolved by choosing better materials. Suitable materials are those containing carbide and nitride precipitates. Next, alloyed steels with molybdenum are recommended, but the elements (Cr, Ni, V, Nb, and N) are also suitable. However, for a good flange joint design, only the right choice of material is not enough. There is a need for creep deformation area research.

3 Examples of Temperature Effect from Technical Practice

Nearly every critical flange connection is affected by high temperature in technical practice. Each critical flange connection requires an individual approach to the solution, due to complex load conditions. Two examples of the negative influence of high temperature on the flange joints are presented in this chapter.

Figure 3 depicts the temperatures on the flange joint by thermocamera. Extreme

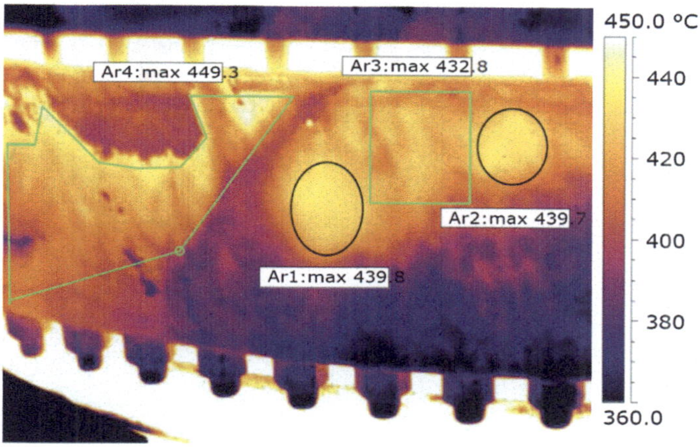

Fig. 3 Thermovision of critical flange joint

temperatures were caused by heat shield failure inside the container.

In this case, the high temperature had several negative effects on the tightness of the flange joint. The first failure occurred at the beginning of the operating state, during the heating of the joint.

Due to the improper heat transfer, the large temperature differences between the flange and the bolt occur. The flange can have up to 100 °C higher temperature. As a result, the bolt working load increases. Even though this phase is only short-term, plastic deformation of bolt has occurred and, consequently, loss of tightness of the gasket joint. From thermovision view, it was obvious that the temperature of the flange is different in circumference. In this case, it is not possible to guarantee any leak rate requirement, due to variable gasket pressure.

To mitigate improper heating effect of the flange connection components, it is possible to use insulation. However, in this case, isolation has caused degradation of the joint. The temperatures could then increase to 600 °C. Such high temperatures are also limiting for heat-resistant materials, especially with respect to relaxation and creep of material. In this case, the heat shield improvement was necessary even though the connection is constantly monitored by heat sensors.

Another example in Fig. 4 does not concern the problem of increased bolt and flange temperatures, but high temperature of the material of the piping systems. Arrows in the middle Fig. 4 indicate a site where leakage has been repeatedly lost. The gasket has been damaged from both sides in the direction of the upper pipe connection. The reason was the defect in compensating the thermal length expansion in the direction of the arrows (see Fig. 4). The thermal expansion of the upper inlet pipe causes an external bending moment to occur. The floating joint type of flange is not able to withstand such a large external load (see Sect. 2.1). It was necessary to adjust the joint by adding a steel ring next to the gasket. Special bolts with narrow neck

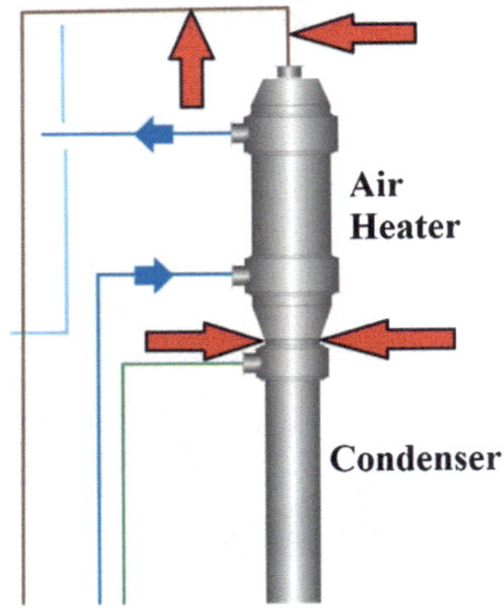

Fig. 4 A pressure vessel with long inlet pipeline

were also used. This solution for eliminating the negative influence of temperature is described later in Chap. 5.

4 Analysis of Temperature Effect by FEM

Based on experience, a computational model was developed in ANSYS Workbench 18.2 Academic Research, which includes the influence of thermal expansion of the flange joints and the influence of the thermal expansion of the inlet pipe. The model must be considerably reduced because it is not advisable to use the symmetry tool when analyzing the influence of the inlet pipe. The purpose of the calculation was to demonstrate how the temperature can affect the operating force in the bolt.

In first step, a transient thermal analysis was developed. The internal media temperature was set at 120 ° C. The analysis was divided into several load conditions. The first loading state is the assembly condition ($I = 0$) where only the ambient temperature is 22 °C. The second load state ($I = 1$) considers the heating of the flange connection only (see Fig. 5). In the third load state ($I = 2$), the supply pipe is also heated to 120 ° C. The next and last load condition ($I = 3$) takes into account the heating of the supply pipe to 200 °C; the other parts of the pressure vessel have ambient temperature. The results further enter the static structure analysis, so it is necessary to select the critical source time from the thermal analysis only. It is the time with maximum temperature differences between flanges and bolts (see Fig. 5) or the time, when the inlet pipe is heated over the whole length, etc. The static structure analysis intentionally simulates the effect of nonfunctional linear thermal expansion compensation. From one side of the pipeline, fixed support was used. The pressure vessel is not loaded by any internal pressure. The M36 screws are used for analysis. The clamping length of the flanges is 200 mm. The inlet pipe axis is located 1500 mm

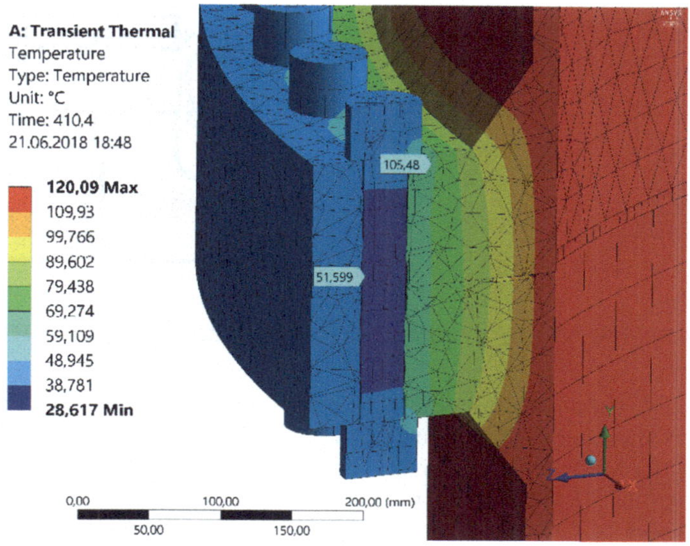

Fig. 5 Transient thermal analyses

Table 1 Bolt working load with temperature dependence

Load conditions	Bolt working load [kN]
$I = 0$	159.54
$I = 1$	180.65
$I = 2$	184.71
$I = 3$	180.52

above the flange connection plane. The pipeline has a DN 170 mm and a length of 1500 mm. The results of the analysis are depicted in Table 1.

Even though this analysis is only simplified and used load conditions do not reach as critical values as in practice (see Chap. 3), it can be concluded from the results that the influence of thermal expansion on the force in the bolt is significant. Although the difference in flange and screw temperatures is only 54 °C in the $I = 1$ state (see Fig. 5), the force increased by 13.2%. Upon heating of the supply pipe in the $I = 2$ state, the force increase is up to 16%. In the operating state $I = 3$, when higher temperature affected inlet pipe only, the increase of bolt pretension is 13.1%.

It should be considered that in practice the internal pressure in the vessel also acts. In addition, the bolt connections are often tightened to the yield strength. Therefore, all critical thermal load cases and interfaces must be included in the calculation of critical flange joints. Knowledge of in situ states may not always be sufficient.

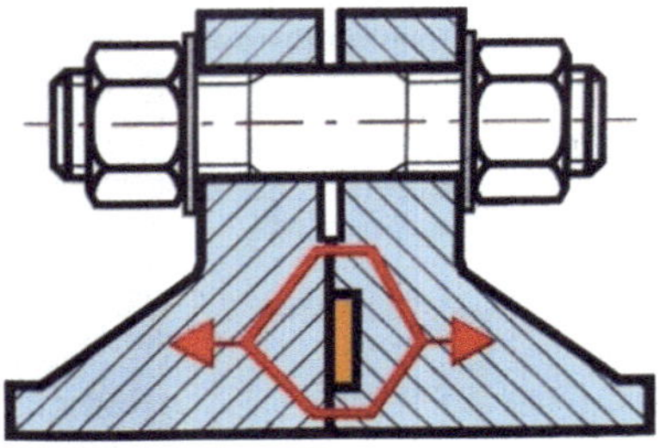

Fig. 6 MMC joint with gap

5 Elimination of the Negative Effect of High Temperatures

Negative effects of temperature can be partially eliminated. First, another type of flange can be used (see Fig. 6). When fully tightened by bolt connection, MMC type of flanges [5] achieves a metal-to-metal contact around the sealing surface. The gap between the flanges ensures a better joint elasticity.

Another way of eliminating the effect of temperature is to allow heat transfer between a warmer flange and a cooler bolt (e.g., by screwing the bolt into one of the flanges) [1]. Next, the overall stiffness of the joint can be changed. Lower stiffness of the bolt and flange (e.g., using special long bolts with narrow neck, inserting the tubular washer) will improve flange joint elasticity. Due to effect of elasticity, additional bending of the bolt occurs. This can be eliminated by using more number of narrow bolts instead of a smaller number of larger bolt diameters. In such a case, it is suitable to use under the nuts also high-carbon washers that will not deform during assembly but will allow plastic deformation in critical operating conditions. This plastic deformation relieves the stress of the bolt. For heat-stressed flange joints, it is also preferable to use high-strength and heat-resistant bolt material. If the preload is expected to increase due to the thermal expansion of the flanges, it is advisable to reduce the bolt pretension in assembly conditions already. The case where the bolt temperature is higher than the flange temperature is in practice unique. In such a case, it is necessary to achieve maximum bolt pretension already in assembly conditions. For joints exposed to dynamic load and high operating temperatures, special loose flange constructions are demonstrated in [1, p. 226].

6 Conclusion

Calculations of heat-affected machine parts can still be considered very complex even nowadays. Even though the effect of the temperature is considerable, attention in practice is very poor in this area. This often leads to unnecessary defects in flange joints. The article describes the basic theory of temperature influence on the flange joint with bolt pretension. It is also described what effect the temperature may have on the gasket and its properties. In the next section of the article are present real examples of extreme temperature effects on flange joints from technical practice. FEM analysis follows, which demonstrates the effect of thermal expansion of flange joints and connected pipes on a simplified model. This is a simplified simulation of the examples from practice, which were presented in Chap. 3. At the end of the paper are presented ways of eliminating the negative effects of temperature on the joints with bolt pretension. These examples are based on experimental measurements and practical experience.

References

1. Pospišil F (1968) Threaded and bolt connection. SNTL, Publishers of technical literature, Prague
2. Začal J (2016) Problem solving of bolted connection of a pressure vessel. Master Thesis. http://hdl.handle.net/10084/115089. Last accessed 22 May 2018
3. CSN, EN 1591-1 (2015) CSN EN 1591-1, Flange and their joints—design rules for gasket circular flange connections—part 1: calculation. Prague, UNMZ, Class number 97212, p 67, May 2015
4. CSN, EN 13 555 (2005) Flanges and their joints—gasket parameters and test procedures relevant to the design rules for gasketed circular flange connections. Prague, CNI (Czech Standards Institute), Aug 2005
5. Schaaf M, Bartoniček J (2003) Calculation of bolted flange connections—state of the art (F533). IASMiRT, Prague

Tribology and Nanotechnology

Scuffing Resistance of 16MnCr5 HCR TiAlCN-Coated Gears

Miroslav Bošanský, František Tóth, Ladislav Gulan, Juraj Rusnák and Adam Furstenzeller

Abstract This paper describes options of application of thin hard coatings on a non-standard involute HCR gearing made up of mat, 16MnCr5, in interaction with ecological gear oil OMV biogear S 150 and gear oil PP90H. Based on the analysis of their geometrical characteristic and it's differences from a non-standard C-C gearing, the options of applying thin hard coatings on the surface of gear tooth flank are reviewed. From the previous results obtained at our workplace (Institute of Transport Technology and Designing, Faculty of Mechanical Engineering of STU in Bratislava) a TiAlCN nitride was applied and the article mentions the results obtained from Niemann's stend for scuffing, which indicate that in the cases of lubrication with both OMV Biogear S 150 and PP90H similar results were achieved. The difference between the oils was only visible on the pinion, where scuffing occurred at the 12th load stage with OMV biogear S 150 oil at the dedendum, but with PP90H, no scuffing was observed.

Keywords TiAlCN coating · HCR · C-C gearing · Niemann's stend · 16MnCr5

1 Introduction

For gear transmission lubrication, oils with high viscosity grade VG from 100 to 680 are very often used. Applying them for the lubrication of tribological elements poses a risk of their leakage into the outside environment, having a negative effect, especially with machinery used in agriculture, forestry and building industries. One of the possibilities of how to prevent an extensive soil or water contamination is the usage of ecologically easily degradable lubricants and oils. Some countries including Germany and Sweden even require the usage of biologically decomposable oils in

M. Bošanský (✉) · L. Gulan
Slovak University of Technology in Bratislava, Námestie Slobody 17, 81231 Bratislava, Slovak Republic
e-mail: miroslav.bosansky@stuba.sk

F. Tóth · J. Rusnák · A. Furstenzeller
Slovak University of Agriculture in Nitra, Tr. A. Hlinku 2, 94975 Nitra, Slovak Republic

Š. Medvecký et al. (eds.), *Current Methods of Construction Design*, Lecture Notes in Mechanical Engineering, https://doi.org/10.1007/978-3-030-33146-7_51

all applications working in ecologically sensitive zones, e.g. areas of water source preservation [1].

Ecological lubricants must therefore fulfil not only the performance requirements defined by the machine manufacturer but ecological demands for degradability and toxicity of the lubricant. The eco-toxicity tests used nowadays in accordance with OECD 201, 202, 203 guidelines evaluate the impact of substances in water environments by studying their impact on standard types of algae, water flea and fish representing water environment and furthermore are approved tests of biodegradability, e.g. the OECD 301 series of the CEC L-33.A.93 test, which are available for the evaluation of biodegradability in water. These different methods allow reviewing the impact of chemical substances on water environment [2]. A limiting factor is the specific gear transmission mechanism.

2 Gearbox as a Tribotechnical System

The transmission of an industrial machine can be characterized as a complex tribotechnical system, consisting of various tribological elements, such as a bearing, a shaft seal and a gear (see Fig. 1).

In terms of increasing the load capacity of gears, it is crucial to observe the indicators of friction, lubrication and wear of the gear mesh in interaction with the used lubricant [3]. With increasing gear load also the contact stress in gear teeth, friction and temperature increase. The amount of contact stress plays a significant role in the formation of tooth flank damage like pitting (see Fig. 2) or scuffing (see Fig. 3). Various factors contribute to the formation of particular damage, e.g.

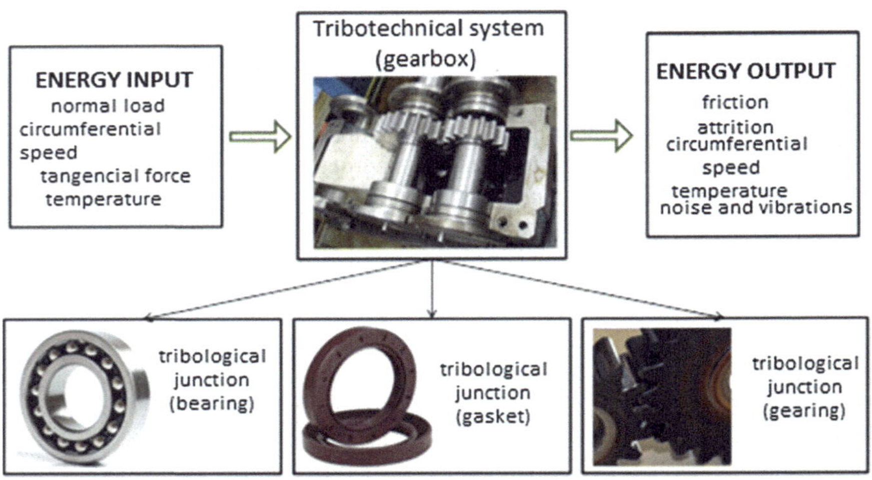

Fig. 1 Gearbox as a tribotechnical system

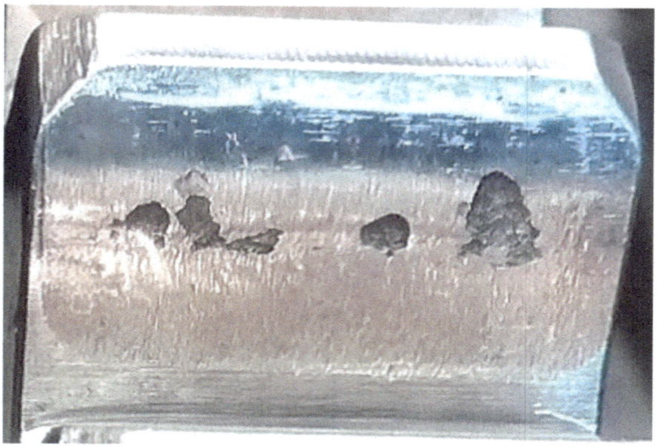

Fig. 2 Tooth flank damaged by pitting

Fig. 3 Scuffing of the tooth flank

operating factors (load size, operating temperature and lubricant), geometry, gear material and surface hardness of the tooth flanks or gear mesh accuracy.

Scuffing mostly occurs on the tooth face and flank, which is in the regions with higher slip velocity and large heat generation. This technical fault mostly occurs with high-speed and highly stressed gears, because of high specific strain energy on the surface of the working profile of the tooth [4].

The analysis of the tooth flank damage [5] implies that the main role in the issue of tooth flank damage in gears is played by the amount of contact stress. Higher resistance to damage can be ensured by increasing surface load capacity of the tooth flank, which can be achieved as follows:

- Change of the geometrical dimensions of the gearing. The disadvantage of this solution is the inevitable increase of overall mass of the gear.
- Using higher quality oils. However, higher oil quality is usually achieved by adding larger concentrations of fitting additives, which are often characterized by considerate amount of eco-toxicity.
- Using higher quality materials. This results in higher costs for the material.
- Increasing the load capacity of the tooth flank surface.
- Changing of the tooth flank geometry—change of the structural shape of the contact surface.

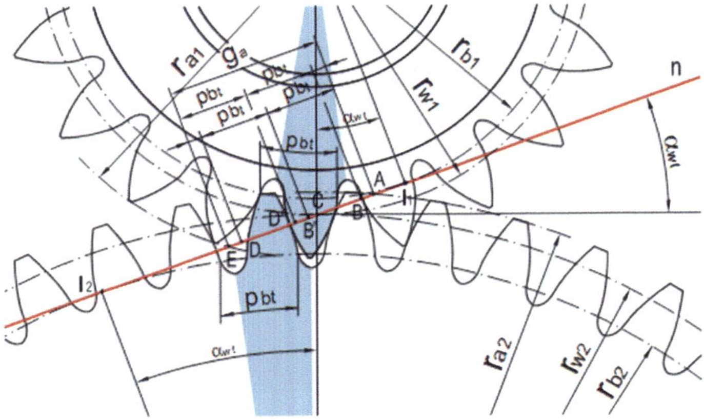

Fig. 4 HCR gearing geometry

In terms of preventing operating failures, the most suitable options out of those mentioned above are increasing the load capacity of the tooth flank surface, or the change of the tooth flank geometry. One of the possibilities of how to increase the surface load capacity of the flank is also the technique of applying various coating layers. Since most degrading processes start as surface damage of the components, the application of coating can be a suitable solution for increasing the load capacity of the surface and therefore prolonging the durability of tooth gears. The formation of coating is possible in a wide range of widths (from a few mm to a few μm). By applying coatings, it is possible to achieve the enhancement of wear resistance, corrosion resistance, fatigue life, etc. Regarding coating, a major role is played by adherence of coating towards the substrate, because it determines the admissible stress of the coated component. HCR gearing (see Fig. 4), where higher slip ratio occurs in gear mesh [5], compared to the standard involute gearing or convexo-concave gearing, but considering continuous meshing of two teeth in a gear set a lower load and therefore lower contact stress occurs (see Fig. 5). Nowadays, this type of gearing is more widely used in automobile industry; therefore, it is interesting to research the options of application of thin hard coating even in this case.

3 Coating on Gears

The processes of application of coating layers on the surface of the base material can be divided into the following categories:

- electrochemical deposition,
- chemical deposition,
- thermal spraying,
- physical and chemical vapour deposition in a vacuum (CVD, PVD) [4].

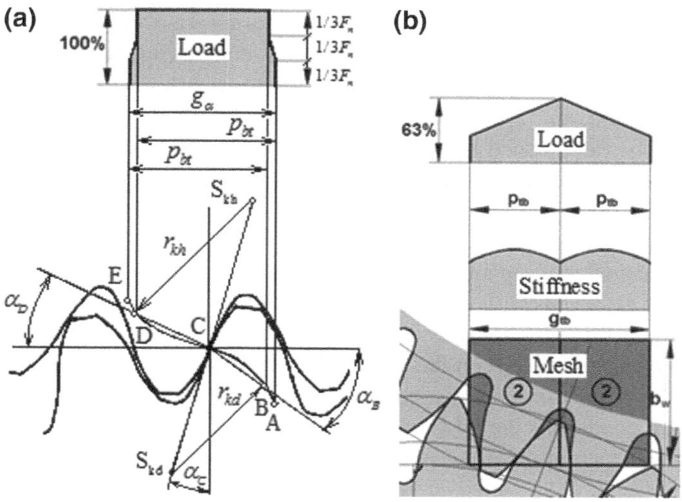

Fig. 5 Load in gear mesh. Left: Allocation F_n along mesh in C-C gearing. Right: Allocation F_n along mesh in HCR gearing

The increasing surface load capacity by applying coating layers with PVD and CVD methods began to be used in the 70s. Titanium carbide coating by CVD method was used at first. Among the first ones to be coated were blades made of cemented carbide. Later, coating materials such as TiN or TiCN started to be used. High deposition temperatures around 1000 °C did not allow for the CVD method to be used for applying a coating on tools made of high-speed steel. Therefore by the end of the 80s, the PVD method started to be enforced [4]. Physical vapour deposition (PVD) is a method of physical metallization from vapour, which consists of the deposited material being transformed into a vapour phase in a vacuum and consequently deposited on the substrate at low temperatures ranging from 150 to 500 °C [6].

In the case of gear transmissions, it is necessary to consider specific conditions for gear mesh while applying thin hard coating; therefore, in this case there are the following requirements:

- resistance towards high stress arising with gear mesh,
- sufficient adhesion of the coating on the gear surface for selected gear transmission materials,
- resistance to temperatures generated by gear mesh up to 450 °C,
- maximum roughness of tooth flanks of gear transmissions Ra 0.6 μm,
- minimum surface stiffness of tooth flanks 60 HRC,
- low value of friction coefficient up to 0.4,
- endurance of the coating to mineral, synthetic or ecological lubricants.

Krantz [7] carried out fatigue tests in the NASA Glenn Research Centre on coated and uncoated spur wheels made of steel AISI 9310 (14NiCrMo13-4) designed for gear manufacturing. The set of gears chosen for coating was modified by blast cleaning before deposition with a water-based medium containing Al_2O_3 particles sized approximately 10 μm. Consequently, they were given a thin adhesive layer of elementary chromium, onto which was alternately applied a layer of Cr and WC and the coating system was finished by a hydrocarbon laser containing wolfram (W-C:H). The layers were created by a magnetron sputtering method with an overall thickness of 2.5–3 μm. The results proved that the fatigue life of coated gears was 5 times higher in comparing with uncoated gears.

The increase of scuffing and pitting resistance is stated by Michalczewski [8] as well, while he also deals with substituting commonly used oils with additives against wear—AW and for extreme pressure—EP, by ecological lubricants and their interaction with coatings applied on the surface of gears made of the material 20MnCr5 (14 221). The results of his research among other things show the fact in extreme load pressure conditions the DLC coating can take on the role of AW and EP additives, allowing to minimize the application of these toxic additives in lubricants and therefore achieve so-called ecological lubrication. At the same time, gears with amorphous C:H:W coating lubricated with ecological oil showed 20% lower oil temperature during operation and also 20% lower friction coefficient compared to gears without coating and lubricated with high-performance gear oil GL-5.

Based on the previous research at the Institute of Transport Technology and Designing, Faculty of Mechanical Engineering of STU in Bratislava, in the field of coated K-K and involute gear sets, we used biodegradable gear oil OMV biogear S 150 and the application of TiAlCN also coating in the case of HCR gearing in Niemann's stend tests. It is a graded nitride coating for universal use with high toughness and stiffness (30 GPa) with low friction coefficient (0.5) and max operation temperature (450 °C) suitable for milling operations. As a comparison, we carried out tests with standard gear oil PP90H. The scuffing test results on Niemann's stend for the OMV biogear S 150 oil are shown in Fig. 6 for a wheel and in Fig. 7 for a pinion. For the PP90H oil, Fig. 8 shows the results for a wheel and Fig. 9 for a pinion.

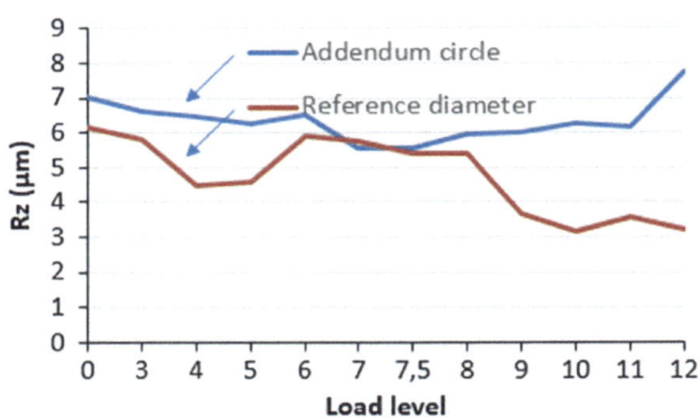

Fig. 6 Change in Rz roughness on a HCR wheel lubricated by OMV biogear S 150

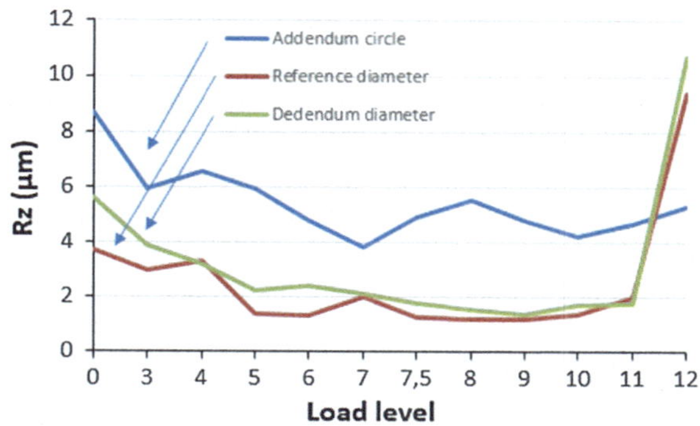

Fig. 7 Change in Rz roughness on a HCR pinion lubricated by OMV biogear S 150

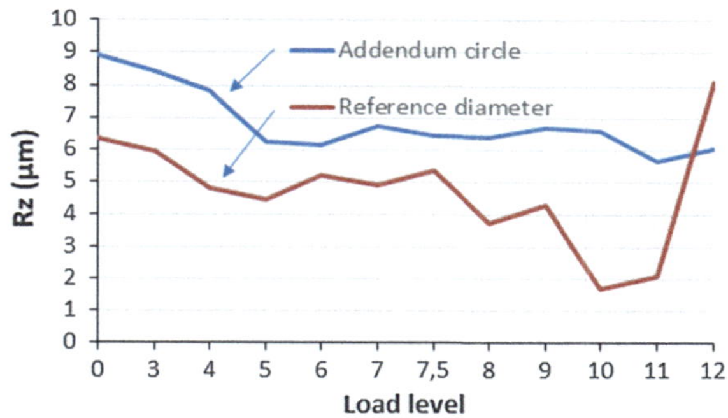

Fig. 8 Change in Rz roughness on a HCR wheel lubricated by PP90H

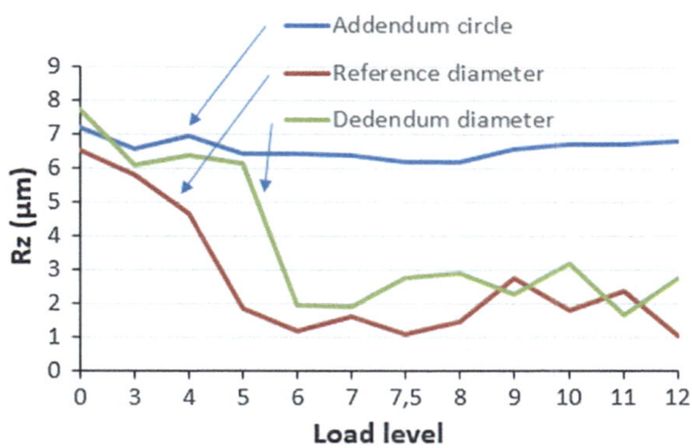

Fig. 9 Change in Rz roughness on a HCR pinion lubricated by PP90H

4 Conclusion

Based on the acquired results, it can be stated that considering the specific operating conditions in HCR gearing mesh (high slip on the dedendum and addendum of gear teeth) and on the contrary lower contact stress resulting from continuous mesh of two gears, in the interaction with the ecological gear oil OMV biogear S 150 in the Niemann's stend scuffing test, better results were achieved than in the cases of standard involute and non-standard C-C gearings [4, 5, 9–12]. Scuffing only occurred on the pinion and at the 11th load stage, as can be seen in Fig. 7. By comparing the two oils, it can be concluded that in the case of the wheel the results were reconcilable until the 11th load stage. At the 12th load stage, an increase in surface roughness occurred with both oils, while with OMV biogear S 150, it was mainly at the wheel's addendum, and in the case of the PP90H oil, it was near the standard pitch diameter of the wheel. In case of the pinion, there is a significant increase of roughness near the pitch diameter at the 12th load stage using the OMV biogear S 150 oil, while an increase of roughness was also recorded in the dedendum. As can be seen in Fig. 9, in the case of a pinion lubricated by the PP90H oil, no increase in roughness occurred even in the 12th load stage.

Acknowledgements The research work reported here was made possible by grant projects VEGA 1/0227/15.

References

1. Staša R (2012) Ekologické maziva na bázi esteru splňují vysoké technologické a ekolgické požadavky. Tribotechnika 3:32–35
2. Cecutti C, Agius D (2008) Ecotoxicity and biodegradability in soil and aqueous media of lubricants used in forestry applications. Bioresour Technol 99(17):8492–8496
3. Vanya A (2012) Návrh štruktúry deponovanej ako systému „povlak-bok zuba" z hľadiska požiadaviek vybraného ozubeného prevodu. Dissertation thesis, STU, Bratislava
4. Fedák M (2008) Povlaky ako možnosť zvýšenia únosnosti ozubených kolies. Dissertation thesis, STU, Bratislava
5. Bošanský M, Vereš M, Tököly P, Vanya A (2012) Neštandardné ozubené prevody. STU, Bratislava
6. Zápotočný J (2014) Určenie kvalitatívnych a kvantitatívnych charakteristík deponovaných povlakov v systéme povlak bok zuba z tribotechnického hľadiska. Dissertation thesis, STU, Bratislava
7. Krantz TL et al (2003) Increased surface fatigue lives of spur gears by application of a coating. NASA/TM, 2003-212463 (ARL-TR-2971). National Aeronautics and Space Administration, Cleveland, Ohio
8. Michalczewski R (2013) New scuffing test methods for the determination of the scuffing resistance of coated gears. In: Tribology, fundamentals and advancements, vol 12, no 4
9. Orokocký R (2004) Zvyšovanie odolnosti ozubených prevodov v interakcii s ekologickými mazivami. Dissertation thesis, STU, Bratislava
10. Bošanský M, Fedák M (2009) Tvrdé povlaky ako alternatíva tepelného spracovania ozubených prevodov. Acta Technologica Agriculturae, Nitra
11. Bošanský M, Vanya A, Čaplovič Ľ, Hudáková M, Sondor J (2013) Evaluation of properties of selected coatings on Aisi Grade 18Ni (250) maraging steel in terms of their use in gears. Adv Mater Res 746:179–185
12. Bošanský M, Vanya A, Vereš M (2013) Evaluation of properties of selected coatings on steel C60E in terms of their use in gearing. Adv Mater Res 716:166–171

Analysis of Tribological Properties of Composite Materials for the Production Designing of Bearing Cage

Jozef Brončk, Ronald Bašt'ovanský, Viera Konstantová and Michal Stupavský

Abstract Some types of bearings, which are used for specific applications of rolling bearings, allow the use of composite materials for cages. Generally, the cages are loaded only with small forces in rolling bearings, but at higher speeds or in disturbed rolling of bodies, the force acting on the cage can significantly increase. In these cases for increasing of elasticity of cages made from technical plastic are preferred, because they are well resistant to bearing lubricants. To improve the dimensional stability of these materials, they are reinforced with glass, carbon fibers, or glass microspheres, and alike. The cages made of composite materials are characterized by their low weight, high elasticity, and very good sliding properties. This paper presents the results of experimental research on the tribological properties of selected composite materials tested in different environments and their mutual comparison. Experimental measurements were realized on the experimental device linear microtribometer working on the principle of method ball on plane. This method according to the standard ISO 7148-2 is standard method for tribological properties testing. The structure analysis of experimental samples was performed because of understanding of potential structure influence on material properties. The aim of experiments was to acquire the knowledge that can be used in designing materials for bearing cages.

Keywords Tribology · Analysis · Composite materials · Structure · Bearing cage

1 Introduce

The rolling bearings are an indispensable part of the traction mechanisms, where they perform multiple tasks associated with the storage of the rotating shafts. One of them is a reduction in friction in the housing, which has a positive effect on the overall efficiency of the device. Solving the surface wear problem of the bearing materials requires knowledge of the wear and how to measure and evaluate it. Growth of knowledge in this area can help us address a wide range of theoretical as well as practical issues encountered by developers and designers of rolling bearings. The

J. Brončk (✉) · R. Bašt'ovanský · V. Konstantová · M. Stupavský
University of Žilina, Univerzitná 8215/1, 01026 Žilina, Slovak Republic
e-mail: jozef.broncek@fstroj.uniza.sk

Š. Medvecký et al. (eds.), *Current Methods of Construction Design*, Lecture Notes in Mechanical Engineering, https://doi.org/10.1007/978-3-030-33146-7_52

constant development of new materials and coatings used for rings, rolling elements, and cages requires their research on experimental devices to evaluate and measure their physical and mechanical properties. On the other hand, comparison of these properties with other materials is equally important. The cages are generally loaded in bearings only by small forces, but at higher speeds or in the case of disturbed rolling of bodies, the force acting on the cage can increase considerably [1–3]. For these cases, a cage made of plastic is preferred for higher elasticity. In order to guarantee their long-lasting functional capabilities, for plastic cages, other material properties and design specificities as applied to classic materials (steel, brass) have to be taken into account: flexibility and strength in the operating temperature range, resistance to fatigue load, tolerance with assumed lubricant, tribotechnical appropriate cage shape and others. Plastic cages are characterized by their low weight, high elasticity, and very good sliding properties. For small- and medium-sized bearings, solid plastic cages made by injection molding made of polyamide PA 6.6, which is also reinforced with glass fibers, respectively, carbon fibers. Polyamide cages (PA 6.6) are used for operating temperatures up to 100 °C, shortly up to 120 °C. For cage material a different types of lubricant according their aggressiveness are used. For large bearings, plastic segments of PEEK polymer with excellent mechanical properties and excellent chemical resistance to organic and inorganic substances at high temperatures are mainly used for production purposes. PEEK material is used for operating temperatures up to 250 °C [4–6].

2 Experimental Materials and Specimens

Two types of composite materials were used for the production of tested specimens. Composite materials with the polyamide basic matrices were reinforced with carbon fibers and with glass fibers, respectively. The specimens had a shape according to Fig. 1, with size 20 × 80 mm and thickness of 4 mm.

Structure analysis was done on longitudinal and cross section of sample extracted from specimens before tests. In this work, only microstructure of longitudinal section in test direction of structure is presented (see Fig. 2).

The samples for metallographic analysis of composite material microstructure were done with conventional method. Image analysis of structure was done on software Quick PHOTO Industrial 3.1; the percentage of phases was measured. In the case of composite material with glass fibers, the phase reinforcement is 25% approx. in the measured area. In the case of composite material with carbon fibers, the reinforcement phase is 65% approx. in the measured area. Photos, shown in Fig. 2, were used as a measurement areas for determination of reinforcement content.

Uniform distribution of reinforcement in matrix was evaluated in the composite material with carbon fibers. However non-uniform distribution of reinforcement in matrix was evaluated in the composite material with glass fiber. It is known that uniformity and density of reinforcement distribution influence mechanical properties of composite materials. Therefore, it is important to perform a structural analysis.

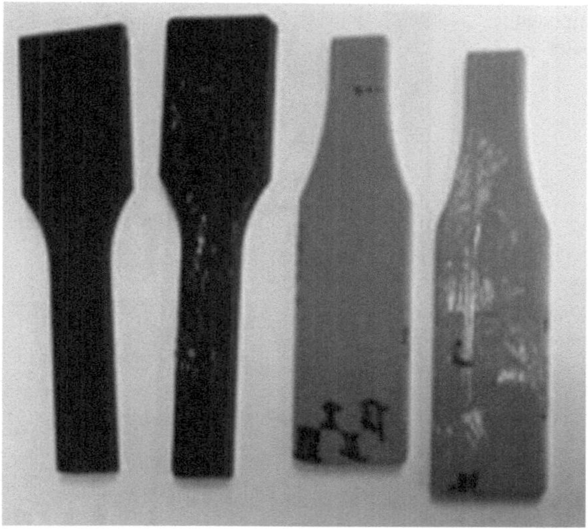

Fig. 1 Photograph of real experimental specimens, left side: composite material with carbon fiber, and right side: composite material with glass fiber

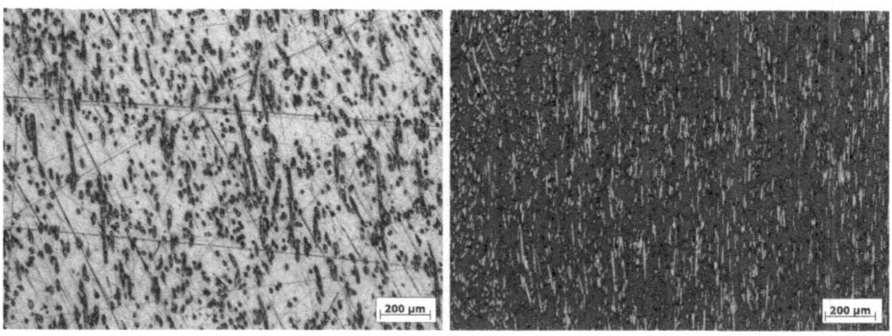

Fig. 2 Microstructure of material, longitudinal section in the test direction, left side: composite material with glass fibers, and right side: composite material with carbon fibers

For micro-innovations, the meaning of the microstructure evaluation is described in [7, 8]. These composite materials were delivered by company without any technical parameters because of private property.

3 Tribological Tests

To evaluate the tribological properties, the testing equipment linear microtribometer, in a tribological laboratory (see Fig. 3) of the department of design and mechanical elements was used. The actual test is based on the ball-on-plate method (see Fig. 4),

Fig. 3 Testing equipment linear microtribometer

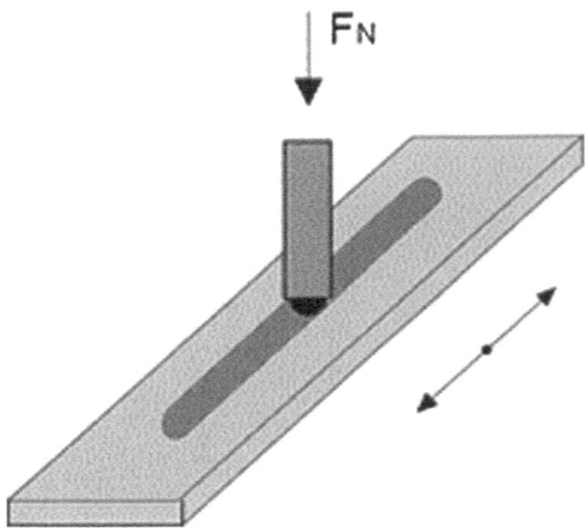

Fig. 4 Principle of tribological test method: ball on plate

which is a standard method of testing the tribological properties of polymer-based bearing materials (ISO 7148-2) [9].

The main part of the microtribometer is the tribological node, which consists of a ball- and plate-shaped specimen. The balls with a diameter of $d = 3.5$ mm made from the 100Cr6 bearing steel and the SiC ceramic material were used. The ball movement is non-uniform, linear, and reversible in each test. The moving ball is pressed onto a static specimen under specific load. The specimen is mounted in the bowl. The bowl is mounted on a loosely constructed plate. The loosely constructed plate is supported at three points and fixed to the strain gauge.

The F_T friction force is determined by tensometric plate positioning. From the F_T and F_N values, friction coefficient values were calculated.

The test methodology was chosen to determine the friction coefficient for the different load values f at the selected sliding velocity v. The duration of the each experimental test was t = 6000 s, at a sliding velocity v = 0–0.02 m s^{-1} and for normal load values of F_N = 5 and 10 N.

In each test, the total sliding distance of the tribological traces was 100 m. The tests were carried out under atmospheric conditions with lubrication and without lubrication.

4 Results and Discussion

The graph of the friction coefficient values in dependence on time for the composite material with glass fiber (specimen Vz1 and specimen Vz2) in different environment: without lubrication (N) and with lubrication (M), and variant condition: loading F_N = 5 and 10 N, with a steel ball (O), respectively, with the ceramic ball (K) is shown in Figs. 5 and 6.

In Table 1, the numerical values of friction coefficient for each specimen in variant conditions are presented for composite material with glass fibers. The graph of the friction coefficient values in dependence on time for the composite material with

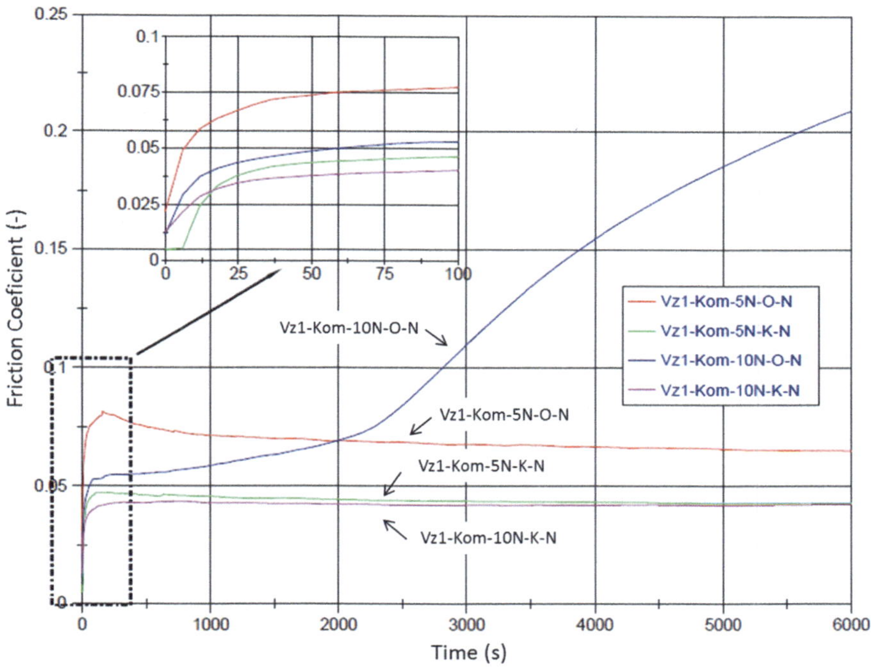

Fig. 5 Graph of friction coefficients of the composite material with glass fibers, specimen Vz1—test environment: without lubricant

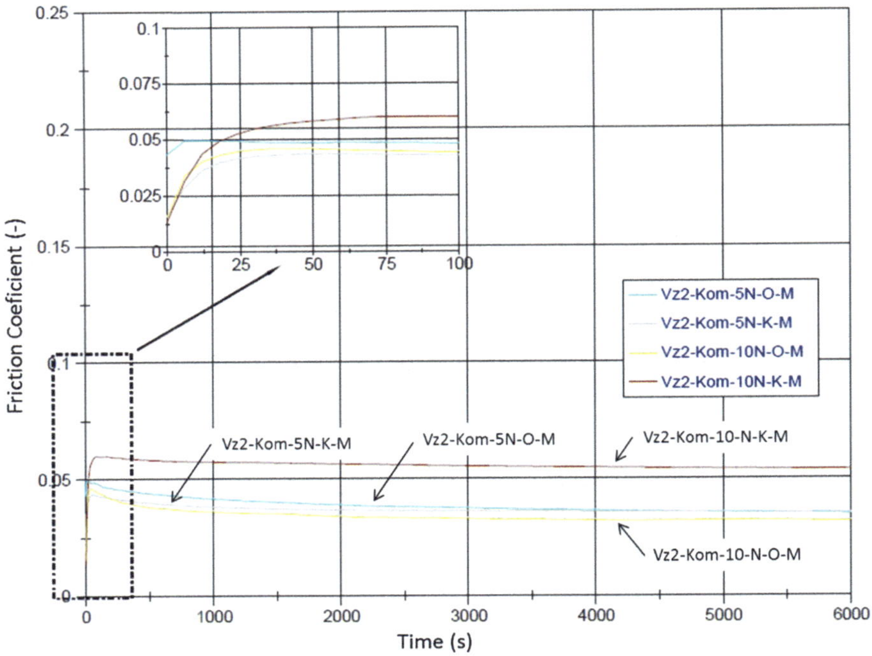

Fig. 6 Graph of friction coefficients of the composite material with glass fibers; specimen Vz2—test environment: with lubricant

Table 1 Friction coefficient values, different environments, composite with glass fibers

Specimen mark	Ball	Lubrication	Load F_N (N)	Friction coefficient (–)
Vz1-Kom-5N-O-N	Steel	No	5	0.075
Vz1-Kom-5N-K-N	SiC	No	5	0.04
Vz1-Kom-10N-O-N	Steel	No	10	0.22
Vz1-Kom-10N-K-N	SiC	No	10	0.04
Vz2-Kom-5N-O-M	Steel	Yes	5	0.035
Vz2-Kom-5N-K-M	SiC	Yes	5	0.035
Vz2-Kom-10N-O-M	Steel	Yes	10	0.03
Vz2-Kom-10N-K-M	SiC	Yes	10	0.06

Legend O—steel ball; K—ceramic ball, SiC; N—without lubricant; M—with lubricant

carbon fibers (specimens Vz3 and Vz4) in two different environment: without lubrication (N) and with lubrication (M), and variant condition: loading $F_N = 5$ and 10 N, with a steel ball (O), respectively, with the ceramic ball (K) is shown in Figs. 7 and 8. In Table 2, the numerical values of friction coefficient for each specimen in variant conditions are presented, for composite material with carbon fibers.

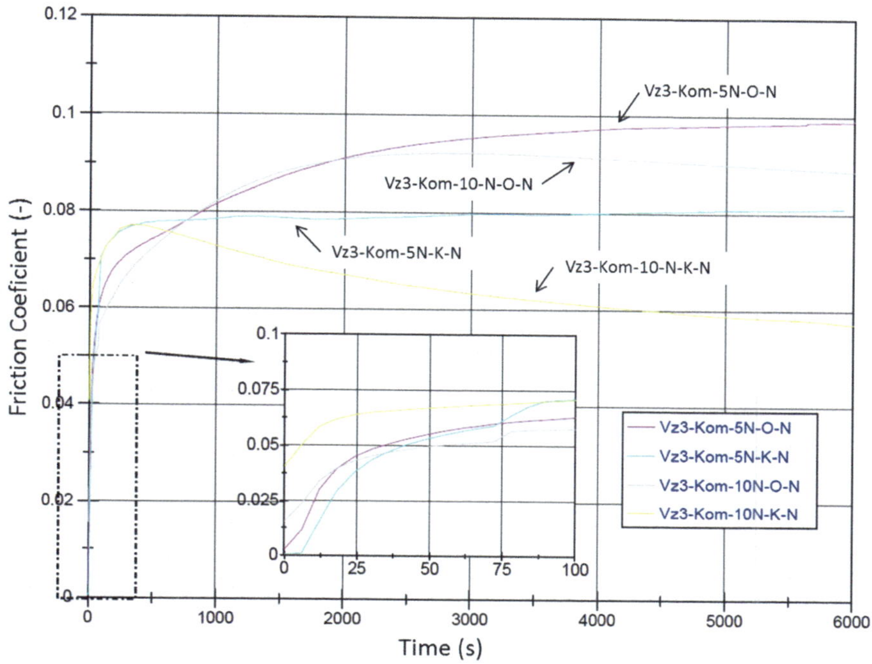

Fig. 7 Graph of friction coefficients of the composite material with carbon fibers, specimen Vz3—test environment: without lubricant

Experimental results of tribological test are possible to discuss in consequence:

1. Comparing the friction coefficient waveforms without lubrication (see Fig. 5, Table 1), it can be stated that the friction coefficient of the glass fiber composite material using a ceramic ball has a lower value than the steel ball at a load of 5 and 10 N.
2. Comparing the friction coefficients under the lubrication conditions (see Fig. 6, Table 1), it can be stated that the course of the friction coefficient for the glass fiber composite material using a ceramic ball is higher than that of the steel ball.
3. Comparing the friction coefficients of the friction coefficients under non-lubrication conditions (see Fig. 7, Table 2), it can be stated that the friction coefficient of the carbon fiber composite material using a ceramic ball has a lower value than the steel ball at 5 and 10 N loads.
4. By comparing the friction coefficients of the friction coefficients under the lubrication conditions (see Fig. 8, Table 2), it can be stated that the frictional coefficients of the carbon fiber composite material using a ceramic ball and a steel ball are similar.

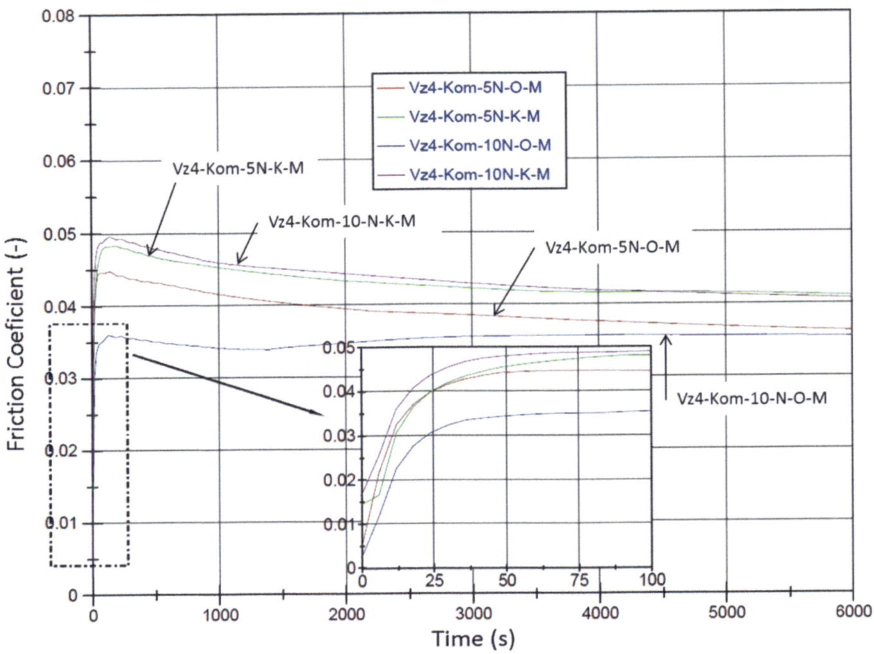

Fig. 8 Graph of friction coefficients of the composite material with carbon fibers, specimen Vz4—test environment: with lubricant

Table 2 Friction coefficient values, different environments, composite with carbon fibers

Specimen mark	Ball	Lubrication	Load F_N (N)	Friction coefficient (–)
Vz3-Kom-5N-O-N	Steel	No	5	0.095
Vz3-Kom-5N-K-N	SiC	No	5	0.082
Vz3-Kom-10N-O-N	Steel	No	10	0.09
Vz3-Kom-10N-K-N	SiC	No	10	0.058
Vz4-Kom-5N-O-M	Steel	Yes	5	0.042
Vz4-Kom-5N-K-M	SiC	Yes	5	0.042
Vz4-Kom-10N-O-M	Steel	Yes	10	0.035
Vz4-Kom-10N-K-M	SiC	Yes	10	0.067

Legend O—steel ball; K—ceramic ball, SiC; N—without lubricant; M—with lubricant

5 Conclusion

The presented work deals with the tribological properties of composite materials and their suitability for use in the production of roller bearing cages. The tribological test analysis done on microscale is also necessary for innovations in this level. The comparison of experimental results of tribological properties at different loads of 5 N, 10 N, and two types of ball (steel ball and ceramic ball) and in different environments allowed monitoring the influence of the load and lubrication on the tested materials. From discussed results the following conclusions are: the values of the friction coefficient varied significantly with the character of the environments, the size of the load force and the type of the body; the values of the friction coefficients detected under the lubrication conditions for both composite materials evaluated and for both bead materials are lower than under non-lubricating conditions.

Acknowledgements This article was written with financial support of Ministry of Education, Science, Research and Sport of the Slovak Republic of KEGA project 031 ŽU-4/2016.

References

1. Brončck J, Fabian P, Radek N (2015) Tribological research of properties of heat—treated cast irons with globular graphite. Mater Sci Forum 818:209–212
2. Martinec T, Mlynec J, Petru M (2015) Calculation of the robot trajectory for the optimum directional orientation of fibre placement in the manufacture of composite profile frames. Robot Comput Integr Manuf 35:42–54
3. Kohár R, Hrček S (2014) Dynamic analysis of a rolling bearing cage with respect to the elastic properties of the cage for the axial and radial load cases. Commun Sci Lett Univ Žilina 16(3A):74–78
4. Brezinová J, Draganovská D, Guzanová A, Balog P, Vňaš J (2016) Influence of the hardfacing welds structure on their wear resistance. Metals 6(2):1–12
5. Brezinová J, Koncz J, Tomáš M, Slota J (2015) Corrosion behaviour of automotive steel sheets depending on the degree of deformation. Eng Mater 635:57–60
6. Antoszewski B, Radek N, Tofil S, Brončck J (2017) Dobór parametrów laserowej mikroobróbki powierzchniowej elementów panewek brązowych. Prz Elektrotech 93(1):41–44
7. Mašín I, Petru M, Novák O (2015) Metódy inivačního inženýrství na mikro-úrovni. Publisher TU v Liberci, Liberec
8. Szczotok A, Pietraszek J, Radek N (2017) Metallographic study and repeatability analysis of gamma phase precipitates in cored, thin-walled castings made from IN713C superalloy. Arch Metall Mater 62(2):595–601
9. ISO 7148-2 (2012) Testing of the tribological behavior of bearing materials, part 2: testing of polymer-based bearing materials

Investigation of the Tribological Properties of the Nitride Layer on Heat-Treated Steel 100Cr6

Jozef Brončék, Martin Vicen, Peter Fabian and Norbert Radek

Abstract The article deals with the tribological properties of the heat-treated steel 100Cr6 and the nitride layer formed on this steel. Steel 100Cr6 is a bearing steel that is used to make bearings' rings and bearings' elements. The reason for the formation of nitride layer on heat-treated steel 100Cr6 is the decrease of the friction coefficient. The reduction of friction coefficient contributes to reduce wear and, at the same time, increases the lifetime of the rings and elements of roller bearings during their work time. Experimental measurements were realized on the experimental device rotary microtribometer working on principle "ball on disc". The aim of tribological tests is to determine the coefficient of friction of material pairs, compare them and then evaluate them. Based on the results, the suitability of the nitride layer for the bearing materials is determined. The results will be suitable not only for bearing materials but also for a wide range of iron-based materials.

Keywords Tribology · Nitride layer · Ball on disc · Steel

1 Introduce

In general, the heat treatment of roller bearings made from bearing steels (steel 100Cr6) serves to change the mechanical properties of the bearings' material. Due to high wear and fatigue resistance requirements, the bearing steels are the most demanding, not only in the production, but also in their shaping and heat treatment [1, 2]. One of the options for obtaining a high surface hardness and tough core of components and reducing the friction coefficient is to form a nitride layer on surface of 100Cr6 bearing steel. It is a requirement that suitable chemical heat treatment processes achieve the desired hardness which will increase the toughness and increase wear resistance of the surface layer which is in contact with other surfaces. Advantage

J. Brončék (✉) · M. Vicen · P. Fabian
University of Žilina, Univerzitná, 8215/1, 01026 Žilina, Slovak Republic
e-mail: jozef.broncek@fstroj.uniza.sk

N. Radek
Kielce University of Technology, al. Tysiaclecia Panstwa Polskiego 7, Kielce, Poland
e-mail: norrad@tu.kielce.pl

Š. Medvecký et al. (eds.), *Current Methods of Construction Design*, Lecture Notes in Mechanical Engineering, https://doi.org/10.1007/978-3-030-33146-7_53

of heat treatment technology used is that nitriding temperature 500–600 °C is lower than for example in cementing. This temperature does not cause large deformations of the shape and geometry of the parts, and surface does not require further heat treatment to achieve the required hardness [3].

Determining the tribological characteristics of real material pairs or tribotechnical systems is currently being realized in many workplaces. It is carried out by extensive experimental research, which provides its own data and possibilities to verify proposed theoretical hypotheses [4–6, 8].

2 Material and Heat Treatment Parameters

As the experimental material was used standard 100Cr6 bearing steel with chemical composition shown in Table 1.

Figure 1 shows sample shape of a circular disc with outer diameter ϕ 46, inner diameter ϕ 25 and thickness 6 mm (the first member of the friction pair). The second member of the friction pair is a ball. Both members are made from 100Cr6 bearing steel.

Samples (circular discs) were heat treated by the recommended mode characteristic of this steel. The austenitizing temperature of samples was 830 °C with a residence time at this temperature of 25 min. Subsequently, samples were quenched into the DURIXOL V71 oil. Tempering of samples was performed at 160 °C for 2 h. As a result of samples, heat treatment has obtained a structure which consisting of the tempered martensite, the evenly distributed carbides $(Fe, Cr)_3C$ and the residual austenite (Fig. 2).

Vickers hardness test was measured on all samples after heat treatment. The hardness results of samples after heat treatment are shown in Table 2. The average

Table 1 Chemical composition of 100Cr6 steel [wg.%]

C	Si	Mn	P	S	Cr	Ni	Fe
1.05	0.35	0.45	0.0029	0.0033	1.55	0.182	All

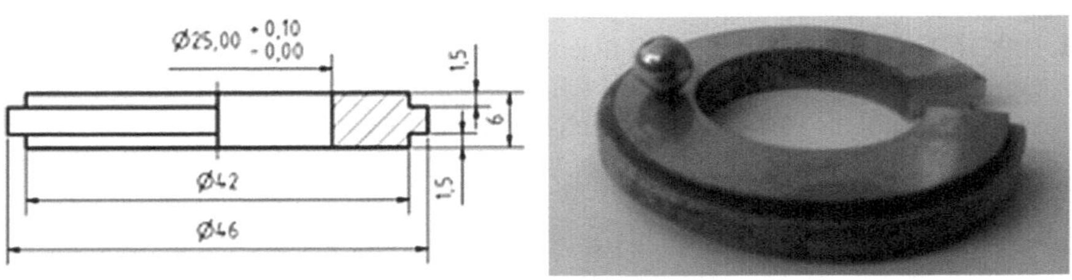

Fig. 1 Dimensions of experimental sample (left). Real view of sample, disc and tribology pair—ball (right)

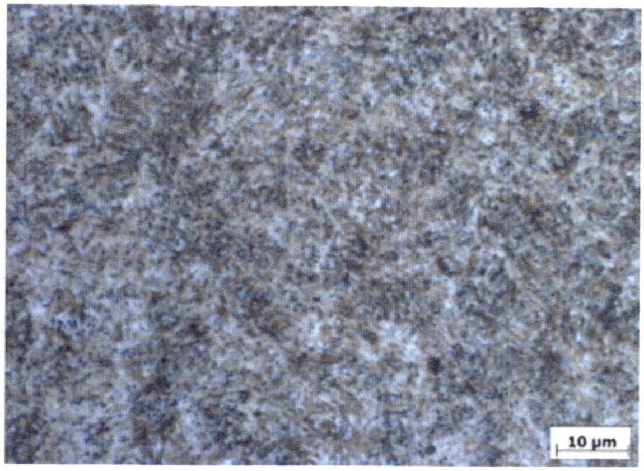

Fig. 2 Microstructure of 100Cr6 steel after heat treatment (etched with 3% Nital)

Table 2 Vickers hardness measurement after heat treatment

No. of samples	No. of measurements				Average hardness
	1	2	3	4	
1	900	896	899	902	899
2	892	897	899	901	897

samples hardness is 898 HV. The nitriding of samples was carried out in a hermetic working chamber in which heating, control and regulation of temperature, pressure and degree of ammonia dissociation were ensured. The degree of dissociation of ammonia was 40% at nitriding temperature of 520–540 °C. Nitriding temperature was determined based on hardness and thickness of the nitride layer. After the nitride layer was formed on heat-treated bearing steel 100Cr6, microhardness was again measured. The results of microhardness measurements of samples after the chemical heat treatment (nitriding) are shown in Table 3.

Figure 3 shows the macrostructure of nitride layer in cross section. On the surface, there is a thin continuous layer ε-phase; in direction into material, there is a diffusion

Table 3 Vickers hardness measurement after chemical heat treatment (nitriding)

No. of samples	No. of measurements				
	1	*2*	*3*	*4*	*5*
1	725	706	690	655	590
2	720	699	685	660	603
	6	*7*	*8*	*9*	*10*
1	542	471	460	395	398
2	549	474	439	410	395

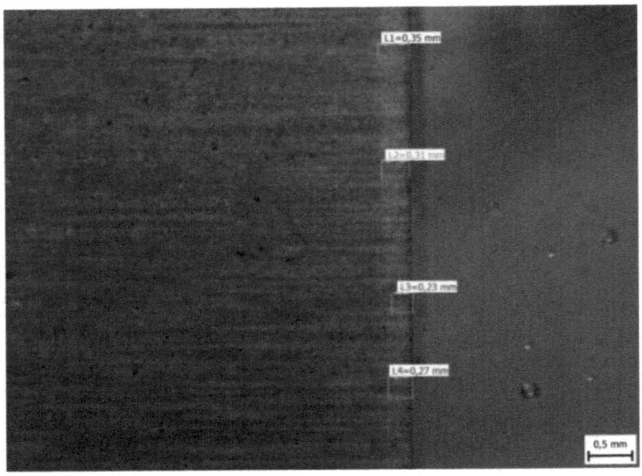

Fig. 3 Cross section of 100Cr6 sample with depth of the nitride layer (etched with 3% Nital)

area. This area consists of γ- and α-phase nitrides. The depth of nitriding layer was measured. The measured values of nitride layer thickness are shown in Fig. 3.

3 Tribological System

From the point of tribology, a real tribological system is explored on a tribological model, which shows a mutual similarity to a real or an ideal friction ball, which enables this mutual description. A rotary microtribometer T–01 M is used for the evaluation of tribological properties [7].

Tribological test is based on the ball-on-disc method, which is a standard method of testing the tribological properties of materials. The main part of the rotary microtribometer is a tribological pair consisting of a ball (fixed at the end of the roller) and a sample that has the shape of a circular disc. The ball is made of standard 100Cr6 bearing steel with a diameter of $d = 6$ mm. The hardness of ball is 750 HV. The ball motion was uniform and rotational, and the velocity was $v = 0.8$ m s^{-1}. Principle of test is that load acts on the ball and at the same time on the face of sample on which nitride layer was applied. Normal load consisted of a set of weights which create the normal force F_N for load values $F_N = 4.9$ and $F_N = 14.7$. The friction force F_T was determined by tensometric measurement of the disc position. Length of test lasted $t = 5000$ s. Friction coefficient values were determined using the software from known F_T and F_N values.

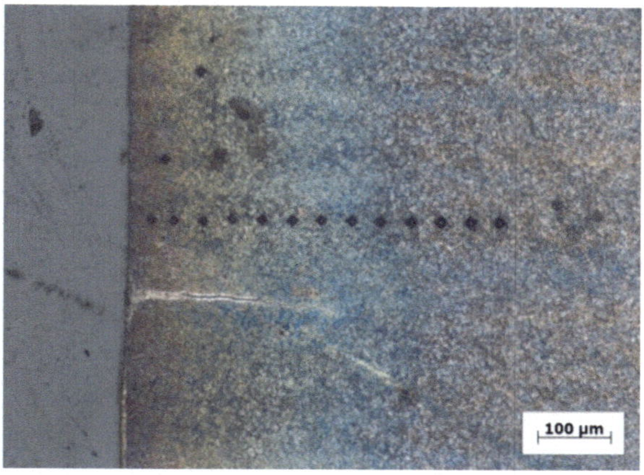

Fig. 4 Position of microhardness HV 0.1 in the cross section of the sample

4 Results and Discussion

4.1 Hardness Measurement Results

Results of measuring hardness of samples after heat treatment are given in Table 2. One of the most important monitored special parameters of nitride components is depth of the nitride layer. The conventional method used to verify the depth of nitride layer is to measure the microhardness of cross sections of nitride components. This method is performed according to DIN 50 190.

Results of substrate hardness evaluation after nitriding are given in Table 3. Figure 4 shows the position of imprints in the cross section of sample from surface to the core of sample.

As can be seen from Fig. 5, microhardness of the nitrided sample (bottom curve) recorded a downward trend from the surface to the core. Figure 5 also shows microhardness behaviour of the sample after heat treatment (upper curve). This curve has a constant straight line resulting from the homogeneity of the material after cross section.

4.2 Evaluation Coefficient of Friction and Wear

In the experiment, the course of friction coefficient values of pairs steel–steel and steel–FeN layer versus time ($v = 0.8$ m s^{-1}) was monitored. The specimens were cleaned by ethanol before test. The variations of the friction coefficients from test beginning to the finish are presented in Table 4, graphically illustrated in Figs. 6 and 7.

Course of friction coefficients in various (a) loading 4.9 N and (b) loading 14.7 N.

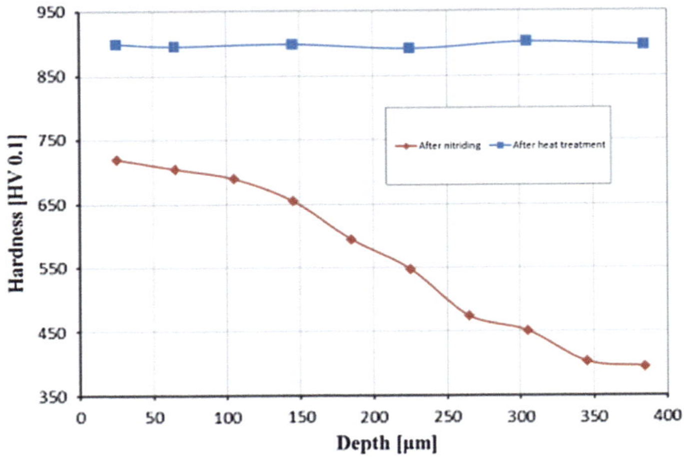

Fig. 5 Behaviour of microhardness HV 0.1 from the surface to the core of sample (µm)

Table 4 Results of measurement and evaluation of material pairs

Material tribological pairs	Steel–steel		Steel–FeN	
Load F_N (N)	4.9	14.7	4.9	14.7
Coefficient of friction (–)	0.85	0.75	0.7	0.7
Ball wear diameter D (mm)	1.32	2.22	1.18	1.99
Ball wear area $A1$ (mm^2)	1.37	3.87	1.09	3.09
Calculated pressure value p (MPa)	3.6	3.8	4.5	4.76

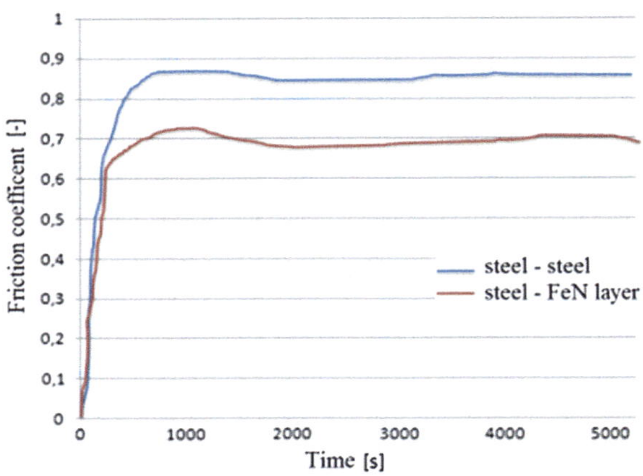

Fig. 6 Course of friction coefficients in loading 4.9 N

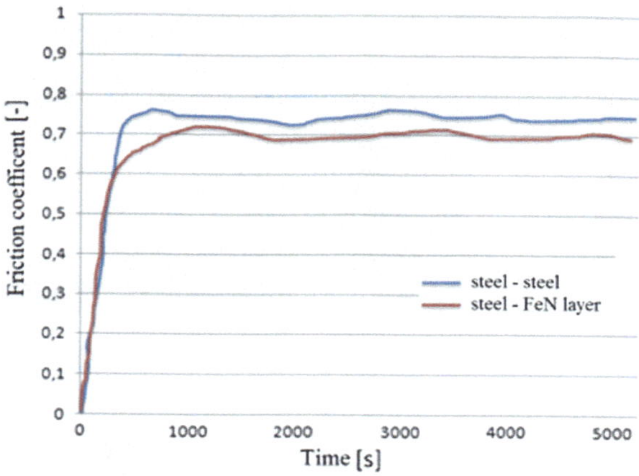

Fig. 7 Course of friction coefficients in loading 14.7 N

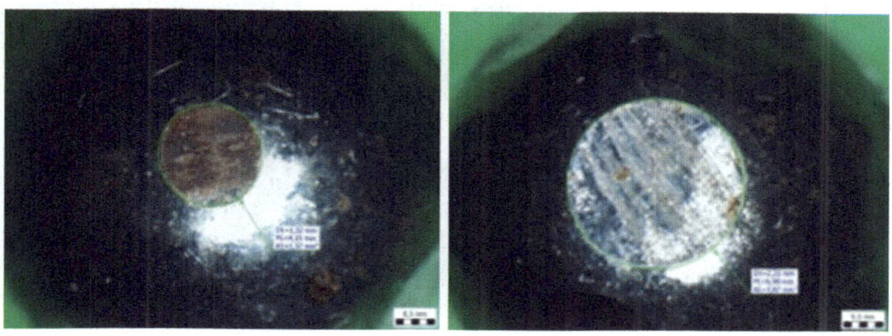

Fig. 8 Wear on the quenching steel ball surface in contact with the steel, left: load $F_N = 4.9$ N, $D = \phi\ 1.32$ mm, right: load $F_N = 14.7$ N, $D = \phi\ 2.22$ mm

Figures 8 and 9 show wear area on the ball after the tribological test. It was found that the friction coefficient of the friction pair steel–FeN layer is less than that of the steel–steel pair by comparison at every load values. Wear diameter as well as the wear area of quenching steel ball was greater at the steel–steel contact than the contact surface steel–FeN layer at all three load values.

The size of the wear surface $A1$ on the ball (in mm^2) is characterized by the intensity of wear on its surface. For the individual samples, the contact pressure values p (MPa) are given in Table 4. Contact pressure values were calculated according to $p = F_N/A1$.

Figures 10 and 11 show microscopic pictures of a tribological trace on the specimen's surface. Figure 10 shows the surface traces of the quenching bearing steel 100Cr6, and Fig. 11 shows traces on quenching bearing steel with a nitrided layer. It is evident from visual observation that on tribological specimens, there is more marked wear on the quenching steel 100Cr6 surface than on the nitride layer.

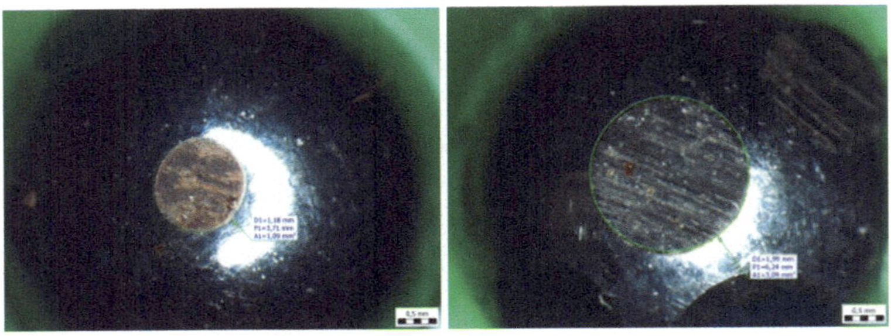

Fig. 9 Wear on the quenching steel ball surface in contact with the FeN layer, left: load $F_N = 4.9$ N, $D = \phi$ 1.18 mm, right: load $F_N = 14.7$ N, $D = \phi$ 1.99 mm

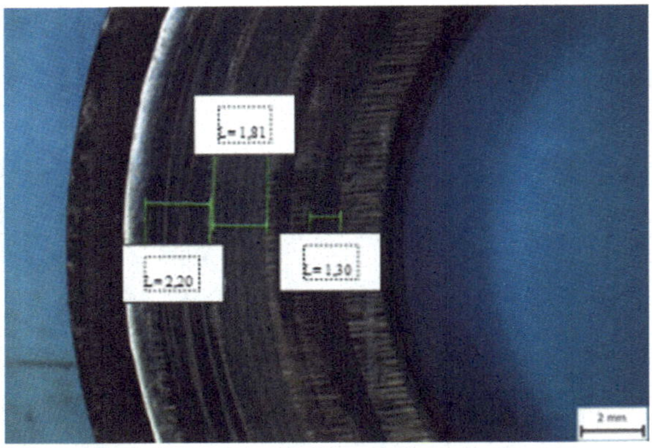

Fig. 10 Tribological trace on the surface of quenching bearing steel 100Cr6

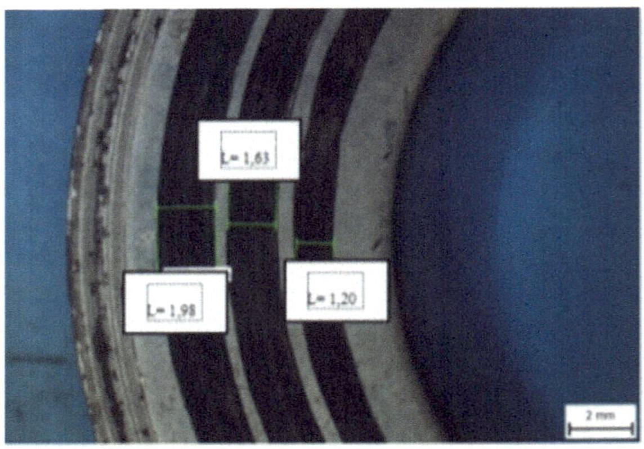

Fig. 11 Tribological trace on the surface of quenching bearing steel with nitriding layer

5 Conclusion

From the achieved results, the conclusion is the following:

- The average value of friction coefficient for the material pair of steel–FeN layer is lower than for steel–steel pair at every load.
- The friction pair wears is smaller for a pair of steel–FeN layer than for a pair of steel–steel.
- Nitriding of 100Cr6 hardened bearing steel causes a significant reduction in hardness, although the FeN layer in contact with steel achieves better friction properties than steel–steel.

Acknowledgements This article was created as a part of project KEGA with number 031 ŽU-4/2016.

References

1. Bronček J, Fabian P, Radek N (2015) Tribological research of properties of heat—treated cast irons with globular graphite. Mater Sci Forum 818:209–212
2. Martinec T, Mlynek J, Petru M (2015) Calculation of the robot trajectory for the optimum directional orientation of fibre placement in the manufacture of composite profile frames. Robot Comput-Int Manuf 35:42–54
3. Kohar R, Hrncek S (2014) Dynamic analysis of a rolling bearing cage with respect to the elastic properties of the cage for the axial and radial load cases. Communications—Scientific Letters of the University of Žilina 16(3A):74–81
4. Brezinová J, Draganovská D, Guzanová A, Balog P, Vňaš J (2016) Influence of the hardfacing welds structure on their wear resistance. Metals 6(2):1–12
5. Brezinová J, Koncz J, Tomáš M, Slota J (2015) Corrosion behaviour of automotive steel sheets depending on the degree of deformation. Key Eng Mater 635:57–60
6. Antoszewski B, Radek N, Tofil S, Bronček J (2017) Dobór parametrów laserowej mikroobróbki powierzchniowej elementów panewek brązowych. Przeglad Elektrotechniczny 93(1):41–44
7. Szczotok A, Pietraszek J, Radek N (2017) Metallographic study and repeatability analysis of gamma phase precipitates in cored, thin walled coating made from IN713C superlloy. Arch Metall Mater 62(2):595–601
8. Mašín I, Petru M, Novák O (2015) Metódy inivačního inženýrství na mikro-úrovni, 1st edn. TU v Liberci, Liberec

Experimental Verification of Tribological Properties of Thin Coatings for Artificial Human Joints

Marián Dzimko, Matúš Kovaliček, Eva Gajdošová and Yoshinori Takeichi

Abstract Qualitative determination of tribological properties of thin coatings, with the aim to test them for possible modification of friction in tribological for artificial human joints system, was done. Experiments were done at common conditions of laboratory environment and temperatures. Attention was paid to tribological characteristics of coated materials suitable for implants of human joints. Unique test apparatus and advanced measurement techniques AFM for experimental procedures were involved. The plane samples for experiments consist of three types of microstructures of steel 100Cr6 obtained after heat treatment. Samples with martensitic, bainitic, and sorbitic structure were coated with a hard and wear resistance WC/C coating. The coatings were deposited to functional surfaces of all plane samples. Magnetron sputtering method was used. As second friction body, a standard ball made of steel 100Cr6 was chosen. Received results extend the existing knowledge of tribology of thin layers. The friction processes in systems with high contact loads and the surface topography, using AFM microscope, were evaluated. The minimum friction coefficient values reached the level of 0.1–0.15; those are approximately 10 times higher as these for natural human joints. The lowest value of coefficient of friction was reached for WC/C coated samples with bainitic structure of the substrate.

Keywords Thin coatings · Nanoindentation · Hertzian contact · Coefficient of friction

1 Introduction

Tribological processes represent primary material interactions of frictional bodies. These interactions run over in space and time. The contact of individual bodies of the tribological system is a fundamental sign of its behavior. The interactions can be

M. Dzimko (✉) · M. Kovaliček · E. Gajdošová
University of Žilina, Univerzitná 8216/1, 01026 Žilina, Slovak Republic
e-mail: dzimko@uniza.sk

Y. Takeichi
Toyohashi University of Technology, Hibarigaoka, Tempakucho, Toyohashi-shi, Aichi 4418580, Japan

Š. Medvecký et al. (eds.), *Current Methods of Construction Design*, Lecture Notes in Mechanical Engineering, https://doi.org/10.1007/978-3-030-33146-7_54

of material, physical, chemical, or energetically origin. This also applies and is valid for biotribological systems. Modern practice has discovered and acknowledged the fact that it is the real surface that is the most important part of technical elements and systems to provide useful tribotechnical functions [1]. The use of surface treatments, layers, and coatings allows the development of new technology and technological constructions where the desired properties of the contact pair are modified at a very particular location and in a specified quantity and quality [1]. Thin coatings can be imagined as layers with a thickness of same of nanometers to micrometers. The layers are applied to the base bulk material (substrate), which defines the geometric shape, volume, and its subsequent use. The coating prepared on the surface of the substrate is limited by two distances very close to each other, while the interface with the external environment can be called the surface [2, 3].

The paper presents an extension of existing knowledge in the field of tribology of thin friction layers. It deals with systems with varying contact loads and uses experimental testing of WC/C-based coating on hard stainless steel substrate.

2 Experiment

2.1 Sample Preparation

The tested pair was composed of coated steel plate and an uncoated steel ball. This corresponds to the form pair "ball on plate" and to the material model "hard-hard." The WC/C-based coatings are commercially available. Tungsten carbide (WC) is a very interesting material of industrial use in the form of hard-wearing, thanks to excellent mechanical and tribological properties. The carbon is found in the WC/C coatings of a-C [4, 5] and DLC [4]. Its anti-wear properties and biocompatibility have led to an idea for testing its frictional behavior for hip joints.

The flat specimens with martensitic, bainitic, and sorbitic microstructure used in this work are made of 100Cr6 stainless steel with dimensions of $25 \times 80 \times 5$ mm. The sample surface before coating was prepared metallographically with three different heat treatment methods and grinding after hardening and tempering. Attention has been paid to its cleanliness, microgeometry, and actual hardness.

Nanometric WC/C coatings were deposited by magnetron sputtering method to functional surfaces of all flat samples. The sample was chemically cleaned before to be placed into deposition chamber and the surface of the sample was ionic purified.

Deposition process was divided into two steps:

1. Thin layer of chromium was deposited to obtain improved adhesion properties.
2. On already deposited Cr layer, the thin WC/C layer was deposited with the thickness of 0.5 μm.

Microstructures of samples before and after coating for martensitic structures (see Fig. 1) and bainitic structures (see Fig. 2).

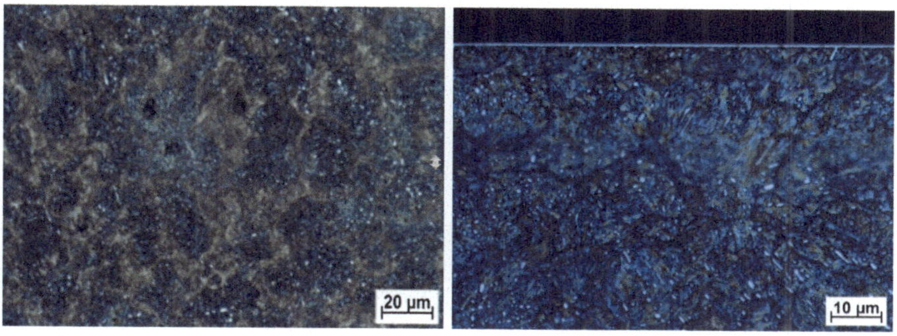

Fig. 1 Microstructures after heat treatment and after coating with visible Cr layer, martensitic 781

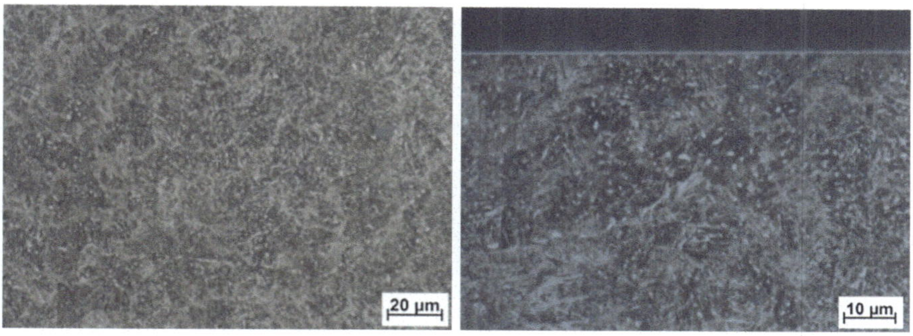

Fig. 2 Microstructures after heat treatment and after coating with visible Cr layer of bainitic 780

The ball specimen are standard ball, made of stainless steel 100Cr6 with the diameter $d = 4$ mm. The ball hardness was measured at 65 HRC.

2.2 *Friction Equipment a Test Conditions*

The friction test was conducted with the "ball on flat" microtribometer with a sliding bracket in which the plate and ball specimens were mounted (see Fig. 3).

The applied load was variable from 1 to 10 N in defined steps and this gives the maximal contact pressures of $p_{\max} = 0.86$ to 1.85 GPa. The environmental conditions were the same for all the types of tested tribological pair.

The horizontal speed of a moving ball on the flat sample respected the sinusoidal curve course within the $v = 0$–20 mm s^{-1}. The length of the frictions way was limited by time ($t = 6\,000$ s).

The value of a friction force in contact is determined by tensometric measurement. The values of friction coefficient are calculated from measured and known values of F_{T} and F_{N}.

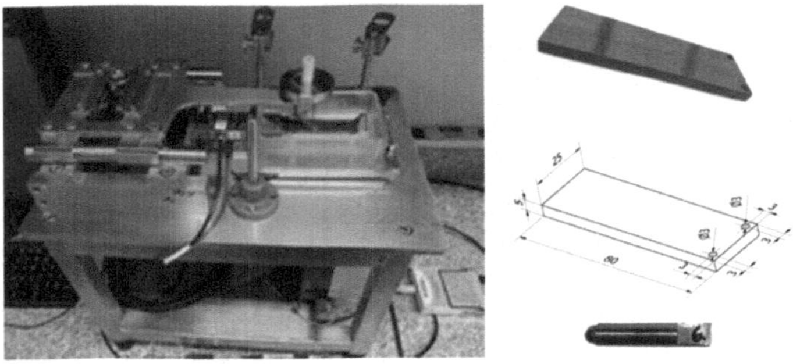

Fig. 3 Microtribometer, flat specimens and specimen mounted in holder

The result of experiment was recorded as the change of the coefficient of friction over time.

3 Results

3.1 *Friction Behavior*

The vertical axis shows the value of the coefficient of friction of all six samples with WC/C coatings on martensitic, sorbitic, and bainitic substrate and the horizontal axis represents running time (see Fig. 4).

Initial value of the coefficient of friction for all samples with was over 0.3 with a maximum of 0.45 for martensite 781. Other samples (bainitic and sorbitic) reached initial values from 0.31 to 0.38. Three different fields of friction tendencies depending on running time have been noted. In the first field, up to 1500 s, the value of the coefficient of friction for all tested samples shows rapidly decreasing tendency. The second field in between 1500 and 3000 s shows slightly increase in the coefficient of friction. For the third field after 3000 s is characteristic low and almost constant value of friction coefficient. The lowest value of the coefficient of friction 0.113 was reached by the sample with bainitic structure 781 at the end of the experiments. Also, martensitic and sorbitic structures reached only slightly higher value of the coefficient of friction. Detailed behavior of the coefficient of friction can be seen in magnification part in Fig. 4.

3.2 *Surface Observations*

Topography of WC/C coating type 780 and 781 on plate samples Coating topography was evaluated on the Solver Next AFM microscope. For both samples, the

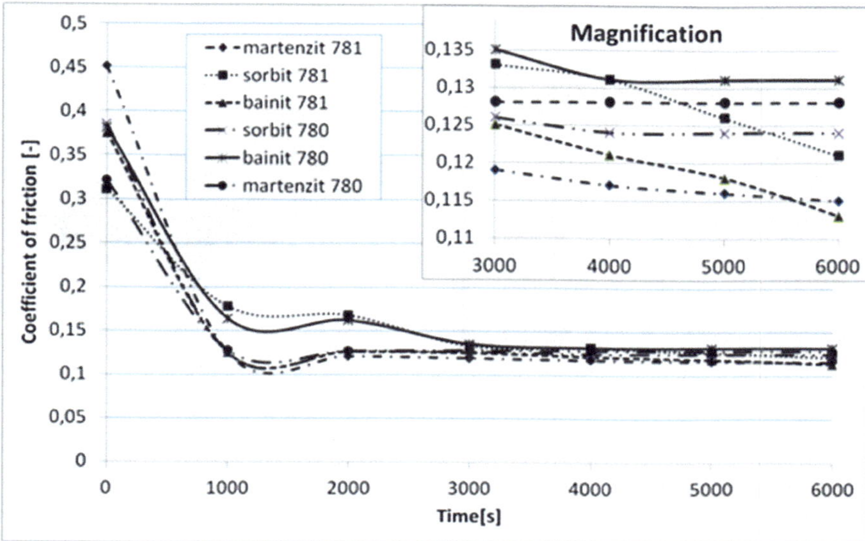

Fig. 4 Coefficient of friction for different heat treated 100Cr6 substrates with deposited WC/C layer

same coating surface topography scanning parameters were used. Surface rasterization occurred in semi-contact mode using the NSG 10 tip. Scanning speed was set to 0.4 Hz and the evaluated scan area was 5 μm × 5 μm. After the surface was scanned, images were displayed showing the actual surface of the samples in 2D and 3D views along with the surface profile with its parameters at the selected location on the sample.

The type 780 coated samples' surface showed low roughness parameter values. The mean arithmetic deviation was $R_a = 3.417$ nm and the maximum profile height R_t was 27.358 nm (see Fig. 5).

The type 781 coated samples surface exhibited low roughness parameter values similar to the previous type 780 coating. The mean arithmetic deviation $R_a =$ 4.748 nm was with slightly higher and the maximum Rt profile was also been greater measured 36.568 nm (see Fig. 6). Both surfaces, i.e., 780 and 781 may be consider as smooth one.

Topography and roughness of 100Cr6 ball samples were not measured. Instead of an AFM observation, an optical microscope was used. Typical picture of worn surface is shown in Fig. 7. It is to remark that wear evaluation was not made however it must be noticed that only the surfaces of ball samples showed significant traces of wear. The size of the worn diameter was compared with those of calculated using Hertzian theory.

The Hertzian theory about the contact of solid elastic bodies is used for solving of tasks connected with the tension conditions in the field of contact points of elastic

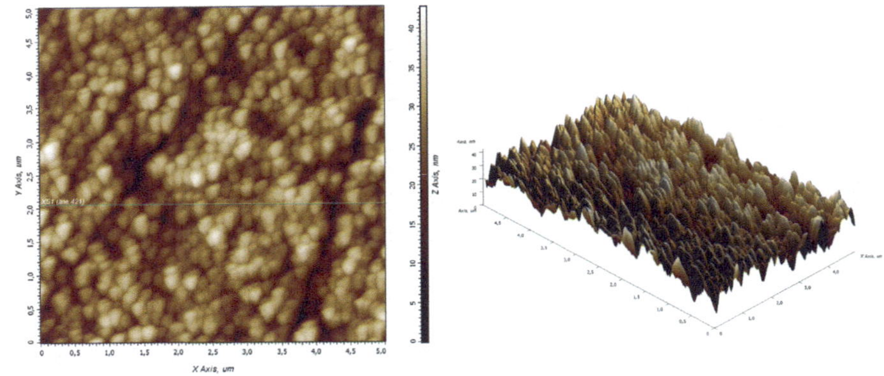

Fig. 5 2D and 3D sample surface after coating with WC/C-780 coating

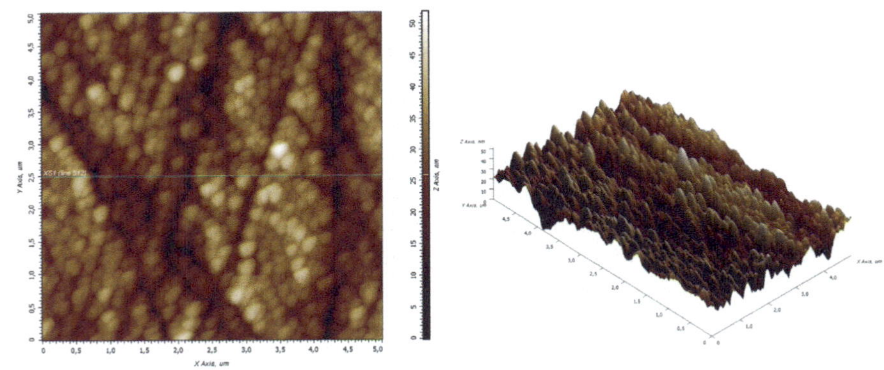

Fig. 6 2D and 3D sample surface after coating with WC/C-781 coating

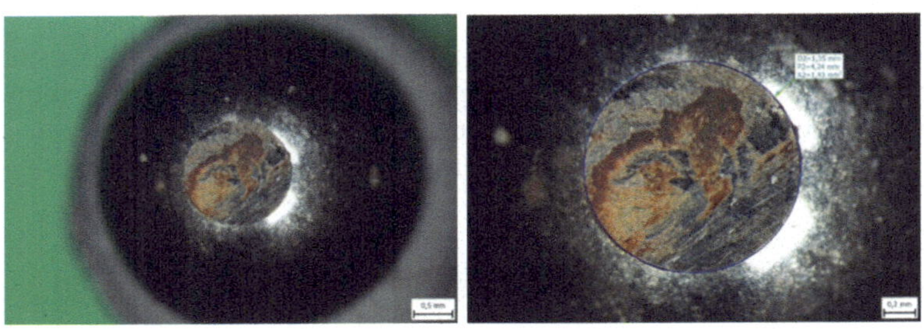

Fig. 7 Photographs of worn contact area on 100Cr6 steel ball

bodies. This is sufficient for setting rules for calculating of a contact tension and a mutual approaching of the bodies [6–8].

For the Hertzian contacts, resulting from applied normal load F_N are the values of maximal pressure p_{max}, and parameters of a semi-contact width r determined by the formulas:

$$P_{\max} = \frac{3F_N}{2\pi r^2} \quad (1)$$

$$r = \left(2R'\delta\right)^{\frac{1}{2}} \quad (2)$$

where the ball with the radius $R_1 = 2$ mm and the plane with the $R_2 = \infty$ are used.

4 Conclusions

The friction properties of two kinds of WC/C coatings on steel with martensitic, sorbitic, and bainitic microstructures were studied under the common laboratory conditions. Received results extend the existing knowledge of tribology of thin layers.

Both types 780 and 781 of WC/C showed lower friction coefficient with increasing tested time. In an unlubricated contact, the final values of friction were around 0.115. This indicates that after the "running in" period, the deformation part of coefficient of friction became low, as the part of steel ball was already worn and the real contact pressure became lower. On the other hand, all coated surface remained nearly unworn as reported in [9, 10]. Slightly lower values of coefficient of friction compared to those of [9, 10] may result from different hardness of ball specimen. The friction processes were evaluated in the systems with high contact loads and the surface topography by using AFM. All of the results indicated WC/C coating on hard substrates, regarding the level of friction (with lower importance of their microstructure) has a potential for the applications of the artificial joints from the understanding of wear resistance. Further investigation regarding biocompatibility and possible simulation of hip prosthesis needs to be conducted. This suggestion is also supported by [11]. Considering the fact that the values of coefficient of friction in unlubricated contacts were up to 10 higher than those in the human joints may lead to continue the tests with the synovial fluids.

References

1. Tribológia tenkých tuhých vrstiev, filmov a povlakov. http://old.uk.fme.vutbr.cz/kestazeni/tribologie/prednasky/prednaska9.pdf. Last Accessed 30 Jan 2014
2. Radek N, Bronček J, Fabian P (2017) Tribological properties of electro-spark deposited coating after laser treatment. In: Herák D (eds) Proceedings of 58th international conference of machine

design departments—ICMD 2017, pp 314–319. Czech University of Life Sciences Prague, Prague
3. Radek N, Brončeк J, Fabian P, Pietraszek J (2015) Properties of the electro-spark deposited coatings—technology and applications. Mater Sci Forum 818:61–64
4. Abad MD, Muñoz-Márquez MA, Mrabet El S, Justo A, Sánchez-López JC (2010) Tailored synthesis of nanostructured WC/a-C coatings by dual magnetron sputtering. Surf CoatS Technol 204:3490–3500
5. Zhou S, Wang L, Wang SC, Xue Q (2011) Comparative study of simplex doped nc-WC/a-C and duplex doped nc-WC/a-C(Al) nanocomposite coatings. Appl Surf Sci 257:6971–6979
6. Ábel M, Brezinová J, Draganovská D (2003) Vlastnosti povrchu predupraveného tryskaním. Transfer inovácií 6
7. Kohar R, Hrcek S (2014) Dynamic analysis of a rolling bearing cage with respect to the elastic properties of the cage for the axial and radial load cases. Commun Sci Lett Univ Žilina 16(3A):74–81
8. Radek N, Szczotok A, Gądek-Moszczak A, Brončeк J, Pietraszek J (2018) The impact of laser processing parameters on the properties of electro-spark deposited coatings. Arch Metall Mater Open Access 63(2):809–816
9. Kovalíček M (2015) Experimental valuation of tribological properties of thin solid lubricant layers. Doctoral Thesis. University of Žilina
10. Broncek, J, Kovalicek M, Fabian P, Bastovansky R (2015) Investigation of tribological properties of WC/C coatings applied on 100Cr6 bearing steel substrate. In: Bošanský M, Bucha J, Kadnár M, Nemčeková M, Rusnák J, Vereš M (eds) Book of proceeding of 56th international conference of machine design departments, ICMD 2015 pp 337–342. Slovak University of Agriculture, Nitra
11. Luo Y, Yang L, Tian M, Ge S, Wang Q (2013) The friction and wear behavior of WC coating on medical grade titanium alloys. Proc Inst Mech Eng, Part J J Eng Tribol 227(8):845–849

Additive Technologies

Modular 3D Printer Concept

Róbert Kohár, Marián Stopka, Peter Weis, Peter Spišák and Ján Šteininger

Abstract The current situation on a field of 3D print does not allow significant changes in the printer's construction. Usually, printer is equipped with definitive construction without ability to change parts. 3D printer's build area is often a limiting factor, so there is a potential need for changing print area dimensions. The goal of this project is the design of modular construction of a 3D printer. In the future, that would mean, that after purchase and after query for increasing build area, it would be necessary just to connect another unified block. All the changes could allow rapid modification of the printer's properties without increasing cost or rebuilding construction. As a result, it was able to develop a construction that is able to expand the build area at one axis. It is able to expand the build area with almost no limit in one direction. Expanding dimensions for more than one axis is a goal for the next research.

Keywords Modularity · 3D print · Engineering

1 Introduction

Modern technologies play a very important role in the engineering and manufacturing process. Undoubtedly, 3D print is such a technology. Most manufacturers are focused on production of prototypes, design parts, and medical implants. Small dimensions and high accuracy requirements characterize all these items. This leads to the construction of single-purpose devices, which are not capable of fast printing without need for precision. That fact shows us new market opportunities that are on a field of big-dimensional printing. There are several printers on the market for printing large products, of course, but they are often 'demonstration only' devices. The problem with such large printers is big usage of space. The customer often purchases a printer that is larger than the really needed to satisfy the potential need for large-scale printing. As a result, space is not used efficiently and causes increased costs. Based on the above, this study was done to design a 3D printer's construction

R. Kohár (✉) · M. Stopka · P. Weis · P. Spišák · J. Šteininger
University of Žilina, Univerzitná 8215/1, 01026 Žilina, Slovak Republic
e-mail: rober.kohar@fstroj.uniza.sk

Š. Medvecký et al. (eds.), *Current Methods of Construction Design*, Lecture Notes in Mechanical Engineering, https://doi.org/10.1007/978-3-030-33146-7_55

that can change its dimensions according to the actual need. The goal was to make the dimensions change simple, fast, cheap, and reversible.

2 Methods and Materials

The first step in the engineering process is the selection of print material. It is important that the material is affordable and inexpensive. It is necessary to avoid complicated processing of material. Therefore, the use of materials such as metals and ceramics, which require the use of a laser and a protective atmosphere, are excluded [1]. Their use places great requirements on the sealing of the workspace (often called build area) as well as the more complicated construction. This could cause problems during engineering a modular system. At this moment, we are able to say that the use of plastic is suitable [2].

Plastic processing does not impose any special requirements for workspace sealing or protective atmosphere. In principle, it is just needed to melt the material and ensure its application to the required place. From this viewpoint, it is clever to use fused deposition modeling (FDM) or fused granular fabrication (FGF) method [3].

The FDM method (also known as FFF—fused filament fabrication) is a 3D printing process that uses a continuous filament (a) of a thermoplastic material. This is fed from a coil through a moving, heated printer extruder head (b). Molten material is forced out of the print head's nozzle and is deposited on the growing workpiece (c). The head is moved, under computer control, to define the printed shape. Usually, the head moves in layers, moving in two dimensions to deposit one horizontal plane at a time, before moving slightly upwards (e) to begin a new slice. Figure 1 shows also workspace (f) and supports (d).

Fused Granular Fabrication (FGF) is an additive 3D printing method in which the 3D model is printed by layer. FGF works by melting granular plastics and feeding the resulting mass at a constant speed through a nozzle onto a platform. By changing the nozzle, more or less plastic can be outputted at a time resulting in higher print speeds or finer details. For each layer printed, the printing platform is lowered (or printhead lifted), allowing for the next layer of the model to be added [6].

Figure 2 shows main parts of the printing system. The granulated material is placed in a reservoir (1) and gravitationally fed into the extruder (3). The screw (2) pushes material downwards into the heated part. Material is molten here and flows through nozzle (4). Molten material is placed on the desired place to be cooled. Cooled material becomes solid and creates final product. The product (6) is placed on the workspace (5).

Overall, it is possible to say that these methods are equivalent and differ only in the way how material is fed. Since the FGF method uses a granular material, its transport is practically uninterrupted. This leads to shorter downtime and simplified operation. According to the mentioned advantages, the FGF method was selected as a starting method for the engineering process [7].

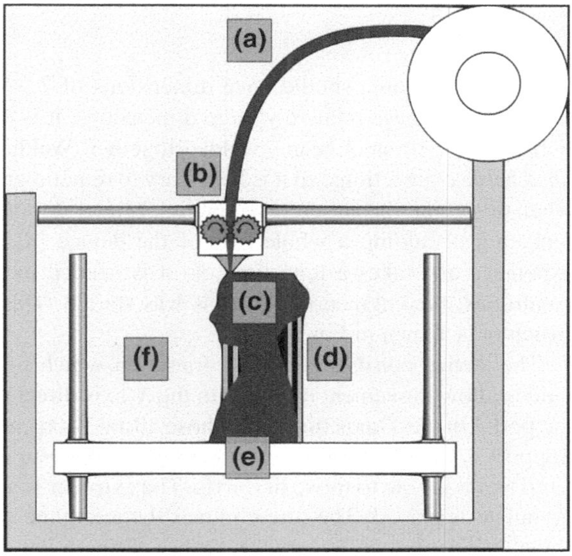

Fig. 1 FDM (FFF) method principle [4]. Legend: **a**—continuous filament, **b**—printer extruder, head, **c**—workpiece, **d**—supports, **e**—upwards, **f**—workspace

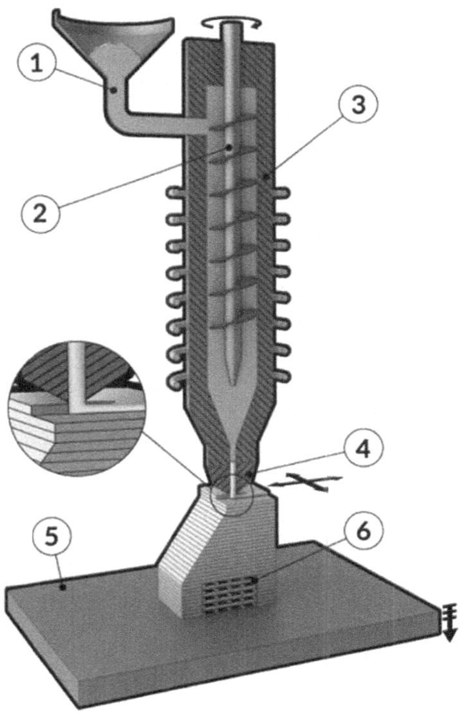

Fig. 2 FGF method principle [5]. Legend 1—reservoir, 2—screw, 3—extruder, 4—nozzle, 5—workplace, 6—product

3 Results

The base build area should have dimensions of 2 × 2 meters to a height of 1 m. According to these relatively large dimensions, it is desirable that the construction will be made of steel beams welded together. Welding technology does not allow detachable connections, so it is necessary to remember the modularity of the solution when designing the structure. Changing printer dimensions needs to be resolved by replacing or adding a whole part of the device [8]. The constructions change is expensive and takes a long time, so it is needed to avoid it. After defining these conditions, the engineering process was started. This resulted in a design of base structure is shown in Fig. 3.

The device consists of a base frame on which linear guides are placed. These guides allow movement of portal in the *X*-axis direction. There are linear guides on the portal in the *Y*-axis direction; those allow *Y*-axis movement of a mounting plate. Support is mounted on the mounting plate. The support allows the extruder to be fixed and it is able to move in *Z*-axis. The extruder is a container in which the plastic granulate is placed. The screw moves the granulate into the heating section where it melts. The molten material is pushed through the nozzle into the required space [10].

A unified extension unit (see Fig. 4) can be used if workspace expansion is needed. This unit is structurally identical to the base frame of 3D printer. It is designed to be able to adjust the frame position using adjusting feet independent to the position of the table. This allows it possible to adjust the plane of the workspace precisely. The extension unit is connected with screw connections after setting a level of the workspace. The portal can be moved from one frame to another on the linear guides then. The only change that needs to be made when expanding the print area is the

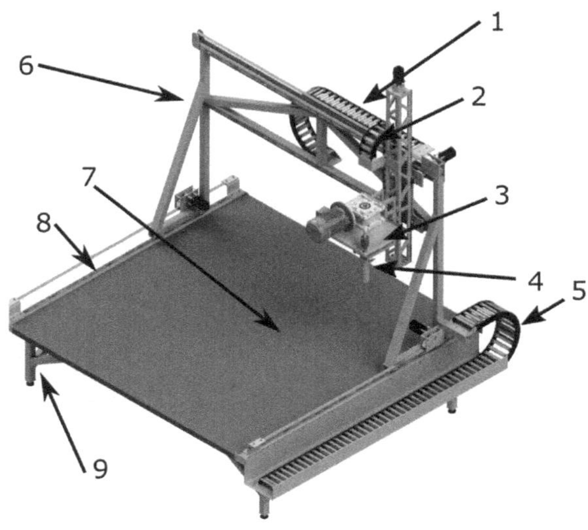

Fig. 3 3D printer's construction overview [9]. Legend: 1—*Y*-axis energy chain, 2—*Z*-axis energy chain, 3—extruder support, 4—extruder/printer head, 5—*X*-axis energy chain, 6—portal, 7—workspace, 8—*X*-axis toothed belt, 9—unified base frame

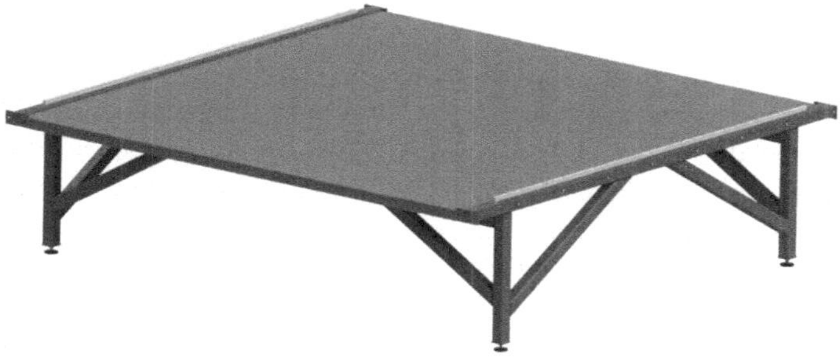

Fig. 4 Unified extension unit equipped by linear guides [9]

use of a longer toothed belt. The servomotor provides movement of each axis. The computer controls all of the servomotors. Energy chains, to transfer energy, realize flexible connection of moving parts.

By adding, one extension unit will result in a workspace with dimensions of 2 × 4 meters. The cable (and also energy chain) length for the *X*-axis is sufficient for use in the basic configuration (2 × 2 meters) even in the extended (2 × 4 meters). If necessary, the area can be further expanded by adding additional unified blocks. In this case, it is necessary to change the length of the cables and also the toothed belt (as mentioned before). Such a solution makes it possible to extend the workspace almost unlimited (Fig. 5).

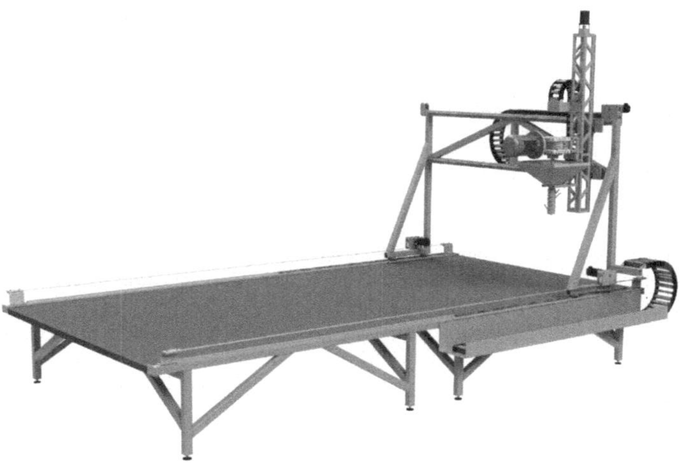

Fig. 5 Extended workspace to 2 × 4 m configuration [9]

4 Conclusion

The goal of this project was design of a 3D printer whose dimensions could be changed as needed. The printer is focused on the fast printing of large-sized objects without a requirement for high accuracy. The printing technology was determined at the beginning and the entire engineering process was then based on it. The proposed solution uses a unified extension unit, which is also the basic framework of the whole device. Using this unit, the basic workspace size 2 × 2 m can be extended in the direction of one axis to almost unlimited length (2 × 4 m, 2 × 6 m …). The use of a unified extension unit requires absolutely no change in construction. In the case of 2 × 4 extensions, it is needed to use a longer toothed belt, in case of bigger extension; the longer cable lengths are needed for the *X*-axis. Extending the workspace in *Y* and *Z* direction requires huge constructional changes and is not possible at this time.

Acknowledgements This study was supported by Cultural and Educational Grant Agency MŠVVaŠ under the contract no. 040ŽU-4/2016.

References

1. Čilík L, Žarnay M (2001) Metodika konštruovania. EDIS, Žilina
2. Kučera L, Gajdosik T (2014) The vibrodiagnostics of gear, in Modern methods of construction design In: Ševčík L, Lepšík P, Petru M, Mašín I, Martonka R (eds) Modern methods of construction design, proceedings of ICMD 2013, pp 113–118. Springer International Publishing, Switzerland
3. Málik L, Medvecký Š, Brumerčík F, Hrček S, Chrzová J, Kohár R, Lukáč M (2013) Konštruovanie II. EDIS, Žilina
4. Scopigno R, Cignoni P, Pietroni N, Callieri N, Dellapiane M (2017) Digital fabrication techniques for cultural heritage: a survey. Comput Graph Forum 36(1):6–21
5. BLB Industries (2017) 3D printers: fused granular fabrication. http://blbindustries.se/3d-printers/fused-granular-fabrication-fgf/, last accessed 10 Dec 2017
6. Jedlinski L, Caban J, Krzywonos L, Wierzbicki S, Brumercik F (2015) Application of vibration signal in the diagnosis of IC engine valve clearance. J Vibro Eng 1:175–187
7. Bastovansky R, Tropp M, Lukac M, Brumercik F (2017) Molybdenum sheet metal test device. Commun, Sci Lett Univ Žilina 1(19):124–127
8. Lukac M, Brumercik F, Krzywonos F, Krzysiak Z (2017) Transmission system power flow model. Commun, Sci Lett Univ Žilina 2:27–31
9. Stopka M (2017) Návrh procesu realizácie veľkorozmerných modelov na báze recyklácie odpadových surovín, Žilina
10. Medvecká I, Binasova V, Kubinec L (2017) Planning and performance evaluation of the manufacturing organizations. Procedia Eng 192:46–51

Optimizing Setting of Open Source Fused Deposition Modeling 3D Printer

Silvester Poljak, Ronald Bašt'ovanský and Pavol Podhora

Abstract This chapter discusses the mechanical properties of samples made using 3D printing using Fused Deposition Modeling. At present, there is a great popularity of cheap RepRap 3D printers. One of these printers is Prusa i3 printers. The advantage of this printer is its low purchase price and low cost of ownership. This 3D printer allows infinitely many options for setting up 3D printing itself. The chapter describes Fused Deposition Modeling (FDM) printing technology, the use of materials, and the effect of setting up padding for printed items. The product is mainly aimed at adjusting the hexagonal charge of the measured samples and then evaluating the strength of these samples. It is known that a hexagonal structure called a honeycomb is the basis for lightweight cores. For usage of this hexagonal structure in conventional application, it is needed to know properties of material with different dimension. Measured values of material properties are evaluated at the end of this chapter.

Keywords 3D printer · Infill · Hexagonal structure

1 Introduction

Nowadays, rapid prototyping technology is experiencing a lot of booming and it is proportionally related to the increasing in materials usable of this technology. Although it exists amount of usable materials, from metals and concrete to biocompatible materials, the most widely used are due to their versatility and price of plastics.

This chapter will describe some of the most commonly used materials for individual forms of production by rapid prototyping, in particular the SLS technologies. The mechanical properties of the samples in different directions of the individual samples will be examined. A brief description of the course of the material tests to be carried out in the practical part will be followed. From the data obtained, the material characteristics are calculated and compared with the values reported by the manufacturer in the material sheet. The results will be analyzed and the reasons for

---

S. Poljak · R. Bašt'ovanský (✉) · P. Podhora
Žilinská Univerzita v Žiline, Univerzitná, 8215/1, 01026 Žilina, Slovak Republic
e-mail: ronald.bastovansky@fstroj.uniza.sk

Š. Medvecký et al. (eds.), *Current Methods of Construction Design*, Lecture Notes in Mechanical Engineering, https://doi.org/10.1007/978-3-030-33146-7_56

possible deviations between measured and reported values will be determined. The most common materials used in 3D printing today include plastics, metals, ceramics, paper, biomaterials, or even food. Each material is uniquely used and is therefore becoming an increasingly frequent combination of multiple materials to guarantee the best mechanical properties of the manufactured part. The most used material, however, is still plastic and mainly due to a wide range of usable types. Each type is characterized by different strength, flexibility, surface finish, or color. Please note that the first paragraph of a section or subsection is not indented. The first paragraphs that follow a table, figure, equation, etc. do not have an indent, either. The drawbacks of printing using this technology are that the parts have different strengths in each direction of printing. This is due to the 3D printing technology itself.

1.1 FDM 3D Printing Technology

FDM technology allows the construction of three-dimensional models directly from 3D CAD software. The technology works on the principle of applying the molten material layer-by-layer using a heat-controlled nozzle. Applying this procedure enables functional testing of prototypes, production of high-quality parts and precise shapes. The advantage of FDM technology is therefore high accuracy, functionality, assembly of prototypes from production materials, durability and service life of parts. FDM technology pushes the model by applying thin layers of molten fiber material. This creates a rigid structure with a laminated surface suitable for models with the use of functional testing, presentation as master models for mold making or directly for the production of final parts. The application of molten plastic (PLA, ABS) or metal wire is deposited in individual layers, the thickness of the individual layers being usually 0.25 mm. The most widely available and most affordable technology we often encounter when presenting prototypes and product samples. The printer produces solid and shape-like models. FDM, or FFF, is the most common and best-known technology for 3D printing. The FDM abbreviation is a registered trademark of Stratasys, so FFF is often used for virtually identical product and prototype production processes. The basic principle of FDM/FFF technology is the extrusion of thermoplastic material by individual layers. This method is similar like process when the molten adhesive material is extruded through the nozzle of hot glue gun. The head of the 3D printer is supplied with thermoplastic material (most often in the form of a string) which is heated to a partially liquid state. The head then extrudes and applies the material in thin layers. As a result of the layering of the solidified material on the previous layer, it is a plastic 3D model. The FDM/FFF process requires the use of support structures for most geometry models (it can not be printed in the air). In general, this means the use of second, water-soluble material that allows relatively easy removal of the support structures once the printing is complete. The advantage of FDM/FFF technology is its affordability and the huge amount of materials that can be used to make 3D models. FDM technology utilizes primary ABS plastics with 40% higher strength than conventional ABS casing, external stability, and high

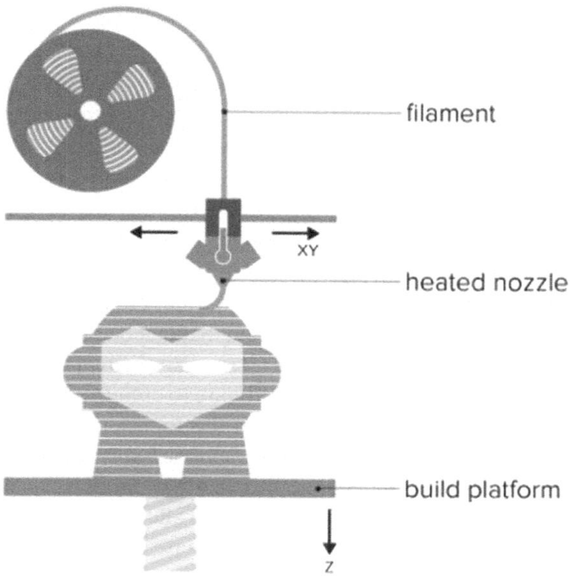

Fig. 1 Fused deposition modeling, 3D printing technology [1]

impact resistance. This material has the necessary features to produce components for final use, functional testing, and presentation with the possibility of using as master models for mold making. At present, biodegradable plastics such as PLA plastics have also been used. PLA plastic is a durable plastic thermoplastic with excellent properties, designed for fast and inexpensive prototyping or finished products. It has a fine structure of material layering. It is not suitable for complex models with complex structure and fine details. PLA is a plastic based on a natural starch base (Fig. 1).

1.2 RepRap Community and Prusa Mendel Printer

RepRap is an international community-based 3D printer project based on open hardware. RepRap is composed mostly of many plastic parts that can be printed on another RepRap. The name itself RepRap stands for replicating rapid prototyper, which means it, is capable of self-replication and rapid prototyping. All documentation needed to build hardware and run RepRap, including firmware and control software, is released under the GNU General Public License under which a range of free software is also released. With total openness and affordability, RepRap has become a very popular project by the worldwide DIY/Community Maker. RepRap consists of a number of components. Besides plastic, which make up the vast majority of the structure, it contains metal rods, screws, stepper motors, bearings, belts, heated glass plate, and jet print head. Step motors are controlled by a Sanguinolol single-board computer. Sanguinolol is the Arduino clone, which is just like RepRap's open hardware. Compared to Arduino, Sanguinolol includes a more powerful microcontroller and electronics for stepper motor control. There are currently four "official" versions of RepRap with the complete manual available and a list of parts needed to build your RepRap. In addition to the official versions, there are many variants on them. Prusa Mendel is based on Original Mendel. Thanks to a simpler building process, it is currently the most widespread modification of RepRap between builders. This is a full-size RepRap that attempts to use cheap and commonly available materials. Currently, there is a third iteration of Prusa Mendel. You can use any CAD or 3D scanner program to create a 3D object model that can be printed on RepRap. The only condition is to support the STL format for storing the resulting model. You can use the FreeCAD or Blender 3D modeling software to create a new model or convert to STL. Models can also be downloaded from the Internet. The most well-known storage site is Thing verse containing models designed specifically for 3D printing and GrabCAD. In addition to manual modeling in CAD, models can be programmed using the CSG solid geometry in OpenSCAD. CSG allows absolute control over the shape of the object and is therefore suitable for precise modeling of parts. The STL format must be trimmed onto individual print layers before printing and calculate the print head movement including the amount of extruded plastic. This is called slicing from English slice. The calculation may take several hours depending on the size and complexity of the model, the performance of the computer processor, and the specific implementation of the slicer. The result is a G-code format file for printing on a specific 3D printer.

1.3 3D Printing Materials Used in FDM Technology

Each print material has general features.

ABS—it is a polymer made from oil, which can be designed with different properties. Generally, however, it is a rigid and slightly flexible plastic having a relatively high melting point. Due to its good mechanical properties, it has a wide use, especially for mechanically stressed components. ABS natural (without dye) has a milky white dye but is also available in different shades. Easily dissolves in acetone, which can also serve as an adhesive for bonding parts or for surface treatment and polishing of the model.

PLA—a natural polymer made from corn, potatoes, or sugar beet. This makes it more environmentally friendly than ABS, even if it does not degrade immediately in your flowerpot or garden. PLA natural naturally releases light but transparent is certainly not. Adding a color creates a wide range of shades. It has worse mechanical properties than ABS, such as worse flexure and lower melting point. For example, a day in a well-heated car can cause a change in shape. Finished models have a nice glossy finish.

Both plastics must be stored in a dry place. Their quality decreases, the longer they are exposed to the environment. Their main disadvantage is that they are able to suck moisture out of the environment, which in turn significantly exacerbates the printing process and reduces the quality of the prints. Absorption of moisture is not very sensitive and almost negligible compared to PLA. For drying, you can simply use a hairdryer or hot air oven. PLA—the moist material tends to produce vapor during the printing at the end of the nozzle of the bubble, which is released by heating from the material. This greatly reduces the accuracy, quality, and strength of the model. When dry, similarly to ABS, it is well served with a hairdryer or oven. Care must be taken, however, that the temperature is not too high and that the material does not deform. In addition to bubbles and precision problems, color fading may also occur. ABS—the biggest disadvantage is that during cooling, the plastic shrinks and changes shape, which is most noticeable especially in larger models. Part of this can be avoided by slow cooling. It is necessary for the model to remain on the hot tray of the printer as long as it is well bonded. Another problem that follows the previous one is the adhesion of the first layer to the mat. This can be ensured by heating the washer to a minimum of 90 °C and forming a thin ABS layer on the pad with a solution that is formed by dissolving the ABS in acetone. (Moreover, we will not forget about this fact.)

PLA—compared to ABS, it exhibits much less deformation, which makes it possible to print more accurate models. Adhesion of the material is also somewhat better but we do not recommend printing without a heated pad, if possible. Sometimes it is enough to spray a layer of hair (very strong) on the mat. One disadvantage, however, is that the material is molten more fluid, which can cause the model to run off when applying layers quickly and inadequate cooling. ABS plastic is a strength and flexible plastic with a high melting point and can be used in a wide range of applications, especially at higher mechanical stresses. Vapors from molten plastic may be unpleasant and deterrent to someone. Its disadvantage is a change of shape when cooling and a worse grip. PLA—a wide range of colors and translucency will appeal to people who prefer interesting toys or household accessories. Also, the natural base and the more acceptable odor of the user get to their side. If it is cool, it is possible to achieve nice sharp edges, lower layer height and higher printing speed. The disadvantage is the worse mechanical properties and the lower melting point.

2 Method for Testing

The tensile test is a basic test of the mechanical properties of the materials [3]. Testing is performed on standardized specimens clamped in jaws of the tearing machine, with the axis of the specimen being aligned with the force axis. A growing force develops on the sample until it breaks down. During the test, the instrument measures the load force and the relative elongation of the rod. By evaluating, the measured data we determine for the material to be tested the slope, the tensile strength, the relative elongation, and the constriction. Test results are used to select the appropriate material for the required application, quality control, and prediction of material behavior under load. For extruded plastic samples, the test is performed according to STN EN ISO 527-1: Plastics. Determination of tensile properties. Part 1: General principles (ISO 527-1: 2012) and STN EN ISO 527-2: Plastics. Determination of tensile properties and Part 2: Test conditions for pressed and extruded plastics (ISO 527-2: 2012).

2.1 Test Specimen for Tensile Strength Testing

The test sample has the most frequent cross-section of square, rectangular, or circular shape. The shape of the samples is defined by the standard due to easily clamped to the jaw of the test equipment. The shape of the measured sample can be seen in Figs. 2 and 4. There is a calibration part of smaller cross-section that ensures initialization of the crack. Test rods are produced in long or short designs [3].

Fig. 2 Fused deposition modeling, 3D print [2]

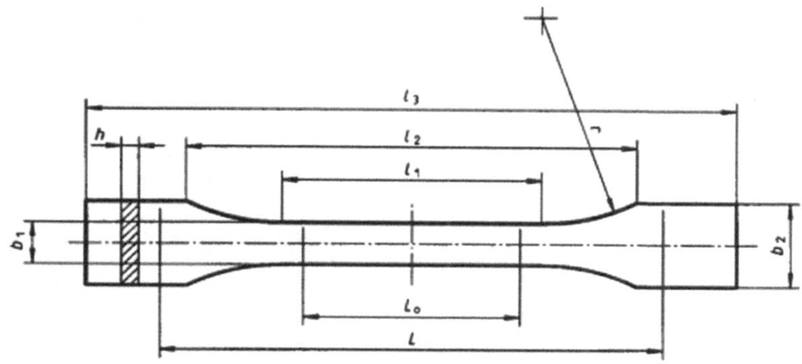

Fig. 3 Test rod with rectangular cross-section

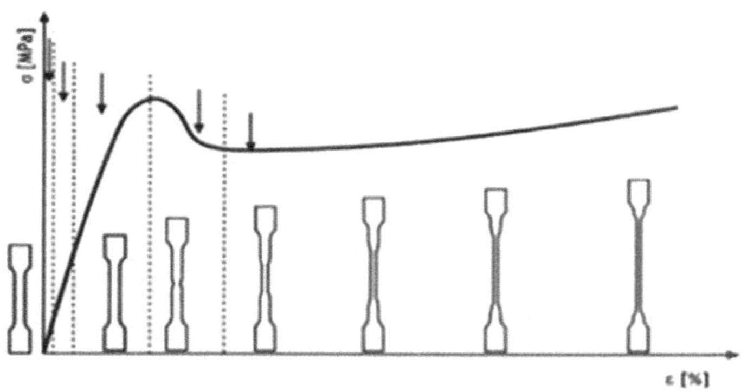

Fig. 4 Deformation curve of thermoplastic materials rectangular cross-section [3]

2.2 *Test Evaluation*

The main result is the Hook diagram, which shows the dependence of the elongation of the bar on the voltage generated. From the graph, in addition to the voltage characteristics, such as the slope Re, the strength limit Rm, the elastic limit Rp, the deformation characteristics determining the plasticity of the material are deformed. Thus, it is possible to determine at what load the material occurs in the area of the elastic deformation, whether it comes into plastic deformation and therefore there is a breakage of the sample. The deformation curve can be seen in Fig. 3. Criteria for assessing deformation characteristics are elongation and contraction [3].

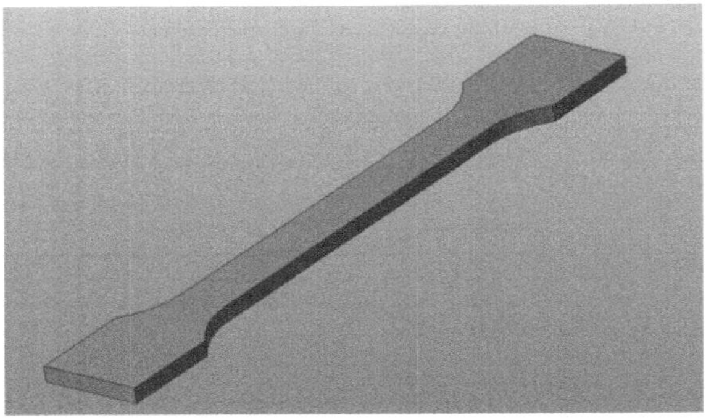

Fig. 5 3D Model of specimen

2.3 Verification of Material Properties by Means of Measuring Instrument

The first step was to model the STD EN ISO 527 in the PTC Creo 3 CAD standardized sample for the tensile test. The sample model has been exported to STL, which is used to work with 3D models in modeling software and 3D printer control. Once the settings have been made in the software, samples were printed on the Prusa i3 printer (Fig. 5).

The sample model has been exported to STL, which is used to work with 3D models in modeling software and 3D printer control. Prusa 3D Slic3r software was used to set different parameters of the tested samples. The percentage of sample filling was changed on samples using the honeycomb structure (hexagon). Percentage of structure ranged from 5 to 100%. After performing the settings in the software, the samples were printed on the Prusa i3 3D printer.P 100 printer. A total of 135 samples were printed out from the PLA material. Samples were printed for 27 settings and 5 pieces of every setting. Due to such decomposition, it is possible to observe the influence of the printing direction on the mechanical properties of the sample. The basic reason for this setting is that we use the different fill settings to influence the strength of the samples as well as the time to be printed. The more the sample is filled, the higher the strength, but of course, the longer the print time is. The less sampled the sample is, the lower the sample strength. The result of these settings is to find the optimum fill for printed samples. These settings apply to the honeycomb infill (hexagonal).

Examples of adjusting the different percentage of honeycomb infill (hexagonal) are shown in Fig. 6.

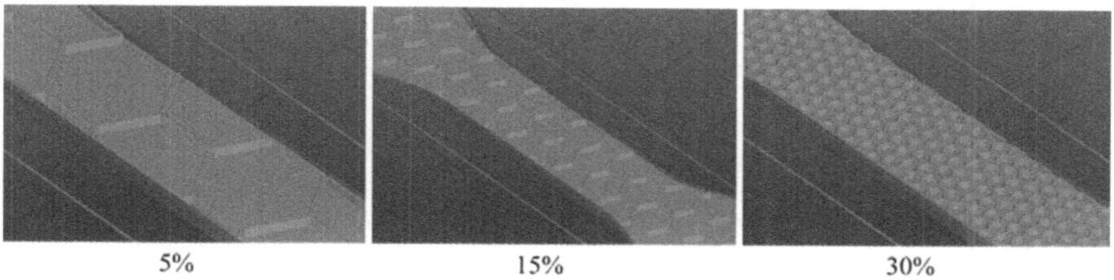

Fig. 6 Samples of specimen with different infill (%)

3 Results and Discussion of the Measured Values

Test was performed on 135 samples. Only the total tensile strength of the individual samples was evaluated. The tension and Young's modulus of elasticity were not judged. This is an overall assessment of the samples, the young modulus of elasticity could be determined only for the filament, which was in our case PLA plastic. For a better explanation of the readings and determine optimum filler, we use evaluative tensile strength. It is the ratio between the total strength of the sample and the percentage of the infill. Only the average measured values are listed in the table. Measured values are in Table 1.

Measured and calculated values are transferred to the graph (see Fig. 5). The graph shows that the best ratio between strength and percent filling has samples with 7% fillings. It can be stated that increasing the percentage of fillers increases the strength of the samples, but also increases the time of sample printing and material consumption (filament). These measurements served as a recommendation for the workers who operate the printer to determine the percentage of fillers to choose for their printed model (Fig. 7).

Table 1 Measured and calculated values

Infill of specimen [%]	0	5	6	7	8	9	10
Force F [N]	300	380	530	630	640	670	685
Ratio F/% infill		7600	8833	9000	6500	7400	6850
Infill of specimen [%]	11	12	13	14	15	16	17
Force F [N]	650	702	770	800	840	860	900
Ratio F/% infill	5909	6416	5923	5714	5600	5375	5294
Infill of specimen [%]	18	19	20	25	30	40	50
Force F [N]	915	1020	1010	1223	1550	1970	2285
Ratio F/% infill	5083	5386	5050	4892	5166	4925	4570

Infill of specimen [%]	60	70	80	90	100
Force F [N]	2660	2855	3340	3580	4000
Ratio F/% infill	4433	4078	4175	43,977	4000

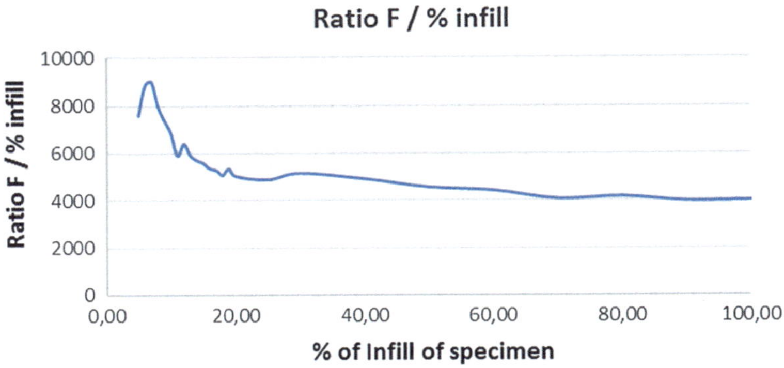

Fig. 7 Graph for explanation of measured values

4 Conclusions

This chapter discusses the suitability of 3D printer settings. This is a cheap 3D printer called RepRap. The measured samples were printed on the Prusa i3 3D printer. The Slic3r program from Prusa Printers was used to generate the G-code for the printer.

When choosing a fill pattern, there are several aspects: object strength, time and material, and personal preferences. It can be deduced that a more complex model will require more strokes and therefore will take more time and material. Slic3r offers several fill patterns, four regular, and three special fill shapes. Some types of models are more suitable for a particular pattern, such as organic or mechanical. From the point of view of the sample types, the six-angular filler structure was chosen which best suit this mechanical part because each hexagon is bonded to the same basic structure of each layer and forms a strong vertical structure. Most models require only low-density fillers, because more than, for example, 50% will produce a very tightly packed model that uses more material than needed. For this reason, the common pattern range is between 5 and 30%, but the model requirements determine which density is best. The figures show how patterns change with respect to density increase.

The topic of this chapter was explained the percentage of filler that has the best properties with respect to the time of printing the samples on the 3D printer and the printing of the material. It can be stated from the chapter that the most suitable setting for solid objects is 7% of the infill. The disadvantage of using a six-angle shield is that it is only uniaxial stress. We can only load the object in one direction. FDM 3D push technology has the disadvantage that objects printed using this technology have low strength, particularly in the z-axis direction, and are concurrently tensile in tensile strength. The strength of the samples in the direction of the x-axis and in the y-axis is comparable. This is because the hexagonal structure is symmetrical about just one axis, not two axes. This effect can be eliminated by setting the fill at an angle of 45° the samples are modeled in the PTC Creo 3 CAD program. Today, manufacturers of cheap 3D printers are working to upgrade their previous versions of devices, including a new 3D hexagonal fill structure that will allow print models to be loaded nearly the same way in all directions.

Acknowledgements This study was supported by Cultural and Educational Grant Agency MŠVVaŠ under the contract no. 040 ŽU-4/2016.

References

1. D Hubs. https://www.3dhubs.com/what is 3d printing. Last Accessed 23 April 2018
2. Prusa Research s.r.o., https://www.prusa3d.cz/. Last Accessed 23 April 2018
3. STN EN ISO 527-2 (2012) Plasty. Stanovenie ťahových vlastností, Časť 2 Skúšobné podmienky pre lisované a vytláčané plasty

The Comparison of Selected Strength Indicators of Manufactured Prototypes Produced by Metal Additive Manufacturing (3D Printing) System

Robert Sásik, Ronald Bašt'ovanský, Michal Hoč, Rudolf Madaj and Peter Spišák

Abstract Laser melting is one of the additive manufacturing methods that uses very strong ytterbium laser ray for the melting of very fine metallic powders to produce fully functioning 3D parts. The production process origins in 3D CAD model that is virtually divided into separate layers (2D curves—25 μm up to 100 μm thin). Sequential application of individual layers of metallic powder and its laser melting within the inert atmosphere creates 3D metallic part. We compared selected strength indicators of samples that were manufactured by the selective laser melting technology on AM 250 machine. The used material was AlSi10 Mg from two different suppliers. The comparison was also important in terms of economic efficiency, since the price difference between the two suppliers was up to 40%. Sample rods were produced by AM 250–200 W laser machine. Samples were cut after the prescribed heat treatment by annealing 300 °C ± 10 °C for 2 h with gradual air cooling.

Keywords Selective laser melting · AM 250 · AlSi10 Mg

1 Introduction

The AM 250 machine from Renishaw company belongs to SLM technology of Rapid Prototyping (renishaw.com). This machine is placed in the Rapid Prototyping Laboratory at the University of Žilina. Laser melting is one of the additive manufacturing methods that uses very strong ytterbium laser ray for the melting of very fine metallic powders to produce fully functioning 3D parts. The production process origins in 3D CAD model that is virtually divided into separate layers (2D curves—25 μm up to 100 μm thin). Sequential application of individual layers of metallic powder and its laser melting within the inert atmosphere creates 3D metallic part. Argon is the gas used for creation of the inert atmosphere. The finished part is ready for the surface post-processing right after the removal from the 3D printer [1, 2].

R. Sásik (✉) · R. Bašt'ovanský · M. Hoč · R. Madaj · P. Spišák
University of Žilina, Univerzitná, 8215/1, 01026 Žilina, Slovak Republic
e-mail: robert.sasik@fstroj.uniza.sk

R. Madaj
e-mail: rudolf.madaj@fstroj.uniza.sk

Š. Medvecký et al. (eds.), *Current Methods of Construction Design*, Lecture Notes in Mechanical Engineering, https://doi.org/10.1007/978-3-030-33146-7_57

Table 1 Technical data of AM 250 machine [1]

DATA	AM 250
Max. dimensions of the Produced part	250 × 250 × 300 mm (*X*, *Y*, *Z*) *Z*-axis is stretchable up to 360 mm
Production speed	5–20 cm^3 per hour
Scan speed	Op to 2000 mm/s
Positioning speed (max.)	7000 mm/s
Layer thickness	25–100 μm
Laser ray diameter	70 μm at the surface of the powder
Laser beam power	200 or 400 W
Outer dimensions of the machine	1700 × 800 × 2025 mm (l, w, h)
Total mass	Brutto 1225 kg, netto 1100 kg
Power supply	230 V, 1 phase, 16 A
Available materials	Stainless steel 316L and 17-4PH, tool steel H13, aluminum Al–Si–12, titan CP, Ti–6Al–4 V and Ti–6Al–7Nb, cobalt–chrome alloy (ASTM75), Inconel 718 a 625

The typical applications of the laser melting technology presented also in [3, 4] are as follows:

- The manufacturing of the prototype parts suitable for functional tests.
- The manufacturing of implanters and very complicated shapes and parts.
- Single piece and small series production of complex parts from specific materials (Table 1).

The production and usage of Al–Si–Mg components: AL–Mg–Si alloys are hardenable alloys for molding, characterized by lower strength, but with higher corrosion resistance and higher toughness compared to, e.g., Al–Cu–Mg alloys.

Industrial alloys can be divided into two groups [5]:

1. Alloys with excess of Mg. As a result of the amount of Mg required to form an appropriate amount of Mg2Si, the corrosion resistance increases, the strength properties are reduced, and the deformability is reduced compared to the alloys where the same amount of Mg2Si is present in intermetallic phase, but there is no excess of Mg in the final alloy
2. Alloys with excess of Si. Because of the Si excess, the strength properties of the alloys are increased without reducing the forming ability and without deterioration of the weld ability. The resistance of these alloys to inter-granular corrosion is reduced in the final part.

Material properties AlSi10 Mg AM 250,200 W 25 micron [1] are as follows:

- Low density (suitable for lightweight components).
- Good specific strength (weight strength).
- Good thermal conductivity.

- Excellent electrical conductivity.
- It responds well to post-processing adjustments.

AM 250 machine allows fast drain out of the air from the working chamber, which is afterward filled with pure argon gas, which ensures clean environment so we can use reactive materials, e.g., titan. In this case, the percentage of oxygen in a working chamber has to be very low. The argon consumption is minimized by the use of perfectly tight and professionally welded chamber, which in a meanwhile supports the robustness of the process. It is also possible to treat non-reactive materials in the atmosphere from nitrogen gas [1].

AM250 uses external cartridge of the metallic powder with flap gate, which allows refilling of the material during the production process. The cartridge with material can be removed when it is necessary to clean up the machine or replace with cartridge, with different materials. The storage cartridges for the material, which drops out during the production process, are placed aside from the working zone, and they have their own flap gates. This way it is possible to reuse the material during the printing process. The manipulation with the powder and the produced part itself is possible by using protective glows which are placed in the door to the working zone. The filter protects the user from the emissions that originate from the printing process [1].

2 Materials and Methods

A tensile test belongs to tests of mechanical properties tests that determine the basis of the mechanical properties used to assess the quality of the material, design calculations, and the general assessment of the suitability of certain technological operation. The basic mechanical properties of the material are thus determined by the tensile test.

STN EN ISO 6892-1 governs the static tensile test in the Slovak Republic: 2010 (42,0310), Metallic materials. Tensile test Part 1 Tensile test at ambient temperature (ISO 6892-1: 2009). The principle of the test shall be the static load of the test specimen of prescribed shapes and dimensions. The test bar is loaded by pre-defined velocity and smooth stroke of the traverse. It deforms usually up to the breakage. The specimen is clamped into the test jaw such that the main axis of sample matches the axis of the applied force. It is usually tested at ambient temperature in the range of 10–35 °C.

In order to define the stress and deformation characteristics of the material by the static tensile test, the sample material needs to be shaped. The shape, dimensions, and preparation of the test rods are made according to STN EN ISO 6892-1. The test bars are divided into two categories: short and long. These can have either a square, retangular, or circular cross section.

We compared the mechanical properties of samples of metallic material delivered from two different suppliers. The composition of AlSi10 Mg was declared to have

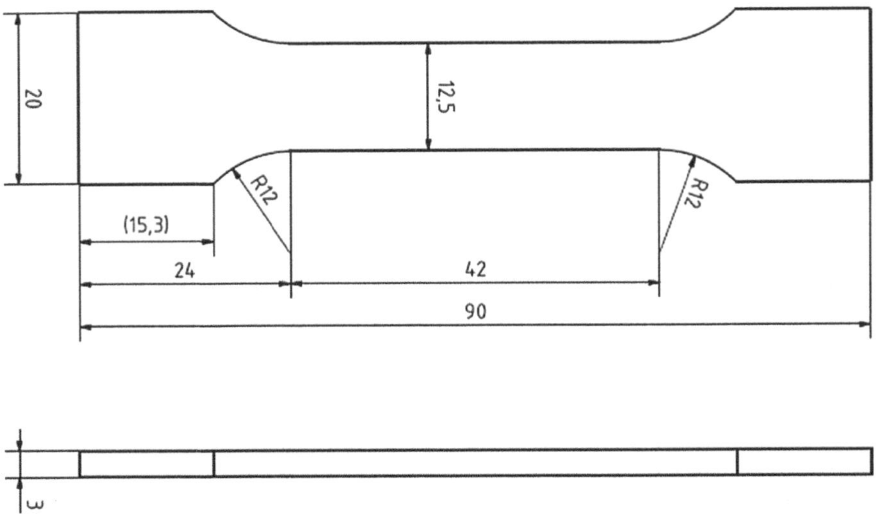

Fig. 1 We checked the size of samples for the blasting tests. Pre-defined size of sample

similar properties for both material samples. The comparison was also important in terms of economic efficiency, since the price difference between the two suppliers was up to 40%. Sample rods were produced by AM 250,200 W laser machine. Samples were cut after the prescribed heat treatment by annealing 300 °C ± 10 °C for 2 h with gradual air cooling (Figs. 1 and 2).

A static tensile test was performed on a LaborTech—LabTest 5.20ST device. The parameter settings for the production of set of samples by laser 3D printer are following:

- Layer thickness—25 μm
- Laser focus diameter—Laser focal length 0.2 mm
- Laser speed consists of the following partial speed parameters:
- Border speed—defines the average laser speed during bending of marginal contours of the model 0.50 m/s.
- Support speed—defines the average laser speed during firing 0.50 m/s support structures.
- Hatch speed—defines the average laser speed while burning the internal volume of the model 0.50 m/s.
- Idle speed—the speed of laser transfer from one point of burning to another, and then the laser does not burn 1.08 m/s.
- Recoating time—the time needed to feed the material, its accumulation—the time needed to prepare the next layer 9.2 s.
- Used Hatch style—Stripes. Layers are divided typically into 10 mm strips. Thus the residual voltage is reduced.

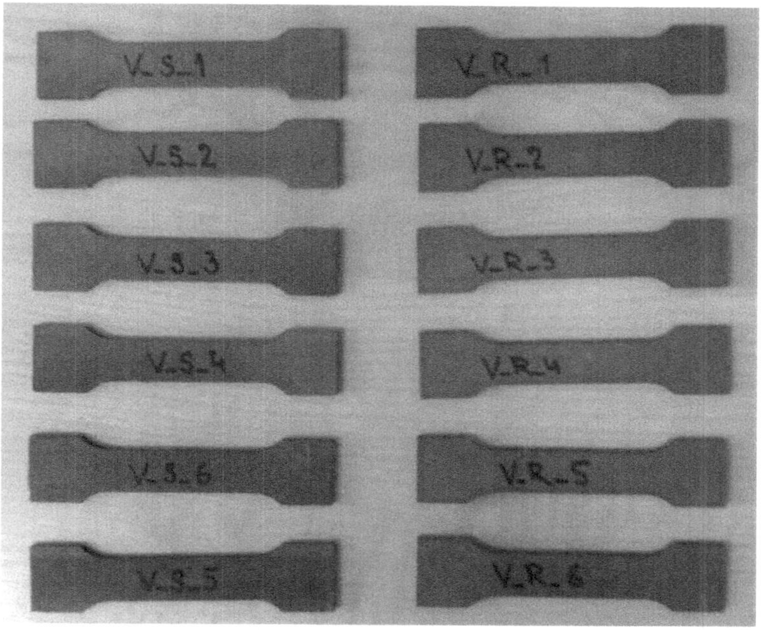

Fig. 2 We designed one set of samples from each manufacturer (V_S and V_R). Each set contains ten pieces of samples

3 Results

Factors affecting the quality of prototypes are as follows:

- Build speed is dependent on alloy properties—and the way laser energy is absorbed.
- The laser energy can be controlled by modulating the laser exposure time and distance between exposures.
- Generally, the more power available, the faster the build rate will be.
- Different parameters can be optimized for the skin (volume border) and core (volume area).
- Sufficient energy must be transferred into the material to achieve the correct melt characteristics; laser power, beam spot diameter, pulse frequency, and exposure time will determine the energy available.
- The material properties—such as reflectivity, conductivity, and powder size will govern energy transfer, and the effect how fast a particular material can be built.
- The geometry of the part will also have an effect on build rate, and as to optimize the desired properties (good surface finish on the skin, and high density when filling) different laser exposure times are used.
- Structural defects—micro and macroporosity.

Table 2 Resulting average values and comparison of the strength limits

The material supplier	Max. Force [N]	Failure strength Rm = F/A [MPa = N/mm^2]
V_R	8461.56	217.00
V_S	9144.93	234.40

The resulting average values and comparison of the strength limits are in Table 2, and in the static test, diagrams are presented in Figs. 3 and 4.

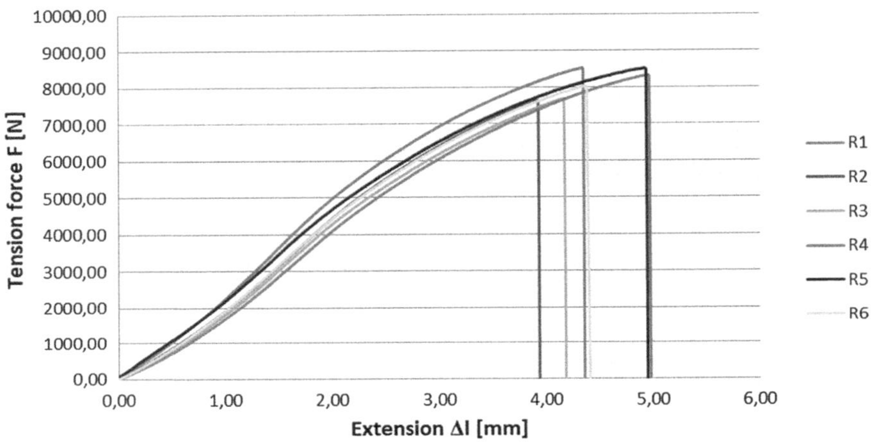

Fig. 3 Resulting values of the V_R samples

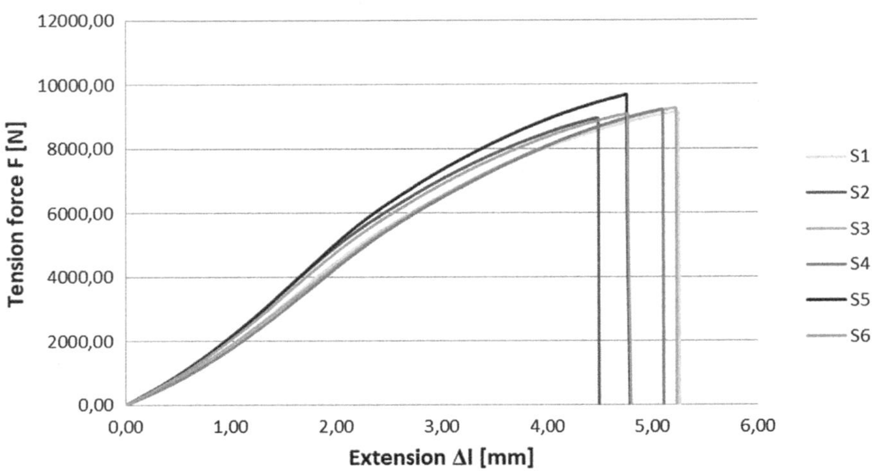

Fig. 4 Resulting values of the V_S samples

The microstructural analysis was done on sample top in cross and longitudinal section. Figures 5 and 6 present the typical microstructure of all analyzed samples, the layers in 25 microns distance approximately, possibility of the porosity, and the structure called as spot contours.

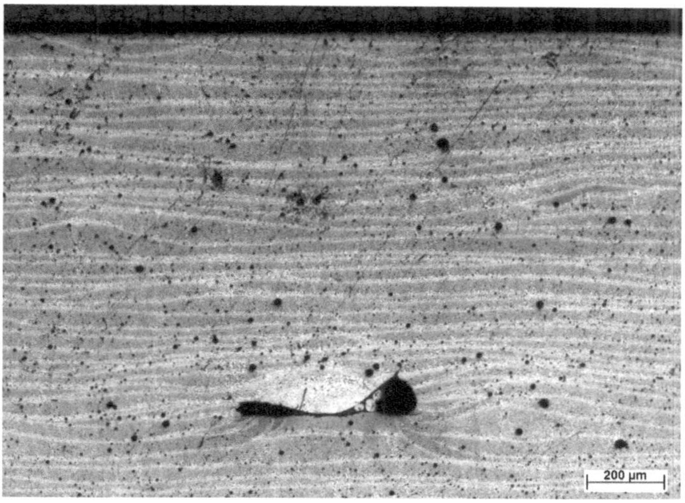

Fig. 5 Microstructure of AlSi10 Mg, sample top in longitudinal section, layers in distance 25 microns approximately and porosity

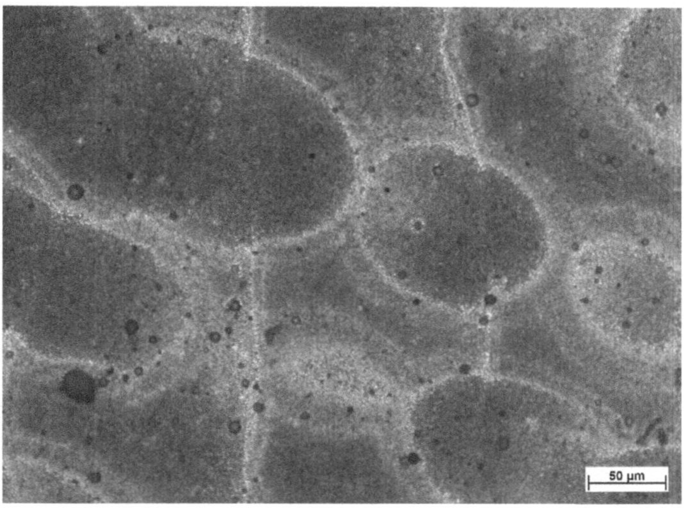

Fig. 6 Microstructure of AlSi10 Mg, sample top in cross section, the structure called as spot contours

4 Conclusion

Since both indicators of strength constants are similar, it is possible to use the AlSi10 Mg material from both suppliers. Even though the material supplier under the V_S label has a 40% better price to streamline prototyping pricing, it has got even more appropriate material in terms of strength indicators.

Additive production of the metallic prototypes in the field of Rapid Prototyping is a progressive method that will be used in many areas of industrial production. Every product, which is produced by this method, is fully functional and individual. The use of this technology is also very promising in the field of biological implants where each patient can obtain its individual implant, which can contribute to the better life of the individual. Automotive industry and customers industry are also interested in the implementation of this technology into their production process [4].

Acknowledgements The research is supported by the Cultural and Educational Grant Agency of the Ministry of Education, Science, Research and Sport of the Slovak Republic under the project No. KEGA 046ŽU-4/2018.

References

1. Renishaw, plc.: Additive manufacturing—laser melting. http://www.renishaw.com/en/laser-melting-metal-3d-printing-systems–15240. Last Accessed 22 May 2018
2. Šarík M, Hudák R, Živčák J (2013) Analýza tepelných procesov pri priamom laserovom sinterovaní titánového prášku. Transfer inovácií 25:249–254
3. Mašín I (2012) Prehled technik využívaných pri Rapid Prototyping. Technická univerzita v Liberci, Czech Republic
4. Osakada K, Shiomi M (2006) Flexible manufacturing of metallic products by selective laser melting of powder. Int J Mach Tools Manuf 46(11):1188–1193
5. Skočovský P, Palček P, Várkoly R, Konečná L (2000) Konštrukcné materiály. Slovak Republic, EDIS

Study of EDSM for Additive Technology in a Smart City

Radek Teply

Abstract Additive technology can cover both manufacturing and user segments. Common usage in households is only in its early stages and far from overshadows of industrial use. Additive technology offers economic benefits as well as simplification of logistics, minimization of inventories, guarantee of time flexibility, availability of spare and complementary parts, which redefine the lifetime of a common equipment. Individuality embedded into the production process, where any part produced by this technology can be customized according to customer's specifications (within the certain limits), allows direct influence on the user's surroundings. Innovative work that promotes uniquely creative design, as well as standardly shared templates, chosen materials, and their recycling, brings a range of options, which also entails certain risks. For the Smart City environment, it is important to secure sustainable efficiency for contained production systems. Utilizing methodologies such as EDSM [1] provides powerful tools for such technology management. To properly reflect the dynamic changes in a modern city, it is necessary to apply the principles of Industry 4.0. By implementing complete life cycle management, it should be possible to provide a secure future for additive technology as an indispensable part of a Smart City.

Keywords Additive technology · Industry 4.0 · EDSM

1 Introduction

Additive technology is more and more shifting from the world of design products, mock-ups, and prototype production into the world of standard production and everyday life. Size and possible placement of these devices make this technology a very flexible tool. They can be placed on a desk in an office or in a production hall. In health care, it is used in the development of artificial bone, vascular prostheses, or organs. For cosmonauts, it could repair and create tools, when supply is out of reach. Thoughts about constructing bases on other planet or spaceships built from stardust

R. Teply (✉)
University of West Bohemia, Univerzitní 22, 30614 Pilsen, Czech Republic
e-mail: teplyrad@kks.zcu.cz

Š. Medvecký et al. (eds.), *Current Methods of Construction Design*, Lecture Notes in Mechanical Engineering, https://doi.org/10.1007/978-3-030-33146-7_58

are mostly theoretical for now. However, the future is in favor of this idea. Nowadays, companies offer printed sand molds for foundry, printed buildings in the form of a rough concrete structure, printed food, clothing, including dresses and footwear, as well as modern jewelry, are all today a reality of additive creation.

2 Principles Used for Additive Technologies

The possibilities of design and principle solutions are quite extensive. Today, additive technologies commonly include technology known as stereolithographic, laminated manufacturing, fused deposition modeling, laser sintering, etc. The diversity of this construction method ranges from manually operated devices for designers and hobbyists, representing the lowest level of automation, when a person both moves and regulates the material flow, up to fully automatic machines (Fig. 1).

Therefore, it is important and appropriate to use a methodology, taking into account the entire life cycle and make it easier to achieve suboptimal solutions in all stages. Methodology can offer synopsis of current or future options. Currently, methods of inserting energy into the materials such as electric resistance, laser, and ultrasound are used. Steppers motors are mainly used as actuators for the movements of such devices. Various sensors can be selected for calibration and measurement of dimensions. Sensors for measurement can be contact or contactless. At a higher level of appliance can be found Laser, Camera, and 3D scanner technology as measuring equipment. By using measurements, the parameters of the manufactured component can be corrected during production. Measured data can indicate inconsistencies with the embedded data and save as from producing the wrong piece. All calibration values can be edited by the user in real time with the use of IoT technologies without interference with correction algorithms.

2.1 Print Materials and Surfaces

An advantage is also the variety of materials for which this technology can be used, e.g., plastics (photopolymers, thermoplastics …), organic materials in the form of paper mixture, metal powders, and string. With the possibility of full-color printing as well as printing for finishing, colors must be added.

Therefore, the logistics and management required to run the production facilities have to guarantee the necessary quantity of raw materials in the connected containers. Today, they are often stored in the form of string, belts, powder, or liquids. At a time when fuel can be created from air and long-distance communication can work without delay, science brings an expectation of efficient recycling technology. Recycling of plastic, such as PET, is often used in domestic conditions. It can be obtained not only in granulate form, but also from the waste materials and from previous creations. Material degradation and transformation processes do not allow a completely closed

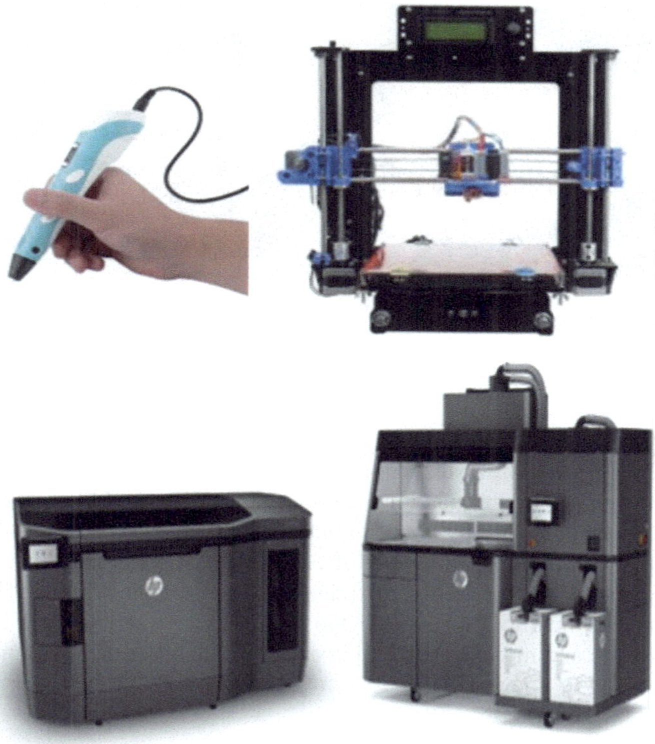

Fig. 1 3D printing tools [2, 3]

loop. However, it shows one of the possible paths that can be taken. Recycling will move more often into our homes.

2.2 Benefits in Food Efficiency

Consumed food has a large influence on the health of a population. Each point of "printed" food can be individually baked or frozen. This offers not only the possibility to choose precise nutritional parameters, but also the consistency and the ratios of the ingredients. The possibility of printing a dessert for coffee will in future be a part of everyday life. Essential for this technology is the possibility to offer more efficient use of resources, significantly reducing the amount of wasted food. Implementations of these technologies will significantly simplify logistics and distribution. It will also lower storage capacity requirements (Fig. 2).

Fig. 2 3D printed food [2]

2.3 *Design*

Additive technologies allow substantial saving of materials, labor, and machine time. They open up new possibilities for creating shapes that would be very difficult or impossible to manufacture by conventional means. The design that future technology can offer will be more limited by the creativity of the creator than the cost of technology that would be needed to create them by standard means. That in itself presents many difficulties especially in the area of security. However, with regard to current software and monitoring technology, there is no reason to assume that such situations would be common.

3 Implications for Smart City

Production through additive technologies is clearly one of the future components of Smart City. This technology serves not only for manufacturing machine parts and components, but also for the production of artworks and for clothing and food. Industry 4.0 and higher support from the Internet of Things (IoT) are the basic tools of a Smart City, where information connects technology with users and gives an overview of the information and material flows. It can take into account storage areas, material flows, machinery implemented in current production, trends, or the product life cycle itself [4, 5]. With the advent of modern technology, complex methodology and lifecycle management will be an absolutely key issue. The component can be updated and produced according to the database and current situation. Theoretically, in specific cases, routine operations, including routine repairs and needs, could be carried out just by using additive technology. Basic repairs prolonging life of product are part of the way to support ecology as well (Fig. 3).

Fig. 3 Repairs with 3D printer [3]

4 Solving Methodology Based on EDSM

Reflecting on the needs and requirements of a user is a function of Industry 4.0 [2]. Properties are according to the methodology given by the design of construction objects [6]. Represent by constructional factors like inputs, outputs, and influences of the environment with the influence of the constructor system operator on the construct process (operands), moral obsolescence the use of knowledge of the transformation process [7] and EDSM [1]. A prerequisite for the good functioning of a system is therefore the communication and knowledge of the state of information, before and after the application of intervention.

The functionality of a system, which represents decentralized domestic production, requires a system (methodology) that must allow functionality and reliability under all foreseeable conditions. This applies for individual dwellings and for a whole city system as well. In this respect, AEDSM [7] and the concept of the Smart City are a very important prerequisite for the sustainable future of urban areas. One of the advantages of these technologies is the potential to greatly reduce the amount of work. The quality of each product is always determined in accordance with the requirements of the product [8]. The term "quality" is largely determined by the point of view from which it is judged. Therefore, the quality can vary considerably to achieve suboptimal results. By means of methodology, freedom of suitable options can be granted. Methodology should not allow create potential safety hazard features in working conditions (like gun parts, explosives, etc.).

5 Example

Constructional design for home-printed appliance focuses mainly on targeted purpose. Not thinking through transformation processes or construction maps. Essential goal is not created to secure living standards or needs, but put such possibility in use. In future, however, it will be mainly standard tool with needs of methodology (Fig. 4).

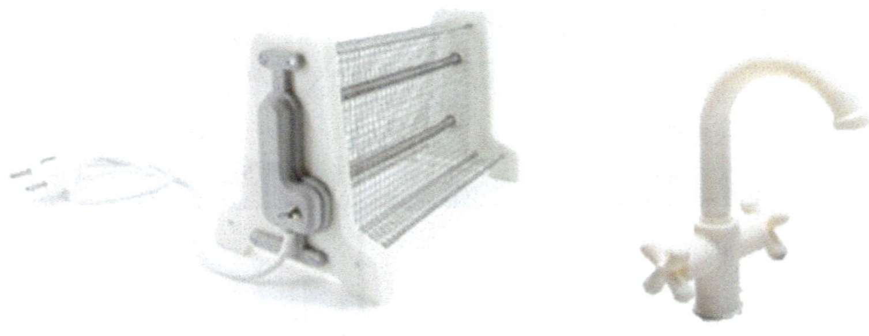

Fig. 4 Example of printed parts: toaster, faucets, [3]

6 Conclusion

Additive technology allows reduction of inventory, utility items, and spare parts. Availability is limited only by production time and the amount of raw material. Further, additive technology allows considerable simplification of logistics. This technology is therefore a suitable tool for reducing the burden of Smart City infrastructure. The methodology contributes to the efficiency and rationalization of the deployment of this technology throughout the life cycle. It is already possible to find companies that are successfully developing individual technologies. In the future, however, these technologies will be even more important. In order to secure individual and complex diversity which are instinctively specific for humanity.

Acknowledgements This article was created within the SGS project.

References

1. Eder WE, Hosnedl S (2010) Introduction to design engineering: systematic creativity and management, 1st edn. CRC Press, Taylor & Francis Group, Boca Raton, Florida
2. MM spectrum. https://www.mmspektrum.com. Last Accessed 12 Jan 2018
3. All3DP 2018. https://all3dp.com. Last Accessed 12 Jan 2018
4. Hubka V, Eder WE (1988) Theory of technical systems, 1st edn. Springer, Berlin Heidelberg
5. Hubka V, Eder WE (1996) Design science. 1st edn. Springer, London
6. Blessing LTM, Chakrabart A (2009) DRM, a design research methodology. Springer, London
7. Hosnedl S (2014) System design engineering of technical products, 2nd edn. University of West Bohemia, Pilsen
8. Eder WE, Hosnedl S (2010) Design engineering: a manual for enhanced creativity, 1st edn. CRC Press, Taylor and Francis group, New York

Development of Process Parameters for SLM Processing of AlSi7Mg Aluminum Alloy

Josef Zvoníček, Daniel Koutný, Libor Pantělejev and David Paloušek

Abstract The paper presents three steps of process parameters development of aluminum alloy AlSi7Mg0.6 for selective laser melting (SLM) additive manufacturing process. Commonly, the components of AlSi7Mg0.6 aluminum alloy are produced by casting. However, to speed up the development phase of new products, often the SLM prototyping is used and thus the processing parameters for this material are required. To develop process parameters of new alloy, single weld tracks were used followed by thin walls and cuboid samples. The cross sections and continuity of single-track welds were evaluated within the range of 200–2000 mm/s laser scanning speed and 175–400 W laser power. The tracks with good size and quality were chosen for further analysis. To minimize the porosity of material, cuboid samples with different overlap of weld tracks were evaluated. The overlap of approximately 80% was found to produce suitable samples with porosity below 0.5%. Finally, the mechanical properties of samples without heat treatment evaluated by tensile testing reached 380 MPa of ultimate tensile strength (UTS), which is slightly higher than cast alloy in T6 state.

Keywords Selective laser melting · Aluminum alloy · Porosity · Mechanical properties

1 Introduction

The selective laser melting is a layer-based technology using metal powder and high power laser beam to form components according to 3D CAD data. Using layer by layer fabrication, it is possible to produce components of complex shape with containing internal cavities or filled with lattice structure which can reduce the weight of the part while maintaining its mechanical properties. This approach has the potential to overcome conventional production technology.

The investigated AlSi7Mg0.6 alloy is specifically used for components in the automotive and aerospace industries. Although it is one of the certified alloys for aviation

J. Zvoníček · D. Koutný (✉) · L. Pantělejev · D. Paloušek
Brno University of Technology, Technická, 2896/2, 61669 Brno, Czech Republic
e-mail: daniel.koutny@vut.cz

Š. Medvecký et al. (eds.), *Current Methods of Construction Design*, Lecture Notes in Mechanical Engineering, https://doi.org/10.1007/978-3-030-33146-7_59

and aerospace, there are not many published information about AlSi7Mg alloy processing by SLM. The most investigated silumin alloys are AlSi12 and AlSi10 Mg [1–7].

Kang et al. [4] tested an AlSi12 alloy made by blending a powder of pure aluminum and pure silicon. They aimed to compare the strength characteristics of this eutectic alloy using a powder made from a mix of pure elements and a powder made from an eutectic alloy. The mixed powder reached a UTS of 273 MPa. Compared to the UTS when using a 325 MPa eutectic powder, this value is about 16% lower. Compared to conventionally cast AlSi12 material, whose UTS is 192 MPa, it is about 42% higher.

Prashanth et al. [5] tested the mechanical properties of the eutectic AlSi12 alloy. The achieved UTS was 380 MPa, which is almost twice as much as that for conventionally cast material. Read et al. [6] examined the strength characteristics of the AlSi10 Mg alloy. Using own optimal process parameters, they achieved a minimal porosity of 0.7%. The measured UTS of a horizontal sample was 335 MPa.

Kimura et al. [7] published a paper about the effect of the percentage of silicon in the Al–Si alloy on the mechanical properties in the tension. The results show that with the increasing percentage of silicon in the Al–Si alloy, the UTS increases, but the elongation decreases. These results apply to all tested alloy compositions except AlSi1 alloy, where cracks appeared in the microstructure and reduce its mechanical resistance.

In different study, Kimura et al. [8] examined similar alloy, the AlSi7Mg0.3 and published mechanical properties with a UTS of 400 MPa, a yield strength (YS) of 200 MPa, and a 13% elongation. Rao et al. [9] examined the AlSi7Mg0.6 alloy and the effect of each process parameter on its relative density. They achieved a relative density of 99.79% $\pm$ 0.43% for the process parameters of laser power (LP) 300 W, laser speed (LS) 2000 mm/s, 0.1 mm of hatch spacing (HS), and a layer thickness (LT) of 0.03 mm. The achieved mechanical properties UTS of 426 MPa, a YS of 280 MPa, and a 10% elongation were over performing those reachable by casting.

The effect of scanning strategy on the resulting relative density of samples was examined by [10–12]. The Aboulkhair et al. [10] achieved the highest relative density of 99.77% using the pre-sinter scanning strategy. This is a one-way strategy that is used on a single layer twice, and the laser power is set to half power for the first time. Thijs et al. [12] used various scanning strategies to produce the AlSi10 Mg alloy. The highest density of 99.4% was achieved with a bidirectional scanning strategy where each layer was scanned twice with a 90° rotation.

2 Materials and Methods

The powder material used in this study is AlSi7Mg0.6 (A357 according to the US Aluminum Association) which was supplied by SLM Solutions (Lübeck, Germany). The chemical composition of the alloy according to EN 1706: 2010 is shown in Table 1 as well as the values measured by supplier. The chemical composition was verified externally before the experiments using ICP-OES Thermo iCAP 6500-ICP spectrometer.

Table 1 Chemical composition of the powder in mass %; e-each; t-total; b-balance

Specification		Al	Si	Mg	Cu	Ti	Fe	Mn	Zn	Others
EN 1706:2010	Min	b	6.5	0.45	0	0	0	0	0	0
	Max		7.5	0.7	0.05	0.25	0.19	0.1	0.07	e 0.03, t 0.1
Supplier		b	6.65	0.47	<0,01	0.01	0.13	<0,01	<0,01	e 0.03, t 0.1
ICP analysis		b	6.45	0.442	<0.01	<0.01	0.1234	<0.01	<0.01	–

The particle size distribution was measured by Horriba LA-960 analyzer. The powder has uniform Gaussian distribution, and the particle size ranges from 13 to 152 μm. The most frequent particle size is 42.1 μm, and the mean particle size is 46.3 μm. The powder contains 10% of the particle smaller than 27 μm and 90% are below 68 μm.

Fabrication of all material samples was carried out on the SLM 280 HL machine (SLM Solutions, Germany). The machine is equipped with Yb fiber laser (maximum power 400 W, beam spot diameter 83 μm). All experiments were carried out in a nitrogen atmosphere with oxygen content below 0.2%. Platform heating was set to 120 °C.

The porosity of samples was analyzed in horizontal planes (perpendicular to build direction) for chosen samples vertical sections were analyzed as well. Samples were ground by a metallographic method, and surface images were captured by light microscope. The porosity evaluation was performed in ImageJ software. The tensile samples were SLM fabricated as the prisms of 13 × 13 × 75 mm and subsequently machined according to DIN 50125 with a gage size of Ø8 and 40 mm length. The tensile test was performed on a Zwick machine with a maximum load of 150 kN.

3 Results and Discussion

3.1 Single-Track Weld Test

For the analysis of the behavior of the individual combinations of process parameters, a single welding test was performed, and their evaluation was processed in the form of a table (see Fig. 1) showing good quality (Group 3) and low quality (Group 1) welds. Within the test, the laser power ranged from 175 to 400 W and a scanning speed changed from 200 to 2000 mm/s. The resulting input energy density was from 3.5 to 20 J/mm^2.

To guarantee a layer thickness of 50 μm, a block of 20 × 50 × 10 mm was fabricated with general aluminum parameters, follow up powder layer was applied,

Energy input (J/mm^2)		Scanning speed (mm/s)																		
		200	300	400	500	600	700	800	900	1000	1100	1200	1300	1400	1500	1600	1700	1800	1900	2000
Laser power (W)	175	17.5	11.7	8.8	7.0	5.8	5.0	4.4	3.9	3.5										
	200	20.0	13.3	10.0	8.0	6.7	5.7	5.0	4.4	4.0	3.6									
	225		15.0	11.3	9.0	7.5	6.4	5.6	5.0	4.5	4.1	3.8								
	250		16.7	12.5	10.0	8.3	7.1	6.3	5.6	5.0	4.5	4.2	3.8							
	275			13.8	11.0	9.2	7.9	6.9	6.1	5.5	5.0	4.6	4.2	3.9						
	300			15.0	12.0	10.0	8.6	7.5	6.7	6.0	5.5	5.0	4.6	4.3	4.0	3.8				
	325				13.0	10.8	9.3	8.1	7.2	6.5	5.9	5.4	5.0	4.6	4.3	4.1	3.8			
	350				14.0	11.7	10.0	8.8	7.8	7.0	6.4	5.8	5.4	5.0	4.7	4.4	4.1	3.9		
	375					12.5	10.7	9.4	8.3	7.5	6.8	6.3	5.8	5.4	5.0	4.7	4.4	4.2	3.9	
	400					13.3	11.4	10.0	8.9	8.0	7.3	6.7	6.2	5.7	5.3	5.0	4.7	4.4	4.2	4.0

Group 1
Group 2
Group 3

Fig. 1 Process map of the single-track welding test with evaluated welds

and single-track welds were fabricated with different parameters onto such block. A total of 120 single-track weld samples were built. By comparing the size and quality of the welds, three groups of welds were specified (see Fig. 2).

The first group included high and low power inputs. For high input energy values, deep depth and wide weld widths were typical. Low input energy curves were characterized by low height, non-continuity, and inappropriate angle of contact with the underlying block. The second group contained welds that met the dimensional condition but was non-continuous. The third group of welds included dimensionally suitable and continuous welds.

The analysis of the results showed that with increasing scanning speed at constant power, the width, and depth of the weld decreases. This is due to the reduction of the energy density flowing into the process and the resulting formation of a smaller molten welding bath. An example of graphs is shown in Fig. 3. For further research, process parameters were selected from the group 3 of Fig. 2 within a LP range of 325–400 W and a LS of 900–1300 mm/s.

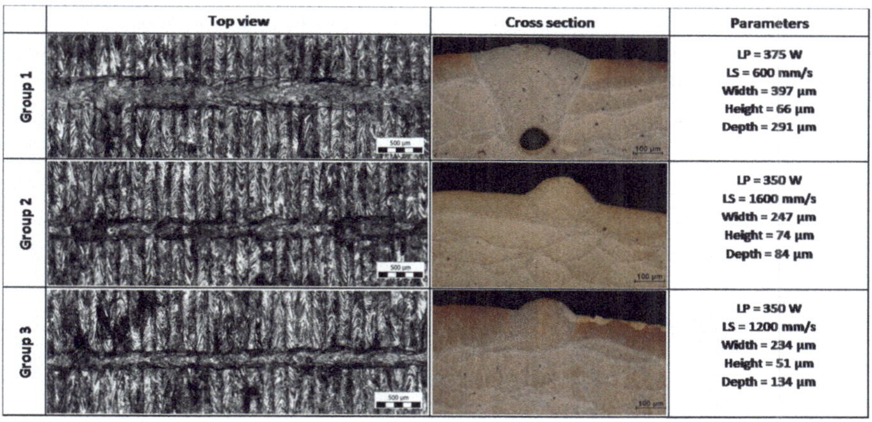

Fig. 2 Sample of individual groups. Top view of the welds and cross-sectional view

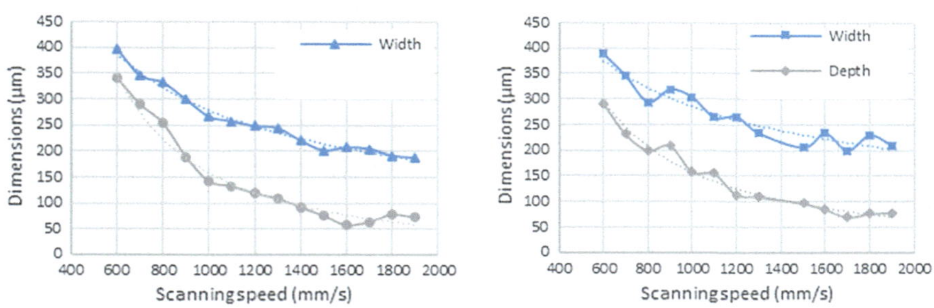

Fig. 3 Dimensions of welds for laser power 375 W depending on the laser scanning speed; direction upstream gas flow (left); direction downstream gas flow (right)

3.2 Thin-Wall Test

Ten thin walls made up of single laser track were built according to the best results of the welding test. The samples had the shape of a 13 × 13 × 10 mm box where only the single outer contour was scanned. The resulting walls were digitized using 3D scanner and subsequently polished in a vertical cross section of the walls. From the scanned data, the thickness of the walls was evaluated in relation to the laser velocity. Figure 4 shows a comparison of measured wall thicknesses from scanned data, polished samples, and welding widths from the welding test. By comparing individual wall thicknesses and weld widths, it can be seen that the wall thickness of the scanned data is 8.5% bigger on average than the width of the welds. This is due to lower heat conduction to surrounding powder during the wall fabrication resulting in the formation of larger molten bath.

The difference between the thickness of the walls obtained from the 3D scanned data and the polished samples is due to the difference of the two measuring approaches. Thickness evaluation of the wall using 3D scanned data is based on 190 points describing the wall surface in chosen virtual cross section of the wall. Thickness evaluation from microscopic images of physical wall sections uses only 16 thickness measurements representing only one transverse plane of the wall.

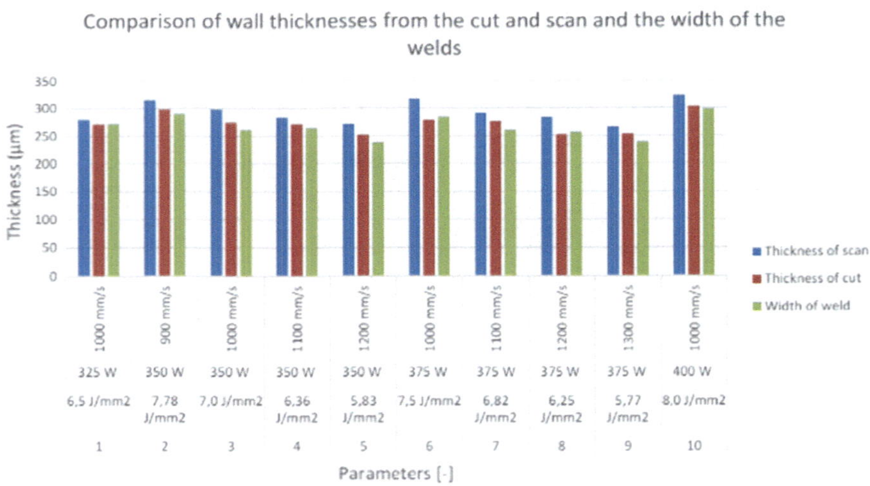

Fig. 4 Comparison of wall thicknesses from scanned data and polished cuts to individual process parameters

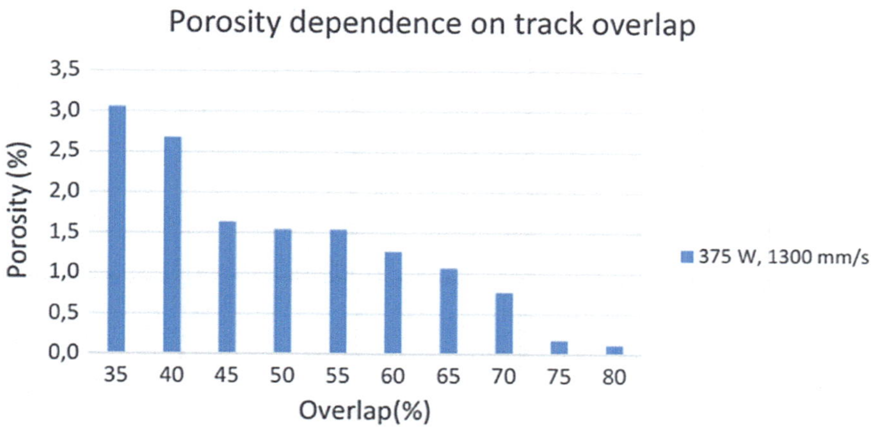

Fig. 5 Diagram of porosity dependence on percent track overlap

3.3 Volumetric Test

The aim of volumetric test was to optimize porosity as a function of the percentage overlap of scan tracks. The best process parameters from the thin-wall test were used in volumetric test. For known wall thicknesses, the hatching distance values for the 20–80% percentage of weld overlaps (defined by hatch distance—HD) were calculated. Samples with a size of 5 × 5 × 10 mm were manufactured. Scanning rotation was set to 0°. With this setup, the welds in each layer will be made in the same place, and it will be possible to evaluate the connection between welds and the resulting porosity for different overlapping (HD) values. Figure 5 shows the decreasing trend of porosity with the increasing overlap.

The lowest porosity value of 0.122% was achieved for process parameters of 375 W, laser velocity 1 300 mm/s and 80% overlap. Figure 6 shows the polished cross sections of individual samples with the appropriate value of their track overlap. For overlapping 25–30%, the occurrence of large discontinuities due to the low overlap can be seen. With increasing weld overlap, a larger part of the already formed weld is repeatedly melted, which results in decrease of porosity.

3.4 Testing of Mechanical Properties

For samples made with 375 W power and laser speeds of 1200 mm/s in volumetric test, microhardness was measured (Fig. 7). The highest average hardness values were obtained with a 50% overlapping sample, i.e., 110.9 HV, and the smallest values reached a 60% overlapping sample, i.e., 108.6 HV. This is the difference of 2.3 HV,

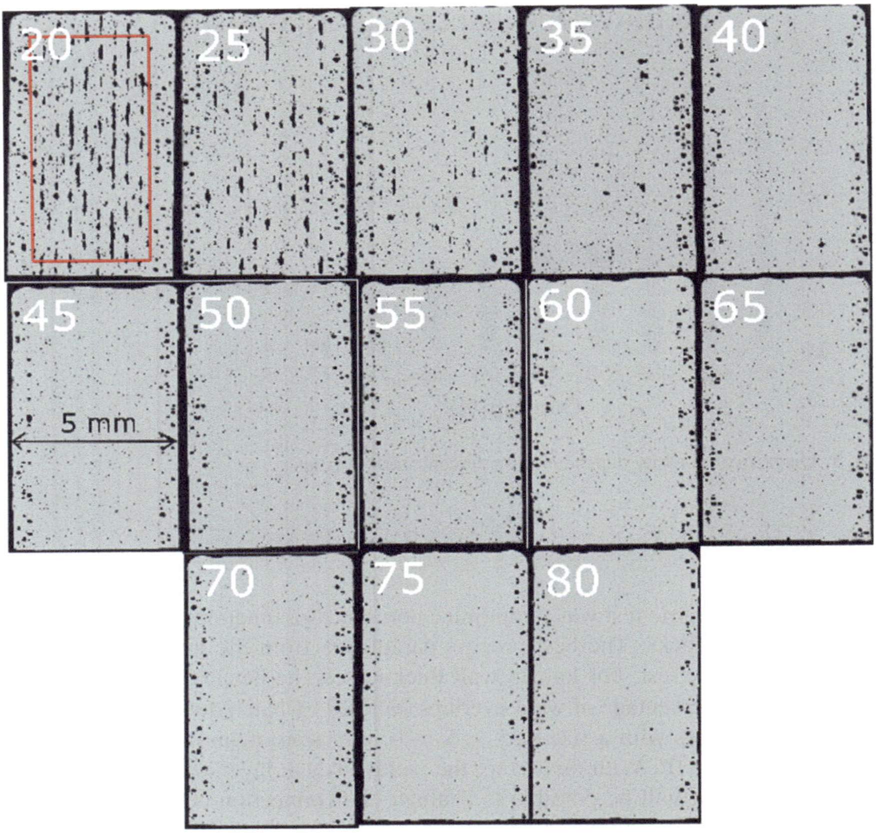

Fig. 6 Example of cuts of samples from volumetric test. The number in the sample indicates its percentage overlap of the tracks. The red field within the first sample shows the area of the volume porosity calculation of the samples

which is within the deviation of this measuring method. Therefore, it is concerned that large overmelting of the track had low influence on mechanical properties.

Mechanical properties of chosen set of parameters were evaluated by a three tensile specimens in as-built state (no heat treatment was used). The tensile test results (Table 2) show low variation. The UTS values of SLM fabricated samples slightly overcoming those of cast alloy in T6 state [9], while the YS values are about 17% lower and the relative elongation is comparable to T6 treated cast alloy.

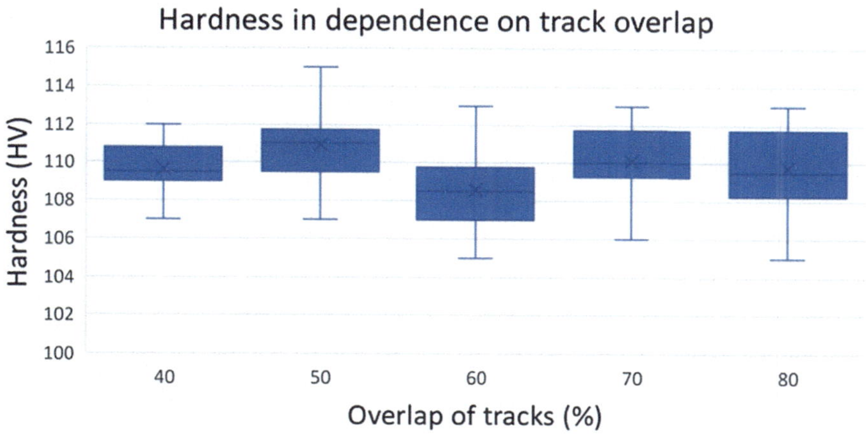

Fig. 7 Hardness in relation to the track overlaps for the LP 350 W and LS 1 200 mm/s

Table 2 Tensile test results of SLM fabricated samples

	Modulus of elasticity	Offset yield strength	Maximal force	Tensile strength	Elongation
	E (GPa)	Rp0.2 (MPa)	F (N)	UTS (MPa)	A (%)
Average value	74.10	229.47	18 983.22	379.87	9.85
Standard deviation	7.71	0.36	19.93	0.81	0.95
Coefficient of variation	10.40	0.16	0.10	0.21	9.61

4 Conclusion

The paper deals with the influence of the process parameters for the porosity of aluminium AlSi7Mg0.6 alloy material processed by SLM technology. The influence of wide range of process parameters on the formation of individual weld tracks was examined. It has been found that with the increasing scanning speed, the width and depth of the welding decrease with the constant laser power. This trend was confirmed in the thin-wall test. Due to the lower heat dissipation during thin-wall fabrication, the wall thickness was about 8.5% higher than the width of individual welds. For volume samples, porosity has been shown to decrease as the track overlap increases. For the 80% overlapping of the hatches, LP 375 W and LS 1300 mm/s the lowest porosity of 0.12% was achieved even without any rotation of the scanning pattern between layers. Mechanical properties showed to be independent of used overlapping with a UTS of 380 MPa, YS of 230 MPa, and a relative elongation of 10%.

Acknowledgements The research leading to these results has received funding from the Technology Agency of the Czech Republic under EPSILON programme Project TH02010514 Development of 3D printing for selected materials and topology optimization of components for aerospace industry.

References

1. Olakanmi EO (2013) Selective laser sintering/melting (SLS/SLM) of pure Al, Al-Mg, and Al-Si powders: Effect of processing conditions and powder properties. J Mater Process Technol 213(8):1387–1405
2. Ilčík J, Koutný D, Paloušek D (2013) Geometrical accuracy of the metal parts produced by selective laser melting: initial tests. In: Book of proceedings of 54th international conference of machine design departments, pp 477–486. Liberec
3. Louvis E, Fox P, Sutcliffe CH (2011) Selective laser melting of aluminium components. J Mater Process Technol 211(2):275–284
4. Kang N, Coddet P, Dembinski L, Liao H, Coddet Ch (2017) Microstructure and strength analysis of eutectic Al-Si alloy in-situ manufactured using selective laser melting from elemental powder mixture. J Alloy Compd 691:316–322
5. Prashanth KG, Scudino S, Klauss HJ, Surreddi KB, Löber L, Wang Z, Chaubey AK, Kühn U, Eckert J (2014) Microstructure and mechanical properties of Al-12Si produced by selective laser melting: Effect of heat treatment. Mater Sci Eng A 590:153–160
6. Read N, Wang AW, Essa K, Attallah MM (2015) Selective laser melting of AlSi10 Mg alloy: process optimisation and mechanical properties development. Mater Des 65:417–424
7. Kimura T, Nakamoto T, Mizuno M, Araki H (2016) Effect of silicon content on densification, mechanical and thermal properties of Al-xSi binary alloys fabricated using selective laser melting. Mater Sci Eng, A 682:593–602
8. Kimura T, Nakamoto T (2016) Microstructures and mechanical properties of A356 (AlSi7Mg0.3) aluminum alloy fabricated by selective laser melting. Mater Des 89:1294–1301
9. Rao H, Giet S, Yang K, Wu X, Davies CH (2016) The influence of processing parameters on aluminium alloy A357 manufactured by selective laser melting. JMADE 109:334–346
10. Aboulkhair NT, Everitt NM, Ashcroft I, Tuck CH (2014) Reducing porosity in AlSi10 Mg parts processed by selective laser melting. Addit Manuf 1:77–86
11. Koutný D, Paloušek D, Pantelejev L, Hoeller CH, Pichler R, Tesicky L, Kaiser J (2018) Influence of scanning strategies on processing of aluminum alloy EN AW 2618 using selective laser melting. Materials 11(2):298
12. Thijs L, Kemplen K, Kruth JP, Van Humbeeck J (2013) Fine-structured aluminium products with controllable texture by selective laser melting of pre-alloyed AlSi10 Mg powder. Acta Mater 61(5):1809–1819

Hydraulics and Fluid Mechanisms

Control System for the Testing Device of Flood Barriers

Aleš Lufinka

Abstract New types of flood barriers must be tested before they are approved for use. One of these tests is the impact of a wood log. Implementation of this test is very difficult because the real wood log is very heavy and large and the test is very space consuming. Therefore, the special test device was built under the project VI20152018005 of the Ministry of the Interior of the Czech Republic. The wood log impact is simulated by a pneumatic engine with impactor head and this engine is controlled so that the impact to the tested flood barrier has the same speed and energy as the impact of the real wood log. It means the impact energy must be evaluated in real time during the test and the engine movement is stopped when the set energy value is reached (these parameters were obtained during impact tests with the real wood log). The design of the system for the pneumatic engine controlling is described in this article.

Keywords Flood barrier · Impact test · Testing device · Control system

1 Introduction

The flood barrier properties must be tested by several different tests prior to its commissioning. One of these tests is the impact of a wood log that simulates the floating object's impact to the barrier. Implementation of this test in the laboratory is not easy because the real wood log is very heavy (400 kg) and long (4 m), its control and handling are difficult and the test is very space consuming. Therefore, the special test device was built and the real wood log impact is replaced by the hit of the impactor, which is accelerated by a linear pneumatic engine (see Fig. 1).

A. Lufinka (✉)
Technical University of Liberec, Studentská 1, 46117 Liberec 1, Czech Republic
e-mail: ales.lufinka@tul.cz

Š. Medvecký et al. (eds.), *Current Methods of Construction Design*, Lecture Notes in Mechanical Engineering, https://doi.org/10.1007/978-3-030-33146-7_60

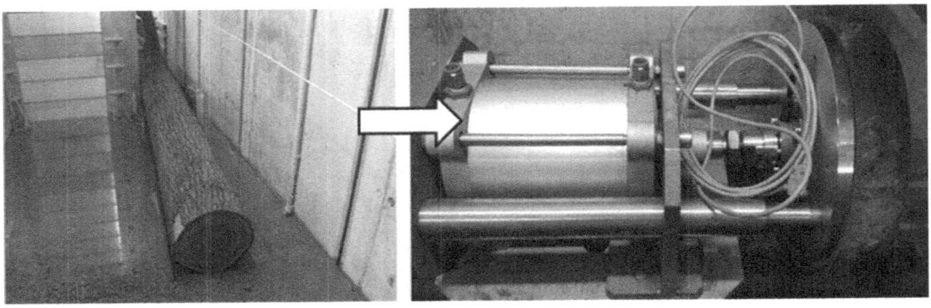

Fig. 1 The real wood log (left) and the pneumatic engine with impactor (right)

2 Testing Methodology

Parameters of impact with a wood log to the flood barrier are described in the methodology of flood barrier testing [1]. The diameter of the log is approximately 300 mm, its weight is about 400 kg and the impact velocity is 2 m s^{-1}. The kinetic energy of the wood log impact can be calculated according to the following simple equation:

$$E_K = \frac{1}{2} m v^2 \quad (1)$$

where E_K—kinetic energy [J], m—mass [kg], v—impact velocity [m s^{-1}].

The principle of replacing the real wood log impact with the impactor is based on achieving the same energy. The kinetic energy of the moving logs will be replaced by the compressed air energy applied to the piston of the pneumatic engine. The impactor with the same diameter as the original log is attached to the engine piston rod.

2.1 *The Real Wood Log Impact Tests*

The tests of the impact of the real wood log into the barrier were performed before the construction of the pneumatic test equipment. A standard lamella barrier with column anchorages was used for testing [2, 3]. Arrangement of the test is shown in Fig. 2.

The wood log speed, acceleration peak and deflection of the barrier lamella were measured during the impact. The impact energy was calculated by Eq. (2) from the curve of the acceleration peak during the wood log impact and from the deformation of the barrier lamella.

$$E_K = m \int_0^{x\max} a(x) \mathrm{d}x \quad (2)$$

where E_K—kinetic energy [J], m—mass [kg], a—acceleration [m s^{-2}], x—displacement [m].

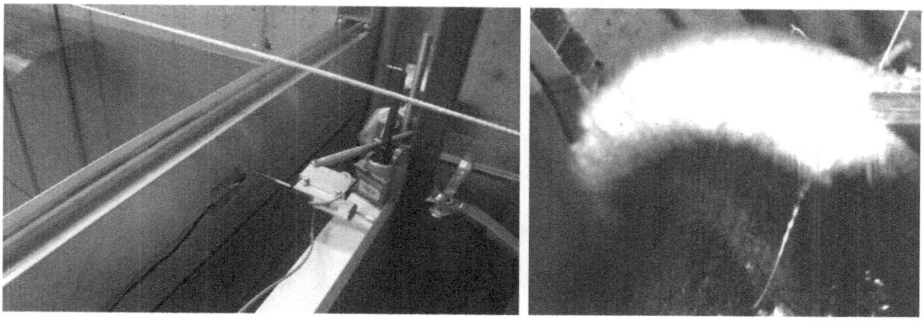

Fig. 2 The real wood log impact test—sensor arrangement left and the real impact right

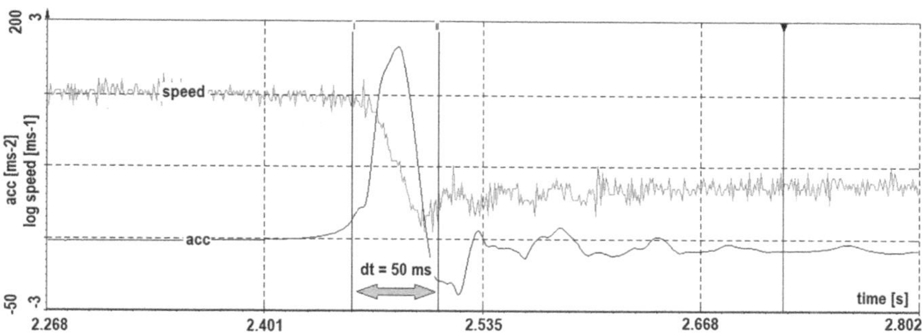

Fig. 3 Example of the real impact measured data

Equation (2) has been calculated to verify the amount of energy because this calculation will be the basic principle of controlling the pneumatic engine. The measured data were further used to verify the duration of the impact process, and the required sampling frequency for data measurement and the timing of the control loop were then determined on the basis of these data. The measured data from these real tests were then also used to verify the functionality of the pneumatic test equipment. An example of the measured waveforms is shown in Fig. 3.

2.2 Principle of the Pneumatic Impactor Control

The principle of replacing a real impact of a log with a pneumatic impactor is to maintain the same impact energy. The calculation of the energy obtained is based on Eq. (2). Because the kinetic energy of the log is replaced by the piston force of the pneumatic engine, the equation is modified to the following shape:

$$E_K = \int_0^{x_{max}} F(x)\mathrm{d}x \tag{3}$$

where E_K—kinetic energy [J], F—force [N], x—displacement [m].

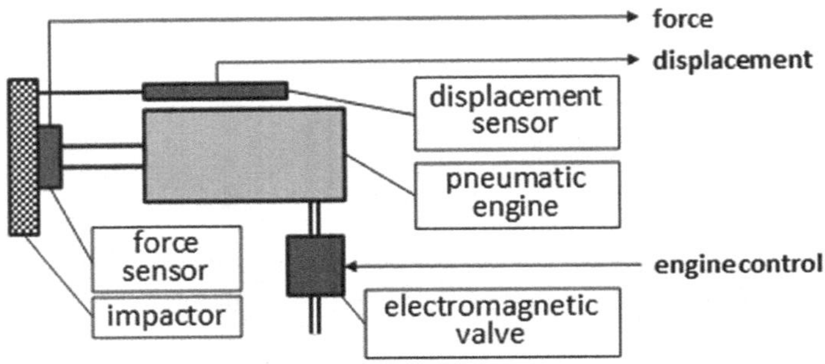

Fig. 4 The pneumatic impactor block diagram

In order for this equation to be computed during impact, the force and displacement of the impactor must be measured. The block diagram of the impactor arrangement is shown in Fig. 4.

The pneumatic engine is supplemented by two sensors—the force sensor is installed between the piston rod and the impactor, and the impactor displacement is measured by the position sensor. Both signals are led to a control computer that calculates the energy in real time in accordance with Eq. (3). After reaching the set value, the computer shuts off the air supply to the engine by the solenoid valve and the impact is thus terminated. From the real impact time analysis (see Fig. 3), it can be seen that the impact time is approximately 50 ms. The sampling frequency for data measurement is therefore set to 5 kHz and the control loop time is 1 ms to stop the motion as accurately as possible after reaching the desired energy. A solenoid valve with a very short reaction time must be used to switch the compressed air in order to prevent the end of the impact from being delayed.

3 The Pneumatic Impactor Building and Testing

3.1 The Pneumatic Impactor Building

The pneumatic impactor test device is built according to the block diagram (see Fig. 4). The linear motor is fitted with a high velocity and high flow electromagnetic valve Festo. The strain gauge force sensor 40 kN from GTM and a 300 mm diameter impactor is attached to the piston rod. The impactor position is measured by the LVDT sensor 150 mm from HBM. Both sensors must be in a special design that allows use below the water level (IP 67) because the pneumatic impactor may be flooded with water during the test (see Fig. 5).

Fig. 5 The pneumatic impactor assembly (left) and its using under the water level (right)

3.2 *The Pneumatic Impactor Testing*

The universal measurement device DEWE5000 from Dewetron was used for the first functional tests. The sensors were connected to the measuring inputs. An optical separator was connected to the TTL output by which the electromagnetic valve was controlled. The DEWE5000 is primarily designed to measure and record data, and the installed version of the software does not allow to count the integral of force from the displacement (see Eq. (3)). So a simpler calculation method was chosen for the tests. Energy was counted as the sum of rectangles of force and the displacement increments $\Delta F.\ \Delta x$. However, this small calculation error did not prevent testing of the impactor functionality.

The first results were not good. The pneumatic engine did not reach the required speeds of 2 m s^{-1}. The real speed was only 0.25 m s^{-1} (see Fig. 6).

The reason for this error was the insufficiently dimensioned intake air pipelines. Their cross-section was too small, the length too large and the airflow was limited.

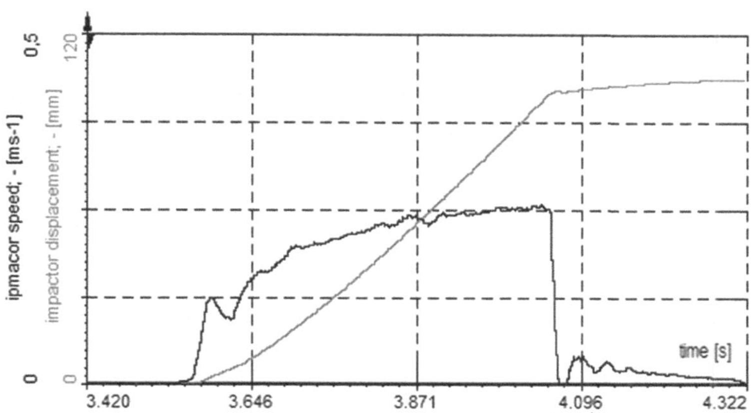

Fig. 6 The first test result example

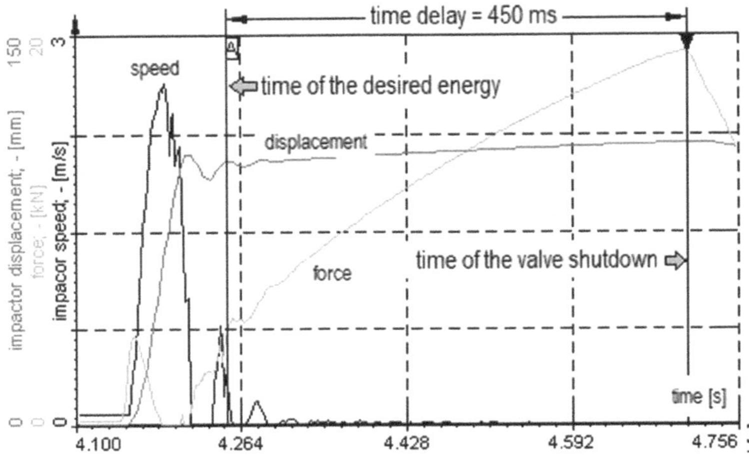

Fig. 7 The new pneumatic circuit measurement result

Another problem was to return the engine piston rod to the starting position, which had to be done manually.

3.3 The Pneumatic Circuit Rebuilding

The pneumatic circuit was rebuilt to eliminate the mentioned errors and the impactor functionality with the newly connected pneumatic circuit was verified again. The maximum speed achieved was 2.6 m s^{-1} (see Fig. 7), so the speed requirement was met after the pneumatic circuit rebuilding.

Stopping motion of the impactor after reaching the set energy could not be verified during these tests because the calculation time of the Dewetron measuring system is too long. The output pulse for stopping the valve was approximately 450 ms delayed from the moment the energy was reached. It means that the final control system must be very fast to stop the impactor movement at the desired time.

New pneumatic circuit block diagram is shown in Fig. 8. The electromagnetic valve was assembled directly on the engine and it was connected through a large cross-section pipeline with an air reservoir near the engine. This arrangement ensures a sufficient volume of compressed air in the vicinity of the engine and the engine function is not affected by the long and thin piping from the compressor. The pneumatic circuit was further equipped with a second electromagnetic valve, so the piston can be automatically returned to the starting position after the test. The remote-controlled pressure regulator was assigned to the inlet pipe of the air reservoir and the actual air pressure measured by the pressure sensor. The air pressure (and thus the maximum force) can be set before the impact test.

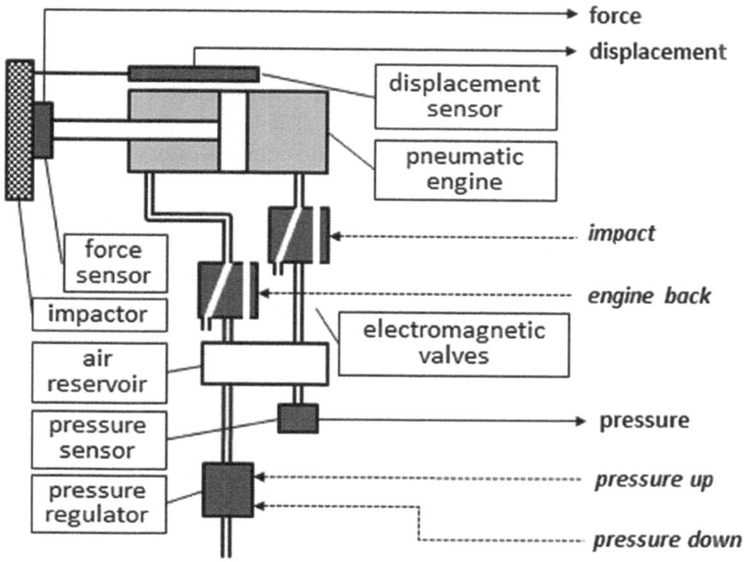

Fig. 8 New configuration of the pneumatic circuit

Table 1 The control system inputs and outputs

Signals	No.	Description
Analog input	3	Force–displacement–pressure
Digital output	4	Impact–engine back–pressure up–pressure down

The necessary inputs and outputs of the control system also came out of the block diagram in Fig. 7 and they are summarized in Table 1.

4 The Control System Design

The pneumatic impactor is located in the test room of flood barriers. Because the barrier may be destroyed during the test, the entire room may be flooded with water. The control system is therefore placed in the control room which is separated from the test room. This spatial arrangement, however, leads to a long distance between the impactor device and the control system, which can reach up to 40 meters.

The customer also required the possibility of simplified control of the impactor by manual control, where the impact energy is not calculated and the impact force is set only by the air pressure.

4.1 The Control System Hardware

The following solution (shown in Fig. 9) resulted from the above requirements.

The control system is divided into three parts. The first is located in a waterproof case (IP 67) directly at the pneumatic impactor and contains only signal amplifier for the force and displacement sensors. The special integrated amplifiers for the strain gauge sensors and LVDT sensors from Analog Devices are used for its construction. The pressure sensor has an integrated amplifier so its output no longer needs to be amplified. Thus three analog signals in the 0–10 V range are connected via a connection line on which a shielded twisted pair cable is used. This solution guarantees a sufficient signal-to-noise ratio. Pneumatic valves and pressure regulator are controlled by 24 V DC signals. These are connected by a separate cable to avoid interference from analog signals from the sensors.

The manual control box is the second part of the control system. It is located in the control room and it allows simplified manual control of the impactor without calculating the impact energy. The 24 V DC power supply for the electromagnetic valves, the pressure regulator and the electronic switches for their control are located in this box. The electronic switches can be either manually operated by push buttons on the front panel of the box or electronically from the parent computer system. The display for the air pressure value in the air reservoir is also on the front panel. The manual control box thus contains all the necessary components and it can be used for simplified pneumatic impactor control without the computer system connection. Its front panel is shown in Fig. 10.

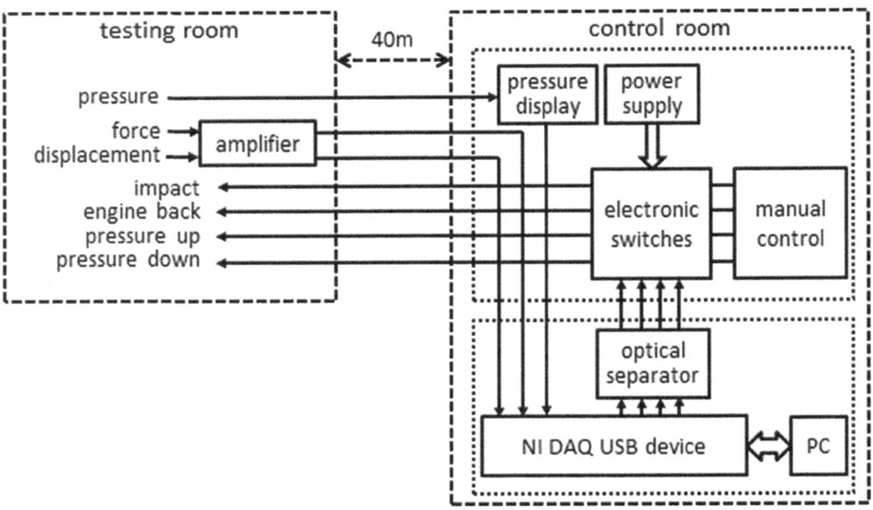

Fig. 9 Block diagram of the control system hardware

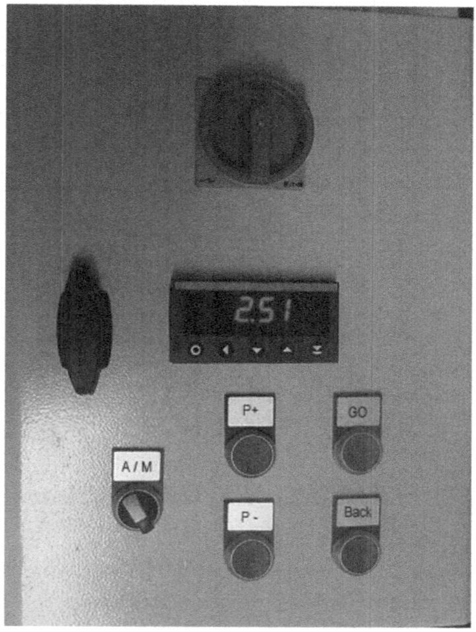

Fig. 10 The manual control box front panel

Force and displacement signals are not used for manual control and they are therefore only connected to the parent computer system. Of course the pressure signal is connected to the computer system as well (see Fig. 9).

The NI USB-6211 universal measuring unit from National Instruments [4] forms the basis of a computer control system. This device has a sufficient number of analog inputs and logic outputs to control the pneumatic impactor. Sensor signals are connected directly to NI USB analog inputs. The digital outputs are connected to the electronic valve switches via the optical separator. This first protects outputs from NI USB and secondly converts their 5 V level to 24 V level for the electronic switches control.

The NI USB-6211 has 16 analog inputs with 16 bits resolution and sampling frequency can be up to 250 kS s^{-1}. It contains four hi-speed TTL digital outputs too. These values ensure sufficient measurement and control speeds needed to stop the piston movement when the set energy is reached. Small delay is due only to the solenoid valve response time.

The NI USB drive is connected via a USB cable to the computer on which the control application is installed. Any computer running Windows can be used.

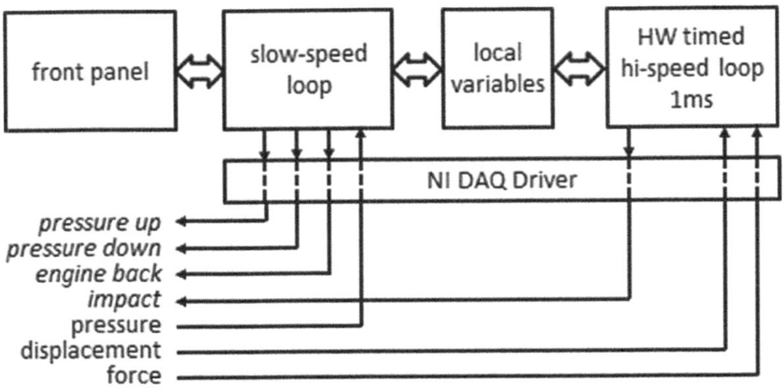

Fig. 11 LabVIEW application block scheme

4.2 *The Control System Software*

Application created in LabVIEW software is used to control the pneumatic impactor. Each LabVIEW application has two parts—a front panel that forms a user interface and a block diagram, which is a program application code. The block diagram can be formed by one or more independent loops. If a measuring device is connected to the application, the NI-DAQ driver is used to control it. Of course, the application for the impactor control is created according to this structure and it is shown in Fig. 11.

The application block diagram contents two independent loops. The first one is the slow-speed loop that is designed for the front panel control. Its repeat period is 250 ms, which is enough for a very good response to user control, while the PC processor load is very small. This loop also directly controls the "slow" actions, i.e., measuring and setting the air pressure and returning the engine to the starting position.

The second loop is the core of the application. It is designed for the analog signal measurement, energy integrity calculation and the impact pneumatic valve control. This loop utilizes the hardware timing of the period, which ensures very accurate loop timing. If no special real-time hardware (which is expensive) is used, the minimum period of HW timed loop is 1 ms. This value is also set in this case. This means that every millisecond the force is measured, the displacement values with the integral of the energy are calculated, the result is compared with the requested value and the electromagnetic valve is controlled according to the result. The valve is therefore stopped with a maximum of 1 ms delay when the set energy is reached.

Hardware of the NI USB unit is maximally used to measure the analog signals using the NI-DAQ driver. Input buffers are used to avoid a loss of data samples and the unit's hardware is further used to convert the measured signals directly into the physical units, i.e., kN for the force and mm for the displacement. So the amount of computational operations in the time-consuming loop is minimized by this efficient use of the unit's capabilities.

Both loops used in the application must run completely independently (the precise timing of the HW timing loop must not be disturbed) but data and command transmission between the two loops is necessary. Therefore, the local variable system is used for this communication. Both loops can store and read data in this system at the moment determined by their timing. The information is passed and loop timing is not disturbed.

5 Conclusion

The new test device for the flood barrier impact tests was built. Complicated and space-intensive real wood log impact tests were effectively replaced by the compact pneumatic impactor. The control system built on the basis of the results of the actual tests allows complete control of the pneumatic impactor and the impact test has the same parameters as the original test with the wood log. So the effectiveness of the impact tests of the flood control systems has increased significantly.

Acknowledgements The results of this project VI20152018005 were obtained with co-funding from the Ministry of the Interior of the Czech Republic as part of targeted support from the project "Development of flood protection systems to enhance protection of the population and infrastructure."

References

1. Approval Standard for Flood Abatement Equipment, Class Number 2510, FM Approvals LLC (2013) https://www.fmglobal.com/assets/pdf/fmapprovals/2510.pdf, Last accessed 15 Jan 2018
2. Petru M, Sevcik L, Lufinka A, Martinec T (2016) Analysis of stress and deformation of the flood barrier. In: EAN 2016—54th international conference experimental stress analysis proceedings, pp 256–295. University of West Bohemia
3. Petru M, Srb P, Sevcik L, Lufinka A (2017) Testing and numerical analysis of mechanical stress mobile flood barriers. In: EAN 2017—55th conference on experimental stress analysis 2017 proceedings, pp 350–353. Technical University of Kosice
4. USB-6211 Specifications. http://www.ni.com/pdf/manuals/375195d.pdf. Last Accessed 15 Jan 2018

Numerical Simulation of Flow Through Porous Media

Ondřej Novák and Michal Petrů

Abstract The filtration process using porous structures (foams, textile structures, granular materials…) is very complicated and its understanding is important for the production of filter materials for a maximum efficiency, low pressure drop, and filter lifetime. To obtain comprehensive information systematic analysis of the structure of the filter media by image analysis of images obtained by optical and electron microscopy, and through micro-tomography were conducted. In particular, it was the characterization of the morphology and structure incl. directional arrangement, fiber diameters, and pore shape as well as porosity. Based on this information, 2D CAD models were created representing the filter medium with the corresponding parameters. Real filter medium was tested on the experimental test device in order to determine pressure drops at different flow rates. These values determined the boundary conditions of the numerical model and the control values for the verification of the model. The result is a numeric model depicting the water flow depending on the depth and the complexity of the filter. The model shows a very good match with the experimental results. It could significantly help to design the structure of the filter media.

Keywords Filter · Flow · Numerical model

1 Introduction

The problem of filtration in porous structures is very complicated because it is very difficult to obtain sufficient information about pore shape. The distribution of pore diameters, lengths, and segmentation plays a major role in this problematic. This is mainly due to the fact that the pore size of the filter materials is in the order of hundreds of microns, but is not limited to materials with pore size in micrometers. However, there are also materials with pore size in nanometers, for example, membranes. An important factor is also the type of filtration that can be depth or surface. The surface filtration is more common in an industrial filtration because it allows repeated

O. Novák (✉) · M. Petrů
Technical University of Liberec, Studentská, 1402/2, 46001 Liberec, Czech Republic
e-mail: novak.ondra1@seznam.cz

Š. Medvecký et al. (eds.), *Current Methods of Construction Design*, Lecture Notes in Mechanical Engineering, https://doi.org/10.1007/978-3-030-33146-7_61

cleaning of filters and thus longer life. The filtering process is typically characterized by two parameters. The first is efficiency, which describes the ability to capture a solid fraction from aerosol with respect to particle size. The second is the pressure drop and its increase during filter clogging. For the modeling, it is necessary to design the geometry of the structural units and for setting of the boundary conditions the basic parameters of the filtration process. The geometry was based on scanning electron microscopy (SEM) and micro-tomography images. Parameters of filtration process were measured with selected filters on designed experimental test device. The model considers fluid flow with particles that clog the filter.

2 Theory

Modeling of flow of fluid [1] containing solid particles can be described by the following relationships. If the liquid contains particles, the calculation of the particle trajectory applies the Newtonian law in the form according to (1)

$$\frac{d}{dt}(m_{\mathrm{p}} \times v) = m_{\mathrm{p}} \times F_{\mathrm{d}}(v - v_{\mathrm{p}}) \tag{1}$$

where v_{p} is the velocity vector of the particle, F_{d} is the force per unit of mass that is

$$F_d = 18\mu/\rho_d d_p^2 \tag{2}$$

where m_{d} is particle mass. These equations are always dealt together with the continuity equation

$$\frac{\partial \rho}{\partial t} + \nabla \times (\rho u) = 0 \tag{3}$$

where the Navier–Stokes equations represent momentum conservation, while the continuity equation represents the conservation of mass [2]. The solution is for a certain set of boundary conditions (such as inputs, outputs, and walls) to predict fluid velocity and its pressure in a given geometry. The principle of the problem is schematically shown in Fig. 1, which represents an idealized formation of a filter cake and filing of the filter media pores with dirt. The particle arrangement is characterized by the type of particles (dimension, shape, and diameter distribution) that are streamed to the geometry of the pores.

To study the pressure loss of flow volume at a fluid flow that clogs pores can be based on Darcy's law [3] according to the Eq. (4)

$$\dot{V} = \frac{A}{\mu}\frac{\Delta p}{\Delta L} \times P \tag{4}$$

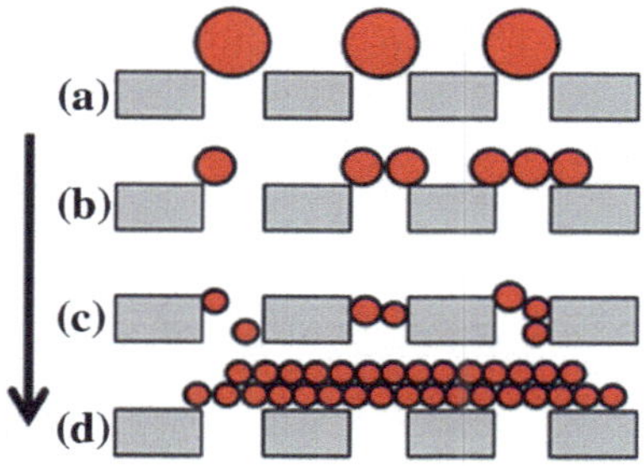

Fig. 1 Scheme of filter media clogging principle: **a** complete blocking, **b** partial blocking, **c** internal pore blocking, and **d** cake filtration

where A is the area of cake filtration, P is the specific permeability coefficient (permeability), which can be expressed as (5)

$$\dot{P} = \frac{\varepsilon^3}{K_2 q^2 a_v^2 (1-\varepsilon)^2} \tag{5}$$

where a_v is the intrinsic specific surface of the particle. For spherical particles is valid Eq. (6)

$$a_v = 6/d_\mathrm{p} \tag{6}$$

3 Experiment and FE Model

The images were prepared from samples of a non-woven filter made of polyester. An electron microscope is used VEGA 3 a micro-CT SKYSCAN 1272 (Fig. 2 left). By the help of image analysis NIS elements images were treated (Fig. 2 right).

Based on the images, a 2D model of the filter unit was created. Unlike the previous [4], a two-layer filter was created. The first layer is a coarse fiber pre-filter; the second

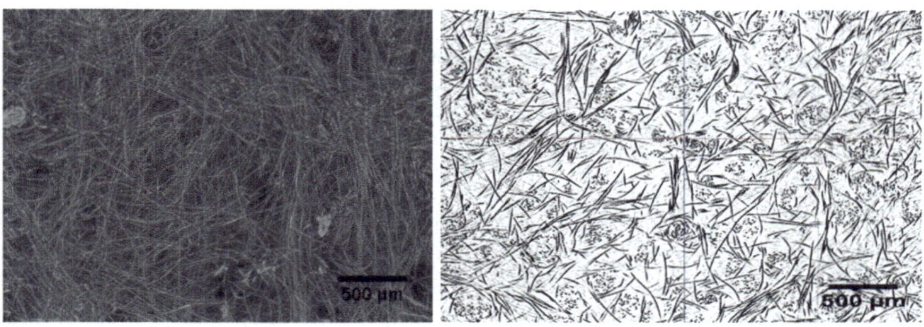

Fig. 2 Micro-CT images (left). Detail of black and white structure (right)

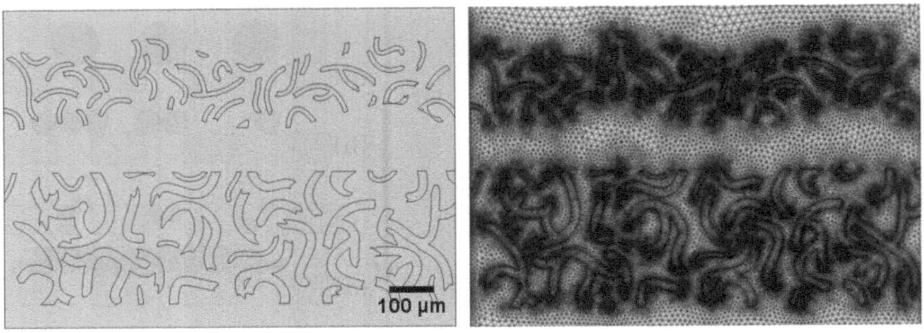

Fig. 3 2D CAD model of structure of two-layer filter (left), meshed model (right)

Table 1 Geometric model parameters

Geometry	Dimension (μm)	Geometry	Dimension (μm)
Thickness of first layer	300	Length (thickness)	750
Thickness of second layer	200	Fiber diameter—first layer	20
Gap between layers	100	Fiber diameter—second layer	10
Width	1 000	Particle size	1

layer is of finer fibers and serves as a filter. The geometrical and physical properties of FE model (Fig. 3) are presented in Table 1.

Boundary conditions are following. Shorter sides of model create walls of system, bottom part side is inlet and upper side is outlet. Inlet pressure was 9.6 Bar, pressure drop 2.2 Bar. The FE mesh is swept and created from triangular and quadrilateral elements. Total number of elements is 145,016. Applied software was COMSOL Mutiphysics. The model uses laminar flow analysis with particle tracing for fluid. Applied medium is water.

The model was assembled with initial conditions, where first in simulation step 1 at time $\tau = 0.1$ were particles with a diameter greater than small pore diameters but at the same time smaller than the large pore diameters of the filter medium by the stream being shifted to the surface of the filtering structure. This created an incomplete filter cake where some particles passed through the structure. Subsequently, in the simulation step 2 at time $\tau > 0.1$ a flow of pure liquid was applied for a prediction the pressure drop.

4 Results

The simulation shows both flow, the pressure drop and the way of the filter clogging. Figure 4 shows the movement of the particles and flow velocity in the filter medium.

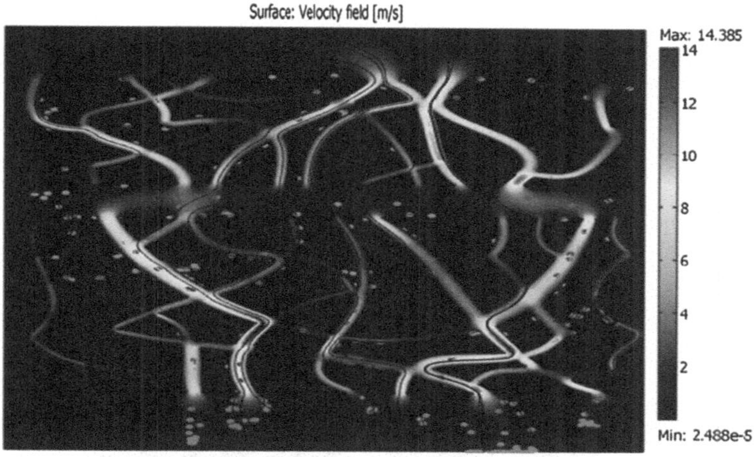

Fig. 4 Filtration at $\tau > 0.1$, velocity of fluid flow

The velocity of water 5.97 ms^{-1} was experimentally determined and set as model input. Obviously, a larger pore filter part catches less particles and, of course, it indicates lower pressure drop. Clogged pores show higher pressure and therefore higher velocity. The output velocity is same, but it differs in individual pores depend on their clogging. From model results, it is seen that velocity in certain parts of filter is almost three times higher (5.97 vs. 14.39 ms^{-1}). The higher pressure and fast flow due to high kinetic energy can move with particles, and it can also change pore size or destroy the filter. Increasing of pore size leads to particle penetration through the filter structure, therefore, a filter efficiency decreasing. In Fig. 5 is shown, the pressure in individual parts of the geometry. It is clear that higher pressure and pressure drop indicates the first part of the filter, especially near to inlet. It corresponds with the fact that pressure decreasing toward the outlet. Also, it is seen decreasing of the pressure between layers. Compare with standard filters, where the layers are put together without the gap, it can bring different behavior of filtration characteristics. But it can be complicated for manufacturing. Two-layered filter with the gap has different behavior compare to filter with same thickness. By this structural change, filtration characteristics can be improved. The problem consists in a requirement of the gap with relatively low dimension. It could be done by the thin rough mesh adding. The mesh with large pores is required. It can also bring reinforcing function to the filter and improving of filter mechanical properties.

5 Conclusion

A two-step simulation was performed on the 2D filter FE model, whose structure is based on a real filter medium. The model uses not only flow, but also particles that simulate filter clogging. The model is unusual with its gap between the layers, which gives to the filtering process different characteristics. Such type of model can

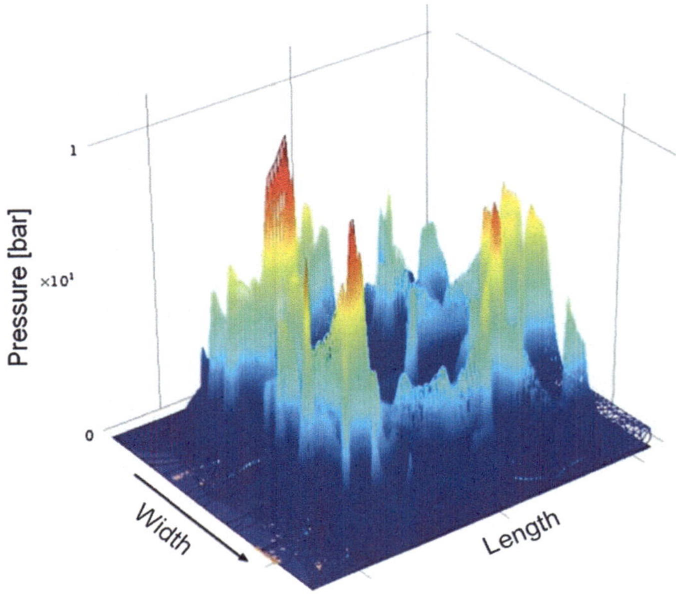

Fig. 5 Pressure in individual layers of filter media

facilitate the design and optimization of the filter medium to achieve both optimal morphology and filter structure, but also a way how to arrange the layers. In the future, it is intended to replace the 2D model with a 3D model that will directly use the scanned filter structures. This will ensure greater accuracy and reproducibility of obtained results. This will allow studying other effects, such as pore size change due to pressure, flow, and particle impact.

Acknowledgements The results of this project LO1201 were obtained with co-funding from the Ministry of Education, Youth, and Sports as part of targeted support from the program "Národní program udržitelnosti I".

References

1. Novák O, Petrů M, Vosáhlo J (2018) Numerical modeling of liquid filtration process. In: Lufinka A, Petríková I, Sivčák L (eds) Experimental stress analysis: 56th international scientific conference, Harrachov, Czech Republic. Book of extended abstracts, pp 103–104. Czech Society of Mechanics, Praha
2. Acheson DJ (1990) Elementary fluid dynamics. Oxford University Press, UK
3. Carman PC (1937) Fluid flow through granular beds. Trans, Inst Chem Eng, Lond 15:150–166
4. Van der Geer J, Hanraads JAJ, Lupton RA (2010) The art of writing a scientific article. J Sci Commun 163:51–59

Modern Material and Technology

Principles of Winding Elbows from Pre-impregnated Carbon Fibers

Petr Kulhavý and Vítězslav Pfliegel

Abstract Nowadays, technologies like winding, braiding or wrapping of high-strength fibers used for manufacturing composite materials are used by producers in almost all sectors. This thesis describes a method of the prepreg (pre-impregnated) carbon filament winding with regards to manufacturing and effective designing of curved rods (elbows). The process of winding fibers is known for a long time. However, until recently was this technology used especially in the field of textile engineering, braiding of ropes, hoses and similar parts from atypical sectors. The basic elements of traditional filament winding are a non-bearing mandrel and spools. In the studied case of winding, the mandrel stands on a place and the spools with fibers rotating around and simultaneously moving parallel to it. The fiber can be wound in helical, circumferential and polar pattern with geodesic or non-geodesic behavior. However, when winding profiles with curved shapes, there could be significant deviations between the real and desired winding angle, caused by the irregular distribution of fibers on the inner and outer tube radius. Therefore, the aim of this work is to describe this phenomenon and find out an optimal solution on how to assure the constant mechanical parameters through the entire shape. The process has been described analytically and also simulated with using specialized software CADWIND.

Keywords Winding · Prepreg · Elbow · Composite · CADWIND

1 Introduction

Nowadays, the uses of composite materials have grown exponentially, thanks to the very good specific strength of the materials and the possibilities of customization the final properties [1]. Compared to conventional metallic materials, it is possible to obtain several times higher specific strength. The high-strength carbon fibers have approximately 2–3 times higher elastic modulus than the best steel and more than five times than the Ti-based alloys used in aeronautics. Composite materials offer some

P. Kulhavý (✉) · V. Pfliegel
Technical University of Liberec, Studentska 2, 46117 Liberec 1, Czech Republic
e-mail: Petr.kulhavy@tul.cz

Š. Medvecký et al. (eds.), *Current Methods of Construction Design*, Lecture Notes in Mechanical Engineering, https://doi.org/10.1007/978-3-030-33146-7_62

unique engineering properties and challenging problems for analysts and designers [2], achieved through the possibilities of significantly varying and tailoring their mechanical parameters. The abnormal strength of the fibers is based on planar bonds, the arrangement of hexagonal meshes and arrangement of the inner layers parallel to the fiber axis. However, with using the conventional manufacturing methods like layering in molds, vacuuming, etc., it is not possible to cover all of the requirements on today's technical devices. Therefore, the so-called advanced composites created by using winding, wrapping or braiding technologies becoming still more important. With using those techniques, it is possible to create some thin-walled, straight or curved hollow tubes or parts of profiled frame structures.

2 The Used Technology and Materials

The general disadvantage of the above-mentioned "wet" methods is their relatively high complexity and especially the high cost of auxiliary devices (e.g., molds), predestines their use primarily for series production. Therefore, technologies using pre-saturated fibers known as prepregs are actually one of the most emerging methods (see Fig. 1) [3, 4]. The pre-impregnated fibers have reached an irreplaceable position between long-fiber composites. Prepregs consist of unidirectional carbon fibers pre-impregnated with partially cured resins (thermosets) and protected by a silicone paper.

They are designed as a basic material for the production of advanced laminates [5]. The main advantage of Prepregs is the high proportion of fiber reinforcement (30% of fibers), the uniformity and smoothness of the finished parts. The reason for using prepregs instead of classic filaments is simplifications of manufacturing processes. Especially, the subsequent resin saturation and curing in form it is not required. As a disadvantage, it is possible to mention higher price of the raw material and requirement to high, ideally constant preloading of the fibers. This is caused by the fact, that the tapes are extremely sticky and this problem significantly arising with the increasing temperature.

The further described and optimized method, based on epicyclic wrapping of pre-impregnated tapes used for manufacturing of thin-walled composite parts with circular or oval profiles, is the so-called winding (see Fig. 2). Winding the composite

Fig. 1 Schematic principle of the two-directional, multifilament winding of straight tube

Fig. 2 Process of winding prepreg tapes on a non-bearing PUR mandrel

manufacturing process that keeps the continuous fibers aligned along the entire length of the component with the declared angle—the so-called winding angle [6].

The global problem of this method arising during the winding mandrels with curved shapes when there are changes of the winding angle on the outer and inner sides of the tube.

Because it is not possible to ensure a constant covering on the both sides, it is necessary to locally adjust the defined winding angle determined by numerical calculation with respect to real boundary conditions. As could be seen on the results of the parametric process simulation—carried in the software CADWIND (Fig. 3), without any local adjustment of the winding angle the coverage of the outer side is poorer.

3 Mathematical Models

Generally, there is no reliable method for non-axisymmetric winding design. Wang [7] was dealt with numerical description of the manufacturing parameters for winding of complicated and multiple curved shapes (so-called combined elbows). The basis of its parametric analysis was to divide the problem solved as a 2D mesh. The winding angle has to be adjusted by changing the doffing points to keep the pattern stable and realize the full coverage. Finally, the problem of combined elbows

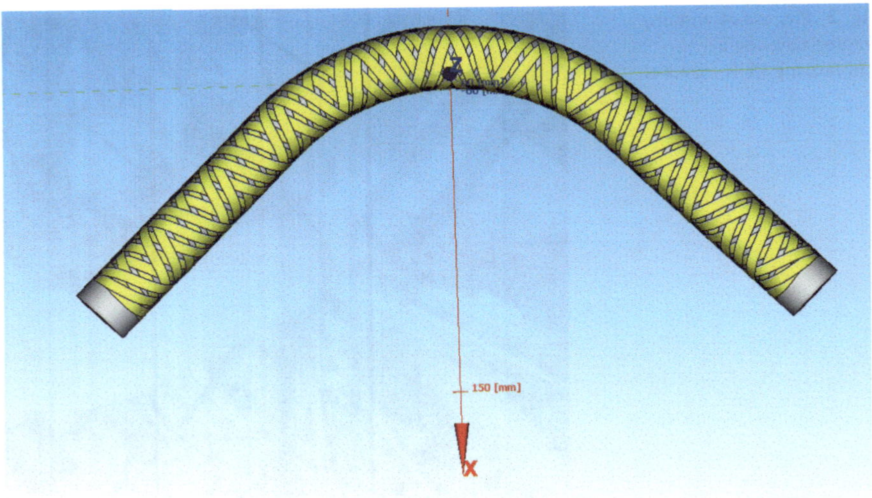

Fig. 3 Parametrical illustration of the elbow winding (created in CADWIND)

winding was solved by a simple table, which contains basic parameter configurations when the individual values for setting the local winding angles when passing through the curved parts could be found.

The design and calculation of fiber paths require considerable mathematical model. The CADFIL and CADWIND are the only two useable software for the elbow and T-shaped winding [8]. Li [9] set the non-axisymmetric filament winding pattern (NFWP). This model is suitable also for 3D models. The main principle of NFWP consists of (1), and the main dimension could be seen in the scheme in Fig. 4.

$$T(\gamma, \varphi) = \begin{matrix} (R + r\ \cos\varphi)\cos\gamma \\ r\ \sin\varphi \\ (R + r\ \cos\varphi)\sin\gamma \end{matrix} \tag{1}$$

where $T(\gamma, \varphi)$ is the vector describing the elbow γ; it is the main angle of, φ is the angle of the rotation of the fiber around the mandrel with diameter r [mm].

For 3D body in space, the so-called geodesic curvature k_s is according to Li [9] using the Liouville formula possible to use Eq. (2):

$$k_s = \frac{\mathrm{d}\alpha}{\mathrm{d}s} + \frac{\sin\varphi}{(R + r\ \cos\varphi)}\cos\alpha \tag{2}$$

If the curvature is set to equal to zero $k_s = 0$, then it is possible to construct simple differential Eqs. (3, 4 and 5):

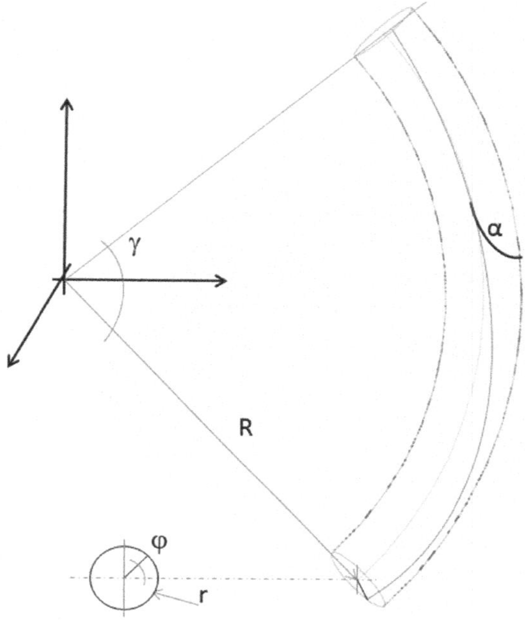

Fig. 4 Scheme of the dimensions in the 3D curved rod

$$\frac{d\alpha}{ds} = -\frac{\sin\varphi}{(R + r\cos\varphi)}\cos\alpha \tag{3}$$

$$\frac{d\gamma}{ds} = -\frac{\cos\alpha}{(R + r\cos\varphi)} \tag{4}$$

$$\frac{d\varphi}{ds} = -\frac{\sin\alpha}{r} \tag{5}$$

With a subsequent modification of the given equations, it is possible to obtain expressions (6 and 7):

$$\frac{d\alpha}{d\varphi} = -\frac{r\sin\varphi}{(R + r\cos\varphi)}\frac{\cos\alpha}{\sin\alpha} \tag{6}$$

$$\frac{d\gamma}{d\varphi} = -\frac{r}{(R + r\cos\varphi)}\frac{\cos\alpha}{\sin\alpha} \tag{7}$$

Finally, using the so-called Clairaut's equation, it is possible to set (8):

$$(R + r\cos\varphi)\cos\alpha = C. \tag{8}$$

In this case, the constant C is defined by the initial value of winding angle α_0. The elbow is not always convex, even though the geodesics may bridge and may not cover the elbow unless the initial winding angle α meets certain condition. The stable winding condition was given in [9, 10] as follows (9):

$$0 < \cos \alpha_0 \leq \frac{R-r}{R+r}\sqrt{\frac{R-r}{R}}. \tag{9}$$

All these conditions, however, describe only geometric dependencies, but they do not take into account the very necessary parameter that [CAD, Geodesic and Haj] point to, and it is the real friction between the fibers and the core.

When introducing normal curvature k_n (10), it is based on normal and geodetic curvature, possible to determine the relative slip resistance λ (11) that could prevent the slipping of the fibers from the sides of the mandrel.

$$k_n = \frac{\sin^2 \alpha}{r} \tag{10}$$

$$\frac{k_g}{k_n} = \lambda \tag{11}$$

To ensure optimal technological parameters, it is recommended the $|\lambda| \leq \mu$, when the μ means friction coefficient between fiber and mandrel.

Generally, the winding angle = 0° corresponds to the longitudinal orientation of the fibers, i.e., the tangent vector is at each moment parallel to the axis of rotation, and the 90° angle corresponds to the circumferential (industry called "Hoop") winding.

Another important feature for optimization the winding structure of curved parts is to set whether the approach will be geodetic or non-geodetic. In modern mathematical meaning, the geodetic path is the shortest way on a spherical surface between two points [11].

Then, the geodetic winding is the path that the fiber can stretch firmly between two points on a curved surface without relying on friction and avoiding slippage.

For this reason, there are the methods of non-geodetic winding, which consider also the friction between the fibers and the mandrel (loxodromes) [12].

Very sticky fibers (e.g., prepreg) and rubber core will be considerably different from the ideal geodetic tracks. However, some very slippery filaments (e.g., lubricated fibers) on smooth cores (e.g., polished stainless steel) should follow a path that is very similar to the geodetic path, otherwise they will slip down. According to CADWIND [13] for the dry fibers wound onto a polyurethane core, the coefficient of friction is approximately 0.2 and for prepreg 0.35.

3.1 Problem of the 2D Curved Shape

The idea of solution in the 2D space is based on the division of the straight and curved parts into one-dimensional sections. The length those sections on the outer radius must be constant, thus guaranteeing a constant winding angle (Fig. 5).

For the solved case, for the optimization of the winding angles and for a 2D curved elbow, it is possible to find the resulting geometric relationship between the angles α_1 a α_2 (according to Fig. 6).

For this scheme, the basic geometrical relationships can be derived. Basically, the length a is equal (12).

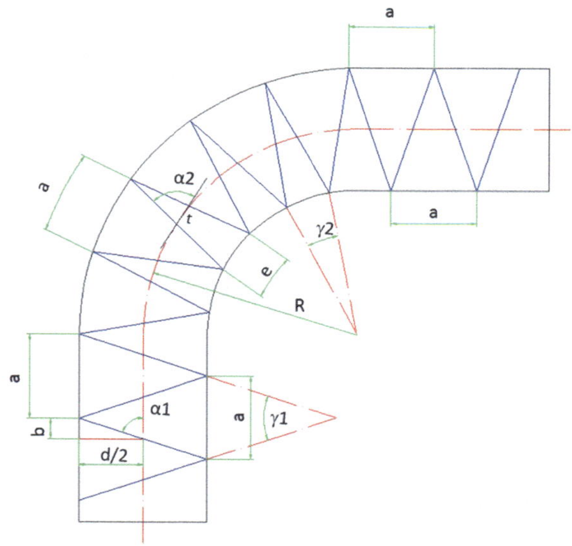

Fig. 5 Scheme of the basic dimensions and the required angle adjustment

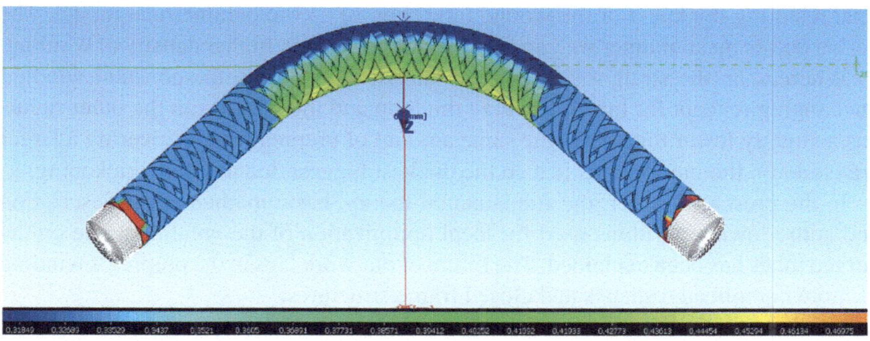

Fig. 6 Model of the various thickness caused by the local thickening in the curved section

$$a = 4b = \frac{2d}{tg \propto_1} \tag{12}$$

Then, the section *e* (13) and (14) depends on the length of the arcs:

$$e = \frac{\pi\left(R - \frac{d}{2}\right)}{180}\gamma_2 \tag{13}$$

$$a = \frac{\pi\left(R + \frac{d}{2}\right)}{180}\gamma_1 \tag{14}$$

And the dependency of winding angle on the fundamental parameters of the elbow is described by (15, 16).

$$\frac{2d}{tg \propto_1} = \frac{\pi\left(R + \frac{d}{2}\right)}{180}\gamma_1 \tag{15}$$

$$\gamma_1 = \frac{2d}{tg \propto_1}\frac{180}{\pi\left(R + \frac{d}{2}\right)} = \frac{d}{tg \propto_1}\frac{360}{\pi\left(R + \frac{d}{2}\right)} \tag{16}$$

Based on these assumptions and with a slight approximation, it is possible to use a simplified but sufficiently accurate relationship (17) to determine the value of the local winding angle when passing through an arc:

$$\propto_2 = \propto_1 + d/2 \propto_1 \ 1/R \cot g \propto_1 = \propto_1 (1 + d/2\, 1/R \cot g \propto_1) \tag{17}$$

4 Conclusion

Probably, the best way how to demonstrate the described problem is to illustrate the final resulting thickness of the wound tube (Fig. 6) on the parametrical model. The darker colors mean thinner wall and the brighter—thicker, higher density of winding.

Whereas on the straight sections, the thickness is due to the constant winding unchanging (except for the ends where the loop and the return), on the outer radius has a slightly lower thickness (the same amount of prepreg tape is covering a larger area) and on the inner radius is then the thickest biggest due to fibers thickening.

In the presented paper, the fundamental theory, basic mathematical description and author own algorithms used for local optimization of the winding angle on the curved tubes has been explained. The future of our work lays in the prepreg's winding of more complicated shapes and closed frame structures.

Acknowledgements This publication was written at the Technical University of Liberec as part of the project "Innovation of technical systems structures with the use of composite materials" with the support of the Specific University Research Grant, as provided by the Ministry of Education,

Youth and Sports of the Czech Republic in the year 2018. Our great thanks belong to Mr. Laval, for the opportunity to verify our models in the provided CADWIND software.

References

1. Cawley PR, Adams DJ (1978) A vibration technique for non-destructively assessing the integrity of structures. Compos Mater 12:336–347
2. Kherredine L, Gouasmi S, Laissaoui R, Zeghib NE (2012) Evaluation and measurement of the damping properties of laminated CFRP composite plates. Materiaux 28
3. Kulhavy P, Syrovatkova M, Srb P, Petru M, Samkova A (2017) Irregular winding of pre-preg fibres aimed at the local improvement of flexural properties. Tekstilec 60(5):310–316
4. Kulhavy P, Lepsik P (2018) Comparison of modal characteristic of carbon prepreg tubes manufactured by wrapping and winding. J Vibro Eng 20(1):519–529
5. Harry NE (2012) Clean filament winding: process optimisation. The University of Birmingham, United Kingdom, Birmingham
6. Allen AK (2004) A method for winding advanced composites of unconventional shapes using continuous and aligned fibers. In: All theses and dissertations. 346. Brigham Young University, Provo
7. Wang X, Xiao J, Wen L (2010) Winding pattern design and simulation of s-elbow. Chin J Aeronaut 23(5):573–577
8. Laval Ch (1991) Process simulation in filament winding. Reinf Plast 32(11):40–42
9. Li H, Liang Y (2002) Computer aided filament winding for elbows. J Softw 13(4):518–525
10. Wu M, Liang Y, Yu Y (2001) Stability of geodesic on torus. Appl Math J Chin Univ 16(4):480–485
11. Harlow JB (2003) Astrophysics. University of Toronto, Toronto
12. James A (2004) Loxodromes: a Rhumb way to go. Math Mag 77(5):349–356
13. CADWIND (2013) User manual. Material SA

Principles of Increasing the Winding Effectivity of Composite Prepregs to Construction Application

Michal Petrů and Josef Vosáhlo

Abstract The development of composite materials for weight reducing is a trend in transport and other industrial areas. The composition of the composite material generates a synergistic effect that provides their advantages. The composite consists of a dispersive and continuous phase. For composites, different fibers (carbon, glass, basalt, textile and natural fibers) and plastic matrices with specific properties can be applied. The type, quantity and arrangement of fibers in the composite are designed for strength characteristics depending on the maximum applied load. Samples of the composite material were made by fiber winding technology on non-bearing core with required geometry. An image analysis of the structure and morphology was performed by scanning electron microscopy (SEM). The study of matrix penetration among fibers has shown that the directional fiber winding production technology in the case of uncontrolled winding significantly affects the resulting ratio between fiber and matrix area. This was reflected in the standard tensile, bending and impact tests by a reduction of mechanical properties of the resulting composite. Insufficiently, saturated areas create the source of defects and crack propagation in the composite. Mechanical properties were determined from numerical models. The nature of strain and stress distribution in individual layers have been identified. The results of the mechanical tests were compared with the numerical model. A comparison of experimental results and numerical models shows that fully controlled winding allows increasing mechanical characteristics of the resulting composite. The cause can be found in the optimal distribution of the fibers in the composite.

Keywords Composite materials · Winding fibers · FEM · Mechanical properties

1 Introduction

The development of composite materials leads to the improvements of their physical and chemical properties and, therefore, spreads into engineering, construction and other industrial applications [1–4]. Improving mechanical properties while reducing

M. Petrů (✉) · J. Vosáhlo
Technical University of Liberec, Studentská 1402/2, 46001 Liberec, Czech Republic
e-mail: michal.petru@tul.cz

Š. Medvecký et al. (eds.), *Current Methods of Construction Design*, Lecture Notes in Mechanical Engineering, https://doi.org/10.1007/978-3-030-33146-7_63

weight and extending the life of parts and frames is one of the directions in which the development of composite materials is oriented [5, 6]. The composite material is comprised of continual phase—matrix (divided into three basic groups: metallic, ceramic and polymer) and a phase of dispersion—reinforcement (particle or continual reinforcement). Mutual interactions of these phases create a synergic effect that is the added value of composite materials compared to classic, for example, metallic materials. For the production of FRP [7] natural, technical and special fibers and reinforcements produced using textile technology (fabrics, multiaxial fabrics, plaited and nonwoven structures) are used. Considering the focus on specific shapes of the final composite parts, it is necessary to develop new technologies of production of composite materials. The classic fibrous structures are in certain cases replaced by pre-impregnated fibrous reinforcements—prepregs [8]. The controlled laying of directionally oriented fibrous layers for the production of spatially formed and closed composite parts the technology of winding [9, 10] was used and modified. The base of spatially formed or closed components (frames) is the core. The core does not have a carrying function; it creates the base of the component mostly due to its shape. The individual directionally oriented layers are put onto the core creating the final composite [11, 12]. A prototype technology of automatic fiber winding is used to create directionally oriented fiber layers onto the surface of the non-bearing core, which is developed in the laboratory of the Institute for Nanomaterials, Advanced Technologies and Innovation at the Technical University of Liberec [13, 14]. The device is capable of simultaneously placing two to three directionally different layers. The combination of angles in consecutive layers is determined based on material tests and mathematical models [14–19], in order to match predicted mechanical properties of the proposed composite [20, 21]. Currently, a third generation of winding head is being used in conjunction with two robots. The winding head of the third-generation winds simultaneously two layers of fibers under various angles as was stated in [22]. The device consists of two independently rotatable robot-controlled carrier rings, carrying the coils with material. Each coil has a separately adjustable break for adjusting the desired tension during winding. The device is carried by the robot for controlled manipulation in space, and the robot simultaneously controls the drives of carrier rings, which are configured as external robot axis. The second robot carries the core. Manipulation with the core was originally performed manually and so was the synchronization of the rotation of carrier rings. This method, however, did not provide the exact fiber-laying angle or the production reproducibility of the molded parts. A version with robots was, therefore, developed linked up into the RoboTeam configuration whose activities are possible to coordinate together. One of the robots controls the movement of the core through the winding device, for the fibers to be placed under the desired angle and in the desired amount. The second robot holds the winding device. The solution is easy and fast to implement.

2 Materials and Methods

The development of composite parts is divided into several basic directions which when put together lead to the merge of all the acquired knowledge and the manufacture of the part with required attributes.

2.1 Mathematical Modeling of Fiber Reinforcement Winding onto the Core Geometry of the Composite Part

A fibrous strand (fibrous reinforcement), which is wounded longitudinally or at an angle onto the core, will be assumed to have a regular rectangular shape clearly determined by the given thickness h and width b. The strand wounded onto the core in the layer j either moves longitudinally (vertically) along the circumference of the core with the length m, therefore, the center of the axis becomes a straight line (longitudinally winded fibers), or the strand is placed spirally onto the circumference of the core. Therefore, the center axis becomes a helix, which rotates around the axis z at an angle of ascent α_j, while α_j, is the angle of ascent of the fibrous strand in tth layer. These geometric formations (line, helix) form surfaces given by the cross section of the fibrous strand as well as create volume geometry by their path of the desired length of winding. Next, we define point P (r_j, 0, 0, 1), which will be the center of the axis of the fiber strand in jth layer, and r_j will be the radius of the axis center distance of the fiber strand (see Fig. 1).

In a single layer of the fiber strand with respect to the design of the optimal strength and stiffness, it can be assumed that there exist a number of fibrous strands, whose maximum possible amount will be dependent on the desired number of layers

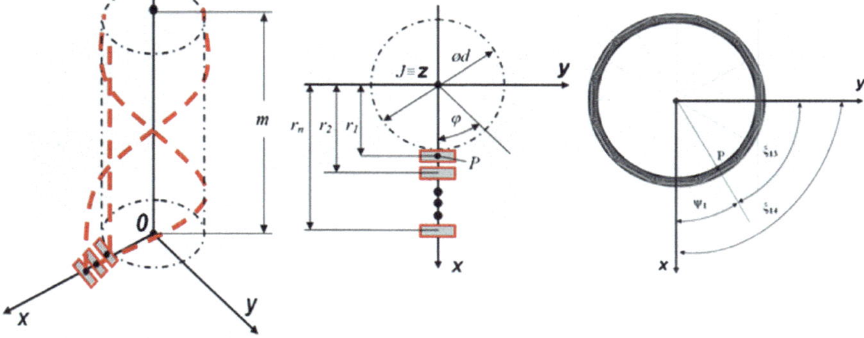

Fig. 1 Principle of winding fibrous strands onto the geometrical shape of the core composite component

for the final frame. To generate a precisely defined amount of fiber strands, we then introduce an angle allowing the rotation of the axis center of the point that will belong to any fibrous strand. Modeling of the winding of the fiber strands amount will be described according to the relationship (1):

$$\left.\begin{array}{l} x(\varphi) = r_j \cos\left(\psi_j + \xi_{ji} + q\varphi\right) \\ y(\varphi) = r_j \cos\left(\psi_j + \xi_{ji} + q_\varphi\right) \\ z(\varphi) = \varphi \frac{r_j}{tg\varphi_j} \end{array}\right\} = \left.\begin{array}{l} x(\varphi) = r_j \cos(\psi\_j + (i-1)2\pi/m\_j + q\varphi) \\ y(\varphi) = r_j \cos(\psi\_j + (i-1)2\pi/m\_j + q\varphi) \\ z(\varphi) = \varphi \frac{r_j}{tg\varphi_j} \end{array}\right\} \quad (1)$$

where ψ_j is the angle of rotation of the first fibrous strand in *j*th layer, *mj* is the number of fibrous strands in *j*—té layer ($j = 1, 2, \ldots n_j$), *q*, defines the direction of rotation, i.e., $q = 1$ for the positive direction or $q = -1$ for the negative direction.

The parametric Eq. (1) allows to mathematically defining geometry of the core shape and geometry as well as position of any fibrous strand that winds the core of the future composite component (see Fig. 2).

For the load causing tensile, compressive or bending stresses in the frame, it will be most appropriate for the fibrous strands to have the direction of winding 0° for maximum stiffness and strength. Since the maximum stress of the fibers in the direction of force load is used, as is apparent in Fig. 3, fibers arranged at an angle of 45° will not sufficiently transfer the load under tensile stress.

This, however, changes for the torsion stress of the frame where the best direction of winding of the fibrous strand is 45° for optimal stiffness and strength. When loading the frame by internal or surface pressure, the most optimal orientation for optimal stiffness and strength will be angular orientation. It is desirable to design the composite frames with at least three layers $\nu = 3$ in combination with fibrous strand

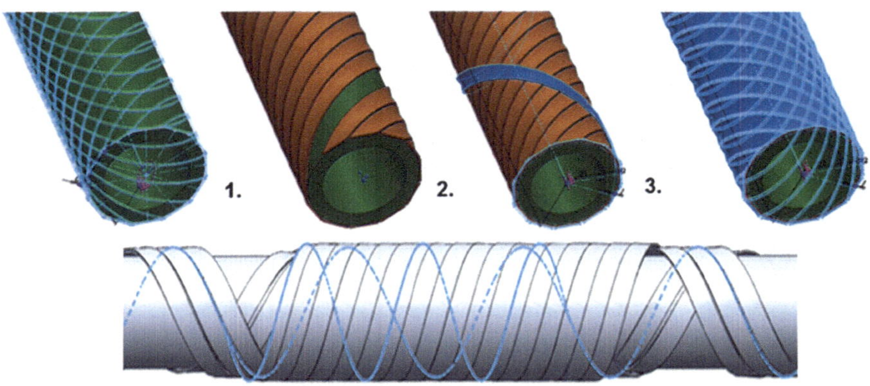

Fig. 2 Model simulation for controlled fiber winding optimization

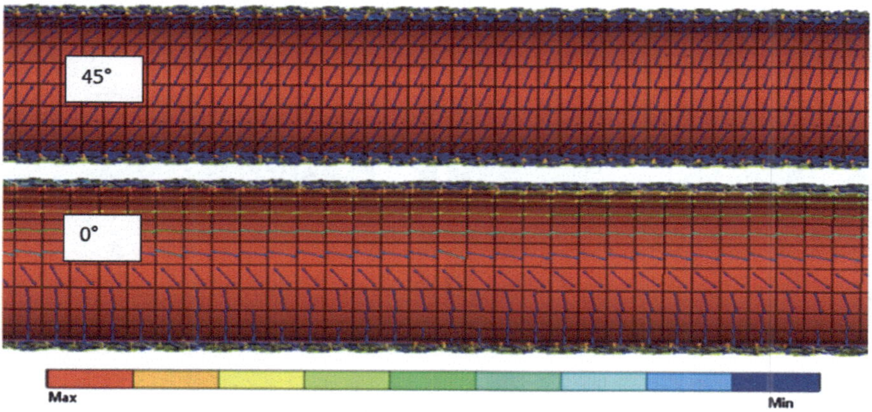

Fig. 3 Strength of lamina comparison **a** CF/thermosetic matrix, **b** CF/thermoplastic matrix

layers placed at an angle of $\alpha^{\nu=3} = 0°; +45°; -45°$ for random loads (pseudo-loads, combined loads, etc.) and due to orthotropic properties of individual layers.

2.2 *Material Research, Laboratory Preparation of Composites and Their Testing*

Composite materials are differentiated according to the matrix type and their reinforcement. The large group is composite materials with polymer matrix reinforced by long fibers. Long fibers for technical applications namely carbon, glass, basalt, aramid and others are processed into longitudinal and flat textiles by various technologies. The type of reinforcement and the arrangement/interconnection of fibers are related to the production technology; properties of a composite are directly dependent on the way fibers are deposited. The primary task is always to determine the appropriate material composition and the technology of composite material production. The selected chemical-mechanical material structures are continuously tested. ASTM standards [23–26] and corresponding experimental preparations were selected to cover the possibilities of verifying material models of composed composite structures for various levels of homogenization of material models of composite materials. Prepregs are pre-impregnated fibrous reinforcements (UD—Unidirectional Layers, Different Bonded Fabrics). Their advantage is the relatively high-weight/volume portion of fibers. The disadvantage is the storage at very low temperatures, the possibility of contamination by impurities during normal temperature handling, final curing under precisely defined condition in the autoclave and a relatively high price. For experiment was prepared composite specimens. Carbon and glass fibers were wounded onto the non-bearing core in six layers (+45°/0°/−45°/+ 45°/0°/−45°).

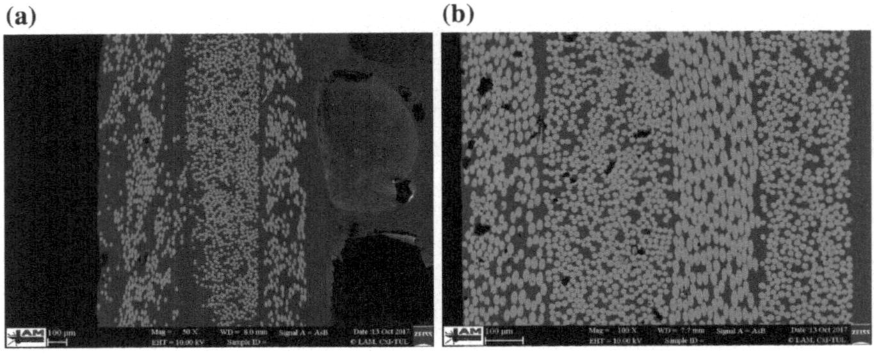

Fig. 4 GFRP—arranged distribution of fibers in composite skin

Part of the material research is the evaluation of the quality of the produced composite by checking the internal structure. In particular, the presence of defects and cracks in individual laminas and between layers. Samples were taken after the manufacture of the final composite to check the quality of saturation of glass fiber layer matrix. The test was conducted in LAM at the Technical University of Liberec using the scanning electron microscope. Figure 3 shows a comparison of the composite skins of two functional samples of composite materials.

Figure 4a shows the composite skin including the base from non-bearing core from porous polyurethane foam which was produced without checking the angle of fiber reinforcement winding. The composite skin itself is a gray matter in which light-gray shapes are deployed—glass fibers. Figure 4b shows the detail of the composite skin, which has been produced with a camera check of the fiber reinforcement winding angle. As can be seen from the comparison of these two results of the image analysis, the controlled winding produced better results in terms of volume fraction of fibers in the matrix.

2.3 The Technology of Production of Composite Parts

The knowledge gained from mathematical modeling and experimental testing allows us to lay the foundation for technology of composite component production. The technology of winding is a process that keeps the continual fibers mutually arranged along the whole length of the produced part. In order to produce directionally oriented layers of composite reinforcement, a prototype technology of automatic fiber winding was used. The concept of the winding head is a modified prototype of an originally patented technology used to winding the fibers to produce closed frames of non-constant cross section. The materials used can be both classical dry roving and pre-impregnated fibers, which in their size and arrangement copy dry roving. The trajectory, motion and tilting of the head were optimized and controlled using mathematical algorithms based on the shape of the control curve of core geometry being the core of the composite. The change of the winding angle and its related composite parameters are possible to directly affect by changing the ratio of the feed rate and the rotation of the impellers. The winding head construction is also adapted to insert a closed core shape.

3 Results and Discussion

This work presents a method of using innovative winding technology for closed 3D formations, which allow simultaneous local optimization of the winding angle in individual layers. At the same time, it presents an option of automated device for controlled placement and feedback angle verification of fibrous reinforcement using PLC. The device was tested both for composites made from conventional fiber roving and pre-impregnated unidirectional oriented fibers. An integral part of the work was also the acquisition of knowledge about the properties and behavior of prepregs. The article represents only a part of the measurements and results obtained. The mathematical models using measurement results to determine elastic constants are also an inherent part. These are the basis for modeling of most tests which have destructive character and would otherwise prolong the work and make it more expensive.

Acknowledgements This work is partially supported by the Ministry of Education, Youth and Sports of the Czech Republic and the European Union in the frames of the project "Modular platform for autonomous chassis of specialized electric vehicles for freight and equipment transportation", Reg. No. CZ.02.1.01/0.0/0.0/16_025/0007293.

References

1. Kulhavy P, Syrovatkova M, Srb P, Petru M, Samkova A (2017) Irregular winding of pre-preg fibres aimed at the local improvement of flexural properties. Tekstilec 60(4):310–316
2. Petru M, Mlynek J, Martinec T, Broncek J (2016) Mathematical modeling of fibre winding process for composite frames. Komunikacie 18(4):103–111
3. Agarwal BD, Broutman LJ, Chandrashekhara K (2006) Analysis and performance of fiber composites 1:576
4. Gay D, Hoa SV (2007) Composite materials. 2nd edn. Press Taylor & Francis Group
5. Laval Ch (2003) Composites design in the real world. Reinf Plast 47(8):50–53
6. Chand S (2000) Carbon fibers for composites. J Mater Sci 35(6):1303–1313
7. Militký J (2012) Textile fibres, classic and special. Mater Sci Forum 919:152–159
8. Petrů M, Syrovátková M, Martinec T, Lepšík P (2015) Analysis of changes in the surface quality of a UD prepregs composite due to mechanical loading. Mat Sci Forum 818:109–112
9. Allen AK (2004) Method for winding advanced composites of unconventional shapes using continuous and aligned fibers. Master thesis. Brigham Young University, Provo-Utah
10. Harry NE (2012) Filament winding, process optimisation. Doctoral thesis. The University of Birmingham, Birmingham

11. Neckar B, Das D (2012) Modelling of fibre orientation in fibrous materials. J Text Inst 103(3):330–340
12. Petru M, Broncek J, Lepsik P, Novak O (2014) Experimental and numerical analysis of crack propagation in light composite materials under dynamic fracturing. Komunikacie 16(3a):82–89
13. Ševčík L, Tuma P, Petru M, Martinec T, Kovar R (2013) Composite reinforcement. Patent application: 1066–2013
14. Ševčík L, Petru M, Kovar R, Srb P, Kulhavy P (2016) Prototype of winding device for composite frames reinforced by long fibers
15. Petru M, Mlynek J, Martinec T (2018) Manufact Technol 1:10
16. Petru M, Novak J (2018) Finite element method—simulation. Numer Anal Solution Tech 1:322
17. Liu D, Fleck N, Sutcliffe M (2004) Compressive strength of fibre composites with random fibre waviness. J Mech Phys Solids 52(7):1481–1505
18. Petrů M, Novák O (2017) FEM Analysis of Mechanical and Structural Properties of Long Fiber-Reinforced Composites. Intechopen, UK
19. Rektorys K (1980) Varitional methods in mathematics, science and engineering, 2nd edn. D. Reidel Publishing Company, Boston
20. Smith PA (2000) Carbon fiber reinforced plastics—properties. Compr Compos Mater 2:107–150
21. Ghamarian N, Hanim MA, Penjumras P, Majid DL (2016) Effect of fiber orientation on the mechanical properties of laminated polymer composites. In: Reference module in materials science and materials engineering, Elsevier
22. Petru M, Mlynek J, Martinec T (2017) Composite production and industrial robot trajectory calculation. Mech Mach Sci 44:271–276
23. ASTM D3039—Standard test method for determining tensile properties of fiber reinforced polymer matrix composites used for strengthening of civil structures
24. ASTM D3410—Standard test method for compressive properties of polymer matrix composite materials with unsupported gage section by Shear loading
25. ASTM D7078—Standard test method for shear properties of composite materials by V-Notched Rail Shear method
26. ASTM D6272—Standard test method for flexural properties of unreinforced and reinforced plastics and electrical insulating materials by four-point bending

Boiling Heat Flux of Multilayer Mesh Microstructures

Norbert Radek, Łukasz J. Orman and Jozef Brončck

Abstract The paper discusses the issue of enhanced boiling heat transfer with the use of metal mesh microstructures. Such heat exchangers can significantly increase heat flux value dissipated during pool boiling. It is especially vital for the production of efficient phase-change heat exchangers used e.g. in refrigeration or electronic cooling. The article provides special focus on a selected heat flux determination method and presents the comparison of the chosen calculation technique with the experimental data of nucleate boiling heat transfer on copper meshed surfaces under ambient pressure. The considered boiling agents and distilled water and ethyl alcohol. The comparison has proven that the congruence between experimental and calculation results may be satisfactory, but not for all the data. Thus, a more reliable model or a correlation that could more successfully determine the heat flux value based on material and geometrical parameters of the microstructures is still needed to be developed.

Keywords Boiling heat transfer · Meshes · Microstructures

1 Introduction

Boiling is a phase-change phenomenon during which significant heat fluxes can be exchanged at small temperature differences. Additional coatings applied on the heater surface can further increase dissipated heat fluxes and rise the efficiency of the process. Such coverings occur in different shapes and are made with technologies (sintering, thermal spraying, etc.). The experiments of Li et al. [1] were focused on water boiling on a horizontal copper heater covered with mesh layers. The structure was sintered at 1030 °C in the gas mixture of nitrogen and hydrogen. Five different

N. Radek (✉) · Ł. J. Orman
Kielce University of Technology, Al. 1000-Lecia Państwa Polskiego 7, 25314 Kielce, Poland
e-mail: norrad@tu.kielce.pl

J. Brončck
University of Žilina, Univerzitná 8215/1, 01026 Žilina, Slovak Republic
e-mail: jozef.broncek@fstroj.uniza.sk

Š. Medvecký et al. (eds.), *Current Methods of Construction Design*, Lecture Notes in Mechanical Engineering, https://doi.org/10.1007/978-3-030-33146-7_64

mesh types were tested. The number of layers ranged from 2 to 9. Heat transfer coefficients of samples with four, six and eight layers were almost identical, only the sample with two layers provided lower heat flux, which was attributed to the flooding of the structure with the liquid. All meshed surfaces enhanced boiling in comparison with the smooth surface. Also an increase in the critical heat flux with increasing height was observed. This heat flux reached 367.9 W/cm^2 and was three times higher than for the smooth surface. Four reasons of the heat transfer enhancement in the porous layers were given in the paper: reduction of heat flux on the heater through additional surface extension, the presence of contact points which join the microstructure with the heater (it reduces the process of vapour film forming on the surface), increased active nucleation sites density and extended surface for evaporation, better liquid transport caused by the flow induced by capillary forces. Li and Peterson [2]—while investigating water boiling at atmospheric pressure on a horizontal copper surface with a copper mesh coating—found out that the application of meshes enhanced boiling heat transfer in comparison with the smooth surface. It was observed that for smaller apertures (distance between the wires in the mesh), higher heat fluxes could be dissipated. In order to determine the impact of volumetric porosity, samples of six mesh layers were produced and compacted to a different porosity, but only a small influence of porosity was recorded. In their paper, Wong and Kao [3] experimentally analysed heat pipes performance with a two-layered mesh wick coating underwater evaporation/boiling conditions. The fine mesh provided more nucleation sites. It needs to be noted that the number of nucleation sites (locations where bubbles are created and grown) influences the boiling performance of the heater.

In the presented paper, mesh microstructures are analysed due to their low cost of production, reliability and repeatability in the production process and regular geometry (of open porosity).

2 Material and Method

The copper mesh microstructures have been produced with the sintering method. The meshes were applied onto the heated, clamped mechanically and inserted into the oven. Sintering took place at the temperature of ca. 940 °C. Figure 1 presents the image of the cross-section of the produced multilayer sample. Copper meshes of aperture 0.50 mm were used for the samples preparation.

The boiling tests have been performed with two and three mesh layers with distilled water and ethyl alcohol as the boiling liquids. For the increased heat flux value provided to the sample with the electric heater, temperature difference (superheat) between the heater surface and the saturation temperatures of the liquid was recorded. Thus, boiling curves (as presented later in the paper) could be drawn. The experiments were conducted under ambient pressure. The vapour produced in the process was condensed and returned to the boiling vessel. Heat flux provided to the sample was determined with the conduction method based on temperature readings within the heater. Temperature values were recorded with K-type thermocouples. Location

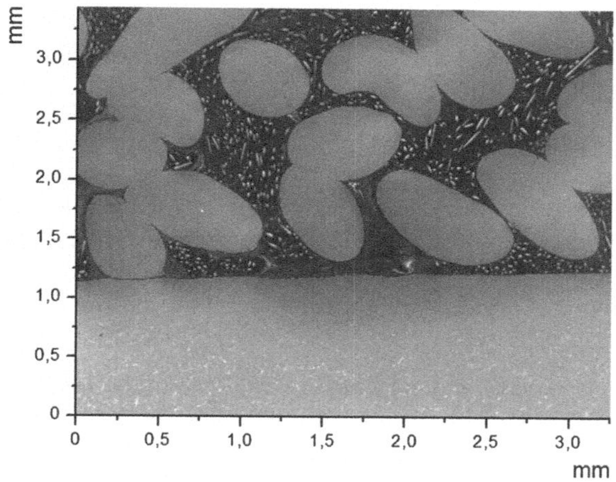

Fig. 1 SEM image of the mulitmesh microstructure, magnification 50x

Fig. 2 Main unit of the experimental stand: 1, 2—thermocouples under the sample, 3, 4, 5—thermocouples in the axis, 6—electric heater, 7—copper block, 8—soldered sample

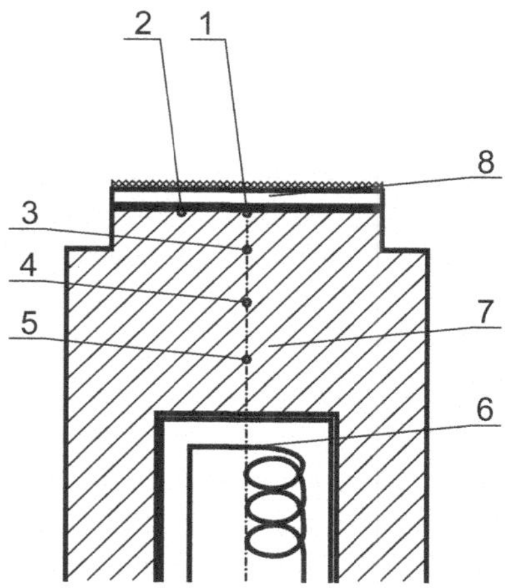

of the thermocouples has been presented in Fig. 2. The additional thermocouple was placed within the boiling liquid in order to read the saturation temperature of the boiling liquid.

3 Heat Flux Determination Model

For the proper design of heat exchangers and solving other engineering problems, it is often necessary to precisely determine the heat flux value which is dissipated by a system. It might be done using dependencies of heat flux on physical and chemical properties provided by different authors across the years in the literature. According to Webb [4], correlations and models developed for microstructure coated surfaces can be divided into:

- exponential correlations—produced with data fitting procedures basing on dimensional and non-dimensional variables;
- asymptotic correlations—developed between minimal and maximal values;
- analytical models that take into account heat transfer and fluid movement mechanisms;
- numerical solutions of complex equations of fundamental laws governing the process.

There are many models and correlations available in the literature for different coating types. Researchers mathematically described boiling heat transfer to suit their own ideas of the physical phenomena of boiling. Smirnov and co-workers [5, 6] experimentally and theoretically analysed boiling heat transfer on meshed surfaces. The model considers that vaporisation occurs in each elementary cell of the microstructure. The heat flux (q) can be determined with the following formulae:

$$q = C\sqrt[6]{\frac{r\sigma\lambda_1^3}{\nu_1}}(1-\varepsilon)^{\frac{5}{6}}\sqrt[3]{\lambda_{\text{eff}}}\left(\frac{\Delta T - \Delta T^*}{L_0}\right)^{\frac{5}{6}} \tag{1}$$

where C—constant, r—heat of vaporisation, λ—thermal conductivity, ε—porosity, σ—surface tension, ν—kinematic viscosity, ΔT—temperature difference (superheat), l—liquid, eff—effective. The specific dimension L_0 is expressed as

$$L_0 = \left\{a\left[\frac{(a+d)^2 - a^2}{4a}\right]^4\right\}^{0.2} \tag{2}$$

where 'a' is the mesh aperture (distance between the wires) and 'd' is the wire diameter on the considered mesh.

The model presented above enables to provide heat flux values for the given parameters of the sample and the temperature difference. It is relatively simple; however, it is worth verifying if it provides proper heat flux values of samples other than those tested by the authors of this model.

4 Results and Discussion

Based on the experimental results, the comparison of the obtained data with the calculations in accordance with the model presented above has been done. Figures 3 and 4 present the results for pool boiling on multilayer structures for distilled water and ethyl alcohol, respectively. The measuring points on the graphs were recorded for increased value of heat flux.

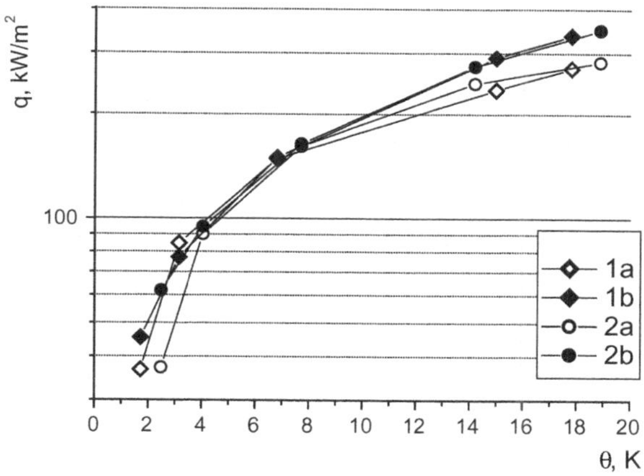

Fig. 3 Comparison of experimental data and calculation results according to Smirnov et al. [5, 6] model, boiling liquid: distilled water; 1a—two-layer mesh coating (experimental data as in [7]); 1b—two-layer mesh coating (calculation results), 2a—three-layer mesh coating (experimental data as in [7]); 2b—three-layer mesh coating (calculation results)

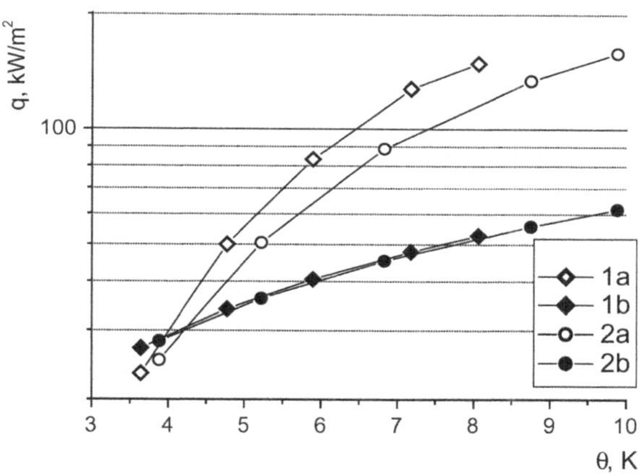

Fig. 4 Comparison of experimental data and calculation results according to Smirnov et al. [5, 6] model, boiling liquid: ethyl alcohol; 1a—two-layer mesh coating (experimental data as in [8]); 1b—two-layer mesh coating (calculation results), 2a—three-layer mesh coating (experimental data as in [8]); 2b—three-layer mesh coating (calculation results)

As can be seen in Fig. 3, congruence between the experimental and calculation results is very satisfactory in the case of water boiling. However, if the working fluid is ethyl alcohol, the determination of heat flux can be considered satisfactory only for low superheats.

Development of a model or a correlation that could successfully predict the performance of microstructure coated heaters (not only mesh coated surfaces) is undoubtedly challenging; however, it could lead to better design possibilities of heat exchangers used in refrigeration, electronic cooling and other areas of engineering. Especially challenging could be the development of the universal model that could be successful for the use with irregular microstructural coatings as well.

5 Conclusions

Meshes can increase the heat flux value exchanged during pool boiling. However, a reliable model of enhanced heat transfer phenomenon is still not available. Undoubtedly, more work needs to be done in this area since the knowledge of phase-change processes is crucial for correct design of heat exchangers used, for example, in refrigeration systems or cooling of electronic devices. It needs to be noted that microstructures can also be used in flow boiling applications [9, 10]. Although the present paper deals with pool boiling, laser-treated surfaces can also be highly effective in the flow boiling heat transfer mode as indicated in [11, 12]. Future experimental and theoretical work of the authors would also cover boiling heat transfer of new and more efficient cooling agents such as nano fluids.

References

1. Li C, Peterson GP, Wang Y (2006) Evaporation/boiling in thin capillary wicks (I)—wick thickness effects. J Heat Transf 128:1312–1319
2. Li C, Peterson GP (2006) Evaporation/boiling in thin capillary wicks (II)—effects of volumetric porosity and mesh size. J Heat Transf 128:1320–1328
3. Wong SC, Kao YH (2008) Visualization and performance measurement of operating mesh—wicked heat pipes. Int J Heat Mass Transf 51:4249–4259
4. Webb RL (1994) Advances in modeling enhanced heat transfer surfaces. In: Proceeding of X international heat transfer conference, Brighton, 445–459
5. Smirnov GF (1977) Približennaja teorija teploobmena pri kipenii na poverchnostjach pokrytych kapilljarno—poristymi strukturami. Teplo energetika 9:77–80
6. Smirnov GF, Afanasiev BA (1982) Investigation of vaporisation in screen wick—capillary structures. In: Proceeding of VI international heat pipe conference advances in heat pipe technology, London, 405–413
7. Dąbek L, Kapjor A, Orman ŁJ (2016) Multilayer mesh structures for efficient phase change heat exchangers. In: Proceedings of XX the application of experimental and numerical methods in fluid mechanics and energy, University of Žilina, Slovakia, 17–23
8. Dąbek L, Kapjor A, Orman ŁJ (2016) Ethyl alcohol boiling heat transfer on multilayer meshed surfaces. EPJ web of conferences 1745
9. Piasecka M (2014) Heat transfer research on enhanced heating surfaces in flow boiling in a minichannel and pool boiling. Ann Nucl Energy 73:282–293
10. Pastuszko R (2010) Pool boiling on micro—fin array with mesh structures. Int J Therm Sci 49(12):2289–2298
11. Pietraszek J, Radek N, Bartkowiak K (2013) Advanced statistical refinement of surface layer's discretization in the case of electro-spark deposited carbide-ceramic coatings modified by a laser beam. Solid State Phenom 197:198–202
12. Radek N, Wajs E, Luchka M (2008) The WC-Co electrospark alloying coatings modified by laser treatment. Powder Metall Met C 47:197–201

Properties of Steel Welded with CO_2 Laser

Norbert Radek, Jacek Pietraszek, Jozef Bronček and Peter Fabian

Abstract This paper discusses the effect of selected parameters of laser welding on the mechanical properties of welds. Two parameters were analysed: the welding speed and the laser power. The properties of the material in the fusion zone and the heat-affected zone were determined by performing static tensile tests, hardness tests and microscopic analysis. The welding was carried out using a CO_2 laser (wavelength $\lambda = 10.6\ \mu m$). The specimens were 0.8 mm in thickness, and they were made of DC04 steel. The results indicate that welds produced at different welding parameters have similar mechanical properties. The experiment was conducted and analysed according to requirements of the design of experiment methodology. Laser beam welding allows us to join various dissimilar metals, which have not been possible with conventional methods. It should be emphasized; however, that not all metals are suitable for laser welding; it all depends on their thermal and structural properties. For example, tin, aluminium, lead, tantalum and tungsten are generally difficult to weld.

Keywords Laser welding · Fusion zone · Properties · OFAT · Main effects · Effects analysis

1 Introduction

We are currently observing the dynamic development of laser techniques in materials processing. The laser beam can be used for cutting, welding, surface modification, etc. [1–5].

N. Radek (✉)
Kielce University of Technology, Al. 1000-Lecia Państwa Polskiego 7, 25314 Kielce, Poland
e-mail: norrad@tu.kielce.pl

J. Pietraszek
Cracow University of Technology, Al. Jana Pawła II 37, 31-864 Cracow, Poland

J. Bronček · P. Fabian
University of Žilina, Univerzitná 8215/1, 01026 Žilina, Slovak Republic

Š. Medvecký et al. (eds.), *Current Methods of Construction Design*, Lecture Notes in Mechanical Engineering, https://doi.org/10.1007/978-3-030-33146-7_65

The process of laser beam welding involves melting the edges of the metals to be joined with a focused beam of radiation at a power density of $10^4 \div 10^6$ W/mm^2 [6]. The development of lasers with a high input power made it possible to join thick plates by directly melting the adjacent edges with no need to chamfer them or use time-consuming filling of the bevel groove. This method of fusion is also used in electron beam welding, but the process has numerous disadvantages, for example, the necessity to place the elements to be welded in vacuum, the necessity to remove random magnetic fields as well as the necessity to protect the workpiece against X-ray radiation occurring during high-voltage electron beam welding [7].

The process of laser welding has the following advantages:

- high purity of the process (dependent on the surface preparation and gas purity),
- joining difficult to weld materials,
- easy automation,
- welding with high precision (e.g. joining thin and thick elements),
- high speed of the welding process,
- one source of radiation for several welding stations,
- welding performed under atmospheric pressure (by contrast, electron beam welding is performed in vacuum),
- high power density (in the case of deep welding), and as a result, small distortions,
- narrow heat-affected zone.

The work discusses the properties of laser-welded joints. Properties were determined based on the results of microstructure analysis, hardness tests and static tensile tests with statistical analysis support.

2 Materials and Methods

2.1 Materials

The specimens to be tested were cut from a steel sheet 0.8 mm in thickness. As a material resistant to ageing, DC04 steel is used extensively in high-performance applications including deep drawing and drawing and in the transport industry. DC04 steel is characterized by good weldability which means no special preparation of the work is needed. Table 1 shows the chemical composition of DC04 steel, respectively.

Table 1 Chemical composition of DC04 steel

Steel grade	Maximum content [%]				
	C	Mn	P	S	Ti
DC04	0.08	0.40	0.030	0.030	0

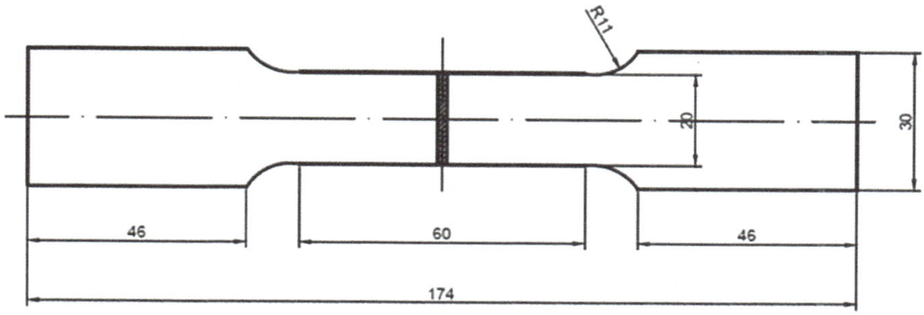

Fig. 1 Shape and dimensions of the specimens used in the tests

The specimens were prepared by cutting them with a laser from a steel sheet. They were then cut into two equal parts to be laser welded. The dimensions and shape of a single specimen are presented in Fig. 1.

2.2 *Welding*

Laser cutting and welding were performed by means of a Lasercell 1005 CO_2 laser (wavelength $\lambda = 10.6\ \mu m$). A photograph of the Lasercell 1005 is shown in Fig. 2.

Fig. 2 Trumpf Lasercell 1005 CO_2 laser

Table 2 Values of the main process parameters

Number of the laser pass line	Laser power P [kW]	Welding speed v [m/min]
A	1.5	3
B	1.8	3
C	2	3
D	2.5	3
E	2	2
F	2	2.5
G	2	3.5
H	2	4
I	2	5

The welding tests were conducted according to one-factor-at-time (OFAT) scheme [8] i.e. selectively changing one factor, while others are fixed. It allows to investigate main effects with negligible/ignored interactions. Two parameters were changed: the laser power at four levels and the welding speed at six levels. According to the literature [9–11], these two parameters have the greatest influence on the quality of welds.

Table 2 provides the parameters of laser welding used during the tests.

3 Analytics

The analysis of the properties of the joints produced by laser welding involved examining their microstructure, measuring their hardness and performing static tensile tests.

The microstructural observations were conducted using a Joel JSM-5400 scanning electron microscope. The hardness was measured with a NEXUS 4304 tester at a load of 1 kg applied for 10 s. The static tensile tests were carried out using a LabTest 5.20SP1 universal testing machine.

4 Results

The macroscopic analysis of the laser pass lines and the phenomena accompanying the welding process helped select the main parameters of laser welding: the laser power and the welding speed.

The laser pass lines A–D were formed at a welding speed of 3 m/min, and an initial power of 1.5 kW gradually increased to 2.5 kW. When the pass line was produced at a power of 1.5 kW, plasma was not present. At a power of 1.8 kW, a small plasma cloud was observed, but its influence on the welding process was negligible. When the laser pass line was formed at a power of 2 kW or 2.5 kW, a positive effect of the plasma cloud was observed.

After the macroscopic examinations of the laser pass lines, it was assumed that the tests would be conducted at a laser power of 2 kW. During the next five tests, the laser pass lines E–I were produced by changing the welding speed from 2 to 5 m/min, with the laser power being constant ($P = 2$ kW). It was found that when the welding speed was 5 m/min, the weld penetration was incomplete (pass line I).

The experimental data were analysed to select the laser welding parameters:

- laser power $P = 2$ kW;
- spot diameter $d = \phi 2$ mm;
- welding speed v: 2; 2.5; 3; 3.5; 4 m/min;
- shielding gas: argon $Q = 10$ l/min;
- preheat time $t = 5$ s;
- pulse repetition rate $f = 30{,}000$ Hz;
- nozzle-workpiece distance $\Delta f = 0$ mm.

The tests were conducted for five series of laser-welded specimens and one series of unwelded specimens, where the base metal was DC04 steel. There were three specimens in each series numbered from 0 (base metal) to 5 (where $v = 4$ m/min). Figure 3 shows examples of the specimens from series 0–5.

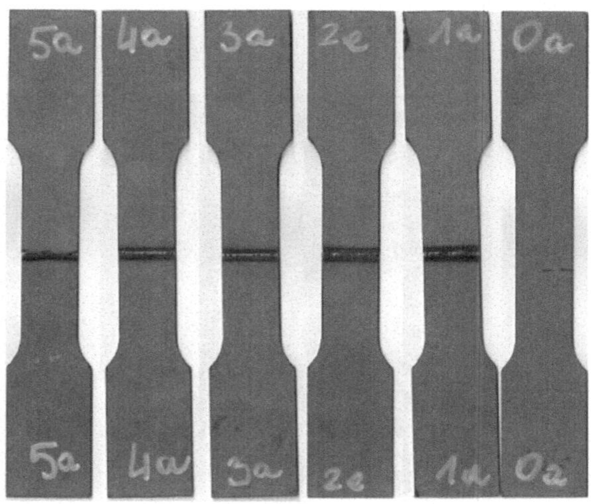

Fig. 3 Examples of specimens from series 0–5

5 Discussion of Results

5.1 Microstructure Analysis

The microstructure of the welded joints was analysed using a Joel JSM-5400 scanning electron microscope.

The microscopic examinations of the welded joints were performed on polished metallographic specimens in the plane perpendicular to the weld, which enabled observation of the fusion zone and the heat-affected zone. It was also possible to analyse the structures formed and to measure the width of the heat-affected zone (HAZ). The aim of the analysis was to compare the resulting microstructures of the welded joints and to determine how variable values of the welding speed affect the shape of the weld.

The laser welding process is characterized by very high heating and cooling rates, which lead to the narrowing of the width of both the fusion zone and the heat-affected zone.

Figure 4 shows an image of the microstructure of the base metal. The results suggest that the base metal had a coarse-grained ferrite–pearlite structure. From Fig. 4, it is clear that the grains are arranged in bands, which indicate that the production of steel sheets (DC04 steel) involved rolling.

The analysis of a specimen of series 1 showed that the material in the HAZ had a fine-grained ferritic–pearlitic microstructure with visible metallic precipitates in the pearlite grains.

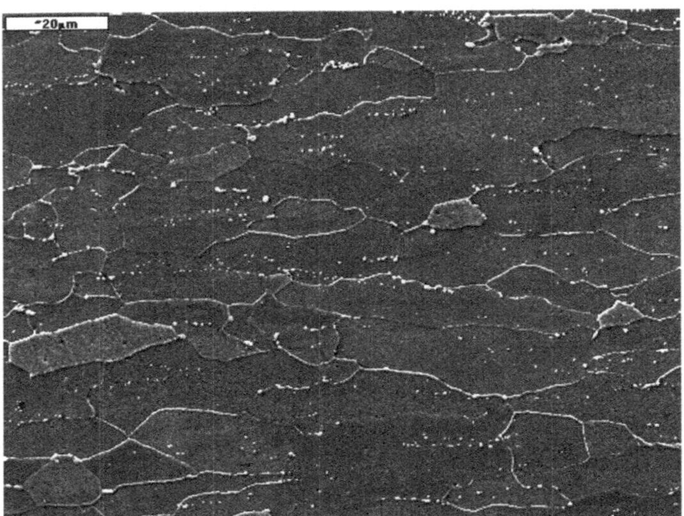

Fig. 4 Ferritic–pearlitic microstructure of the base metal (× 1000 magnification)

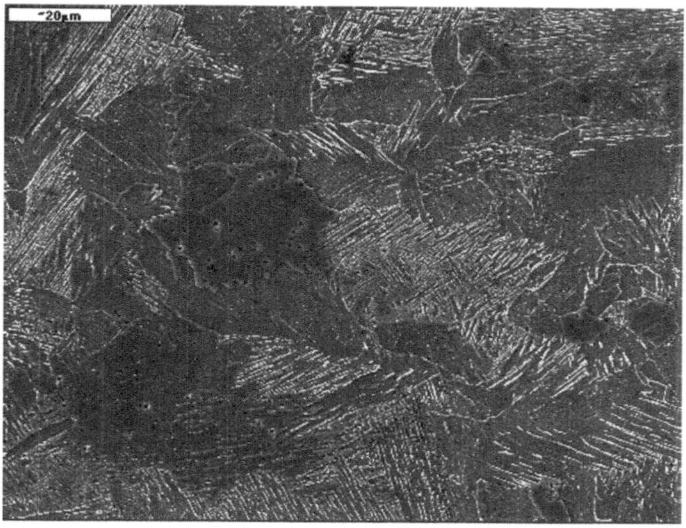

Fig. 5 Widmanstätten structure in the fusion zone (× 1000 magnification)

For a specimen of series 1, a Widmanstätten pattern was observed in the fusion zone (Fig. 5). It contains plate-shaped ferrite precipitates running at 60° and 120°. The occurrence of the Widmanstätten structure indicates that the steel was overheated, which caused a decrease in the mechanical properties of the weld.

The microscopic examinations were performed to determine the influence of different welding speeds on the microstructure and shape of the welds. It was found that the welding speed affected the weld shape and the penetration depth.

The higher the welding speed, the smaller the penetration depth and the more narrow the fusion zone; there was also a change in the direction of the fusion line resulting from a change in the weld shape from mushroom-like to triangular.

When the welding speed was too low, the width of the fusion zone and the width of the heat-affected zone increased.

5.2 *Hardness Tests*

The hardness of the material was measured using the Vickers method. The indentations were made in all the specimens of series 1–5 prepared as metallographic specimens. The measurements were taken on surfaces perpendicular to the three zones: the fusion zone (at the face), the heat-affected zone and the base metal.

The values of the hardness in the base metal zone obtained for all the specimens were comparable. It can be assumed that the specimens had a similar structure with no microdefects that would contribute to the weakening of the material. The average hardness in the base metal zone for all the specimens of series 1–5 was 105 HV1.

The analysis of the hardness of the material in the heat-affected zone indicates that the highest values were reported for a welding speed of 3.5 m/min (a series 4 specimen). The average hardness in the HAZ obtained for that specimen was 184 HV1. For the specimens of series 1–5, the average hardness in the HAZ was 38% higher than the average hardness in the base metal zone.

The highest hardness at the weld face was reported for a specimen of series 4; the average value was 256 HV1. The lowest hardness in the fusion zone was reported for a specimen of series 3 ($v = 3$ m/min); the average hardness for that specimen was 236 HV1. The phase transitions that occurred in the heated material during its rapid cooling contributed to the material hardening both in the fusion zone and the heat-affected zone. The phase transitions were responsible for the formation of martensite-like and ferrite–bainite structures, which improved the material hardness. The average hardness in the fusion zone was 31% higher than that in the heat-affected zone.

5.3 *Static Tensile Tests*

All the specimens (series 0–5) were subjected to tensile testing. The aim of the static tensile tests was to analyse the behaviour of DC04 steel after welding at different process parameters (welding speed).

Before tests, the specimens were measured to determine their width and thickness. Each measurement was performed three times to minimize measurement errors. Then, the arithmetic mean width and thickness were calculated separately for each specimen. The data loaded into the test system program were: the dimensions of the specimens (a constant thickness of 0.78 mm, a constant gauge length of 60 mm and a variable width), their mass (assumed to be 13 g) and the test speed (first, $v_0 = 10$ mm/min, then, after a force F_0 of 100 N was reached, $v_1 = 5$ mm/min, and finally, after four tests, the speed was changed into $v_2 = 10$ mm/min because the duration of a single test was too long). A reference file created on the basis of these parameters was used to automatically calculate the results for each specimen. Initially, the deformation criterion was 30 mm, and the drop in the tensile force was 95%. After two tests, the deformation criterion was changed into 40 mm because of the high plasticity of the material. The tests were carried out at a temperature of 20 °C without the use of an extensometer. The first test was performed for a reference specimen in order to set the reference ranges for the parameters and check whether the assumptions were reasonable. The static tensile tests were conducted for the particular specimens starting from series 0 and ending with series 5. Before each test, the values of displacement and force were set to zero.

The analysis of the tensile test results shows that the unwelded specimens (series 0) had lower values of the offset yield strength ($R_{0.2}$) and the ultimate tensile strength (R_m) but better plastic properties than the welded specimens. The values of the uniform elongation (elongation before the ultimate tensile strength R_m is reached) and the total elongation (elongation at break) are the highest for the unwelded specimens. It is thus evident that the presence of a welded joint improves the tensile strength but reduces the plastic properties of the material.

Figures 6 and 7 illustrate the cumulative results concerning the ultimate tensile strength and the offset yield strength reported for all the specimens under study.

Analysing the values of the offset yield strength and the tensile strength for all the welded specimens (Figs. 6 and 7), we can conclude that, because of the randomness of the results, it is difficult to specify whether there is an upward or downward trend. The highest stability of results was reported for the series 1 and 4 welded specimens.

The differences in the strength and plasticity of the specimens may be due to plasma instability, the related changes in the laser radiation absorption conditions and the heat transfer from the plasma to the material. This phenomenon can be prevented, for instance, by applying a laser beam to the surface of the material with a burning plasma (in argon) or to the surface of an unwelded section of the metal.

6 Conclusions

The shape and thickness of the welds as well as the width of the heat-affected zone are dependent on the welding speed.

Comparing the mechanical properties of the unwelded specimens (series 0) with those of the welded specimens, we can see that the laser welding process slightly improved the strength properties of the material but significantly reduced its plastic properties. The worsening of the plastic properties caused that the total elongation of the welded specimens was smaller than that of the unwelded specimens.

The hardness measurement results confirm that the welded specimens are not homogeneous but very complex in structure.

An increase in the welding speed caused visible, irreversible structural changes in the fusion zone and the heat-affected zone, which were attributable to high heating

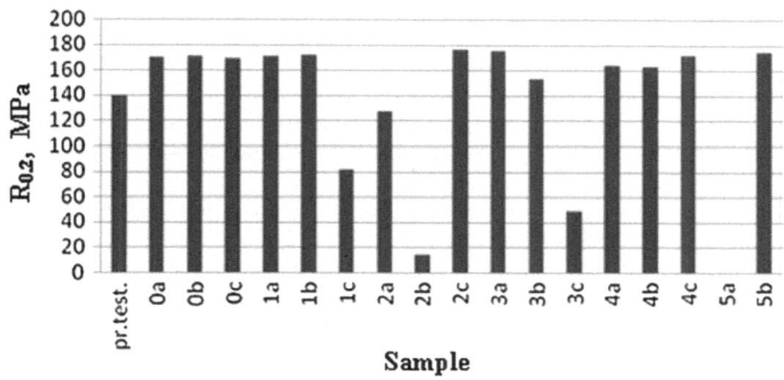

Fig. 6 Cumulative results showing the offset yield strength for all the specimens studied

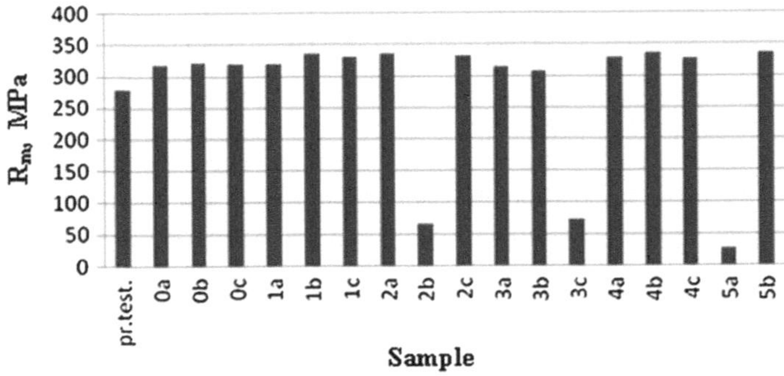

Fig. 7 Cumulative results showing the tensile strength for all the specimens studied

and cooling rates. The changes included refinement of the grain structure and higher hardness.

The obtained results are characterized by a large influence of interfering factors, which are probably caused by ignoring the interaction between the laser power and the feed rate. It strongly suggests to use in future investigation the factorial approach.

References

1. Radek N, Wajs E, Luchka M (2008) The WC-Co electrospark alloying coatings modified by laser treatment. Powder Metall Met C 47:197–201
2. Pietraszek J, Radek N, Bartkowiak K (2013) Advanced statistical refinement of surface layer's discretization in the case of electro-spark deposited carbide-ceramic coatings modified by a laser beam. Solid State Phenom 197:198–202
3. Scendo M, Radek N, Trela J (2013) Influence of laser treatment on the corrosive resistance of WC-Cu coating produced by electrospark deposition. Int J Electrochem Sci 8:9264–9277
4. Antoszewski B (2014) Influence of laser surface texturing on scuffing resistance of sliding pairs. Adv Mater Res Switz 874:51–55
5. Radek N, Pietraszek J, Antoszewski B (2014) The average friction coefficient of laser textured surfaces of silicon carbide identified by RSM methodology. Adv Mater Res Switz 874:29–34
6. Kusinski J (2000) Lasers and their application in material engineering. Akapit, Krakow
7. Szczecinski Z (1985) Spawanie wiazka promieniowania laserowego/welding with laser beam (in Polish). Prz Spaw/Pol Weld Rev 57:12–17
8. Montgomery DC (2008) Design and analysis of experiments. Wiley, Inc, Hoboken
9. Costa A, Miranda RM, Quintino L (2006) Materials behavior in laser welding of hardmetals to steel. Mater Manuf Process 21:459–465
10. Zowczak W (2014) On the possibility of the estimation of the depth of a keyhole formed during laser welding. Adv Mater Res Switz 874:163–168
11. Torkamany MJ, Ghaini FM, Poursalehi R (2016) An insight to the mechanism of weld penetration in dissimilar pulsed laser welding of niobium and Ti-6Al-4V. Opt Laser Technol 79:100–107

Determination of Mechanical Properties of FRP by Bending Test

Martina Ryvolová and Josef Vosáhlo

Abstract The determination of the mechanical properties of the designing materials is carried out in various ways by using standard procedures. The basic test is the tensile test, from which it is possible to determine directly Young's tensile modulus-E [Pa]. Some materials can not be tested by this procedure. These are mainly building materials (concrete), some plastics and composites, glass and ceramics. The biggest problem of these materials is the clamping of the ends of the test specimens in the jaws of the tearing machine. The tensile test is replaced for these materials by a bending test to determine the mechanical properties. The bending test can be performed with a simple supported beam, which is loaded with alone force—it is a three-point bend test or a beam that is loaded with a pair of forces—it is a four-point bend test. The bending modulus and the stiffness characteristics of the material are results of measurement. Bending modulus can be recalculated using the cross-sectional characteristics of the test specimen to the tensile modulus. The article describes the basic differences between the two variants of the bending test, the course of loading and deformation. The results obtained by measuring on composite specimens were verified using mathematical models and their degree of agreement was determined.

Keywords Four-point bending · Fibers reinforced plastic · FE model

1 Introduction

Short-term static strains are among the most commonly used ways of detecting mechanical properties of materials. The test principle is based on a short-term defined load on the test body/specimen at low speed until its failure or a predefined deformation. The test is conducted under predetermined conditions. Tension, pressure, flexure, torsion and shear are the most common tests.

M. Ryvolová · J. Vosáhlo (✉)
Technical University of Liberec, Studentská 2, 46001 Liberec, Czech Republic
e-mail: josef.vosahlo@tul.cz

M. Ryvolová
e-mail: martina.ryvolova@tul.cz

Š. Medvecký et al. (eds.), *Current Methods of Construction Design*, Lecture Notes in Mechanical Engineering, https://doi.org/10.1007/978-3-030-33146-7_66

The tensile test determines the basic characteristics of the material—tensile strength limit, tensile strength, elongation, and contraction. Quantities are used to assess quality of materials, for design calculations and for a general assessment of the suitability of technological operations. The tensile test determines the yield strength and tensile modulus directly. Specimens of brittle materials (such as cast materials—gray cast iron, building materials—concrete, ceramics, glass and some types of plastics) cannot be clamped into the jaws of the tearing machine. The flexure test replaces the tensile test for these cases.

Bending test serves to evaluate weld joints, provides important information on the behavior of the material under flexural stress, the measurement result allows for determination of bending modulus and tensile modulus. The bending test is carried out in such a way that the beam-shaped body is mounted on two supports at a prescribed distance. Two variants of test are used (Fig. 1); the alone force acts on the body in the middle between the supports—the three-point bend test or a pair of forces acting on the beam at a certain distance—the four-point bend test. The load, which gradually increases, causes the test body deflection. Deflection is measured until damage or irreversible deflection of test body [1, 2].

The synergistic interaction of the fibers and the matrix (definition of fibers reinforced composite material) makes it possible to produce composites with mechanical properties which go beyond properties of conventional construction materials. Thick-nets of the wall are achieved by connecting of individual layers. Lamina is the name for one layer; laminate is consisted of several layers. Material composition and type of reinforcement determine the composite material properties [3]. The reinforcement in the layer may be oriented one-direction (using fibers oriented in one-direction), two-direction (using a woven fabric or biaxial warp knitted fabric), multi-direction (using three to multiaxial warp knitting fabric), at random direction (using non-woven structures) and multi-direction in space (using 3D woven, knitted and braided structures). The mechanical properties of the laminate are achieved by the appropriate orientation of the fibers in the individual layers and the composition of the layers in composite [4].

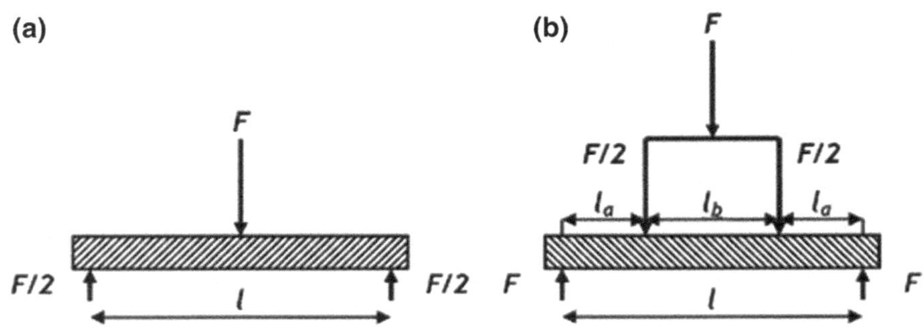

Fig. 1 Arrangement of **a** three-point bending test, **b** four-point bending test

The properties of fibers reinforced laminates are determined by the four-point bend usually. A result of four-point bending test is more accurate and is eliminated damage to the test specimen at the area of action of the test equipment, which in the case of using a three-point bending test cannot be avoided. For the four-point test arrangement, the maximum flexural moment Mo and the maximum flexural stress in the region between the two forces being applied are constant. The test body is stressed the flexure only. The maximal deflection y [mm] and maximal force [N] are usually measured [5].

2 Experiments

2.1 *Material Properties*

Composite tubes were prepared to determine the mechanical properties. Base of test specimens were tubes made by pultrusion with unidirectional-oriented carbon fibers in epoxy matrix. These tubes have good mechanical properties only in the longitudinal direction. The aim of the experiment was to improve the mechanical properties by adding other directionally-oriented layers from two different materials. The four-point bend test was used for the measurement. The mathematical model of the experiment was created for comparison. Two layers of fibers have been added to improve mechanical properties. First variant of material was carbon fibers; second variant of material was flax fibers—UD prepregs with epoxy resin. Improving the mechanical properties of the tube was the task of the experiment. Material composition, laminate composition, dimensions and average specimen weight are shown in Table 1.

Properties of carbon and flax fibers were determined using tensile strength test. One-layer specimen was prepared for test. Orientation of layers and average of results measurement are shown Table 2.

A four-point bend test was performed until the test specimens were broken. Figure 2 shows the graphs of dependence force and deflection for both variants of specimens with carbon and flax fibers.

Table 1 Specimen composition

Signature	Material	Composition	Weight (g)	Length (mm)	Diameter (mm)
C	CF	UD	60	330	12; 14
C_CF	CF/CF	UD/+45°/−45°	108.5	330	12; 15.6
C_FF	CF/flax	UD/+45°/−45°	218	330	12; 16

Table 2 Carbon fibers and flax fibers prepreg properties

Specimen	Orientation (°)	Thickness (mm)	F_{max} (N)	R_{max} (MPa)	E (GPa)
C0	0	0.5	5272.2	1076.5	37.11
C90	90	0.5	106.91	27.07	8.2
C45	45	0.5	46.67	11.81	4.9
F0	0	0.5	1598.28	159.83	9.05
F90	90	0.5	66.08	6.61	1.25
F45	45	0.5	101.65	10.16	1.37

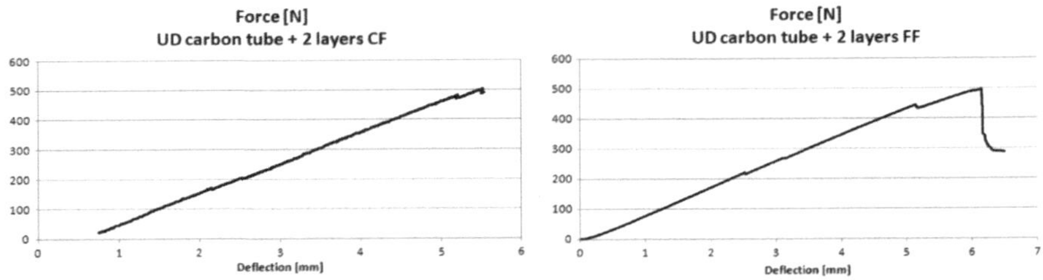

Fig. 2 Graphs (left) dependence F—w_{max} for UD carbon tube/CF, (right) dependence F—w_{max} for UD carbon tube/FF

2.2 *Mathematical Model*

The results obtained from the tensile tests on the UD specimens and from four-point bending test on composite tubes were the input data to the SolidWorks program for creating the numerical model [1]. The created models were imported into the Ansys software, where model topology was modified. The material model was created as orthotropic elastic model. Linear elasticity in an orthotropic material is most easily defined by giving the materials constants (Young's modulus, Poisson's ratios and shear modulus) associated with the material's principal directions. The model of tested specimen was created as a conventional shell composite layup. Contacts between support and specimens were set to non-friction type with Asymmetric behavior and normal Lagrange definition [2]. Sample models, supports, and test equipment have embedded mapped networks combined with 2 mm Face Sizing. For mesh creating was used the total number of 18,792 nodes and 14,192 elements [6]. Supports have been set to fixation boundary conditions that prevent movement and rotation to all sides. Perpendicular to the model of specimen, a sliding boundary condition with average values of deflection was set in the test equipment [7, 8] (Fig. 3).

Fig. 3 Four-point bend geometry and embedded boundary conditions

3 Results

Four-point bending test was realized for three groups of specimens. Average values of experimental results (maximum force and deflection) are shown in Table 3. The value flexural stress σ_o [MPa] was calculated using Eq. (1) [9]:

$$\sigma_0 = \frac{16\,\mathrm{FLD}}{3\pi\left(D^4 - d^4\right)} \quad (1)$$

where F-Force [N], L-Support span [mm], D-External diameter [mm], d-Internal diameter [mm].

Detection of cracks and defects was performed inside the tubes in location of the action of test equipment. Figure 4 shows rupture of fibers inside C_CF composite tube. Figure 5 shows damage of C_FF composite structure. In this case, fibers were not ruptured.

Table 3 Result of measurement

Specimens	$F_{\max}$ (N)	$y_{\max}$ (mm)	σ_o (MPa)
C	212.83	3.21	172.40
C_CF	570.83	6.18	234.04
C_FF	561.25	7.79	202.05

Fig. 4 Specimens with carbon fibers; damage to the fibers in the location of the action of test equipment

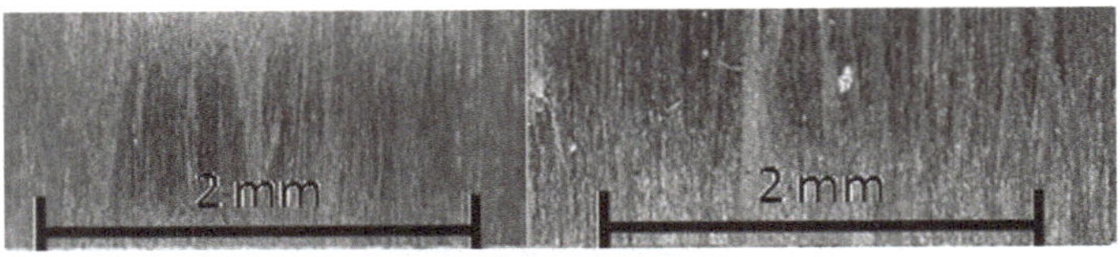

Fig. 5 Specimens with flax fibers; damage to the fibers in the location of the action of test equipment

The output of the simulation was set to have maximum deflections and stresses, depending on the action of the test equipment. The first group of specimens, where the 0° orientation of the fibers was used, was a maximal deflection of 3.1 mm (average of measurement values) and von Mieses stress 397 MPa, see Fig. 6. For the C-CF group specimens were used deflection 6.2 mm, and for the C-FF group specimen was used deflection 7.8 mm. The values of von Mieses stress were determined (sample in Fig. 7).

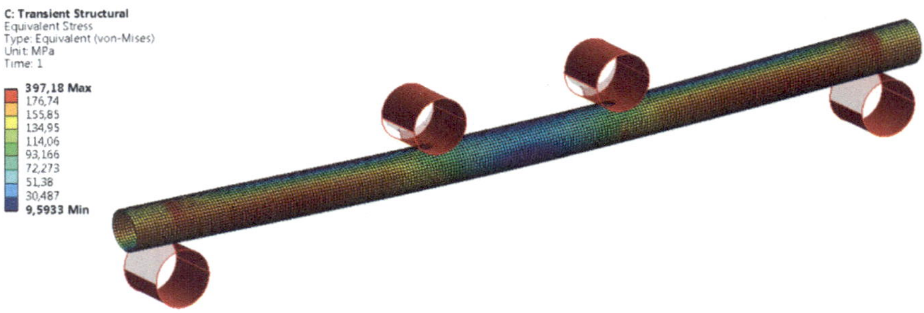

Fig. 6 Equivalent stress (von Mieses) for model C

Fig. 7 Equivalent stress (von Mieses) for model C-CF

4 Conclusions

Mechanical tests of composite specimens from carbon and flax fibers have been performed. Four-point bending test was performed for tubes specimen with specific layers orientation and carbon and flax fibers. The data obtained from the measurements was further used to creation of numerical model. The results of numerical simulations show a relatively good agreement with the performed tests.

Acknowledgements This work is partially supported by the Ministry of Education, Youth and Sports of the Czech Republic and the European Union in the frames of the project "Modular platform for autonomous chassis of specialized electric vehicles for freight and equipment transportation," Reg. No. CZ.02.1.01/0.0/0.0/16_025/0007293.

References

1. Kulhavý P, Kovalova N, Martonka R, Petrik J (2014) Biaxial stress of the textile car seat cover tested in a climatic chamber. In: Dynybyl V, Berka O, Petr K, Lopot F, Dub M (eds) The latest methods of construction design, ICMD, Springer, Heidelberg, 437–442
2. Quaglini V, Corazza C, Poggi C (2008) Experimental characterization of orthotropic technical textiles under uniaxial and biaxial loading. Compos A Appl Sci Manuf 39(8):1331–1342
3. Laš V (2008) Mechanika kompozitních materiálů, 2nd edn. Západočeská univerzita v Plzni, Plzeň
4. Militký J (2012) Classic and special textile fibers. Technical University of Liberec, Liberec
5. ČSN EN ISO 14125 Vlákny vyztužené plastové kompozity—Stanovení ohybových vlastností (1999)
6. Tran Huu Nam (2004) Mechanical properties of the composite material with elastomeric matrix reinforced by textile cords. Doctoral thesis, Technical University of Liberec, Liberec
7. Bittnar Z, Šejnoha J (1992) Numerical methods of mechanics 2. České vysoké účení technické, Praha
8. ASTM D7264 Standard test method for flexural properties of polymer matrix composite materials
9. Leinveber J, Řasa J, Vávra J (1999) Strojnické tabulky, 3rd edn. Scientia, Prague

Design of Belt Wheels Using the 3D Printing Method

Patrik Sniehotta

Abstract This article deals with the design a manufacturing of belt wheel for bicycles, with the help of 3D printing. The market with bicycles, that use belt instead of a traditional chain, is growing steadily. In the design process of a new bicycle model, you need to quick check, if the proposed belt wheels fit into the frame. Even in time of 3D CAD applications, a physical check is necessary to confirm the design. Also, a physical model is helpful to tray the assembly process. For small companies, it is expensive to buy multiple belt wheels, with different tooth numbers, just to confirm the proposed design. On the other hand, some companies want to use her own belt model, for which is no belt wheel for bicycles. Here comes 3D printing handy. This method if fast and flexible, and in small series can be cheaper then belt wheel made by traditional methods. This article deals with the problem of designing a belt wheel from a given belt model and different 3D printing methods for their production.

Keywords Belt wheel · 3D printing · Bicycle

1 Introduction

The impulse for this article was a bachelor's thesis that dealt with the measurement a right installation of a belt drive for bicycles. Belt drive is for a bicycle an unusual drivetrain. Normally bike manufacturers use a chain and chain wheels to transfer the power from the bicyclist's legs to the back wheel. It is a proven, reliable and cheap solution, which has some drawbacks. The biggest disadvantage is that a chain needs to be lubricated to ensure a smooth and quiet ride. But most lubricants are sensitive to water and they get washout with time. Also, from the lubricant you can get dirty. For cyclist that commutes in a city or uses his bicycle for daily movement around the city, can be the need of regular maintenance a big impracticality.

On the other hand, a belt drive is almost maintenance-free, no lubricant is used, is dirt resistant and you can clean it with normal water. These are big benefits for the user.

P. Sniehotta (✉)
Technical University of Ostrava, 17. Listopadu 15/2172, 70833 Ostrava, Poruba, Czech Republic
e-mail: Patrik.sniehotta@vsb.cz

Š. Medvecký et al. (eds.), *Current Methods of Construction Design*, Lecture Notes in Mechanical Engineering, https://doi.org/10.1007/978-3-030-33146-7_67

Fig. 1 Example of a belt drive dirty by mud [1]

For the bicycle, designer creates the use of belt drive some significant issues. The major drawback of belt drive is, that the chainstay must be split into two parts, that it is possible to install the belt. Unlike by a chain that can be uncoupled a belt is made out of one endless piece. Also, a belt drive needs a tensioner to provide the required tension of the belt. Last but not least, the frame and components for the belt drive require much more accurate tolerances (Fig. 1).

All these problems require the designer a bigger attention to the design of the frame. Also, in the prototype phases it is the bigger need to test the design belt drive, if, for example, the selected gear ration is correct or the belt wheels are collinear. For these reasons, there is a bigger need for belt wheel prototypes. This prototyping is 3D printing method very useful but has his rules [2, 3].

2 Methods and Materials

The first step by designing a new belt drive is the right belt choice. Actually, there can be used any toothed belt. The only problem is the strength of the belt. Typical for bicycle use, it is a belt with half-round profile. We choose the CBD belt from Continental that was special designed for cycling. This belt is reinforced with carbon fiber, so it can be narrow and light (Fig. 2).

At the beginning, we also did buy the CBD belt sprocket also from Continental. The special about this sprocket was that it was the first bicycle belt sprocket made completely out of plastic using injection molding. This method dramatically dropped the price of bicycle belt system (Fig. 3).

After then we modeled both belt and sprocket in Autodesk Inventor and Dassault Systèmes Solidworks to see in which application it will be easier to design the sprockets. In the end, we found that for sprocket design is Inventor the better application (Fig. 4).

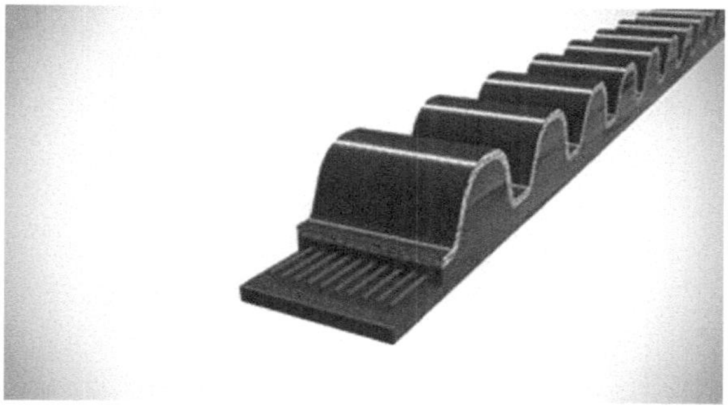

Fig. 2 Continental CBD belt [4]

Fig. 3 Continental CBD sprocket [5]

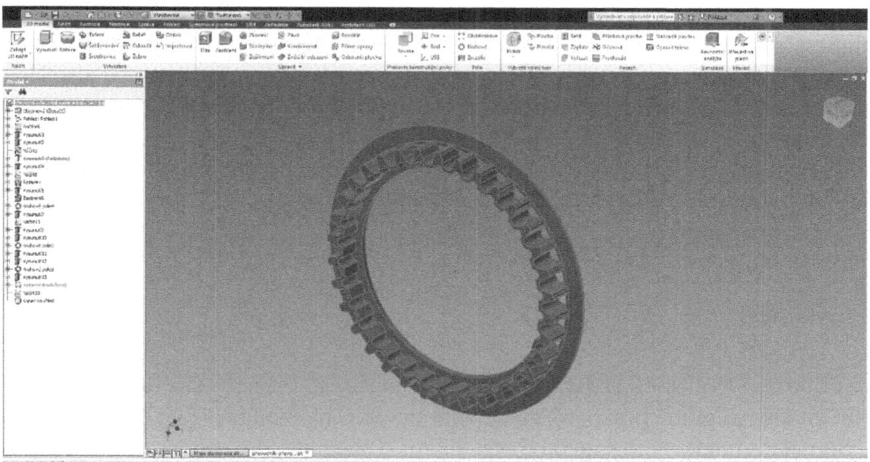

Fig. 4 Sprocket designed in autodesk inventor

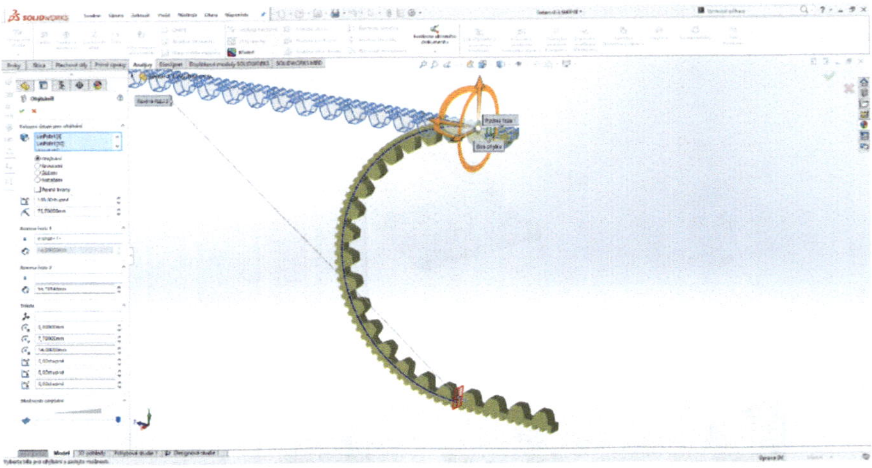

Fig. 5 Bended belt model in solidworks

But for model the bended belt is better Solidworks, because it has better function for bending straight part. The bended belt model is needed to check if the sprocket model has the right geometry and pitch (Fig. 5).

For the 3D printing, we choose three 3D printers. Two were printers for home use, which are cheap to purchase and for use. These printers print from a polylactic acid (PLA) wire. Specifically, it was:

- FELIX Pro 2.
- PRUSA I3 MK3.

The third printer was a professional Laser-Sinter-System EOS P 396 printer that prints selective laser sintering (SLS) method. The material was polyamide PA2200. It creates a more homogeneous structure, that is much stronger and can be also later machined. On the other hand, parts from this type of printer are more than ten times expensive [6, 7].

Due to this price difference, there was the question if there is the need to use always the better but more expensive method.

3 Results

We printed several sprockets with all three printers and from the site of the resulting print quality were the best parts made with the SLS method [8] (Fig. 6).

Sprockets made by this method had the most accurate dimensions, it was easy to program and it could be machined by a grinder. This is very useful in the prototyping time where you need, for example, grind an edge or a hole.

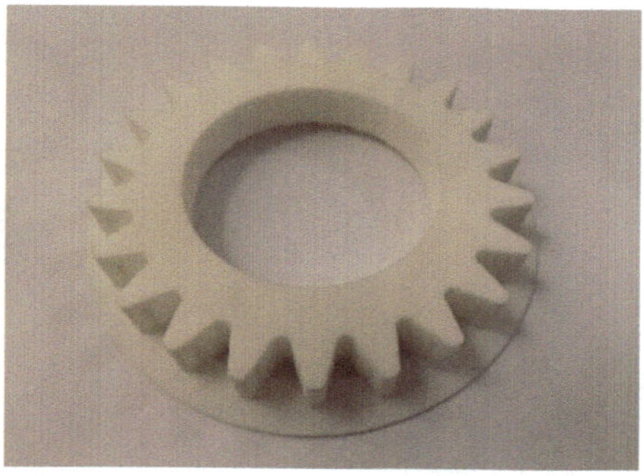

Fig. 6 Sprocket made by SLS method

The sprocket made out of the PLA variate significantly in the resulting quality. But it was not the fault of the method. The problem was the operator of the printer, which programs the trajectory, how the part is printed. The result was fibers hanging from the part and an inaccurate geometry. These printers make first an "accurate shell" of each layer and then fill the space between. The operator can choose how thick or thin the shell would be. This initial setting fundamentally affects the quality of the resulting part [8] (Fig. 7).

When the operator has a good knowledge of printing such complex parts, like these belt sprockets are, then geometric accuracy of the resulting parts can be compared with parts made with the SLS method.

However, there are of course some drawbacks that even a good operator can not compensate, compared to the SLS method. The layers of the material are thicker, so the resulting surface quality is rougher. Also machining (e.g., grinding) is much more difficult. The reason is that the material is actually not homogeneous. The PLA wire has a round profile and when then fibers are lying between each other, they create small gaps. Therefore, you do not know how much resistance the material will make (Fig. 8).

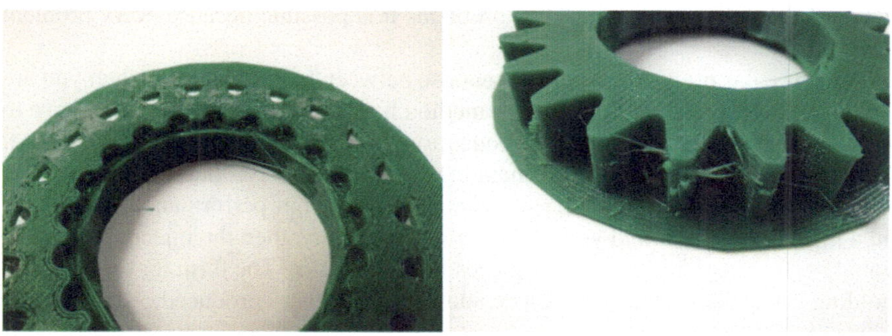

Fig. 7 Example of bad printing

Fig. 8 Example of good printed sprockets, made out of two parts

Also, when the part is to complex, it has to be split into several parts and then glued together. This brings into the process a potential inaccuracy. The gluing requires some additional time and work.

4 Discussion

We found that both methods are useable for belt drive sprockets, but each is better for another purpose. The PLA method is powerful in the preliminary design when you have to check if, for example, the sprockets do not collide with the frame. You can then very fast and cheap test different sizes of sprocket or different mounts.

When all the basic problems are sort out, the SLS method is to recommend for the functional testing of the whole bike. The material is then stronger and shapes are more accurate. That ensures minimum of problems with the sprocket during testing of the bicycle. The testing phases of a new bike demanding and expensive, so as bicycle manufacturer you so few problems it is possible because every problem costs money.

In the regular production, you can choose between both methods. When you are a small bicycle manufacture, the PLA method has the lowest costs, but you have to optimize the method for the regular production and check the quality with every part. The sprockets made with the SLS method are more expensive, but not so expensive like with injection molding. The sprockets offer the high performance. When you are a big manufacturer and you assume a large quantity, then the injection molding would be, from a certain number of pieces, a better choice. The form for the injection molding is expensive, but then costs come back with every produced piece.

5 Conclusions

The 3D printing method is a good choice for testing the belt drive in the early stages of bicycle development. In development phases, both methods find their use. In the regular production, also both methods can be uses, but the SLS is recommended, because it has a better overall quality, performance and has lower risk of complaints. For large series is the best choice after the development phases to invest into an injection molding form. This method has then the lowest production costs.

This process can be used not only for belt drives, but also, for example, for small gearboxes (e.g., for drones, toys, etc.) or other application where is a power transfer. Especially for gearboxes has this method a big potential. The conventional prototyping methods can be then replaced with 3D printing. This will then speed up the whole process.

Acknowledgements This study was supported by SP2018/45 project.

References

1. Naked bicycles (2018) Titanium Belt Drive X Fighter, location [online]. 26.11.201/26.11.2014 [13.6.2018]. Available from https://nakedbicycles.com/gallery/titanium-belt-drive-x-fighter/2014/11/26
2. Novakova-Marcincinova L, Fecova V, Novak-Marcincin J, Janak M., Barna J (2011) Intelligent design of experimental gearbox with rapid prototyping technology support. In: 15th IEEE international conference on intelligent engineering systems, IEEE Xplore Digital Library, Poprad, Slovakia, USA, 77–80
3. Novakova-Marcincinova L, Fecova V, Novak-Marcincin J, Janak M, Barna J (2012) Effective utilization of rapid prototyping technology. In: AIP conference proceedings, vol 1431(57). AIP, 834–841
4. Continental bicycle systems (2018) CBD-Belt, location [online]. 10.5.2018/10.5.2018 [13.6.2018]. Available from http://www.continental-bicycle-systems.com/en-gl/eBikes/Belt-Drive/CBD-Belt
5. Continental bicycle systems (2018) Technical advices, location [online]. 10.5.2018/10.5.2018 [13.6.2018]. Available from http://www.continental-bicycle-systems.com/en-gl/Ebikes/Belt-Drive/Technical-advice
6. Hernandez DD (2015) Factors affecting dimensional precision of consumer 3D printing. Int J Aviation Aeronaut Aerosp 2(4):1–43
7. Kaziunas A (2013) Make: 3D printing: the essential guide to 3D printers, 1st edn. O'Reilly and Associates, Sebastopol
8. Horvath J (2014) Mastering 3D printing, 1st edn. Springer, Berlin

Biomechanics, Biomimicry and Innovation

Performance of a Manually Fed Pelleting Machine with a Horizontal Rotating Matrix

Gurkan Alp Kagan Gürdil, Bahadır Demirel, David Herak and Yunus Özcan Baz

Abstract This study analyzed the performance of a pelleting machine with a manual feeder and with a horizontal rotating matrix. Hazelnut husk residue obtained after hazelnut harvesting is used as a material. The residues were pelleted at M10 moisture content and with 6 mm particle sizes. Physical properties (bulk density, pellet density, mechanical duration and pellet firmness, moisture and equivalent humidity contents) and thermal properties (gas emission values after combustion, ash content and calorific values) of the pellets were measured. Pellets bulk density was 584.44 kg m^{-3}. Particle density, mechanical durability and firmness values were 1238.20 kg m^{-3}, 93.26% and 1146.00 N, respectively. Regarding the emission values, the pellets' flue gases were within the legal limits and had a heating value of 4196 cal g^{-1}, which is close to the heating value of wood. In conclusion, the pelleting machine with a horizontal rotating matrix was found to be very suitable for pelleting hazelnut husks, and this means that this agro-based residue material can be readily used as a solid fuel.

Keywords Bioenergy · Hazelnut · Pellet · Residue

1 Introduction

Biomass defined as different materials of biological origin mainly plant material and animal wastes [1, 2] used primarily as a domestic energy source is naturally abundant and present a renewable energy opportunity that could serve as an alternative to fossil fuel. Behind coal and oil, biomass is the third largest energy resource in the world [3] having dominated the world energy consumption until the mid-nineteenth century [4]. Utilization of agricultural residues is often difficult due to their uneven and troublesome characteristics. Compacting technologies have been in operation for over 150 years [5]. The process of compaction of residues into a product of higher

G. A. K. Gürdil · B. Demirel · Y. Ö. Baz
Ondokuz Mayis University, Körfez Mahallesi, 19 Mayıs Ünv, 55270 Atakum/Samsun, Turkey

D. Herak (✉)
Czech University of Life Sciences Prague, Kamýcká 129, 16500 Prague, Czech Republic
e-mail: herak@tf.czu.cz

Š. Medvecký et al. (eds.), *Current Methods of Construction Design*, Lecture Notes in Mechanical Engineering, https://doi.org/10.1007/978-3-030-33146-7_68

density than the original raw material is known as densification. Densification has aroused a great deal of interest in developing countries all over the world lately as a technique for upgrading residues as an energy source [6]. Densifying the agricultural residues can be done either in the form of briquettes or in pellets. Both methods increase the specific density (gravity) of biomass to more than 1000 kg/m^3 [7, 8]. Worldwide using refined material in recent times is mostly in the form of pellets. The main advantage of pellets is the possibility of fully automated delivery in the process of combustion, along with very precise regulation of the combustion process, even for low-performance furnaces [9]. Hazelnut production is one of the major agricultural products for Turkey, especially in Black Sea Region. The husk of hazelnut is the outer part of the shell which binds it to the branch. The shelled hazelnut is taken out of husk during harvesting. It is reported that Turkey has produced more than 70% of the world's hazelnut need with 675,000 tons [10]. The study used the husk of hazelnut as a material. The objective of the study is to analyze the performance of a pelleting machine with manual feeder and with a horizontal rotating matrix. The residues are pelleted at M10 moisture content and with 6 mm particle sizes (PS). Physical properties (bulk density, pellet density, mechanical duration pellet firmness and moisture contents) and thermal properties (gas emission values after combustion, ash content and calorific values) of the pellets were measured.

2 Materials and Methods

This study was carried out in the workshop and labs of Agricultural Machines and Technologies Engineering Department at Ondokuz Mayis University. Up-to-date European standard EN 14961-2 (Solid biofuels—Fuel specifications and classes—Part 2: Wood pellets for non-industrial use) [11] was taken as a reference for this study. Hazelnut husk agricultural residue which has a big potential in the Black Sea Region of Turkey was used a material for pelleting. The husks obtained after harvesting are brought to labs, and they are left under sun during the day for natural drying. The moisture content of the material is decreased down to 10% and the dried materials ground by a 3 kW powered hammer mill having eight blades and with two 850 r min^{-1} rotation speed. For the homogenization of particle size, a sieve having 6 mm sieve diameter is used during grinding. Laboratory-type pelleting machine is used for pelleting (see Fig. 1).

Pelleting of material is done with a manually fed pellet machine with horizontal rotating matrix. Figure 2 shows the produced pellets. Tests for determining physical–mechanical and thermal properties of pellets, flue gas emission values are measured for determining the quality of pellets. Volume density of pellets is calculated according to the EN 15103 EU norms [12] in kg m^{-3}, where the pellets are poured in a pot, and the ratio of mass to the volume of the filled pot is calculated. Pellet density of pellets is calculated by stereometric method according to EN 15150 EU standards in kg m^{-3}, as well [13]. Mechanical durability of pellets is calculated according to EN 15210-1 EU standards [14]. For this, 500 g sample is poured into the testing chamber

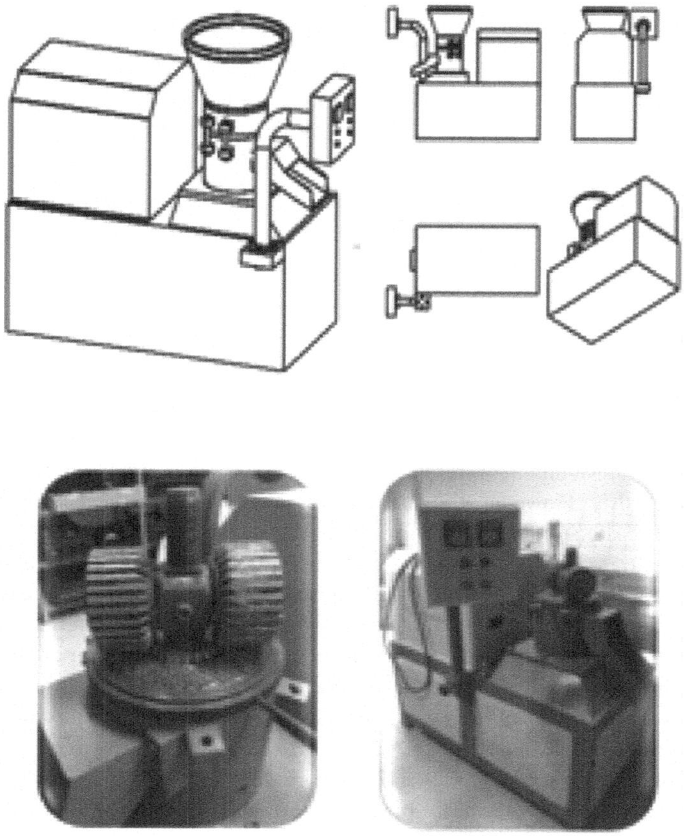

Fig. 1 Manually fed laboratory-type pelleting machine with horizontal rotating matrix

Fig. 2 Hazelnut husk pellets

and rotated with 50 r min^{-1} speed for 10 min. Then, the pellets are sifted out by a sieve having 3.15 mm sieve holes, the oversized pellets are weighed, and the durability of pellets is calculated. Moisture content of pellets is defined as in EN 14961-2 EU standards [11] by using a desiccator. Pellets are burnt in a pellet stove, and the flue gas emission values such as O_2, CO, CO_2, NO and NO_x are measured by a gas emission device. Lower heating values of pellets are determined by a calorimeter device according to EN 14918 EU standards [15]. Ash contents are then calculated according to EN 14775 EU standards [16].

3 Results and Discussion

Tables 1 and 2 show some physical properties of grinded material.

In [17, 18], it is reported that the pellet quality is higher when the pellet durability is 80% or higher. Mechanical durability of hazelnut husk pellets indicated that they are in high quality. However, the value of it was lower than the values suggested in EN 14961-2 EU standards [11] for EN plus-A1, EN plus-A2 (≥97.5%) and for EN-B classes (≥96.5%). This is because of the material used in standards. They used wood chips only but we used hazelnut husk agricultural residue. This material is mostly plant-based material rather than wood-based. The difference comes out due to this original differentiation. But, nevertheless, the produced pellets are good enough for transportation and storage purposes regarding their mechanical durability and firmness values. Table 3 shows heating value, ash content and gas emissions of hazelnut husk pellets.

Ash content values in EN 14961-2 EU standard [11] are given as ≤0.7% for EN plus-A1 class, ≤1.5% for EN plus-A2 class and ≤3.0% for EN-B class. Ash content of pellets made from hazel husk residue was more than the values reported in the particular standard. The reason for this is the pellets given in EU standard

Table 1 Some physical properties of grinded husk material

Particle size (mm)	Bulk density of material (kg m^{-3})	Geometrical mean diameter of material (mm)
6	220.18	1.64

Table 2 Some physical–mechanical properties of hazelnut husk pellets

Properties	Value
Bulk density (kg m^{-3})	584.44 ± 3.88
Particle density (kg m^{-3})	1238.20 ± 16.08
Mechanical durability (%)	93.26 ± 0.13
Pellet firmness (N)	1146.00 ± 34.08
Moisture content of pellet (%)	878 ± 0.04

Table 3 Heating value, ash content and gas emissions of hazelnut husk pellets

Ash content dry (%)	Heating value dry (cal g^{-1})	CO (ppm)	CO_2 (%)	O_2 (%)	NO (ppm)	NOx (ppm)
7.97 ± 0	4196 ± 0	1383.67	0.90	19.17	121.0	61.67

Table 4 Heating values of some agricultural materials

Residue	Heating value (cal g^{-1})
Wood	4200
Pruning residues of other fruits and vineyard	4300
Hazelnut pruning residue	4494

are made from wood chips. However, the heating value of hazelnut husk pellets was conformable to the value ($Q \leq 19$ MJ kg^{-1}) given in that standard. As seen from Table 4, the heating value of hazelnut husk pellets (4196 cal g^{-1}) is very close to those of agricultural materials especially, to the heating value of wood (4196 cal g^{-1}). This means that this agro-based residue material can be readily used as a solid fuel.

4 Conclusions

In this study, a laboratory-type manually fed pelleting machine with horizontal rotating matrix is designed and developed for utilization of hazelnut husk agricultural residues to be used as solid biofuel. Pellets with M10 and with 6 mm PS are produced. To decide the appropriateness of this developed machine, some physical and thermal properties of the pellets were investigated.

After all the tests, it is found that the developed pelleting machine was suitable for pelleting agricultural residues. These kinds of researches will help to improve the design and function of pelleting machines for the future and by this way for the energy deficiency of the world by converting agricultural residues to energy sources.

Acknowledgements This study was supported by TÜBİTAK (THE SCIENTIFIC AND TECHNOLOGICAL RESEARCH COUNCIL OF TURKEY) with the project number 2140652 and by CULS FE IGA 2018 project number 31130/1312/3110.

References

1. Sampson RN, Wright LL, Winjum JK, Kinsman JD, Benneman J, Kursten E, Scurlock JMO (1993) Biomass management and energy. Water Air Soil Pollut 70(1–4):139–159

2. Trebbi G (1993) Power-production options from biomass: the vision of a southern Euorpean utility. Bioresour Technol 46:23–29
3. Bapat DW, Kulkarni SV, Bhandarkar VP (1997) Design and operating experience on fluidized bed boiler burning biomass fuels with high alkali ash. In: Proceedings of the 14th international conference on fluidized bed combustion, ASME, New York, Vancouver, 165–174
4. Tumuluru JS, Wright CT, Kenny KL, Hess JR (2010) A review of biomass densification technologies for energy application. A report prepared for the U. S. department of energy by the Idaho National Laboratory, biofuels and renewable energy technologies department, US
5. Strehler A (2000) Technologies of wood combustion. Ecol Eng 16:25–40
6. Bhattacharya SC, Leon MA, Rahman MM (2002) A study on improved biomass briquetting. Energy Sustain Dev 6(2):67–71
7. Lehtikangas P (2001) Quality properties of pelletized sawdust, logging residues and bark. Biomass Bioenerg 20(5):351–360
8. Mani S, Tabil LG, Sokhansanj S (2004) Evaluation of compaction equations applied to four biomass species. Can Biosyst Eng 46(1):3.55–3.61
9. Šooš L, Matúš M, Beniak J, Križan P (2015) Development of the compaction machine for the production of new shapes of pressed biofuels. IOP Conf Ser Mater Sci Eng 297(012008):1–9
10. Tuik (2017) Hazelnut Production Data. Turkish Statistical Institute, Ankara, Turkey
11. EN 14961-2 (2010) Solid biofuels-fuel specifications and classes-part 2: wood pellets for non-industrial use European Committee for Standardization: Management Centre, Avenue Marnix 17, B-1000 Brussels
12. EN 15103 (2009) Solid biofuels—determination of bulk density. European Committee for Standardization: Management Centre, Avenue Marnix 17, B-1000 Brussels
13. EN 15150 (2011) Solid biofuels—determination of particle density. European Committee for Standardization: Management Centre, Avenue Marnix 17, B-1000 Brussels
14. EN 15210-1 (2009) Solid biofuels—determination of mechanical durability of pellets and briquettes—Part 1: pellets. European Committee for Standardization: Management Centre, Avenue Marnix 17, B-1000 Brussels
15. EN 14918 (2009) Solid biofuels—determination of calorific value. European Committee for Standardization: Management Centre, Avenue Marnix 17, B-1000 Brussels
16. EN 14775 (2009) Solid biofuels—determination of ash content. European Committee for Standardization: Management Centre, Avenue Marnix 17, B-1000 Brussels
17. Tabil LG, Sokhansnj S (1996) Process conditions affecting the physical quality of alfalfa pellets. Appl Eng Agric 12(3):345–350
18. Tabil LG, Sokhansanj S (1997) Bulk properties of alfalfa grind in relation to its compaction characteristics. Appl Eng Agric 13(4):499–505

Innovation of Device for Nanofibers Production Using TRIZ

Petr Lepšík

Abstract The paper deals with the innovation of the device for nanofibers producing from a free surface of polymer solution. The collector (negative electrode) is one of the main parts of the device. The original flat collector did not provide sufficient results at producing 3D nanostructure. The production of 3D nanostructure demands high intensity of electrostatic field (big force) for obtaining sufficient productivity but at the same time low intensity of electrostatic field (small force) for obtaining required 3D nanostructure at the collector. This problem can be solved by the use of the method Theory of Inventive Problem Solving (TRIZ) and its tools for physical contradiction solving. The defining of the physical contradiction, resolving the contradiction by the use of separation and also inventive principles (especially principle nr. 9 Preliminary anti-action) are described in the paper. The idea of collector innovation has been confirmed by the numerical simulation FEM and verified by the experimental way.

Keywords Electrospinning · 3D nanostructure · TRIZ · Physical contradiction · Separation principles

1 Introduction

The electrospinning is a fiber production technology, which uses electric force to draw charged threads of polymer solutions, or polymer melts up to fiber diameters in the order of some hundred nanometers. The technology allows the production of nanofibers from polymers solved in water, acids or bipolar solvents and is suitable for the production of organic high-quality fibers. The numbers of fibers per machine width are given by the distance of the Taylor cones. The free liquid surface electrospinning lets natural physics define this distance, rather than using individual needles. This allows higher fiber packing density and thus an increased productivity as well as better fiber homogeneity and more consistent Web morphology. The schema of the needleless electrospinning setup is shown in Fig. 1.

P. Lepšík (✉)
Technical University of Liberec, Studentska 2, 46117 Liberec 1, Czech Republic
e-mail: petr.lepsik@tul.cz

Š. Medvecký et al. (eds.), *Current Methods of Construction Design*, Lecture Notes in Mechanical Engineering, https://doi.org/10.1007/978-3-030-33146-7_69

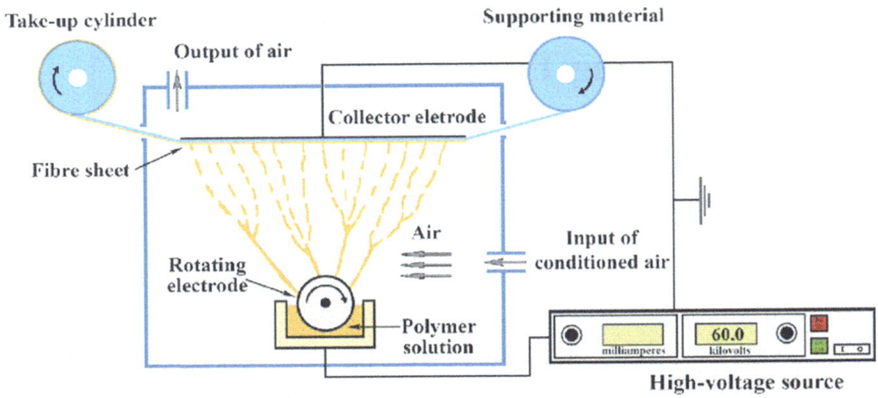

Fig. 1 Schema of the needleless electrospinning setup [1]

The productivity of electrospinning process is dependent on the intensity of electrostatic field. The productivity increases with the intensity of electrostatic field, respectively, with the electric forces. High electric forces cause undesirable flatness of the fibers in the collector plane.

The main goal of this paper is to show how can be this issue solved with the use of TRIZ method and its separation principles for physical contradiction solving.

2 Methods and Materials

TRIZ [2–10] presents a systematic approach for understanding and defining challenging problems: difficult problems require an inventive solution, and TRIZ provides a range of strategies and tools for finding these inventive solutions. One of the earliest findings of the massive research on which the theory is based is that the vast majority of problems that require inventive solutions typically reflect a need to overcome a dilemma or a trade-off between two contradictory elements. The central purpose of TRIZ-based analysis is to systematically apply the strategies and tools to find superior solutions that overcome the need for a compromise or trade-off (contradiction) between the two elements. TRIZ recognizes technical and physical contradictions. The physical contradictions involve those situations in which we desire different properties of a certain parameter. A model of the physical contradiction is shown in Fig. 2.

The physical contradictions can be solved with the use of main following separation principles:

- separation in time,
- separation in space,
- separation in system level (separation between the parts and the whole),

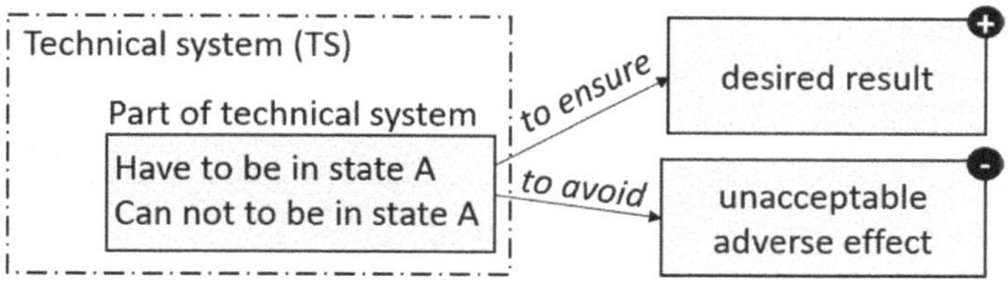

Fig. 2 Model of the physical contradiction

Table 1 Inventive principles recommended for the separation principles

Separation principle	Inventive principles [2]
Separation in time	9, 10, 11, 15, 16, 18, 19, 20, 21
Separation in space	1, 2, 3, 4, 7, 17, 24, 26
Separation in system level	1, 27, 5, 22
Separation upon conditions	13, 28, 32, 35, 36, 38, 39

- separation upon conditions.

To implement the separation principles, it is appropriate to use the so-called inventive principles [2] that are also used to solve technical contradictions (see Table 1).

The described problem of producing 3D nanostructure with high productivity can be solved by the use of separation principles for solving physical contradiction.

3 Results

The physical contradiction has been defined as: the production of 3D nanostructure demands high intensity of electrostatic field (big force) for obtaining sufficient productivity but also low intensity of electrostatic field (small force) for obtaining required 3D nanostructure at the collector (see Fig. 3).

This contradictory demand is demanded on different time. The force should be big in time t_1 (a moment of Taylor´s cones creations), and the force should be small

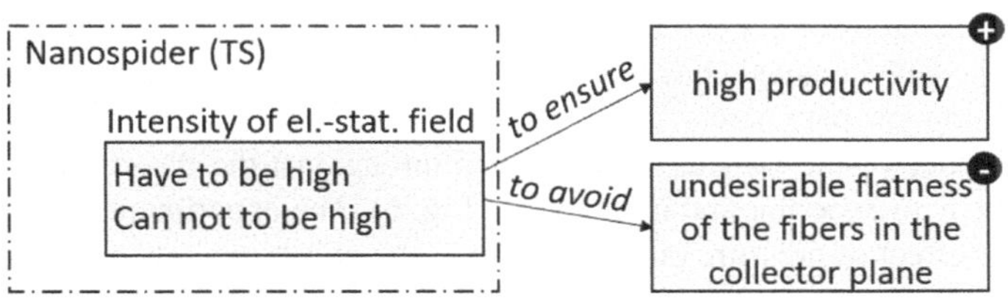

Fig. 3 Defined physical contradiction

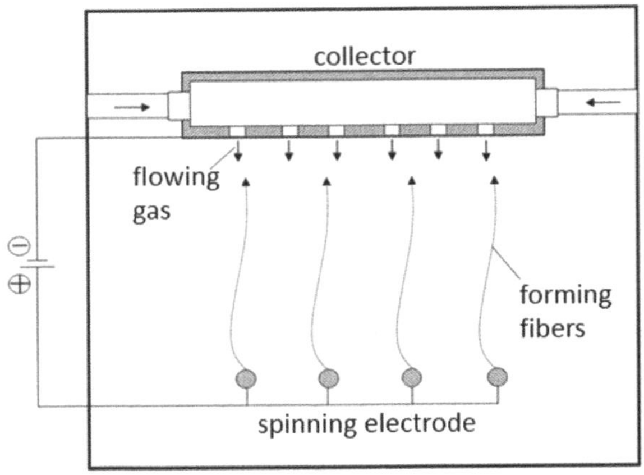

Fig. 4 Idea of the use of flowing gas against the direction of forming fibers [11]

in time t_2 (a moment when flying fibers meet the collector). If contradictory demand is demanded in different time then the principle of separation in space should be used.

The inventive principles (9, 10, 11, 15, 16, 18, 19, 20 and 21) can be used for ensuring separation in time. First of the recommended inventive principle is the principle nr. 9 Preliminary anti-action. This principle says:

- If it will be necessary to do an action with both harmful and useful effects, this action should be replaced with anti-actions to control harmful effects.
- Create beforehand stresses in an object that will oppose known undesirable working stresses later on.

The application of the inventive principle 9 Preliminary anti-action can consist it introducing flowing gas from collector against the direction of fibers creating (see Fig. 4).

4 Discussion

The influence of flowing gas into nanostructure has been studied using FEM analysis. The results show required aeration of the nanostructure and thickness increasing of structure in the areas of collector jets. The structure deformation in time t is shown in Fig. 5. The FEM analysis has been used for optimization of the collector design and the flowing gas parameters.

The influence of the flowing gaseous medium against the direction of spinning was verified in an experimental manner (see Fig. 6). The experiment has shown that the dynamic effect of pressure on the surface of the non-homogeneous nano-fibrous structure causes the structure of the fiber reorganization. The flowing gas penetrates the structure and expands the space between the layers not only at the nozzle area but among them. This is because gas flows also in the pores of the forming nano-fibrous

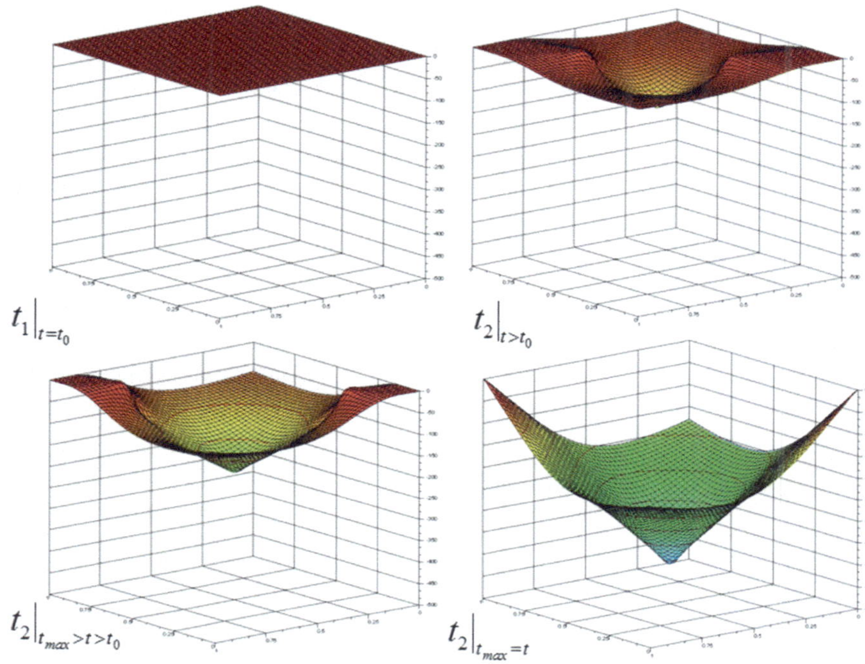

Fig. 5 FEM analysis—the structure deformation in time t [11]

Fig. 6 Experimental verification—electrospinning with the use of flowing gas

structure. The value of the maximum thickness of the real structure formed by the flowing gas was measured microscopically.

The results obtained by the FEM analysis and the experimental method correspond. The FEM analysis and the experiment confirmed the expectation of positive influence of flowing gas for creation of 3D nanostructure.

5 Conclusions

The presented task demonstrated the suitability of using tools of TRIZ method in identifying and solving the problem of fiber arrangement in the collector plane. The problem was defined in the form of physical contradiction when an opposing requirement was placed on one parameter of the technical system. The problem of physical contradiction can be solved by the separation principle of separation in time, using the inventive principle nr. 9 Preliminary anti-action. This principle can be accomplished by blowing the gaseous media from the collector against the direction of spinning. The influence of flowing gas was studied and optimized by numerical simulation of FEM and verified experimentally. The results of the FEM analysis and the experiment showed the suitability of the use of flowing gaseous media, which causes the desired deformation (aeration) of the nano-fibrous structure.

Acknowledgements This publication was written at the Technical University of Liberec as part of the project "Innovation of products and equipment in engineering practice" with the support of the Specific University Research Grant, as provided by the Ministry of Education, Youth and Sports of the Czech Republic in the year 2018.

References

1. Sasithorn N, Martinova L, Horakova J (2016) Fabrication of silk fibroin nanofibres by needleless electrospinning. Electrospinning—material, techniques, and biomedical applications, INTECH
2. Altshuller G (1999) The innovation algorithm: TRIZ, systematic innovation and technical creativity. Technical Innovation Center, Worchester
3. Rantanen K, Domb E (2007) Simplified TRIZ: new problem solving applications for engineers and manufacturing professionals. Auerbach Publications, Boca Raton
4. Fey VR, Rivin EI (2005) Innovation on demand: new product development using TRIZ. Cambridge University Press, Cambridge
5. Bukhman I (2012) TRIZ technology for innovation. Cubic Creativity Company
6. Gadd K (2011) TRIZ for engineers: enabling inventive problem solving. Wiley
7. Orloff, M. A (2012) Modern TRIZ a practical course with easytriz technology. Springer
8. San YT, Jin YT, Li SC (2009) TRIZ systematic innovation in manufacturing. Firstfruits
9. Belski I (2007) Improve your thinking: substance-field analysis, TRIZ4U, Melbourne
10. Cameron G (2010) TRIZICS—teach yourself TRIZ, how to invent, innovate and solve impossible technical problems systematically
11. Petrů M, Novák O, Vejrych D, Lepšík P (2013) FEM study of the strain kinematics in the 3D nanofibrous structure prepared by the electrospinning process. Appli Math 4(5):80–90

The manufacturer's authorised representative in the EU is Springer Nature Customer Service Centre GmbH, Europaplatz 3, 69115 Heidelberg, Germany. If you have any concerns regarding our products, please contact ProductSafety@springernature.com

Printed and bound by CPI Group (UK) Ltd, Croydon, CR0 4YY
22/09/2026
02220979-0011